STUDENT SOLUTIONS MANUAL
to accompany

CHEMISTRY
The Molecular Science

MOORE ▲ STANITSKI ▲ JURS

Judy Ozment Payne
The Pennsylvania State University, Abington College

Harcourt College Publishers

Fort Worth   Philadelphia   San Diego   New York   Orlando   Austin
San Antonio   Toronto   Montreal   London   Sydney   Tokyo

Printed in the United States of America

ISBN 0-03-032384-3

123 202 7654321

Table of Contents

Introduction

This solutions manual was written specifically to accompany the first edition of *Chemistry – The Molecular Science*, by Moore, Stanitski, and Jurs. It presents detailed solutions for the bold numbered Questions for Review and Thought at the end of each chapter.

Many of these solutions are presented using the same format described in Chapter 1 and Appendix A. This technique uses a four-stage process. Define the problem, develop a plan, execute the plan, and check your answer. Following these stages should help to make the methods you learn more readily applicable to similar problems, or the same kind problem in a different context.

Using this Book

It is important to try to answer any question for yourself, before looking at this book. When you find it necessary to use this book to give you hints or direction, try reading only the first segment "define the problem," to see if that information jives with how you read the problem. If you find that your interpretation or evaluation of the problem is routinely incorrect or incomplete, then you might seek help from your instructor, a teaching assistant, a learning center, or a tutor in reading word problems for comprehension. Sometimes, the best help for these kinds of problems can be gained from a math tutor, since they often have experience helping students with reading word problems.

If you find that you usually defined the problem in a similar fashion as described here, but still need help understanding how to solve the problem, read the second segment called "develop a plan." This will give general step by step instructions for how the problem is solved. Once you can solve problems before actually looking at the solution, shown in the segment called "execute the plan," you have made a big step toward actually learning.

It is always important to look at your solution to make sure that it makes sense. Often a quick check will help you confirm the correctness of an answer, or exhibit its flaws. For example, if you just determined that an atom of gold weighs four times the mass of the entire earth, the "check your answer" stage of the problem solving method could help you see that you made an error. Once you have the answer to a question, take a moment to think about whether it makes sense. Is it the right size and sign? Is it what you expected? Check the units and the significant figures at this point, also.

A true sign that you are learning how to do chemistry problems is if you find yourself relying on this solutions book less and less. Set goals for yourself whenever you use this book to limit how often you consult it and how extensively. It is easy to use this as a crutch, and crutches keep you from walking on your own.

Cautions

Resist the temptation to just read the solutions in this book without doing any work on your own. There are two very obvious reasons for this: (1) Recognition is easier than recall. It is far easier to look over a solution done correctly and feel like you understand it, than to stare at the same question followed by a blank space (as on a test) and recall how to do things. (2) Most teachers will not let you use this book during a test. These solutions will not be there when you need to know how to do things.

There are many ways to solve problems. Often times the right answer can be derived in several ways that are equally valid. Your instructors may have completely different ways of describing how to work these problems than are described here. All good methods have three things in common: they always give the right answer, they demonstrate how the answer was achieved, and they make sense. If any one of these three is missing, the method is flawed.

The first edition of any textbook has a few more errors than anyone would wish. As a result, a few questions in this textbook have inconsistent or incomplete information. In these cases, the solution in this solutions book provides a note about the missing or erroneous information in italics, and shows you the solution with this correction. If your instructors assign you any of these questions, you might also make them aware that these errors or omissions exist.

Most of the answers in this solutions book correspond to the answers in the back of the textbook. In the cases when they disagree, the answers in this solutions book are more reliable. The same person who wrote those answers wrote this book. Some information, figures, and diagrams that were needed to answer the questions were more readily available during the production of this book than during the production of the answers in the back of the textbook. Again, if your instructors assign you these questions, you might also make them aware that these discrepancies exist.

Acknowledgments: I want to especially thank Ms. Leslie N. Kinsland for her help, advice, patience, experience, expediency, and companionship during the production of this book. Perhaps someday we'll have a chance to meet in person. It was much easier to accomplish this task with a mentor like Leslie. I also want to thank my husband, Jeffrey S. Payne, for putting up with my very long work hours and serious distractions during the many months I worked on this book.

Harcourt College Publishers

Where Learning Comes to Life

TECHNOLOGY

Technology is changing the learning experience, by increasing the power of your textbook and other learning materials; by allowing you to access more information, more quickly; and by bringing a wider array of choices in your course and content information sources.

Harcourt College Publishers has developed the most comprehensive Web sites, e-books, and electronic learning materials on the market to help you use technology to achieve your goals.

PARTNERS IN LEARNING

Harcourt partners with other companies to make technology work for you and to supply the learning resources you want and need. More importantly, Harcourt and its partners provide avenues to help you reduce your research time of numerous information sources.

Harcourt College Publishers and its partners offer increased opportunities to enhance your learning resources and address your learning style. With quick access to chapter-specific Web sites and e-books . . . from interactive study materials to quizzing, testing, and career advice . . . Harcourt and its partners bring learning to life.

Harcourt's partnership with Digital:Convergence™ brings :CRQ™ technology and the :CueCat™ reader to you and allows Harcourt to provide you with a complete and dynamic list of resources designed to help you achieve your learning goals. Just swipe the cue to view a list of Harcourt's partners and Harcourt's print and electronic learning solutions.

http://www.harcourtcollege.com/partners/

Chapter 1: The Nature of Chemistry

How Science is Done

14. Some qualitative observations: The object is solid and roughly cubic in shape. It appears to have cubic sub-units. It appears to be a slightly gold or silvery shiny metal.

Some quantitative observations: The side of the cube is about 1.5 inches or a little more than 4 centimeters across.

16. (a) The atomic mass of an element (12.011 amu) is **quantitative** information. Information about what element it is (carbon) is **qualitative**.

(b) The purity, the details of the appearance of a substance (silvery-white), the lack of magnetic capabilities (nonmagnetic), the relative density (low), the inability to produce sparks when struck, and information about the specific element it contains (aluminum) are all **qualitative**.

(c) The density of a substance (0.968 g/mL) is **quantitative** information. Information about what element it is (sodium) is **qualitative**.

(d) The primary location of an element in the animal body (extracellular), the element's ionic form (cation), the biological importance of an element (nerve function), and what element it is (sodium) are all **qualitative**.

Physical and Chemical Properties of Matter

18. Many Americans only remember the human body temperature in the Fahrenheit scale. That is 98.6 °F. If that is the case, we can quickly apply the °F to °C conversion equation, so we can compare to the 29.8 °C melting point.

$$°C = \frac{5}{9} \times \left(°F - 32 \right) = \frac{5}{9} \times \left(98.6 - 32 \right) = 37.0°C$$

Gallium is a solid at room temperature (22 °C), but it melts at 29.8 °C, so it will melt into the **liquid** state when you hold it in your hand.

19. One reason for these questions about temperature using different scales is to help us get familiar with the Celsius scale. So, it is important to try estimating the answer before using equations. For a while it is useful to have a few connection points to a familiar scale. For that reasons, let's make some key connections between the Celsius scale and the Fahrenheit scale

Let's look at two thermometers, one calibrated with a Celsius scale and one calibrated with a Fahrenheit scale:

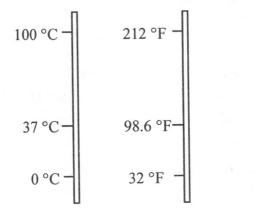

(a) 20 °C (above freezing) is higher than 20 °F (below freezing).

(b) 100 °C (at boiling) is higher than 180 °F (below boiling)

(c) 100 °F is close to body temperature, which is around 40 °C. Therefore 60 °C is higher than 100 °F.

(d) – 12 °C and 20 °F are both below freezing, so let's use the conversion equation to figure this one out:

$$°C = \frac{5}{9} \times \left(°F - 32 \right) = \frac{5}{9} \times \left(20\ °F - 32 \right) = -6.7\ °C$$

– 6.7 °C is warmer than –12 °C, so 20 °F is a higher temperature than – 12 °C.

21. Use the side-by-side thermometers again, using body temperature and the freezing point of water as reference points:

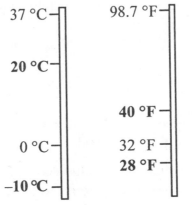

This comparison shows that Charlotte (at 20 °C) is the warmest city and Montreal (at – 10 °C) is the coldest. To be quantitative, convert 40 °F to °C to make sure it's lower than 20 °C, and convert 28 °F to °C to make sure it's higher than –10 °C.

$$°C = \frac{5}{9} \times \left(°F - 32\right)$$

$$\frac{5}{9} \times \left(40 \ °F - 32\right) = 4 \ °C \text{ lower than } 20 \ °C \text{ (confirms estimate)}$$

$$\frac{5}{9} \times \left(28 \ °F - 32\right) = -2 \ °C \text{ higher than } -10 \ °C \text{ (confirms estimate)}$$

22. *Define the problem*: We have the mass of the metal and some volume information. We need to determine the density.

Develop a plan: Use the initial and final volumes to find the volume of the metal piece, then use the mass and the volume to get the density.

Execute the plan: The metal piece displaces the water when it sinks, making the volume level in the graduated cylinder rise. The difference between the starting volume and the final volume, must be the volume of the metal piece:

$$V_{metal} = V_{final} - V_{initial} = (37.2 \text{ mL}) - (25.4 \text{ mL}) = 11.8 \text{ mL}$$

$$d = \frac{m}{V} = \frac{105.5 \text{ g}}{11.8 \text{ mL}} = 8.94 \ \frac{\text{g}}{\text{mL}}$$

Check your answer: The metal piece sinks, so the density of the metal piece needed to be higher than water. (Table 1.1 gives water density as 0.998 g/mL.) So, the answer seems right.

24. *Define the problem*: We have the three linear dimensions of a regularly shaped piece of metal.

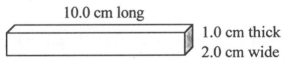
10.0 cm long
1.0 cm thick
2.0 cm wide

We also have its mass. We have a table of densities (Table 1.1). We need to determine the identity of the metal.

Develop a plan: Use the three linear dimensions to find the volume of the metal piece. Use the volume and the mass to find the density. Use the table of densities to find the identity of the metal.

Execute the plan:

$$V = (\text{thickness}) \times (\text{width}) \times (\text{length}) = (1.0 \text{ cm}) \times (2.0 \text{ cm}) \times (10.0 \text{ cm}) = 20. \text{ cm}^3$$

Using dimensional analysis, find the volume in mL:

$$20.\ cm^3 \times \frac{1\ mL}{1\ cm^3} = 20.\ mL$$

Find the density: $$d = \frac{m}{V} = \frac{54.0\ g}{20.\ mL} = 2.7\ \frac{g}{mL}$$

According to Table 1.1, the density that most closely matches this one is **aluminum** (density = 2.70 g/mL).

Check your answer: An object whose mass is larger than its volume will have a density larger than one. The mass of this object is between two and three times larger than the volume. This answer looks right.

26. *Define the problem*: We have the three linear dimensions of a regularly-shaped sodium chloride crystal:

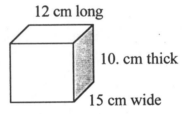

12 cm long

10. cm thick

15 cm wide

We have a table of densities (Table 1.1). We need to determine the mass of the crystal.

Develop a plan: Use the three linear dimensions to find the volume of the crystal. Use the volume and density (Table 1.1) to find the mass.

Execute the plan:

V = (thickness) × (width) × (length) = (10. cm) × (15 cm) × (12 cm)

$$= 1800\ cm^3 = 1.8 \times 10^3\ cm^3$$

Using dimensional analysis, find the volume in mL:

$$1.8 \times 10^3\ cm^3 \times \frac{1\ mL}{1\ cm^3} = 1.8 \times 10^3\ mL$$

Find the mass using the density: $$1.8 \times 10^3\ mL \times \frac{2.16\ g}{1\ mL} = 3.9 \times 10^3\ g$$

Note: We're carrying two significant figures since the length data was only that precise.

Check your answer: This crystal is pretty big. So, while the mass calculated is a large number, the volume is still about half the mass, in keeping with a density around two.

Chemical Changes and Chemical Properties

27. (a) The normal color of bromine is a **physical** property. Determining the color of a substance does not change its chemical form.

    (b) The fact that iron can be transformed into rust is a **chemical** property. Iron, originally in the elemental metallic state, is incorporated into a compound, rust, when it is observed to undergo this transformation.

    (c) The fact that dynamite can explode is a **chemical** property. The dynamite is chemically changed when it is observed to explode.

    (d) Observing the shininess of aluminum does not change it, so this is a **physical** property. Melting aluminum does not change it to a different substance, though it does change its physical state. It is still aluminum, so melting at 387 °C is a physical property of aluminum.

29. (a) Bleaching clothes from purple to pink is a **chemical** change. The purple substance in the clothing reacts with the bleach to make a pink substance. The purple color cannot be brought back nor can the bleach.

    (b) The burning of fuel in the space shuttle (hydrogen and oxygen) to form water and energy is a **chemical** change. The two elements react to form a compound.

    (c) The ice cube melting in the lemonade is a **physical** change. The H_2O molecules do not change to a different form in the physical state change.

31. (a) The conversion of excess food into fat molecules is the body's way of storing energy for doing work later. So, this represents the outside source of energy (from the food we eat) is forcing a chemical reaction to occur (the production of fat).

    (b) Sodium reacts with water rather violently. It produces a lot of heat and causes work to be done.

    (c) Sodium azide in an automobile's airbag decomposes causing the bag to inflate. This uses a chemical reaction to release energy and cause work to be done (inflation of the air bag).

    (d) The process of hard-boiling an egg on your stove uses energy from the stove to cause a chemical reaction to occur (the coagulation of the white and yolk of the egg).

Substances, Mixtures, and Separation

32. It is clear by visual inspection that the mixture is non-uniform (**heterogeneous**) at the macroscopic level. The iron could be separated from the sand **using a magnet**, since iron is attracted to magnets and the sand is not.

34. The terminology "heterogeneous" and "homogeneous" is somewhat subjective. In Section 1.5, the terms are described. The heterogeneous mixture is described as one whose uneven texture can be seen without magnification or with a microscope. The homogeneous mixture is defined as one that is completely uniform, wherein no amount of optical magnification will reveal variations in the mixture's properties. Note that there is a "gap" between these two terms. The question "how close do I look?" comes to mind.

    The bottom line is this: These terms were designed to HELP us classify things, not to create trick questions. If you explain your answer with a valid defense, the answer ought to be right; however, don't go out of your way to imagine unusual circumstances that make the question more difficult to classify. Think about what you would SEE. Identify whether what you see has variations, then make a case for the proper term.

    (a) Vodka is classified as a **homogenous** mixture. It is a clear, colorless solution of alcohol, water, and probably some other minor ingredients.

    (b) Blood appears smooth in texture and to our eyes most likely appears to be homogeneous. Upon closer examination it is found to have various particles within the liquid, and might, for that reason, be called **heterogeneous**.

    (c) Cowhide is **heterogeneous**. Even folks who have never seen a cowhide might be able to imagine that brown cows, white cows, black cows and spotted cows probably have different coloration to their hide. There are probably pores where the hair grows out, making it rough in texture. The hide of a young cow is probably softer and more pliable than the hide of an old cow. Unless presented with a sample that visually changed our perception of what constitutes a cowhide, it is quite safe to call the cowhide heterogeneous.

    (d) Bread is **heterogeneous**. The crust is a different color. Some parts of the bread have bigger bubbles than other parts. Some breads have whole grains in them. Some breads are composed of different colors (like the rye/white swirl breads). In general, most samples of bread have identifiable regions that are different from other regions.

Elements and Compounds

37. (a) A blue powder turns white and loses mass. The loss of mass is most likely due to the creation of a gaseous product. That suggests that the original material was a compound that decomposed into the white substance (a compound or an element) and a gas (a compound or an element).

   (b) If three different gases were formed, that suggests that the original material was a compound that decomposed into three compounds or gases.

38. (a) A reddish metal is placed in a flame. It turns black and the black material has a higher mass than the original reddish material. That suggests that the material produced is a compound, and either a compound or an element combined with something in air (oxygen or nitrogen) to make the product.

   (b) A white solid is heated in oxygen and forms two gases. The mass of the gases produced is the same as the solid and the oxygen. Oxygen is an element. That suggests that the product gases are compounds and the elemental oxygen combined with another element or compound to make one or more compounds.

39. (a) A piece of newspaper is a **heterogeneous mixture**. Paper and ink are distributed in a non-uniform fashion at the macroscopic level.

   (b) Solid granulated sugar is a **pure compound**. Sugar is made of two or more elements.

   (c) Fresh squeezed orange juice is a heterogeneous mixture. The presence of unfiltered pulp makes part of the mixture solid and part of the mixture liquid.

   (d) Gold jewelry is a **homogeneous mixture**, most of the time. Most jewelry is less than 24 carat – pure gold. If it is 18 carat, 14 carat, 10 carat gold, that means that other metals are mixed in with the gold (usually to make it more durable and cheaper). The combining of metals in a fashion that prevents us from seeing variations in the texture or properties of the metal qualifies it as a homogeneous mixture.

41. (a) Chunky peanut butter is definitely a **heterogeneous mixture**. The uncrushed peanut chunks do not have the same properties as the smooth, sweetened part of the mixture.

   (b) Distilled water is a **pure compound**. The distillation process removes other minerals and substances from water, leaving it just water.

   (c) Platinum is an **element** with the symbol "Pt".

(d) Air is usually considered to be a **homogeneous mixture**. Now, sometimes air has enough variable properties to qualify as heterogeneous, such as near the tailpipe of a diesel truck. However, most of the time, the gases in a sample of air are sufficiently well mixed such that there is no visible difference in the properties of various regions of that air sample.

Nanoscale Models: Solids, Liquids, Gases

43. The crystal of halite pictured is in the **macroscopic world**. Its shape is cubic. It could be expected that the arrangement of particles (atoms and/or ions) in the nanoscale world are also arranged in a cubic fashion.

45. The bacterium is in the **microscale world**. We need some enhanced magnification to see.

47. When we open a can of a soft drink, the carbon dioxide gas expands rapidly as it rushes out of the can. At the nanoscale, this can be explained as large number of carbon dioxide molecules crowded into the unopened can. When the can is opened, the molecules that were about to hit the surface where the hole was made continue forward through the hole. A large number of the carbon dioxide particles that were contained within the can very quickly escape through the same hole.

Atomic Theory

50. Conservation of mass is easy to see from the point of view of atomic theory. A chemical change is described as the rearrangement of atoms. Since the atoms in the starting materials must all be accounted for in the substances produced, then there would be no change in the mass.

52. Four Postulates of Modern Atomic Theory

    (I)   All matter is composed of atoms, which are extremely tiny.

    (II) All atoms of a given element have the same chemical properties.

    (III) Compounds are formed by the chemical combination of two or more different kinds of atoms.

    (IV) A chemical reaction involves joining separating or rearranging atoms.

54. The law of multiple proportions says that if two compounds contain the same elements and samples of those two compounds both contain the same mass of one element, then the ratio of the masses of the other elements will be a small whole number.

The Chemical Elements

56. Many responses are equally valid here. Below are a few common examples. These lists are not comprehensive; many other answers are also right. The periodic table on the inside cover of your text is color coded to indicate metals, non-metals and metalloids.

(a) Common metallic elements: Fe, iron; Au, gold; Pb, lead; Cu, copper; Al, aluminum.

(b) Common non-metallic elements: C, carbon; H, hydrogen; O, oxygen; N, nitrogen

(c) Metalloids: B, boron; Si, silicon; Ge, germanium; As, arsenic; Sb, antimony; Te, tellurium.

(d) Elements that are diatomic molecules: nitrogen, N_2; oxygen, O_2; hydrogen, H_2; fluorine, F_2; chlorine, Cl_2; bromine, Br_2; iodine, I_2

Chemical Symbolism

58. Formula for each substance and nanoscale picture:

(a) Water H_2O

(b) Nitrogen N_2

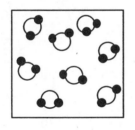

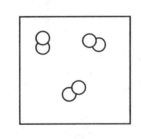

(c) Neon   Ne

(d) Chlorine Cl_2

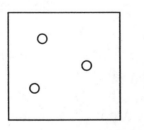

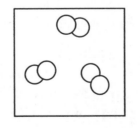

60. $2 H_2 (g) + O_2 (g) \longrightarrow 2 H_2O (g)$

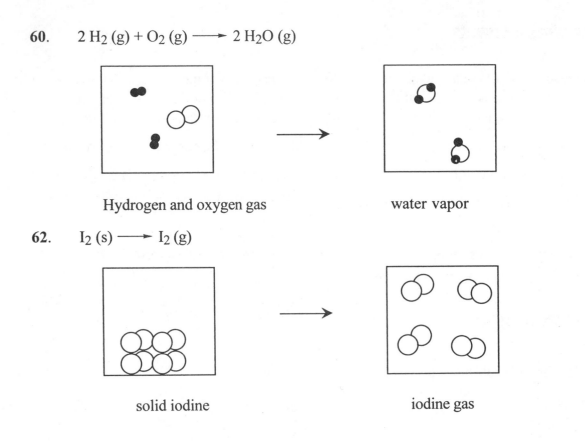

Hydrogen and oxygen gas   water vapor

62. $I_2 (s) \longrightarrow I_2 (g)$

solid iodine   iodine gas

Risks and Benefits

64. The answers to these questions are very subjective. The answers given here may not reflect your own opinion.

(a) The benefits of a new drug outweigh the risks. The FDA is charged with the responsibility of insuring that there are no unwanted (or unmanageable) side effects of the drugs. Those side effects would occur only in those people taking the drugs. The benefit of prolonged life would outweigh the disadvantage of most side effects. The drug companies and their stockholders and the HIV-positive patients would benefit. The HIV-positive individuals are also at risk.

(b) The benefits of better health through reduced caloric and fat intake outweigh the risks of having diarrhea that Olestra might cause. Those individuals that are subject to these side effects do not have to eat chips cooked in Olestra. The risks are not life threatening. The manufacturer of Olestra, through increased profits, and the general chip-eating public, through better health, would be the main benefactors. People with "sensitive" digestive systems are at risk.

Why Care About Chemistry

66. Fourteen questions correlating to chemistry and science phenomena are found in Section 1.12 of your textbook.

General Questions

68. (a) The mass of the compound (1.456 grams) is quantitative and relates to a physical property. The color (white), the fact that it reacts with a dye, and the color change in the dye (red to colorless) are all qualitative. The colors are related to physical properties. The reaction with the dye is related to a chemical property.

(b) The mass of the metal (0.6 grams) is quantitative and related to a physical property. The identity of the metal (lithium) and the identities of the chemicals it reacts with and produces (water, lithium hydroxide, and hydrogen) are all qualitative information. The fact that a chemical reaction occurs when the metal is added to water is qualitative information and related to a chemical property.

70. If the density of solid calcium is almost twice that of solid potassium, but their masses are approximately the same size, then the volume must account for the difference. This suggests that the atoms of calcium are smaller than the atoms of potassium:

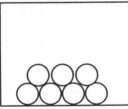

solid calcium
smaller atoms
closer packed
smaller volume

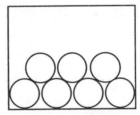

solid potassium
larger atoms
less closely packed
larger volume

Applying Concepts

73. (a) (Table 1.1) density of butane = 0.579 g/mL; density of bromobenzene = 1.49 g/mL

1 mL butane weighs less than 1 mL of bromobenzene so, 20 mL butane weighs less than 20 mL of bromobenzene. The **bromobenzene sample has a larger mass.**

(b) (Table 1.1) density of benzene = 0.880 g/mL; density of gold = 19.32 g/mL

There are 0.880 grams of benzene in 1 mL of benzene, so there are 8.80 grams of benzene in 10 mL of benzene. Since 1.0 mL of gold has a mass of 19.32 grams that means the **gold sample has a larger mass**.

(c) (Table 1.1) density of copper = 8.93 g/mL; density of lead = 11.34 g/mL

Any volume of lead has a larger mass than the same volume of copper. That means the **lead sample has a larger mass**.

75. (a) *Define the problem*: Use the volume of the bottle and the densities of water and ice to determine the volume of ice formed from a fixed amount of water.

Develop a plan: Use the volume of the bottle and the density of water to figure out the mass of water frozen, then calculate the volume of the ice.

Execute the plan: At 25 °C, density of water is 0.997 g/mL.

At 0 °C, density of ice = 0.917 g/mL

$$250 \text{ mL water} \times \frac{0.997 \text{ g H}_2\text{O}(\ell)}{1 \text{ mL water}} \times \frac{1 \text{ g H}_2\text{O(s)}}{1 \text{ g H}_2\text{O}(\ell)} \times \frac{1 \text{ mL ice}}{0.917 \text{ g H}_2\text{O(s)}} = 2.7 \times 10^2 \text{ mL ice}$$

Check your answer: The density of water is larger than the density of ice. It makes sense that the volume of the ice produced is larger than the volume of water.

(b) If the bottle is made of flexible plastic, it might be deformed and bulging if not cracked and leaking ice. If the bottle is made of glass and the top came off, there would be ice (approximately 20 mL of it) oozing out of the top of the bottle. Worst case scenario would be if the bottle was glass and the top did not come off, so it broke.

78. (a) (Table 1.1) density of water = 0.998 g/mL, density of bromobenzene = 1.49 g/mL.

Since water does not dissolve in bromobenzene, the lower density water will be the top layer of the two imiscible layers.

(b) (Table 1.1) density of benzene = 0.880 g/mL. Benzene doesn't dissolve in water and since it has a lower density, it will sit on top of the water layer.

(c) (Table 1.1) density of ethanol = 0.789 g/mL. Ethanol doesn't dissolve in benzene, and since it has a lower density, it will sit on top of the benzene layer.

(d) The ethanol and water will mix and make one phase. The benzene and bromobenzene will also mix into one phase. Assuming the new mixtures have the average density of the original liquids, the water/ethanol layer (average density is 0.894 g/mL) will sit on top of the benzene/bromobenzene layer. (average density is 1.185 g/mL)

79. The melting point is the temperature at which a solid changes into a liquid. In the nanoscale picture, the liquid molecules are moving faster and they are farther apart. That motion spreads out the molecules and makes the volume larger, even though the collection of molecules have the same mass. Hence, the density of the liquid is smaller than the density of the solid.

The boiling point is the temperature at which the liquid molecules change into gas-phase molecules. In the nanoscale picture, the gas molecules are moving faster and they are farther apart. That motion spreads out the molecules and makes the volume larger, even though the collection of molecules have the same mass. Hence, the density of the gas is smaller than the density of the liquid.

14

Chapter 2: Atoms and Elements

Units and Unit Conversions

9. *Define the problem*: If the nucleus were scaled to a diameter of 6 cm, determine the diameter of the atom.

   Develop a plan: Find the accepted relationship between the size of the nucleus and the size of the atom. Use size relationships to get the diameter of the "artificially large" atom.

   Execute the plan: The atom is about 10,000 times bigger than the nucleus.

   $$10,000 \times 6 \text{ cm} = 60,000 \text{ cm}$$

   Check your answer: A much larger nucleus means a much larger atom. This large atomic diameter result looks right.

11. *Define the problem*: The pole vault record is a height of 6.14 m. Use conversion factors to change the units to centimeters, feet, and inches.

    Develop a plan: Use the metric relationship between m and cm to convert m into cm. Then, use the metric relationship between cm and inches to convert cm into inches. Then, use the relationship between inches and feet to convert from inches to feet.

    Execute the plan: If any one of these questions were asked separately, we would start with the given information and apply the appropriate conversion factors.

    $$6.14 \text{ m} \times \frac{100 \text{ cm}}{1 \text{ m}} = 614 \text{ cm}$$

    $$6.14 \text{ m} \times \frac{100 \text{ cm}}{1 \text{ m}} \times \frac{1 \text{ in}}{2.54 \text{ cm}} = 242 \text{ in}$$

    $$6.14 \text{ m} \times \frac{100 \text{ cm}}{1 \text{ m}} \times \frac{1 \text{ in}}{2.54 \text{ cm}} \times \frac{1 \text{ ft}}{12 \text{ in}} = 20.1 \text{ ft}$$

    When answering all three questions, we can use the results of the previous calculation to make the next calculation faster:

    $$614 \text{ cm} \times \frac{1 \text{ in}}{2.54 \text{ cm}} = 242 \text{ in}$$

    $$242 \text{ in} \times \frac{1 \text{ ft}}{12 \text{ in}} = 20.1 \text{ ft}$$

Check your answers: The number of centimeters should be larger than the number of meters, since "centimeter" is a smaller unit of measure. The number of inches should be smaller than the number of centimeters, since "inch" is a larger unit of measure. The number of feet should be smaller than the number of inches, since "foot" is a larger unit of measure. These answers make sense.

13. *Define the problem:* Given the distance to the fence in feet, determine the distance in meters.

Develop a plan: Use conversion factors between feet and inches, then inches and centimeters, then centimeters and meters to change the units from feet to meters.

Execute the plan: $\quad 404 \text{ ft} \times \dfrac{12 \text{ in}}{1 \text{ ft}} \times \dfrac{2.54 \text{ cm}}{1 \text{ in}} \times \dfrac{1 \text{ m}}{100 \text{ cm}} = 123 \text{ m}$

Check your answer: Feet are shorter than meters, so the number of meters should be smaller than the number of feet. This answer looks right.

15. *Define the problem:* Given displacement volume of 250. in^3, determine the volume in cm^3 and liters.

Develop a plan: Use conversion factor between inches and centimeters to change the units from cubic inches to cubic centimeters. Then use conversion factors between cubic centimeters and milliliters, then milliliters and liters to change the units from cubic centimeters to liters.

Execute the plan: $\quad 250. \text{ in}^3 \times \left(\dfrac{2.54 \text{ cm}}{1 \text{ in}} \right)^3 = 4.10 \times 10^3 \text{ cm}^3$

$$250. \text{ in}^3 \times \left(\dfrac{2.54 \text{ cm}}{1 \text{ in}} \right)^3 \times \dfrac{1 \text{ mL}}{1 \text{ cm}^3} \times \dfrac{1 \text{ L}}{1000 \text{ mL}} = 4.10 \text{ L}$$

Notice: The conversion between liters and milliliters used here indicates that a larger number (1000) of small things (mL) are equal to a smaller number (1) of large things (L). Conversely, you can use the conversion relating a smaller number (10^{-3}) of larger things (L) to a larger number (1) of small things.

$$250. \text{ in}^3 \times \left(\dfrac{2.54 \text{ cm}}{1 \text{ in}} \right)^3 \times \dfrac{1 \text{ mL}}{1 \text{ cm}^3} \times \dfrac{10^{-3} \text{ L}}{1 \text{ mL}} = 4.10 \text{ L}$$

Check your answers: Centimeters are smaller than inches, so a cubic centimeter is much smaller than a cubic inch. It makes sense that the number of cubic centimeters is larger than the number of cubic inches. A liter is larger than a cubic centimeter, so it makes sense that the number of liters is smaller than the number of cubic inches. These answers look right.

Percent

18. *Define the problem*: Given a 17.6-gram bracelet that contains 14.1 grams of silver and the rest copper, determine the percentage silver and the percentage copper.

Develop a plan: Determine the percentage of silver by dividing the mass of silver by the total mass and multiplying by 100 %. Since the metal is made up of only silver and copper, determine the percentage of copper by subtracting the percentage of silver from 100 %.

Execute the plan: $$\frac{14.1 \text{ g silver}}{17.6 \text{ g bracelet}} \times 100 \text{ \%} = 80.1 \text{ \% silver}$$

$$100 \text{ \% total} - 80.1 \text{ \% silver} = 19.9 \text{ \% copper}$$

NOTICE: When you are using grams of different substances, be careful to carry enough information in the units so you don't confuse one mass with another.

Check your answers: A significant majority of the metal in the bracelet is silver, so it makes sense that the percentage of silver is larger than the percentage of copper.

20. *Define the problem*: Given the volume of a battery acid sample, the density and the percentage of sulfuric acid in the battery acid by mass, determine the mass of acid in the battery.

Develop a plan: Always start with the sample. Use the density of the solution to create a conversion factor between milliliters and grams, so you can determine the mass of the battery acid in grams.   Then use the mass percentage of sulfuric acid in the battery acid as a conversion factor to determine the mass of sulfuric acid in the sample.

Execute the plan:

Every 1.000 mL of battery acid solution weighs 1.285 grams.

Every 100.00 g of battery acid solution contains 38.08 grams of sulfuric acid.

$$500. \text{ mL solution} \times \frac{1.285 \text{ g solution}}{1.000 \text{ mL solution}} \times \frac{38.08 \text{ g sulfuric acid}}{100.00 \text{ g solution}} = 245 \text{ g sulfuric acid}$$

NOTICE: When you are using grams of different substances, be careful to carry enough information in the units so you don't confuse one mass with another.

Check your answer: Only about a third of the sample is sulfuric acid, so the mass of sulfuric acid should be smaller than the volume of the solution. This looks right.

Isotopes

24. *Define the problem*: Given the identity of an element (americium) and the atom's mass number (241), determine the number of electrons, protons, and neutrons in the atom.

    Develop a plan: Look up the symbol for americium and find that symbol on the periodic table. The periodic table gives the atomic number. The atomic number is the number of protons. The number of electrons is equal to the number of protons since the atom has no charge. The number of neutrons is the difference between the mass number and the atomic number.

    Execute the plan: The element americium has the symbol Am. On the periodic table, we find it listed with the atomic number 95. So, the atom has **95 protons, 95 electrons** and (241 – 95 =) **146 neutrons**.

    Check your answer: The number protons and electrons must be the same (95 = 95). The sum of the protons and neutrons is the mass number (146 + 95 = 241). This is correct.

26. Atoms of the same element have the same number of protons in the nucleus, and therefore all the atoms for any given element would have the same atomic **number**.

28. To estimate an atom's mass, one only has to add up the number of protons and neutrons in the nucleus. This estimate, which is a whole number, refers to the **mass** number for a given atom. The mass of an electron is 1800 times smaller than the mass of a proton or neutron. That makes its contribution to the atom's mass negligible.

30. Mass number is the sum of the number of protons and neutrons. Atomic number is the number of protons. So, when you subtract the atomic number from the mass number, you obtain the number of neutrons.

32. *Define the problem*: Given the identity of an element and the number of neutrons in the atom, determine the mass number of the atom.

    Develop a plan: Look up the symbol for the element and find that symbol on the periodic table. The periodic table gives the atomic number, which represents the number of protons. Add the number of neutrons to the number of protons to get the mass number.

    Execute the plan:

    (a) The element iron has the symbol Fe. On the periodic table, we find it listed with the atomic number 26. The given number of neutrons is 30. So, (26 + 30 =) 56 is the mass number for this iron atom.

(b) The element americium has the symbol Am. On the periodic table, we find it listed with the atomic number 95. The given number of neutrons is 148. So, (95 + 148 =) 243 is the mass number for this americium atom.

(c) The element tungsten has the symbol W. On the periodic table, we find it listed with the atomic number 74. The given number of neutrons is 110. So, (74 + 110 =) 184 is the mass number for this tungsten atom.

Check your answers: Mass number should be close to (but not always the same as) the atomic weight given on the periodic table. Iron's atomic weight (55.85) is close to its 56 mass number. Americium's atomic weight (243) is the same as its 243 mass number. Tungsten's atomic weight (183.85) is close to its 184 mass number. These look right.

34. *Define the problem*: Given the identity of an element and the number of neutrons in the atom, determine the atomic symbol $^A_Z X$.

Develop a plan: Look up the symbol for the element and find that symbol (X) on the periodic table. The periodic table gives the atomic number (Z), which represents the number of protons. Add the number of neutrons to the number of protons to get the mass number (A).

Execute the plan:

(a) The element nitrogen has the symbol N. On the periodic table, we find it listed with the atomic number 7. The given number of neutrons is 8. So, (7 + 8 =) 15 is the mass number for this nitrogen atom. Its atomic symbol looks like this: $^{15}_{7} N$.

(b) The element zinc has the symbol Zn. On the periodic table, we find it listed with the atomic number 30. The given number of neutrons is 34. So, (30 + 34 =) 64 is the mass number for this zinc atom. Its atomic symbol looks like this: $^{64}_{30} Zn$.

(c) The element xenon has the symbol Xe. On the periodic table, we find it listed with the atomic number 54. The given number of neutrons is 75. So, (54 + 75 =) 129 is the mass number for this xenon atom. Its atomic symbol looks like this: $^{129}_{54} Xe$.

Check your answers: Mass number should be close to (but not exactly the same as) the atomic weight also given on the periodic table. Nitrogen's atomic weight (14.01) is close to the 15 mass number. Zinc's atomic weight (65.38) is close to the 64 mass number. Xenon's atomic weight (131.3) is close to the 129 mass number. These numbers seem reasonable.

36. *Define the problem*: Given the atomic symbol $^{A}_{Z}X$ of the isotope, determine the number of electrons, protons, and neutrons.

Develop a plan: The atomic number (Z) represents the number of protons. In neutral atoms, the number of electrons are equal to the number of protons. To get the number of neutrons, subtract the number of protons from the mass number (A).

Execute the plan:

(a) The isotope given is $^{13}_{6}C$. That means A = 13 and Z = 6. So, the number of protons is 6, the number of electrons is 6, and the number of neutrons is (13 – 6 =) 7.

(b) The isotope given is $^{50}_{24}Cr$. That means A = 50 and Z = 24. So, the number of protons is 24, the number of electrons is 24, and the number of neutrons is (50 – 24 =) 26.

(c) The isotope given is $^{205}_{83}Bi$. That means A = 205 and Z = 83. So, the number of protons is 83, the number of electrons is 83, and the number of neutrons is (205 – 83 =) 122.

Check your answers: The number of protons and electrons must be equal in neutral atoms. The mass number must be the sum of the protons and neutrons. These answers look okay.

40. *Define the problem*: Knowing the identity of the element and the number of neutrons, determine the atomic symbol $^{A}_{Z}X$.

Develop a plan: The element's identity can be used to get the atomic number (Z). The atomic number represents the number of protons. The mass number (A) is the number of neutrons and protons. Determine the atomic number and mass number of each isotope and construct the atomic symbol.

Execute the plan: Look up the symbol for cobalt (Co), and find Co on the periodic table: Z = 27.

$$A = Z + \text{number of neutrons} = 27 + 30 = 57 \qquad ^{57}_{27}Cr$$

$$A = Z + \text{number of neutrons} = 27 + 31 = 58 \qquad ^{58}_{27}Cr$$

$$A = Z + \text{number of neutrons} = 27 + 33 = 60 \qquad ^{60}_{27}Cr$$

Check your answers: Isotopes must have the same symbols and the same Z values, but different A values. These answers look okay.

Atomic Weight

42. *Define the problem*: Using the exact mass and the percent abundance of several isotopes of an element, determine the atomic weight.

Develop a plan: Calculate the weighted average of the isotope masses.

Execute the plan:

Every 10000 atoms of magnesium contains 7899 atoms of the ^{24}Mg isotope, 1000 atoms of the ^{25}Mg isotope, and 1101 atoms of the ^{26}Mg isotope.

$$\frac{7899 \text{ atoms } ^{24}Mg}{10000 \text{ Mg atoms}} \times \left(\frac{23.985042 \text{ amu}}{1 \text{ atom } ^{24}Mg} \right) + \frac{1000 \text{ atoms } ^{25}Mg}{10000 \text{ Mg atoms}} \times \left(\frac{24.98537 \text{ amu}}{1 \text{ atom } ^{25}Mg} \right)$$

$$+ \frac{1101 \text{ atoms } ^{26}Mg}{10000 \text{ Mg atoms}} \times \left(\frac{25.982593 \text{ amu}}{1 \text{ atom } ^{26}Mg} \right) = 24.31 \frac{\text{amu}}{\text{Mg atom}}$$

Note: The given percentages limit each term to four significant figures, therefore the first term has only two decimal places. So, this answer is rounded off significantly.

Check your answer: The periodic table value for the atomic weight is the same this calculated value. This answer looks right.

44. *Define the problem*: Using the exact mass of several isotopes and the atomic weight, determine the abundance of the isotopes.

Develop a plan: Establish variables describing the isotope percentages. Set up two relationships between these variables. The sum of the percents must be 100 %, and the weighted average of the isotope masses must be equal to the reported atomic weight.

Execute the plan: X % ^{63}Cu and Y % ^{65}Cu, This means: X + Y = 100 %

That gives us one equation with two unknowns.

Every 100 atoms of copper contains X atoms of the ^{63}Cu isotope and Y atoms of the ^{65}Cu isotope. The atomic weight for Cu from the periodic table is 63.546 amu/atom.

$$\frac{X \text{ atoms } ^{63}Cu}{100 \text{ Cu atoms}} \times \left(\frac{62.939598 \text{ amu}}{1 \text{ atom } ^{63}Cu} \right) + \frac{Y \text{ atoms } ^{65}Cu}{100 \text{ Cu atoms}} \times \left(\frac{64.927793 \text{ amu}}{1 \text{ atom } ^{65}Cu} \right)$$

$$= 63.546 \frac{\text{amu}}{\text{Cu atom}}$$

That gives us a second equation with the same two unknowns.

$$\frac{X}{100} \times 62.939598 + \frac{Y}{100} \times 64.927793 = 63.546$$

Solve the first equation for Y in terms of X, then substitute back into the second equation:

$$Y = 100 - X$$

$$\frac{X}{100} \times 62.939598 + \frac{(100-X)}{100} \times 64.927793 = 63.546$$

Now, solve for X then Y:

$$0.62939598\,X + 64.927793 - 0.64927793\,X = 63.546$$

$$64.927793 - 63.546 = 0.64927793\,X - 0.62939598\,X = (0.64927793 - 0.62939598)X$$

$$1.382 = (0.01988195)X$$

$$X = 69.51 \text{ and } Y = 100 - X = 100 - 69.50 = 30.51$$

Therefore the abundances for these isotopes are: 69.51 % $^6_3$Cu   and   30.49 % ^{65}Cu

Check your answers: The periodic table gives the atomic weight as closer to 62.939598 than it is to 64.927793, so it makes sense that the percentage of ^{63}Cu is larger than ^{65}Cu. This answer looks right.

46. Knowing that almost all of the argon in nature is ^{40}Ar, a good estimate for the atomic weight of argon is a little less than 40 amu/atom.

Define the problem: Using the exact mass and the percent abundance of several isotopes of an element, determine the atomic weight.

Develop a plan: Calculate the weighted average of the isotope masses.

Execute the plan: Every 100000 atoms of argon contains 337 atoms of the ^{36}Ar isotope, 63 atoms of the ^{38}Ar isotope, and 99600 atoms of the ^{40}Ar isotope.

$$\frac{337 \text{ atoms } ^{36}\text{Ar}}{100000 \text{ Ar atoms}} \times \left(\frac{35.968 \text{ amu}}{1 \text{ atom } ^{36}\text{Ar}} \right) + \frac{63 \text{ atoms } ^{38}\text{Ar}}{100000 \text{ Ar atoms}} \times \left(\frac{37.963 \text{ amu}}{1 \text{ atom } ^{38}\text{Ar}} \right)$$

$$+ \frac{99600 \text{ atoms } ^{40}\text{Ar}}{100000 \text{ Ar atoms}} \times \left(\frac{39.962 \text{ amu}}{1 \text{ atom } ^{40}\text{Ar}} \right) = 39.95 \frac{\text{amu}}{\text{Ar atom}}$$

Check your answer: This calculated answer matches the estimate. Also, the periodic table value for the atomic weight is the same as this calculated value.

The Mole

47. A few common counting units are: pair (2), dozen (12), gross (144), hundred (100), million (1,000,000), billion (1,000,000,000), etc.

51. *Define the problem*: Determine the mass in grams from given quantity in moles.

Develop a plan: Look up the elements on the periodic table to get the atomic weight (with at least four significant figures). Use that number for the molar mass (with units of grams per mole) as a conversion factor between moles and grams.

Note: Whenever you use physical constants that you look up, it is important to use more *significant figures than the rest of the numbers, to prevent causing inappropriate round-off errors.*

Execute the plan:

(a) Gold (Au) has atomic number 79 on the periodic table. Its atomic weight is 197.0, so the molar mass is 197.0 g/mol.

$$6.03 \text{ mol Au} \times \frac{197.0 \text{ g Au}}{1 \text{ mol Au}} = 1.19 \times 10^3 \text{ g Au}$$

(b) Uranium (U) has atomic number 92 on the periodic table. Its atomic weight is 238.0, so the molar mass is 238.0 g/mol.

$$0.045 \text{ mol U} \times \frac{238.0 \text{ g U}}{1 \text{ mol U}} = 11 \text{ g U}$$

(c) Neon (Ne) has atomic number 10 on the periodic table. Its atomic weight is 20.18, so the molar mass is 20.18 g/mol.

$$15.6 \text{ mol Ne} \times \frac{20.18 \text{ g Ne}}{1 \text{ mol Ne}} = 315 \text{ g Ne}$$

(d) Radioactive plutonium (Pu) has atomic number 94 on the periodic table. The atomic weight given on the periodic table is the weight of its most stable isotope 244, so the molar mass is 244 g/mol.

$$3.63 \times 10^{-4} \text{ mol Pu} \times \frac{244 \text{ g Pu}}{1 \text{ mol Pu}} = 0.0886 \text{ g Pu}$$

Check your answers: Notice that the "moles" units cancel when the factor is multiplied, leaving the answer in grams.

53. *Define the problem*: Determine the quantity in moles from given mass in grams.

Develop a plan: Look up the elements on the periodic table to get the atomic weight (with an appropriate number of significant figures). Use that number for the molar mass (with units of grams per mole) as a conversion factor between grams and moles.

Execute the plan:

(a) Sodium (Na) has atomic number 11 on the periodic table. Its atomic weight is 22.99, so the molar mass is 22.99 g/mol.

$$16.0 \text{ g Na} \times \frac{1 \text{mol Na}}{22.99 \text{ g Na}} = 0.696 \text{ mol Na}$$

(b) Platinum (Pt) has atomic number 78 on the periodic table. Its atomic weight is 195.1, so the molar mass is 195.1 g/mol.

$$0.0034 \text{ g Pt} \times \frac{1 \text{ mol Pt}}{195.1 \text{ g Pt}} = 1.7 \times 10^{-5} \text{ mol Pt}$$

(c) Phosphorus (P) has atomic number 15 on the periodic table. Its atomic weight is 30.97, so the molar mass is 30.97 g/mol.

$$1.54 \text{ g P} \times \frac{1 \text{ mol P}}{30.97 \text{ g P}} = 0.0497 \text{ mol P}$$

(d) Arsenic (As) has atomic number 33 on the periodic table. Its atomic weight is 74.92, so the molar mass is 74.92 g/mol.

$$0.876 \text{ g As} \times \frac{1 \text{ mol As}}{74.92 \text{ g As}} = 0.0117 \text{ mol As}$$

(e) Xenon (Xe) has atomic number 54 on the periodic table. Its atomic weight is 131.3, so the molar mass is 131.3 g/mol.

$$0.983 \text{ g Xe} \times \frac{1 \text{ mol Xe}}{131.3 \text{ g Xe}} = 7.49 \times 10^{-3} \text{ mol Xe}$$

Check your answers: Notice that the "grams" units cancel when the factor is multiplied, leaving the answer in moles.

55. *Define the problem*: Determine the quantity in moles from given mass in grams.

Develop a plan: Look up krypton on the periodic table to get the atomic weight (with an appropriate number of significant figures). Use that number for the molar mass (with units of grams per mole) as a conversion factor between grams and moles.

Execute the plan:

Krypton (Kr) has atomic number 36 on the periodic table. Its atomic weight is 83.80, so the molar mass is 83.80 g/mol.

$$0.00789 \text{ g Kr} \times \frac{1 \text{ mol Kr}}{83.80 \text{ g Kr}} = 9.42 \times 10^{-5} \text{ mol Kr}$$

Check your answer: Notice that the "grams" units cancel when the factor is multiplied, leaving the answer in moles.

57. *Define the problem*: A sample of sodium has a given density. Determine the volume of a cube of sodium and the length of each side.

Develop a plan: Always start with the sample; in this problem, start with the given moles. Use the molar mass of sodium as a conversion factor between moles and grams. Then use the density as a conversion factor between grams and cubic centimeters. Then take the cube-root of the volume to find the length of each side of the cube.

Execute the plan:

$$0.125 \text{ mol Na} \times \frac{22.99 \text{ g Na}}{1 \text{ mol Na}} \times \frac{1 \text{ cm}^3}{0.968 \text{ g Na}} = 2.97 \text{ cm}^3$$

$$\text{length} = \sqrt[3]{\text{volume}} = \sqrt[3]{2.97 \text{ cm}^3} = 1.45 \text{ cm}$$

Check your answer: The problem makes it sound like this cube will be manageable by a normal strength chemist with a simple knife. This cube's size doesn't look too big or too small. This answer looks right.

59. *Define the problem*: A ring of gold has a known mass. Determine the number of atoms in the sample.

Develop a plan: Start with the mass. Use the molar mass of gold as a conversion factor between grams and moles. Then use Avogadro's number as a conversion factor between moles of gold atoms and the actual number of gold atoms.

Execute the plan:

$$1.94 \text{ g Au} \times \frac{1 \text{ mol Au atoms}}{197.0 \text{ g Au}} \times \frac{6.022 \times 10^{23} \text{ Au atoms}}{1 \text{ mol Au atoms}} = 5.93 \times 10^{21} \text{ Au atoms}$$

Check your answer: A ring of gold is something that a person can see and hold. It will contain a very large number of atoms. This answer seems right.

61. *Define the problem*: Given the number of titanium atoms, determine the mass.

Develop a plan: Start with the quantity. Use Avogadro's number as a conversion factor between the number of titanium atoms and moles of titanium atoms. Then use the molar mass of titanium as a conversion factor between moles and grams.

Execute the plan:

$$1 \text{ Ti atom} \times \frac{1 \text{ mol Ti atoms}}{6.022 \times 10^{23} \text{ Ti atoms}} \times \frac{47.88 \text{ g Ti}}{1 \text{ mol Ti atoms}} = 7.951 \times 10^{-23} \text{ g Ti}$$

Check your answer: An atom of titanium is NOT something that a person can see or hold. It will have a very small mass. This answer seems right.

The Periodic Table

62. A group on the periodic table is the collection of elements that share the same vertical column, whereas a period on the periodic table is the collection of elements that share the same horizontal row.

70. The fourth period of the periodic table has eighteen elements. They are the following metals: potassium (K), calcium (Ca), scandium (Sc), titanium (Ti), vanadium (V), chromium (Cr), manganese (Mn), iron (Fe), cobalt (Co), nickel (Ni), copper (Cu), zinc (Zn), gallium (Ga); the following metalloids: germanium (Ge) and arsenic (As); and the following non-metals selenium (Se), bromine (Br), and krypton (Kr).

72. Two periods of the periodic table have eight elements (periods 2 and 3). Two periods of the periodic table have 18 elements (Periods 4 and 5). One period of the current periodic table has 32 elements (Period 6). Period 7 is currently six elements short of the total of 32 elements – this is not because there aren't 32, but because no one has yet isolated and characterized them.

74. There are many answers to this question. These are examples.

    (a) An element in Group 2B is zinc (Zn).

    (b) An element in the fifth period is xenon (Xe) (or any other element whose atomic number is within the range 37–54).

    (c) The element in the sixth period in Group 4A is lead (Pb).

    (d) The element in the third period in Group 6A is sulfur (S).

    (e) The alkali metal in the third period is sodium (Na).

(f) The noble gas in the fifth period is xenon (Xe).

(g) The element in Group 6A and the fourth period is selenium (Se). It is a non-metal.

(h) A metalloid in the fifth period is antimony (Sb) or tellurium (Te).

76. The chart showing the plot of relative abundance of the first 36 elements is given in Question 75. There is a general trend showing that the lighter elements are more abundant than the heavier ones. Looking more closely, even numbered elements are fairly consistently more abundant than the odd numbered elements on either side of them. (Both of these general trends have exceptions.)

General Questions

79. *Define the problem*: A distance is given in angstroms (Å), which are defined. Determine the distance in nanometers and picometers.

Develop a plan: Use the given relationship between angstroms and meters as a conversion factor to get from angstroms to meters. Then use the metric relationships between meters and the other two units to find the distance in nanometers and picometers.

Execute the plan:

$$1.97 \text{ Å} \times \frac{1 \times 10^{-10} \text{ m}}{1 \text{ Å}} \times \frac{1 \text{ nm}}{1 \times 10^{-9} \text{ m}} = 0.197 \text{ nm}$$

$$1.97 \text{ Å} \times \frac{1 \times 10^{-10} \text{ m}}{1 \text{ Å}} \times \frac{1 \text{ pm}}{1 \times 10^{-12} \text{ m}} = 197 \text{ pm}$$

Check your answers: The unit nanometer is larger than an angstrom, so the distance in nanometers should be a smaller number. The unit picometer is smaller than an angstrom, so the distance in picometers should be a larger number. These answers look right.

81. *Define the problem*: The edge length of a cube is given nanometers. Determine the volume of the cube in cubic nanometers and in cubic centimeters.

Develop a plan: Cube the edge length in nanometers to get the volume of the cube in cubic nanometers. Use metric relationships to convert nanometers into meters, then meters into centimeters. Cube the edge length in centimeters to get the volume of the cube in cubic centimeters.

Execute the plan:

$$V = (\text{edge length})^3 = (0.563 \text{ nm})^3 = 0.178 \text{ nm}^3$$

$$0.563 \text{ nm} \times \frac{1 \times 10^{-9} \text{ m}}{1 \text{ nm}} \times \frac{100 \text{ cm}}{1 \text{ m}} = 5.63 \times 10^{-8} \text{ cm}$$

$$V = (\text{edge length})^3 = \left(5.63 \times 10^{-8} \text{ cm}\right)^3 = 1.78 \times 10^{-22} \text{ cm}^3$$

Check your answers: Cubing fractional quantities makes the number smaller. The first volume makes sense. The unit centimeter is larger than an nanometer, so the volume in cubic centimeter should be a very small number. These answers look right.

83. (a) *Define the problem*: Given the volume of a bottle containing water, the density of water and ice (at different temperatures), determine the volume of ice.

    Develop a plan: Always start with the sample. Convert the volume of water to grams using the density of water. The entire mass of water freezes at the new temperature. Use the density of water as a conversion factor to determine the volume of ice.

    Execute the plan:

    $$250 \text{ mL water} \times \frac{0.997 \text{ g water}}{1 \text{ mL water}} \times \frac{1 \text{ g ice}}{1 \text{ g water}} \times \frac{1 \text{ mL ice}}{0.917 \text{ g ice}} = 272 \text{ mL ice}$$

    Check your answer: The density of ice is lower than the density of water, so the volume should be greater once the water freezes.

    (b) The ice could not be contained in the bottle. The volume of ice exceeds the capacity of the bottle.

85. *Define the problem*: Given the number of people in the city who use water, the volume of water each person needs per day, the number of days in a year, the mass of one gallon of water, the fluoride concentration, and the mass percentage of fluoride in sodium fluoride, determine the number of tons of sodium fluoride needed per year.

    Develop a plan: Always start with the sample. Given the number of people in the city, use the volume of water each person needs per day, then calculate everyone's water needs for the day. Using the number of days in a year calculate the total volume of water used per year. Then using the mass of one gallon of water as a conversion factor, determine the grams of water. Convert the grams to tons. Then, using the fluoride concentration as a conversion factor, determine the number of tons of fluoride, and using the mass percentage of fluoride in sodium fluoride to determine the number of tons.

Execute the plan:

$$150,000 \text{ people} \times \frac{\dfrac{175 \text{ gal water}}{1 \text{ day}} \times \dfrac{365 \text{ days}}{1 \text{ year}}}{1 \text{ person}} \times \frac{8.34 \text{ lb water}}{1 \text{ gal water}} \times \frac{1 \text{ ton water}}{2000 \text{ lb water}}$$

$$\times \frac{1 \text{ ton fluoride}}{1,000,000 \text{ tons water}} \times \frac{100 \text{ tons sodium fluoride}}{45.0 \text{ tons flouride}} = 89 \ \frac{\text{tons sodium fluoride}}{\text{year}}$$

Check your answer: The significant figures are limited to two by the 150,000 figure. The mass units are appropriately labeled. The units cancel appropriately to give tons per year. This is a large number of people using a large amount of water so the large quantity of sodium fluoride makes sense.

87. Potassium's atomic weight is 39.0983. The isotopes that contribute most to this mass are ^{39}K and ^{41}K, since the problem tells us that ^{40}K has a very low abundance. Since the atomic mass is closer to 39 than 41, that confirms that the ^{39}K isotope is more abundant.

89. (a) Ti has atomic number = 22 and atomic weight = 47.88.

    (b) Titanium is in Period 4 and Group 4B. The other elements in its group are zirconium (Zr), hafnium (Hf), and rutherfordium (Rf).

    (c) Titanium is light-weight and strong, making it a good choice for something that needs to be sturdy and small.

    (d) According to the New College Edition of the American Heritage Dictionary of the English Language (©1978): **Titanium**: A strong, low-density, highly corrosion resistant, lustrous white metallic element that occurs widely in igneous rocks and is used to alloy aircraft metals for low weight, strength, and high-temperature stability. (page 1348)

90. 1-2. Metal used in ancient times (Sn)

    3-4. Metal that burns in air and is found in Group 5A. (Bi)

    1-3. Two letter symbol for a metalloid. (Sb)

    2-4. Two letter symbol for metal used in US coins (Ni)

    1. A colorful non-metal (S)

    2. A colorless gaseous nonmetal (N)

    3. An element that makes fireworks green (B)

    4. An element that has medicinal uses (I)

    1-4. A two-letter symbol for an element used in electronics (Si)

2-3. A two-letter symbol for a metal used with Zr to make wires for superconducting magnets. (Nb)

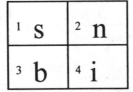

92. *Define the problem*: Given the carat mass of a diamond and the relationship between carat and milligrams, determine how many moles of carbon are in the diamond.

Develop a plan: Always start with the sample. Diamond is an allotropic form of pure carbon. Given the carats of the diamond, use the relationship between carats and milligrams as a conversion factor to determine milligrams of carbon. Then using metric relationships to determine grams of carbon, and the molar mass of carbon to determine the moles of carbon.

Execute the plan:

$$2.3 \text{ carats C} \times \frac{200 \text{ mg C}}{1 \text{ carat C}} \times \frac{1 \text{ g C}}{1000 \text{ mg C}} \times \frac{1 \text{ mol C}}{12.01 \text{ g C}} = 0.038 \text{ mol C}$$

Check your answer: The significant figures are limited to two by the 2.3 figure. The units cancel appropriately to give moles of carbon. This looks right.

94. *Define the problem*: Given the moles of gold you want to buy, the relationship between troy ounces and grams, and the price of a troy ounce of gold, determine the amount of money you must spend.

Develop a plan: The sample is the 1.00 mole of gold. Use the molar mass of gold to determine the grams of gold and the given relationship between grams and troy ounces to determine the troy ounces of gold, then use the price per troy ounce to determine how much that would cost.

Execute the plan:

$$1.00 \text{ mol Au} \times \frac{197.0 \text{ g Au}}{1 \text{ mol Au}} \times \frac{1 \text{ troy ounce Au}}{31.1 \text{ g Au}} \times \frac{\$338.70}{1 \text{ troy ounce Au}} = \$2,145 \approx \$2,150$$

Check your answer: The number of moles has three significant figures, so the answer must be reported with three significant figures. The units cancel properly. A mole of gold is pretty heavy, so this seems like a reasonable amount of money to spend.

96. *Define the problem*: Given the dimensions of a piece of copper wire and the density of copper, determine the moles of copper and the number of atoms of copper in the wire.

Develop a plan: Use the metric and English length relationships to convert the wire's dimensions to centimeters. Then use those dimensions to find its volume. Then use the density of copper to determine the mass of the wire. Then use the molar mass of copper to determine the moles of copper and Avogadro's number to determine the actual number of copper atoms.

Execute the plan:

$$\text{wire length in centimeters} = L = 25 \text{ ft long} \times \frac{12 \text{ in}}{1 \text{ ft}} \times \frac{2.54 \text{ cm}}{1 \text{ in}} = 7.6 \times 10^2 \text{ cm long}$$

$$\text{wire diameter in centimeters} = d = 2.0 \text{ mm diameter} \times \frac{1 \text{ m}}{1000 \text{ mm}} \times \frac{100 \text{ cm}}{1 \text{ m}} = 0.20 \text{ cm}$$

$$\text{wire radius in centimeters} = r = \frac{d}{2} = \frac{0.20 \text{ cm}}{2} = 0.10 \text{ cm}$$

$$\text{cylindrical wire's volume} = V = A \times L = (\pi r^2) \times L$$

$$V = (3.14159) \times (0.10 \text{ cm})^2 \times (7.6 \times 10^2 \text{ cm}) = 24 \text{ cm}^3$$

$$24 \text{ cm}^3 \text{ Cu} \times \frac{8.92 \text{ g Cu}}{1 \text{ cm}^3 \text{ Cu}} \times \frac{1 \text{ mol Cu atoms}}{63.55 \text{ g Cu}} = 3.4 \text{ mol Cu atoms}$$

$$3.4 \text{ mol Cu atoms} \times \frac{6.022 \times 10^{23} \text{ Cu atoms}}{1 \text{ mol Cu atoms}} = 2.0 \times 10^{24} \text{ Cu atoms}$$

Check your answers: The length and diameter of the wire both have two significant figures, so the answer must be reported with two significant figures. The units cancel properly. The number of atoms in a wire you can see and hold is conveniently represented in the quantity unit "moles". The actual number of atoms is very large. These answers make sense.

Applying Concepts

98. A particle is an atom or a molecule. The unit "mole" is a convenient way of describing a large quantity of particles. It is also important to keep in mind that 1 mol of particles contains 6.022×10^{23} particles and the molar mass describes the mass of 1 mol of particles.

(a) A sample containing 1 mol of Cl has **the same number** of particles as a sample containing 1 mol Cl_2, since each sample contains 1 mole of particles. (The masses of the samples are different, the numbers of atoms contained in the samples are different, but those statements do not answer this question.)

(b) A sample containing 1 mol of O_2 contains 6.022×10^{23} molecules of O_2. This 1-mol sample has **more** particles than a sample containing just 1 molecule of O_2.

(c) Each sample contains only one particle. These samples have **the same number** of particles.

(d) A sample containing 6.022×10^{23} molecules of F_2 contains 1 mol of F_2 molecules. This sample has **the same number** of particles as 6.022×10^{23} molecules of F_2.

(e) The molar mass of Ne is 20.18 g/mol, so a sample of 20.2 grams of neon contains 1 mol of neon. Note that, with the degree of certainty limited to one decimal place, the slightly smaller molar mass is indistinguishable from the sample mass. This 20.2-gram sample has **the same number** of particles as the sample with 1 mol of neon.

(f) The molar mass of bromine (Br_2) is (79.9 g/mol Br)$\times$(2 mol Br) = 159.8 g/mol, so a sample of 159.8 grams of bromine contains 1 mol bromine which equals 6.022×10^{23} molecules. This 159.8-gram sample has **more** particles than a sample containing just 1 molecule of Br_2.

(g) The molar mass of Ag is 107.9 g/mol, so a sample of 107.9 grams of Ag contains 1 mol of Ag. The molar mass of Li is 6.9 g/mol, so a sample of 6.9 grams of Li contains 1 mol of Li. These samples have **the same number** of particles, since each sample contains 1 mole of particles.

(h) The molar mass of Co is 58.9 g/mol, so a sample of 58.9 grams of Co contains 1 mol of Co. The molar mass of Cu is 63.55 g/mol, so a sample of 58.9 grams of Cu contains less than 1 mol of Cu. The 58.9-g sample of Co has **more** particles than the 58.9-g sample of Cu.

(i) A sample containing 6.022×10^{23} atoms of Ca contains 1 mol of Ca atoms. The molar mass of Ca is 40.1 g/mol, so a sample of 1 gram of Co contains less than 1 mol of Co. The 6.022×10^{23}-atoms sample has **more** particles than the 1-gram sample.

(j) A chlorine molecule contains two chlorine atoms. Since chlorine atoms all weigh the same, a chlorine molecule weighs twice as much as a chlorine atom. If the two samples of chlorine both weigh the same, the atomic chlorine (Cl) sample will have **more** particles in it than in the molecular sample (Cl_2).

99. The unit "mole" is a convenient way of describing a large quantity of particles. It is also important to keep in mind that 1 mol of particles contains 6.022×10^{23} particles and that the molar mass gives the grams in one mol of particles.

(a) The molar mass of iron (Fe) is 55.85 g/mol. The molar mass of aluminum (Al) is 27.0 g/mol, so a 1 mol sample of Fe has a **greater** mass than a 1 mol sample of Al.

(b) A sample of 6.022×10^{23} lead atoms contains 1 mol of lead. This sample will have **the same mass** as a sample containing 1 mol of lead.

(c) 1 mol of copper contains 6.022×10^{23} copper atoms. This sample will have a **greater** mass than a sample containing only 1 copper atom.

(d) A Cl_2 molecule has twice the mass of one Cl atom. So, comparing samples with the same number of particles, 1 mol, the Cl_2 sample will have a **greater** mass than the Cl sample.

(e) A "gram" is a unit of mass. If both samples weigh 1 gram, then they have **the same mass**.

(f) The molar mass of magnesium (Mg) is 24.3 g/mol, so a sample weighing 24.3 grams contains 1 mol of Mg. This sample will have **the same mass** as a sample containing 1 mol of Mg.

(g) The molar mass of Na is 23.0 g/mol, so a 1 mol sample of Na will weigh 23.0 grams. This sample has a **greater** mass than a sample containing 1 gram of Na.

(h) The molar mass of He is 4.0 g/mol, so a 1 mol sample of He weighs 4.0 grams. A sample of 6.022×10^{23} He atoms contains 1 mol of He. These two samples have **the same mass**.

(i) A sample with 1 mol of I_2 contains 6.022×10^{23} I_2 molecules. This sample **weighs more** than a sample that only contains 1 I_2 molecule.

(j) An oxygen molecule (O_2) has twice the mass of one O atom, so the O_2 sample will have a **greater** mass than the O sample.

34

Chapter 3: Chemical Compounds

Molecular and Structural Formulas

12.

	Molecular Formula	Condensed Formula	Structural Formula
butanol	$C_4H_{10}O$	$CH_3CH_2CH_2CH_2OH$	H—C—C—C—C—OH with H atoms on each carbon (H H H H on top, H H H H on bottom)
pentanol	$C_5H_{12}O$	$CH_3CH_2CH_2CH_2CH_2OH$	H—C—C—C—C—C—OH with H atoms on each carbon (H H H H H on top, H H H H H on bottom)

14. Sucrose, $C_{12}H_{22}O_{11}$, has eleven oxygen atoms per molecule. Glutathione, $C_{10}H_{17}N_3O_6S$, has only six oxygen atoms per molecule. Therefore sucrose has more oxygen atoms. Sucrose has (12 + 22 + 11=) 45 atoms, total. Glutathione has (10 + 17 + 3 + 6 + 1=) 37 atoms, total. Therefore sucrose has more atoms of all kinds.

17. *Note: Atoms in a formula found inside parentheses that are followed by a subscript get multiplied by that subscript.*

(a) CaC_2O_4 contains one atom of calcium, two atoms of carbon, and four atoms of oxygen.

(b) $C_6H_5CHCH_2$ contains eight atoms of carbon and eight atoms of hydrogen.

(c) $(NH_4)_2SO_4$ contains two (1 × 2) atoms of nitrogen, eight (4 × 2) atoms of hydrogen, one atom of sulfur, and four atoms of oxygen.

(d) $Pt(NH_3)_2Cl_2$ contains one atom of platinum, two (1 × 2) atoms of nitrogen, six (3 × 2) atoms of hydrogen, and two atoms of chlorine.

(e) $K_4Fe(CN)_6$ contains four atoms of potassium, one atom of iron, six (1×6) atoms of carbon, and six (1×6) atoms of nitrogen.

Predicting Ion Charges

21. A general rule for the charge on a metal cation: the group number represents the number of electrons lost. Hence, the group number will be the cation's positive charge.

(a) Lithium (Group 1A) Li^+

(b) Strontium (Group 2A) Sr^{2+}

(c) Aluminum (Group 3A) Al^{3+}

(d) Calcium (Group 2A) Ca^{2+}

(e) Zinc (Group 2B) Zn^{2+}

23. A general rule for the charge on a metal cation: the group number represents the number of electrons lost. Hence, the group number will be the cation's positive charge. For nonmetal elements in Groups 5A-7A, the electrons gained by an atom to form a stable anion are calculated using the formula: 8 – (group number). That means the (group number) – 8 is the negative charge of the anion.

Barium (Group 2A) has a +2 charge. Bromine (Group 7A) has a –1 charge. So, the ions are Ba^{2+} and Br^-.

25. For nonmetal elements in Groups 5A-7A, the electrons gained by an atom to form a stable anion are calculated using the formula: 8 – (group number). That means the (group number) – 8 is the negative charge of the anion. Transition metals often have a +2 charge. Some have +3 and +1 charged ions, as well.

(a) selenium (Group 6A) 6 – 8 = –2 Se^{2-}

(b) fluorine (Group 7A) 7 – 8 = –1 F^-

(c) nickel (a transition metal) Ni^{2+}

(d) nitrogen (Group 5A) 5 – 8 = –3 N^{3-}

27. The two compounds containing lead and chloride are made from the neutral combination of the charged ions:

one Pb^{2+} and two Cl^- [net charge = $+2 + 2 \times (-1) = 0$] $PbCl_2$

one Pb^{4+} and four Cl^- [net charge = $+4 + 4 \times (-1) = 0$] $PbCl_4$

29. (a) Calcium ion (from Group 2A) is Ca^{2+}. Oxide ion (from Group 6A) is O^{2-}.

    Ca_2O is **not** a neutral combination of these two ions. The proper formula would be CaO. [net charge $= +2 + (-2) = 0$]

   (b) Strontium ion (Group 2A) is Sr^{2+}. Chloride ion (from Group 7A) is Cl^-.

    $SrCl_2$ **is** the proper neutral combination of these two ions.

    [net charge $= +2 + 2 \times (-1) = 0$]

   (c) Iron ion (from the transition elements) is Fe^{3+} or Fe^{2+}. Oxide ion (from Group 6A) is O^{2-}. Fe_2O_5 is **not** a neutral combination of these ions. The proper possible formulas would be FeO [net charge $= +2 + (-2) = 0$] or

    Fe_2O_3 [net charge $= 2 \times (+3) + 3 \times (-2) = 0$]

   (d) Potassium ion (from Group 1A) is K^+. Oxide ion (from Group 6A) is O^{2-}.

    K_2O **is** the proper neutral combination of these two ions.

    [net charge $= 2 \times (+1) + (-2) = 0$]

Polyatomic Ions

31. (a) $Ca(CH_3CO_2)_2$ has one ion of calcium (Ca^{2+}) and two ions of acetate ($CH_3CO_2^-$ also written: CH_3COO^-).

   (b) $Co_2(SO_4)_3$ has two ions of cobalt(III) (Co^{3+}) and three ions of sulfate (SO_4^{2-}).

   (c) $Al(OH)_3$ has one ion of aluminum (Al^{3+}) and three ions of hydroxide (OH^-).

   (d) $(NH_4)_2CO_3$ has two ions of ammonium (NH_4^+) and one ion of carbonate (CO_3^{2-}).

33. (a) Nickel(II) nitrate $Ni(NO_3)_2$ (b) sodium bicarbonate $NaHCO_3$

   (c) Lithium hypochlorite $LiClO$ (d) magnesium chlorate $Mg(ClO_3)_2$

   (e) Calcium sulfite $CaSO_3$

Ionic Compounds

35. To tell if a compound is ionic or not, look for metals and nonmetals together, or common cations and anions. If a compound contains only nonmetals or metalloids and nonmetals, it is probably not ionic.

   (a) Methane, CH_4, contains only nonmetals. Not ionic.

(b) Dinitrogen pentoxide, N_2O_5, contains only nonmetals. Not ionic.

(c) Ammonium sulfide, $(NH_4)_2S$, has a common cation and anion together. Ionic.

(d) Hydrogen selenide, H_2Se, contains a metalloid and a nonmetal. Not ionic.

(e) Sodium perchlorate, $NaClO_4$, has a metal and a common anion together. Ionic.

37. (a) Calcium hydrogen carbonate, $Ca(HCO_3)_2$

(b) Potassium permanganate, $KMnO_4$

(c) Magnesium perchlorate, $Mg(ClO_4)_2$

(d) Ammonium hydrogen phosphate, $(NH_4)_2HPO_4$

39. (a) $Ca(CH_3CO_2)_2$ is calcium acetate.

(b) $Co_2(SO_4)_3$ is cobalt(III) sulfate.

(c) $Al(OH)_3$ is aluminum hydroxide.

41. Magnesium oxide is MgO, and it is composed of Mg^{2+} ions and O^{2-} ions. The relatively high melting temperature of MgO compared to NaCl (composed of Na^+ ions and Cl^- ions) is probably due to the higher ionic charges and smaller sizes of the ions. The large opposite charges sitting close together have very strong attractive forces between the ions. Melting requires that these attractive forces be overcome.

Electrolytes

43. An electrolyte is a compound that conducts electricity when dissolved in water. Electricity is conducted when ions are present in the solution. When a strong electrolyte (such as NaCl) dissolves in water it will ionize completely (producing Na^+ and Cl^-). When a weak electrolyte (such as acetic acid, CH_3COOH) dissolves in water it ionizes only a small amount (most of what is in the solution is the molecular form, CH_3COOH, with only small amounts of the ions, CH_3COO^- and H^+)

45. "Molecular compounds are generally non-electrolytes." This general trend is sensible, since the molecular compounds are generally not ionic compounds, and therefore would not ionize in water.

47. (a) The ions present in a solution of KOH are K^+ and OH^-.

(b) The ions present in a solution of K_2SO_4 are K^+ and SO_4^{2-}.

(c) The ions present in a solution of $NaNO_3$ are Na^+ and NO_3^-.

(d) The ions present in a solution of NH_4Cl are NH_4^+ and Cl^-.

Moles of Compounds

49. Consider a sample of 1 mol of methanol (M).

	CH_3OH	Carbon	Hydrogen	Oxygen
No. of moles	1 mol	1 mol	4 mol	1 mol
No. of molecules or atoms	6.022×10^{23} molecules	6.022×10^{23} atoms	2.409×10^{24} atoms	6.022×10^{23} atoms
Molar mass	32.042 g/mol M	12.011 g/mol M	4.0316 g/mol M	15.9994 g/mol M

51. To calculate the molar mass of a compound, we perform a series of steps: First, look up the atomic weight of each element in the compound and identify the atomic weight as the molar mass of each element. Second, determine the number of moles of atoms in one mole of the compound. Third, multiply the number of moles of the element by the molar mass of the element. Last, add up all the individual mass contributions.

(a) 1 mol of Fe_2O_3 contains 2 mol of Fe and 3 mol of O.

$$\left(\frac{2 \text{ mol Fe}}{1 \text{ mol Fe}_2O_3} \times \frac{55.847 \text{ g}}{1 \text{ mol Fe}} \right) + \left(\frac{3 \text{ mol O}}{1 \text{ mol Fe}_2O_3} \times \frac{15.9994 \text{ g}}{1 \text{ mol O}} \right) = 159.692 \frac{\text{g}}{\text{mol Fe}_2O_3}$$

(b) 1 mol of BF_3 contains 1 mol of B and 3 mol of F.

$$\left(\frac{1 \text{ mol B}}{1 \text{ mol BF}_3} \times \frac{10.811 \text{ g}}{1 \text{ mol B}} \right) + \left(\frac{3 \text{ mol F}}{1 \text{ mol BF}_3} \times \frac{18.9984 \text{ g}}{1 \text{ mol F}} \right) = 67.806 \frac{\text{g}}{\text{mol BF}_3}$$

(c) 1 mol of N_2O contains 2 mol of N and 1 mol of O.

$$\left(\frac{2 \text{ mol N}}{1 \text{ mol N}_2O} \times \frac{14.0067 \text{ g}}{1 \text{ mol N}} \right) + \left(\frac{1 \text{ mol O}}{1 \text{ mol N}_2O} \times \frac{15.9994 \text{ g}}{1 \text{ mol O}} \right) = 44.0128 \frac{\text{g}}{\text{mol N}_2O}$$

(d) 1 mol of $MnCl_2 \cdot 4 H_2O$ compound has 1 mol of $MnCl_2$ with 4 mol of H_2O molecules. So, it contains 1 mol Mn, 2 mol Cl, $(4 \times 2 =)$ 8 mol H and $(4 \times 1 =)$ 4 mol O.

$$\left(\frac{1 \text{ mol Mn}}{1 \text{ mol comp}} \times \frac{54.938 \text{ g}}{1 \text{ mol Mn}} \right) + \left(\frac{2 \text{ mol Cl}}{1 \text{ mol comp}} \times \frac{35.4537 \text{ g}}{1 \text{ mol Cl}} \right)$$

$$+ \left(\frac{8 \text{ mol H}}{1 \text{ mol comp}} \times \frac{1.0079 \text{ g}}{1 \text{ mol H}} \right) + \left(\frac{4 \text{ mol O}}{1 \text{ mol comp}} \times \frac{15.9994 \text{ g}}{1 \text{ mol O}} \right) = 197.906 \frac{\text{g}}{\text{mol comp}}$$

(e) 1 mol of $C_6H_8O_6$ compound contains 6 mol of C, 8 mol of H, and 6 mol of O.

$$\left(\frac{6 \text{ mol C}}{1 \text{ mol comp}} \times \frac{12.011 \text{ g}}{1 \text{ mol C}}\right) + \left(\frac{8 \text{ mol H}}{1 \text{ mol comp}} \times \frac{1.0079 \text{ g}}{1 \text{ mol H}}\right) + \left(\frac{6 \text{ mol O}}{1 \text{ mol comp}} \times \frac{15.9994 \text{ g}}{1 \text{ mol O}}\right)$$

$$= 176.126 \frac{\text{g}}{\text{mol comp}}$$

53. *Define the problem:* Determine the number of moles in a given mass of a compound.

Develop a plan: Adapt the method described in the answer to Question 51 to calculate the molar mass for the compound, then use the molar mass as a conversion factor between grams and moles.

Execute the plan:

(a) Molar mass $CH_3OH = (12.011 \text{ g}) + 4 \times (1.0079 \text{ g}) + (15.9994 \text{ g}) = 32.042$ g/mol

$$1.00 \text{ g } CH_3OH \times \frac{1 \text{ mol } CH_3OH}{32.042 \text{ g } CH_3OH} = 0.0312 \text{ mol } CH_3OH$$

(b) Molar mass $Cl_2CO = 2 \times (35.4527 \text{ g}) + (12.011 \text{ g}) + (15.9994 \text{ g}) = 98.916$ g/mol

$$1.00 \text{ g } Cl_2CO \times \frac{1 \text{ mol } Cl_2CO}{98.916 \text{ g } Cl_2CO} = 0.0101 \text{ mol } Cl_2CO$$

(c) Molar mass $NH_4NO_3 = 2 \times (14.0067 \text{ g}) + 4 \times (1.0079 \text{ g}) + 3 \times (15.9994 \text{ g})$

$$= 80.043 \text{ g/mol}$$

$$1.00 \text{ g } NH_4NO_3 \times \frac{1 \text{ mol } NH_4NO_3}{80.043 \text{ g } NH_4NO_3} = 0.0125 \text{ mol } NH_4NO_3$$

(d) Molar mass $MgSO_4 \cdot 7 H_2O$

$$= (24.305 \text{ g}) + (32.066 \text{ g}) + 11 \times (15.9994 \text{ g}) + 14 \times (1.0079 \text{ g}) = 246.475 \text{ g/mol}$$

$$1.00 \text{ g } MgSO_4 \cdot 7H_2O \times \frac{1 \text{ mol } MgSO_4 \cdot 7H_2O}{246.475 \text{ g } MgSO_4 \cdot 7H_2O} = 0.00406 \text{ mol } MgSO_4 \cdot 7H_2O$$

(e) Molar mass $AgCH_3CO_2$

$$= (107.868 \text{ g}) + 2 \times (12.011 \text{ g}) + 3 \times (1.0079 \text{ g}) + 2 \times (15.9994 \text{ g}) = 166.913 \text{ g/mol}$$

$$1.00 \text{ g } AgCH_3CO_2 \times \frac{1 \text{ mol } AgCH_3CO_2}{166.913 \text{ g } AgCH_3CO_2} = 0.00599 \text{ mol } AgCH_3CO_2$$

Check your answers: The quantity in moles is always going to be smaller than the mass in grams. These numbers look right.

57. *Define the problem:* Given the masses of three compounds in a mixture, determine the number of moles of each, then determine the number of molecules of one compound.

Develop a plan: Adapt the method described in the answer to Question 51 to calculate the molar mass for the compounds. Convert milligrams to grams, then use the molar mass as a conversion factor between grams and moles. Use Avogadro's number to determine the actual number of molecules.

Execute the plan:

(a) Molar mass $C_9H_8O_4$ = 9 × (12.011 g) + 8 × (1.0079 g) + 4 × (15.9994 g)

$$= 180.160 \text{ g/mol}$$

Molar mass $NaHCO_3$ = (22.98977 g) + (1.0079 g) + (12.011 g) + 3 × (15.9994 g)

$$= 84.007 \text{ g/mol}$$

Molar mass $C_6H_8O_7$ = 6 × (12.011 g) + 8 × (1.0079 g) + 7 × (15.9994 g)

$$= 192.125 \text{ g/mol}$$

$$324 \text{ mg} \, C_9H_8O_4 \times \frac{1 \text{ g } C_9H_8O_4}{1000 \text{ mg } C_9H_8O_4} \times \frac{1 \text{ mol } C_9H_8O_4}{180.160 \text{ g } C_9H_8O_4}$$

$$= 0.00180 \text{ mol } C_9H_8O_4$$

$$1904 \text{ mg } NaHCO_3 \times \frac{1 \text{ g } NaHCO_3}{1000 \text{ mg } NaHCO_3} \times \frac{1 \text{ mol } NaHCO_3}{84.007 \text{ g } NaHCO_3}$$

$$= 0.02266 \text{ mol } NaHCO_3$$

$$1000. \text{ mg } C_6H_8O_7 \times \frac{1 \text{ g } C_6H_8O_7}{1000 \text{ mg } C_6H_8O_7} \times \frac{1 \text{ mol } C_6H_8O_7}{192.125 \text{ g } C_6H_8O_7}$$

$$= 0.005205 \text{ mol } C_6H_8O_7$$

(b) $$0.00180 \text{ mol } C_9H_8O_4 \times \frac{6.022 \times 10^{23} \, C_9H_8O_4 \text{ molecules}}{1 \text{ mol } C_9H_8O_4}$$

$$= 1.08 \times 10^{21} \, C_9H_8O_4 \text{ molecules}$$

Check your answers: The quantity in moles is always going to be smaller than the mass in grams or milligrams. The number of molecules for a macroscopic sample will be huge. These numbers look right.

59. *Define the problem:* Given the mass of a compound, determine the number of moles of that compound, the number of molecules of that compound, and the number of atoms of both elements.

Develop a plan: Adapt the method described in the answer to Question 51 to calculate the molar mass for the compound. Convert pounds to grams, then use the molar mass as a conversion factor between grams and moles, then use Avogadro's number to determine the actual number of molecules, then use the formula stoichiometry to determine the number of atoms of each type.

Execute the plan:

$$\text{Molar mass } SO_3 = (32.066 \text{ g}) + 3 \times (15.9994 \text{ g}) = 80.064 \text{ g/mol}$$

(a) $$1.00 \text{ lb } SO_3 \times \frac{453.6 \text{ g } SO_3}{1 \text{ lb } SO_3} \times \frac{1 \text{ mol } SO_3}{80.064 \text{ g } SO_3} = 5.67 \text{ mol } SO_3$$

(b) $$5.67 \text{ mol } SO_3 \times \frac{6.022 \times 10^{23} \text{ molecules } SO_3}{1 \text{ mol } SO_3} = 3.41 \times 10^{24} \text{ molecules } SO_3$$

(c) Stoichiometry of the chemical formula: Each molecule of SO_3 contains one atom of S.

$$3.41 \times 10^{24} \text{ molecules } SO_3 \times \frac{1 \text{ S atom}}{1 \, SO_3 \text{ molecule}} = 3.41 \times 10^{24} \text{ S atoms}$$

(d) Stoichiometry of the chemical formula: Each molecule of SO_3 contains three atom of O.

$$3.41 \times 10^{24} \text{ molecules } SO_3 \times \frac{3 \text{ O atom}}{1 \, SO_3 \text{ molecule}} = 1.02 \times 10^{25} \text{ O atoms}$$

Check your answers: The number of molecules for a macroscopic sample will be huge. The atom ratio of sulfur to oxygen in the compound SO_3 is 1:3, so the number of atoms of O will be three times greater than the number of atoms of S. These numbers look right.

60. *Define the problem:* Given the mass of a compound, determine the number of moles of that compound, and the number of atoms of one of the elements.

Develop a plan: Adapt the method described in the answer to Question 51 to calculate the molar mass for the compound. Use the molar mass as a conversion factor between grams and moles, then use Avogadro's number to determine number of molecules, then use the formula stoichiometry to determine number of atoms of each type.

Execute the plan: Molar mass CF_3CH_2F

$$= 2 \times (12.011 \text{ g}) + 4 \times (18.9984 \text{ g}) + 2 \times (1.0079 \text{ g}) = 102.031 \text{ g/mol}$$

(a) $25.5 \text{ g CF}_3\text{CH}_2\text{F} \times \dfrac{1 \text{ mol CF}_3\text{CH}_2\text{F}}{102.031 \text{ g CF}_3\text{CH}_2\text{F}} = 0.250 \text{ mol CF}_3\text{CH}_2\text{F}$

(b) Stoichiometry of the chemical formula: Each mol of CF_3CH_2F contains 4 mol of F atoms.

$$0.250 \text{ mol CF}_3\text{CH}_2\text{F} \times \dfrac{4 \text{ mol F}}{1 \text{ mol CF}_3\text{CH}_2\text{F}} \times \dfrac{6.022 \times 10^{23} \text{ F atoms}}{1 \text{ mol F}}$$

$$= 6.02 \times 10^{23} \text{ F atoms}$$

Check your answers: The number of atoms in a macroscopic sample will be huge. The mass of a substance will always be larger than the number of moles. These numbers look right.

61. *Define the problem:* Given the volume of a compound and its density, determine the number of molecules of the compound.

Develop a plan: Adapt the method described in the answer to Question 51 to calculate the molar mass for the compound. Use the density to convert the volume from milliliters to grams, then use the molar mass as a conversion factor between grams and moles, then use Avogadro's number to determine the number of molecules.

Execute the plan:

$$\text{Molar mass H}_2\text{O} = 2 \times (1.0079 \text{ g}) + (15.9994 \text{ g}) = 18.0152 \text{ g/mol}$$

$$0.050 \text{ mL H}_2\text{O} \times \dfrac{1.0 \text{ g H}_2\text{O}}{1 \text{ mL H}_2\text{O}} \times \dfrac{1 \text{ mol H}_2\text{O}}{18.0152 \text{ g H}_2\text{O}} \times \dfrac{6.022 \times 10^{23} \text{ H}_2\text{O molecules}}{1 \text{ mol H}_2\text{O}}$$

$$= 1.7 \times 10^{21} \text{ H}_2\text{O molecules}$$

Check your answer: The number of atoms in a macroscopic sample will be huge. These numbers look right.

Percent Composition

63. *Define the problem:* Given the formula of a compound, determine the molar mass, and the weight percent of each element

Develop a plan: Calculate the mass of each element in one mole of compound, while calculating the molar mass of the compound. Divide the calculated mass of the element by the molar mass of the compound and multiply by 100 % to get weight percent. To get the last element's weight percent, subtract the other percentages from 100 %.

Execute the plan:

(a)

$$\text{Mass of Pb per mole of PbS} = 207.2 \text{ g Pb}$$

$$\text{Mass of S per mole of PbS} = 32.066 \text{ g S}$$

$$\text{Molar mass PbS} = (207.2 \text{ g}) + (32.066 \text{ g}) = 239.3 \text{ g/mol PbS}$$

$$\% \text{ Pb} = \frac{\text{mass of Pb per mol PbS}}{\text{mass of PbS per mol PbS}} \times 100\% = \frac{207.2 \text{ g Pb}}{239.3 \text{ g PbS}} \times 100\%$$

$$= 86.60 \% \text{ Pb in PbS}$$

$$\% \text{ S} = 100 \% - 86.60 \% \text{ Pb} = 13.40 \% \text{ S in PbS}$$

(b)

$$\text{Mass of C per mole of } C_2H_6 = 2 \times (12.011 \text{ g}) = 24.022 \text{ g C}$$

$$\text{Mass of H per mole of } C_2H_6 = 6 \times (1.0079 \text{ g}) = 6.0474 \text{ g H}$$

$$\text{Molar mass } C_2H_6 = (24.022 \text{ g}) + (6.0474 \text{ g}) = 30.069 \text{ g/mol } C_2H_6$$

$$\% \text{ C} = \frac{\text{mass of C / mol } C_2H_6}{\text{mass of } C_2H_6 / \text{mol } C_2H_6} \times 100 \% = \frac{24.022 \text{ g C}}{30.069 \text{ g } C_2H_6} \times 100 \%$$

$$= 79.889 \% \text{ C in } C_2H_6$$

$$\% \text{ H} = 100 \% - 79.889 \% \text{ C} = 20.111 \% \text{ H in } C_2H_6$$

(c)

$$\text{Mass of C per mole of } CH_3CO_2H = 2 \times (12.011 \text{ g}) = 24.022 \text{ g C}$$

$$\text{Mass of H per mole of } CH_3CO_2H = 4 \times (1.0079 \text{ g}) = 4.0316 \text{ g H}$$

$$\text{Mass of O per mole of } CH_3CO_2H = 2 \times (15.9994 \text{ g}) = 31.9988 \text{ g O}$$

$$\text{Molar mass } CH_3CO_2H = (24.022 \text{ g}) + (4.0316 \text{ g}) + (31.9988 \text{ g})$$

$$= 60.052 \text{ g/mol } CH_3CO_2H$$

$$\% \text{ C} = \frac{\text{mass of C / mol } CH_3CO_2H}{\text{mass of } CH_3CO_2H / \text{mol } CH_3CO_2H} \times 100 \%$$

$$= \frac{24.022 \text{ g C}}{60.052 \text{ g } CH_3CO_2H} \times 100 \% = 40.002 \% \text{ C in } CH_3CO_2H$$

$$\% \text{ H} = \frac{\text{mass of H / mol } CH_3CO_2H}{\text{mass of } CH_3CO_2H / \text{ mol } CH_3CO_2H} \times 100 \%$$

$$= \frac{4.0316 \text{ g H}}{60.052 \text{ g } CH_3CO_2H} \times 100 \% = 6.7135 \% \text{ H in } CH_3CO_2H$$

$\%\ O = 100\ \% - 40.002\ \%\ C - 6.7135\ \%\ H = 53.285\ \%\ O$ in CH_3CO_2H

(d)   Mass of N per mole of $NH_4NO_3 = 2 \times (14.0067\ g) = 28.0134\ g\ N$

Mass of H per mole of $NH_4NO_3 = 4 \times (1.0079\ g) = 4.0316\ g\ H$

Mass of O per mole of $NH_4NO_3 = 3 \times (15.9994\ g) = 47.9982\ g\ O$

Molar mass $NH_4NO_3 = (28.0134\ g) + (4.0316\ g) + (47.9982\ g)$

$$= 80.0432\ g/mol\ NH_4NO_3$$

$$\%\ N = \frac{\text{mass of N/mol } NH_4NO_3}{\text{mass of } NH_4NO_3\ /mol\ NH_4NO_3} \times 100\ \% = \frac{28.0134\ g\ N}{80.0432\ g\ NH_4NO_3} \times 100\ \%$$

$$= 34.9979\ \%\ N\ \text{in } NH_4NO_3$$

$$\%\ H = \frac{\text{mass of H/mol } NH_4NO_3}{\text{mass of } NH_4NO_3\ /mol\ NH_4NO_3} \times 100\ \% = \frac{4.0316\ g\ H}{80.0432\ g\ NH_4NO_3} \times 100\ \%$$

$$= 5.0368\ \%\ H\ \text{in } NH_4NO_3$$

$\%\ O = 100\ \% - 34.9979\ \%\ C - 5.0368\ \%\ H = 59.9654\ \%\ O$ in NH_4NO_3

Note: When masses of different things are used in the same problem, make sure your units clearly specify what each mass refers to.

Check your answers: Calculating the last element's weight percent using the formula gives the same answer as subtracting the other percentages from 100 %. These answers are consistent.

65. *Define the problem:* Given the weight percent of one compound, M_2O, containing one known element, O, and one unknown element, M, calculate the percent by weight of another compound, MO.

Develop a plan: Choose a convenient sample mass of M_2O, such as 100.0 g. Find the mass of M and O in the sample, using the given weight percent. Using the molar mass of oxygen as a conversion factor, determine the number of moles of oxygen, then using the formula stoichiometry of M_2O as a conversion factor determine the number of moles of M. Find the molar mass of M by dividing the mass of M by the moles of M. Use the molar mass of M, and the formula stoichiometry of MO, to determine the weight percent of M in MO.

Execute the plan:

73.4 % M in M_2O means that 100.0 grams of M_2O contains 73.4 grams of M.

$$\text{Mass of O} = 100.0 \text{ g } M_2O - 73.4 \text{ g M} = 26.6 \text{ g O}$$

Formula Stoichiometry: 1 mol of M_2O contains 2 mol M and 1 mol O.

$$26.6 \text{ g O} \times \frac{1 \text{ mol O}}{15.9994 \text{ g O}} \times \frac{2 \text{ mol M}}{1 \text{ mol O}} = 3.33 \text{ mol M}$$

$$\text{Molar Mass of M} = \frac{\text{mass of M in sample}}{\text{mol of M in sample}} = \frac{73.4 \text{ g M}}{3.33 \text{ mol M}} = 22.1 \frac{\text{g}}{\text{mol}}$$

$$\text{Molar mass of MO} = 22.1 \text{ g} + 15.9994 \text{ g} = 38.07 \text{ g/mol}$$

$$\%M = \frac{\text{mass of M / mol MO}}{\text{mass of MO / mol MO}} \times 100\% = \frac{22.1 \text{ g M}}{38.07 \text{ g MO}} \times 100\% = 58.0 \text{ \% M in MO}$$

Check your answer: It makes sense that the compound with more atoms of M has a higher weight percent of M. The closest element to M's atomic mass (22.1) is sodium (atomic mass = 22.99). If M is sodium, the two compounds would probably be sodium oxide (Na_2O) and sodium peroxide (Na_2O_2, a compound made up of two Na^+ ions and one O_2^{2-} ion. The simple ratio of Na and O atoms in this compound is 1:1). The results make sense.

67. *Define the problem:* Given the percent by mass of an element in an enzyme and the number of atoms of that element in one molecule of the enzyme, determine the molar mass of the enzyme.

Develop a plan: Choose a convenient sample of the enzyme, such as 100.0 g. Using the percent by mass, determine the number of grams of Mo in the sample. Use the molar mass of Mo as a conversion factor to get the moles of Mo. Use the formula stoichiometry as a conversion factor to get the moles of enzyme. Determine the molar mass by dividing the grams of enzyme in the sample, by the moles of enzyme in the sample.

Execute the plan: The enzyme contains 0.0872 % Mo by mass, this means that 100.0 g of enzyme contains 0.0872 grams Mo.

Formula stoichiometry: One molecule of enzyme contains 2 atoms of Mo, so 1 mol of enzyme molecules contains 2 mol of Mo atoms.

$$0.0872 \text{ g Mo} \times \frac{1 \text{ mol Mo}}{95.94 \text{ g Mo}} \times \frac{1 \text{ mol enzyme}}{2 \text{ mol Mo}} = 4.54 \times 10^{-4} \text{ mol enzyme}$$

$$\text{Molar Mass of enzyme} = \frac{\text{mass of enzyme in sample}}{\text{moles of enzyme in sample}} = \frac{100.0000 \text{ g enzyme}}{4.54 \times 10^{-4} \text{ mol enzyme}}$$

$$= 2.20 \times 10^5 \text{ g/mol}$$

Check your answer: Enzymes are large molecules with large molar masses. This answer makes sense.

69. *Define the problem:* Given the percent by mass of an element in a compound and the compound's formula with an unknown subscript, determine the value of the unknown subscript.

Develop a plan: Choose a convenient sample of Si_2H_x, such as 100.00 g. Using the percent by mass, determine the number of grams of Si and H in the sample. Use the molar mass of Si as a conversion factor to get the moles of Si. Use the molar mass of H as a conversion factor to get the moles of H. Set up a mole ratio to determine the value of x.

Execute the plan: The compound is 90.28 % Si by mass. This means that 100.00 g of Si_2H_x contains 90.28 grams Si and the rest of the mass is from H.

Mass of H in sample = 100.00 g Si_2H_x − 90.28 g Si = 9.72 g H

$$90.28 \text{ g Si} \times \frac{1 \text{ mol Si}}{28.0855 \text{ g Si}} = 3.214 \text{ mol Si} \qquad 9.72 \text{ g H} \times \frac{1 \text{ mol H}}{1.0079 \text{ g H}} = 9.64 \text{ mol H}$$

$$\text{Molar Ratio} = \frac{\text{moles of H in sample}}{\text{moles of Si in sample}} = \frac{9.64 \text{ mol H}}{3.214 \text{ mol Si}} = \frac{3 \text{ mol H}}{1 \text{ mol Si}} = \frac{6 \text{ mol H}}{2 \text{ mol Si}}$$

Therefore, the formula is Si_2H_6 and x = 6.

Check your answer: The molar ratio is clearly a whole number relationship indicating a sensible number of hydrogen atoms in this molecule. The answer makes sense.

71. Most of the percentages given on food labels are related to percent daily value (%DV) based on a 2000 calorie diet. Food calories vary depending on the types of food substance. For example, on a 2000 calorie/day diet, it is recommended that you eat no more than 300 grams of carbohydrates per day. The masses of additives, vitamin, and mineral also vary in the daily value. For example, on a 2000 calorie/day diet, it is recommended that you eat no more than 2400 milligrams of sodium (in the form of salt) per day.

For example, here is some of the information on a can of Campbell's Tomato Soup:

	Amount per serving	%DV
Total Carb.	19 g	6 %
Fiber	1 g	4 %
Sodium	710 mg	30 %
Vitamin A	-	10 %
Vitamin C	-	10 %
Calcium	-	2 %
Iron	-	4 %

Notice that 19 grams of carbohydrates is 6 % by mass (to one significant figure) of 300 grams, and that 710 milligrams is 30 % by mass (to two significant figures) of 2400 milligrams.

Empirical and Molecular Formula

73. An empirical formula shows the simplest whole number ratio of the elements in a compound. The molecular formula gives the actual number of atoms of each element in one formula unit of the compound. For ethane, C_2H_6 is the molecular formula. CH_3 is the empirical formula, since 6 is exactly divisible by 2 to give the smallest whole number ratio of 1:3.

75. *Define the problem:* Given the empirical formula of a compound and the molar mass, determine the molecular formula.

Develop a plan: Find the mass of 1 mol of the empirical formula. Divide the molar mass of the compound by the calculated empirical mass to get a whole number. Multiply all the subscripts in the empirical formula by this whole number.

Execute the plan: The empirical formula is C_2H_4NO, the molecular formula is $(C_2H_4NO)_n$.

Mass of 1 mol $C_2H_4NO = 2 \times (12.011 \text{ g}) + 4 \times (1.0079 \text{ g}) + 14.0067 \text{ g} + 15.9994 \text{ g}$

$$= 58.060 \text{ g / mol}$$

$$n = \frac{\text{mass of 1 mol of molecule}}{\text{mass of 1 mol of } C_2H_4NO} = \frac{116.4 \text{ g}}{58.060 \text{ g}} = 2.005 \approx 2$$

Molecular Formula is $(C_2H_4NO)_2 = C_4H_8N_2O_2$

Check your answer: The molar mass is about 2 times larger than the mass of one mole of the empirical formula, so the molecular formula $C_4H_8N_2O_2$ makes sense.

77. *Define the problem:* Given the percent by mass of one element in a compound containing two elements with unknown subscripts and a list of possible empirical formulas, determine the empirical formula.

Develop a plan: Choose a convenient sample of B_xH_y, such as 100.0 g. Using the percent by mass, determine the number of grams of B and H in the sample. Use the molar masses of B and H as conversion factors to get the moles of B and H. Set up a mole ratio to determine the y/x ratio. Compare to the list given to select the empirical formula that has the closest mole ratio.

Execute the plan: The compound is 88.5 % B by mass. This means that 100.0 g of B_xH_y contains 88.5 grams B and the rest of the mass is from H.

Mass of H in sample = 100.00 g B_xH_y − 88.5 g B = 11.5 g H

$$88.5 \text{ g B} \times \frac{1 \text{ mol B}}{10.81 \text{ g B}} = 8.19 \text{ mol B} \qquad 11.5 \text{ g H} \times \frac{1 \text{ mol H}}{1.0079 \text{ g H}} = 11.41 \text{ mol H}$$

$$\text{Molar Ratio} = \frac{\text{moles of H in sample}}{\text{moles of B in sample}} = \frac{11.41 \text{ mol H}}{8.19 \text{ mol B}} = 1.39 \approx \frac{7 \text{ mol H}}{5 \text{ mol B}}$$

Therefore, the formula is B_5H_7.

Check your answer: The other possible formulas' mole ratios were 2.5, 2.2, and 2.0, so the 1.4 ratio (from 7/5) was the closest to 1.39. The answer makes sense.

79. *Define the problem:* Given the percent by mass of one element in a compound containing two elements with unknown subscripts and the molar mass, determine the empirical formula and the molecular formula.

Develop a plan: Choose a convenient sample of C_xH_y, such as 100.00 g. Using the percent by mass, determine the number of grams of C and H in the sample. Use the molar masses of C and H as conversion factors to get the moles of C and H. Set up a mole ratio to determine the whole number y/x ratio for the empirical formula. Find the mass of 1 mol of the empirical formula. Divide the molar mass of the compound by the calculated empirical mass to get a whole number. Multiply all the subscripts in the empirical formula by this whole number.

Execute the plan: The compound is 89.94 % C by mass. This means that 100.00 g of C_xH_y contains 89.94 grams C and the rest of the mass is from H.

Mass of H in sample = 100.00 g C_xH_y − 89.94 g C = 10.06 g H

Find moles of C and H in the sample:

$$89.94 \text{ g C} \times \frac{1 \text{ mol C}}{12.011 \text{ g C}} = 7.488 \text{ mol C} \qquad 10.06 \text{ g H} \times \frac{1 \text{ mol H}}{1.0079 \text{ g H}} = 9.981 \text{ mol H}$$

$$\text{Mole ratio} = \frac{9.981 \text{ mol H}}{7.488 \text{ mol C}} = 1.333 = \frac{4 \text{ mol H}}{3 \text{ mol C}}$$

The empirical formula is C_3H_4, so the molecular formula is $(C_3H_4)_n$.

Mass of 1 mol C_3H_4 = 3 × (12.011 g) + 4 × (1.0079 g) = 40.065 g / mol

$$n = \frac{\text{mass of 1 mol of molecule}}{\text{mass of 1 mol of } C_3H_4} = \frac{120.2 \text{ g}}{40.065 \text{ g}} = 3.000 \approx 3$$

Molecular Formula is $(C_3H_4)_3 = C_9H_{12}$

Check your answer: The mole ratio of C and H in the sample is very close to $\frac{4}{3}$, so the empirical formula makes sense. The molar mass is 3 times larger than the mass of one mole of the empirical formula, so the molecular formula C_9H_{12} makes sense.

81. *Define the problem:* Given the percent by mass of all the elements in a compound and the molar mass, determine the empirical formula and the molecular formula.

Develop a plan: Choose a convenient sample of $C_xH_yN_z$, such as 100.00 g. Using the percent by mass, determine the number of grams of C, H, and N in the sample. Use the molar masses of C, H, and N as conversion factors to get the moles of C, H and N. Set up a mole ratio to determine the whole number x:y:z ratio for the empirical formula. Find the mass of 1 mol of the empirical formula. Divide the molar mass of the compound by the calculated empirical mass to get a whole number. Multiply all the subscripts in the empirical formula by this whole number.

Execute the plan:

(a) The compound is 74.0 % C, 8.65 % H, and 17.35 % N by mass. This means that 100.00 g of $C_xH_yN_z$ contains 74.0 grams C, 8.65 grams H, and 17.35 grams N.

Find moles of C, H, and N in the sample:

$$74.0 \text{ g C} \times \frac{1 \text{ mol C}}{12.011 \text{ g C}} = 6.16 \text{ mol C} \qquad 8.65 \text{ g H} \times \frac{1 \text{ mol H}}{1.0079 \text{ g H}} = 8.58 \text{ mol H}$$

$$17.35 \text{ g N} \times \frac{1 \text{ mol N}}{14.0067 \text{ g N}} = 1.239 \text{ mol N}$$

Mole ratio: 6.16 mol C : 8.58 mol H : 1.239 mol N

Divide each term by the smallest number of moles, 1.239 mol

Atom ratio: 5 C : 7 H : 1 N

(b) The molecular formula is $(C_5H_7N)_n$.

Mass of 1 mol $C_5H_7N = 5 \times (12.011 \text{ g}) + 7 \times (1.0079 \text{ g}) + 14.0067 \text{ g} = 81.117 \text{ g / mol}$

$$n = \frac{\text{mass of 1 mol of molecule}}{\text{mass of 1 mol of } C_5H_7N} = \frac{162 \text{ g}}{81.117 \text{ g}} = 2.00 \approx 2$$

Molecular Formula is $(C_5H_7N)_2 = C_{10}H_{14}N_2$

Check your answers: The mole ratios of C, H, and N in the sample are close to integer values, so the empirical formula makes sense. The molar mass is twice the mass of one mole of the empirical formula, so the molecular formula $C_{10}H_{14}N_2$ makes sense.

83. *Define the problem:* Given the percent by mass of all the elements in a compound and the molar mass, determine the molecular formula.

Develop a plan: Choose a convenient sample of $C_xH_yN_z$, such as 100.00 g. Using the percent by mass, determine the number of grams of C, H, and N in the sample. Use the molar masses of C, H, and N as conversion factors to get the moles of C, H and N. Set up a mole ratio to determine the whole number x:y:z ratio for the empirical formula. Find the mass of 1 mol of the empirical formula. Divide the molar mass of the compound by the calculated empirical mass to get a whole number. Multiply all the subscripts in the empirical formula by this whole number.

Execute the plan:

The compound is 58.77 % C, 13.81 % H, and 27.42 % N by mass. This means that 100.00 g of $C_xH_yN_z$ contains 58.77 grams C, 13.81 grams H, and 27.42 grams N.

Find moles of C, H, and N in the sample:

$$58.77 \text{ g C} \times \frac{1 \text{ mol C}}{12.011 \text{ g C}} = 4.893 \text{ mol C} \qquad 13.81 \text{ g H} \times \frac{1 \text{ mol H}}{1.0079 \text{ g H}} = 13.70 \text{ mol H}$$

$$27.42 \text{ g N} \times \frac{1 \text{ mol N}}{14.0067 \text{ g N}} = 1.958 \text{ mol N}$$

Mole ratio: 4.893 mol C : 13.70 mol H : 1.958 mol N

Divide each term by the smallest number of moles, 1.958 mol, to get atom ratio:

Atom ratio: 2.5 C : 7 H : 1 N

Multiply each term by 2 to get whole number atom ratio:

Whole number atom ratio: 5 C : 14 H : 2 N

The empirical formula is $C_5H_{14}N_2$, so the molecular formula is $(C_5H_{14}N_2)_n$.

Mass of 1 mol $C_5H_{14}N_2$ = 5 × (12.011 g) + 14 × (1.0079 g) + 2 × (14.0067 g)

$$= 102.179 \text{ g / mol}$$

$$n = \frac{\text{mass of 1 mol of molecule}}{\text{mass of 1 mol of } C_5H_{14}N_2} = \frac{102.2 \text{ g}}{102.179 \text{ g}} = 1.000 \approx 1$$

Molecular Formula is $(C_5H_{14}N_2)_1 = C_5H_{14}N_2$

Check your answer: The mole ratios of C, H, and N in the sample are close to integer values, so the empirical formula makes sense. The molar mass is almost the same as the mass of one mole of the empirical formula, so the molecular formula $C_5H_{14}N_2$ makes sense.

85. *Define the problem:* The mass of a sample of a hydrate compound is given. The formula of the hydrate is known except for the amount of water in it. All the water is dried out of the sample using high temperature, leaving a mass of completely dehydrated salt. Determine out the number of water molecules in the formula of the hydrate compound.

Develop a plan: Use the molar mass of the dehydrated compound as a conversion factor to convert the mass of the dehydrated compound into moles. Since the only thing lost was water, a mole relationship can be established between the dehydrated and hydrated compound. In addition, the difference between the mass of the hydrated compound and the mass of the dehydrated compound gives the amount of water lost by the sample. Convert that water into moles, using the molar mass of water. Divide the moles of water by the moles of hydrate, to determine how many moles of water are in one mole of hydrate compound.

Execute the plan:

1.687 g of hydrated compound, $MgSO_4 \cdot xH_2O$, is dehydrated to make 0.824 g of the dehydrated compound, $MgSO_4$.

Molar Mass $MgSO_4$ = 24.305 g + 32.066 g + 4 × (15.9994 g) = 120.369 g/mol

Molar Mass H_2O = 2 × (1.0079 g) + 2 × (15.9994 g) = 18.0152 g/mol

Find moles of $MgSO_4 \cdot xH_2O$ in the sample:

$$0.824 \text{ g } MgSO_4 \times \frac{1 \text{ mol } MgSO_4}{120.369 \text{ g } MgSO_4} \times \frac{1 \text{ mol } MgSO_4 \cdot xH_2O}{1 \text{ mol } MgSO_4}$$

$$= 6.85 \times 10^{-3} \text{ mol } MgSO_4 \cdot xH_2O$$

Find mass of water lost by the sample:

$$= (1.687 \text{ g } MgSO_4 \cdot xH_2O) - (0.824 \text{ g } MgSO_4) = 0.863 \text{ g } H_2O$$

Find moles of H_2O lost from sample:

$$0.863 \text{ g } H_2O \times \frac{1 \text{ mol } H_2O}{18.0152 \text{ g } H_2O} = 0.0479 \text{ mol } H_2O$$

$$\text{Mole ratio} = \frac{\text{mol } H_2O \text{ from sample}}{\text{mol hydrate is sample}} = \frac{0.0479 \text{ mol } H_2O}{6.85 \times 10^{-3} \text{ mol hydrate}} = 7.00 \approx 7$$

The proper formula for the hydrate is $MgSO_4 \cdot 7H_2O$ and x = 7.

Check your answer: The formula for epsom salt was given earlier in Chapter 3 (Question 44). It had 7 water molecules then, so it better have 7 water molecules, now!

Biological Periodic Table

87. According to Table 3.11, the ten elements most abundant in the human body are: H, O, C, N, Ca, P, Cl, S, Na, K

89. (a) Metals are found in the body as ions.

(b) Two uses for metals in the body are calcium (Ca^{2+}) in bones and Fe^{2+} in hemoglobin. There are many others.

91. An essential mineral that is toxic at high levels is selenium. (See the "Chemistry in the News" box on page 107). This is the only one mentioned in the text. Most heavy metals such as iron and copper and nonmetals such as iodine are safe and necessary, but they are toxic at high concentrations.

Carbohydrates and Fats

93. (a) The monosaccharide that is most important to humans is glucose, the building block of carbohydrates.

(b) The ring structure and name:

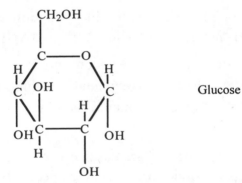

Glucose

(c) Plants make glucose and store it as starch. The glucose is used in plants to form cellulose for structural support. Animals rely on glucose as their major source of energy – we metabolize it to produce energy needed to stay alive.

General Questions

96. (a) Trinitrotoluene, TNT, has seven C atoms, six in the ring and one more in the –CH_3 at the top of the structure. It has five H atoms, two on ring carbons and three more in the –CH_3 at the top of the structure. It has three N atoms (one in each of the three –NO_2 groups attached to the ring carbons). It also has six O atoms (two in each of the three –NO_2 groups attached to the ring carbons). So, the molecular formula is $C_7H_5N_3O_6$.

(b) Serine has three carbon atoms. It has three hydrogen atoms attached to carbons, two H atoms in the two –OH groups, and two H atoms in the –NH$_2$ groups. It has one N atom. It has two O atoms in the –OH groups and one more in the =O. So, the molecular formula is C$_3$H$_7$NO$_3$.

98. (a) Chlorine (Cl) and bromine (Br) are **not** likely to form an ionic compound, since they are both nonmetals in Group 7A.

(b) Lithium (Li) and tellurium (Te) **might** make an ionic compound. Lithium is a metal and tellurium is a metalloid. The likely compound contains ions Li$^+$ (Group 1A cation; charge is +1) and Te^{2-} (Group 6A anion; charge is –2). The compound's formula will be **Li$_2$Te**.

(c) Sodium (Na) and argon (Ar) are **not** likely to form an ionic compound, since argon is in Group 8A. Those elements a very unreactive and do not form ions at all.

(d) Magnesium (Mg) and fluorine (F) **will** make an ionic compound. Magnesium is a metal and fluorine is a nonmetal. The likely compound contains ions Mg^{2+} (Group 2A cation; charge is +2) and F$^-$ (Group 7A anion; charge is –1). The compound's formula will be **MgF$_2$**.

(e) Nitrogen (N) and bromine (Br) are **not** likely to form an ionic compound, since they are both nonmetals in Groups 5A and 7A, respectively.

(f) Indium (In) and sulfur (S) **will** make an ionic compound. Indium is a metal and sulfur is a nonmetal. The likely compound contains ions In^{3+} (Group 3A cation; charge is +3) and S^{2-} (Group 6A anion; charge is –2). The compound's formula will be **In$_2$S$_3$**.

(g) Selenium (Se) and bromine (Br) are **not** likely to form an ionic compound, since they are both nonmetals in Groups 6A and 7A, respectively.

100. (a) sodium hypochlorite (common cation, Na$^+$, and anion, ClO$^-$; ionic): NaClO

(b) aluminum perchlorate (common cation, Al^{3+}, and anion, ClO$_4^-$; ionic): Al(ClO$_4$)$_3$

(c) potassium permanganate (common cation, K$^+$, and anion, MnO$_4^-$; ionic): KMnO$_4$

(d) potassium dihydrogen phosphate (common cation, K$^+$, and anion, H$_2$PO$_4^-$; ionic): KH$_2$PO$_4$

(e) chlorine trifluoride (binary molecular compound since both elements are nonmetals; not ionic). ClF$_3$

(f) boron tribromide (binary molecular compound since both elements are nonmetals; not ionic). BBr$_3$

(g) calcium acetate (common cation, Ca^{2+}, and anion, $CH_3CO_2^-$ or CH_3COO^-; ionic):
$Ca(CH_3CO_2)_2$ or $Ca(CH_3COO)_2$ *Either one is correct.*

(h) sodium sulfite (common cation, Na^+, and anion, SO_3^{2-}; ionic): Na_2SO_3

(i) disulfur tetrachloride (binary molecular compound since both elements are nonmetals; not ionic). S_2Cl_4

(j) phosphorus trifluoride (binary molecular compound since both elements are nonmetals; not ionic). PF_3

102. *Define the problem:* Given the moles of a cube of lithium and the density of lithium, determine the volume of the cube in cubic centimeters and the length of one edge of the cube.

Develop a plan: Start with the sample – the number of moles in the cube. Use the molar mass of lithium (Li) as a conversion factor to get the number of grams of Li. Then use the density as a conversion factor to get the volume in cubic centimeters. Use the relationship between the edges of a cube and the volume of a cube to find the length of one edge of the cubic block.

Execute the plan:

(a) $256 \text{ mol Li} \times \dfrac{6.941 \text{ g Li}}{1 \text{ mol Li}} \times \dfrac{1 \text{ cm}^3 \text{ Li}}{0.534 \text{ g Li}} = 3.33 \times 10^3 \text{ cm}^3 \text{ Li}$

(b) $V = (x \text{ cm})^3$

$3.33 \times 10^3 \text{ cm}^3 = (x \text{ cm})^3$

$x \text{ cm} = \sqrt[3]{3.33 \times 10^3 \text{ cm}^3} = 14.9 \text{ cm}$ is the length of one side

Check your answer: The question made it sound like this cube might not fit on the starship. Even this large number of moles only occupies a cubic space that has 15 cm (about 6 inches) per side. That's not very big. Rechecking the math, though, confirms that these numbers are calculated correctly; therefore, we don't have to write the authors of Star Trek and tell them to redesign their ships.

104. *Define the problem:* Given the percent by mass of carbon in a hydrocarbon compound and the molar mass, determine the empirical formula and the molecular formula.

Develop a plan: A hydrocarbon contains just carbon and hydrogen. Choose a convenient sample of C_xH_y, such as 100.00 g. Using the percent by mass of carbon, determine percent by mass of hydrogen, and the number of grams of C and H in the sample. Use the molar masses of C and H as conversion factors to get the moles of C and H. Set up a mole ratio to determine the whole number x:y ratio for the empirical formula. Find the mass of 1 mol of the empirical formula. Divide the molar mass of the compound by the calculated empirical mass to get a whole number. Multiply all the subscripts in the empirical formula by this whole number.

Execute the plan:

(a) The compound is 93.71 % C and the rest is hydrogen. This means that 100.00 g of C_xH_y contains 93.71 grams C and the mass of hydrogen is calculated like this:

$$100.00 \text{ g } C_xH_y - 93.71 \text{ g C} = 6.29 \text{ g H}$$

Find moles of C and H in the sample:

$$93.71 \text{ g C} \times \frac{1 \text{ mol C}}{12.011 \text{ g C}} = 7.802 \text{ mol C} \qquad 6.29 \text{ g H} \times \frac{1 \text{ mol H}}{1.0079 \text{ g H}} = 6.24 \text{ mol H}$$

Mole ratio: 7.802 mol C : 6.24 mol H

Divide each term by the smallest number of moles, 6.24 mol

Atom ratio: 1.25 C : 1 H

Multiply by 4 (since $4 \times 1.25 = 5$) to get the whole number ratio:

Atom ratio: 5 C : 4 H

The empirical formula is C_5H_4

(b) The molecular formula is $(C_5H_4)_n$.

Mass of 1 mol C_5H_4 = $5 \times (12.011 \text{ g}) + 4 \times (1.0079 \text{ g}) = 64.087$ g/mol

$$n = \frac{\text{mass of 1 mol of molecule}}{\text{mass of 1 mol of } C_5H_4} = \frac{128.16 \text{ g}}{64.087 \text{ g}} = 1.9998 \approx 2$$

Molecular Formula is $(C_5H_4)_2 = C_{10}H_8$

Check your answers: The simplest ratio of C and H in the sample ends up very close to whole number values, so the empirical formula makes sense. The molar mass is very close to double the mass of one mole of the empirical formula, so the molecular formula $C_{10}H_8$ makes sense.

106. *Define the problem:* Given the mass of an element used to make a mass of a compound and the molar mass of the compound, determine the empirical formula and the molecular formula of the compound.

Develop a plan: Use the mass of the compound, I_xCl_y, and the mass of I_2 to determine the mass of Cl_2 used. Use the molar masses of I_2 and Cl_2 and the stoichiometry of their formulas to determine the moles of I and moles of Cl in the sample of the compound. Set up a mole ratio to determine the whole number x:y ratio for the empirical formula. Find the mass of 1 mol of the empirical formula. Divide the molar mass of the compound by the calculated empirical mass to get a whole number. Multiply all the subscripts in the empirical formula by this whole number.

Execute the plan:

(a) The I_xCl_y sample has a mass of 1.246 grams. This sample was produced using 0.678 grams of iodine. The rest of the mass is from chlorine.

$$1.246 \text{ g } I_xCl_y - 0.678 \text{ g iodine} = 0.568 \text{ g chlorine}$$

Molar mass of $I_2 = 2 \times (126.90 \text{ g}) = 253.80$ g/mol

Molar mass of $Cl_2 = 2 \times (35.4527 \text{ g}) = 70.9054$ g/mol

Find moles of I and Cl in the sample:

$$0.678 \text{ g } I_2 \times \frac{1 \text{ mol } I_2}{253.80 \text{ g } I_2} \times \frac{2 \text{ mol I}}{1 \text{ mol } I_2} = 5.34 \times 10^{-3} \text{ mol I}$$

$$0.568 \text{ g } Cl_2 \times \frac{1 \text{ mol } Cl_2}{70.9054 \text{ g } Cl_2} \times \frac{2 \text{ mol Cl}}{1 \text{ mol } Cl_2} = 1.60 \times 10^{-2} \text{ mol Cl}$$

Mole ratio: 5.34×10^{-3} mol I : 1.60×10^{-2} mol Cl

Divide each term by the smallest number of moles, 5.34×10^{-3} mol, and round to whole numbers:

$$1 \text{ I} : 3 \text{ Cl}$$

The empirical formula is ICl_3

(b) The molecular formula is $(ICl_3)_n$.

Mass of 1 mol $ICl_3 = 126.90 \text{ g} + 3 \times (35.4527 \text{ g}) = 233.26$ g/mol

$$n = \frac{\text{mass of 1 mol of molecule}}{\text{mass of 1 mol of } ICl_3} = \frac{467 \text{ g}}{233.26 \text{ g}} = 2.00 \approx 2$$

Molecular Formula is $(ICl_3)_2 = I_2Cl_6$

Check your answer: The simplest ratio of I and Cl in the sample ends up very close to whole number values, so the empirical formula makes sense. The molar mass is twice the mass of one mole of the empirical formula, so the molecular formula I_2Cl_6 makes sense.

108. *Define the problem:* Given the mass of an iron pyrite sample in kilograms and the formula of iron pyrite, determine the mass of one element in the sample.

Develop a plan: Always start with the sample – in this case, the kilograms of the FeS_2. Use metric conversions to determine the sample mass in grams. Then use the molar mass of FeS_2 to determine the moles of FeS_2 in the sample. Then use the formula stoichiometry to get the moles of Fe in the sample. Then use the molar mass of Fe to get the grams of Fe in the sample. Use metric conversions to determine the Fe mass in kilograms.

Execute the plan: The sample is composed of 15.8 kg of FeS_2.

Molar mass of FeS_2 = 55.847 g + 2 × (32.066 g) = 119.979 g/mol

Find grams of Fe in the sample:

$$15.8 \text{ kg FeS}_2 \times \frac{1000 \text{ g FeS}_2}{1 \text{ kg FeS}_2} \times \frac{1 \text{ mol FeS}_2}{119.979 \text{ g FeS}_2} \times \frac{1 \text{ mol Fe}}{1 \text{ mol FeS}_2} \times \frac{55.847 \text{ g Fe}}{1 \text{ mol Fe}}$$

$$\times \frac{1 \text{ kg Fe}}{1000 \text{ g Fe}} = 7.35 \text{ kg Fe}$$

Notice that the molar mass and stoichiometry can be interpreted in kilograms and kilomoles, removing the redundant metric conversions and making the problem shorter:

$$15.8 \text{ kg FeS}_2 \times \frac{1 \text{ kmol FeS}_2}{119.979 \text{ kg FeS}_2} \times \frac{1 \text{ kmol Fe}}{1 \text{ kmol FeS}_2} \times \frac{55.847 \text{ kg Fe}}{1 \text{ kmol Fe}} = 7.35 \text{ kg Fe}$$

Check your answer: The mass percentage of Fe in FeS_2 is around 50 %. So, it makes sense that the mass of Fe in the sample is about half of the sample mass. This number looks right.

110. *Define the problem:* Given the mass of a sample of an ore, the mass percent of antimony in the ore, the fact that the ore contains the compound stibnite, and the formula of stibnite (Sb_2S_3), determine the mass in grams of Sb_2S_3 in the sample of ore.

Develop a plan: Always start with the sample – in this case, 1.00 pound of ore. Use English to metric conversion factor to determine the sample mass in grams. Then use the mass percent of Sb as a conversion factor to determine the mass of Sb in the sample. Then use the molar mass of Sb to find the moles of Sb in the sample. Then use the formula stoichiometry of stibnite to determine the moles of stibnite, Sb_2S_3. Then use the molar mass of Sb_2S_3 to get the grams of Sb_2S_3 in the sample.

Execute the plan: It is not clear how precisely the sample's mass is measured. Mass measurements are pretty easy to do precisely, so we'll use three significant figures, the limit of the other given information.

The ore is 10.6 % Sb. That means, 100.0 grams of ore contains 10.6 grams Sb.

Formula Stoichiometry: 2 mol Sb is contained in 1 mol Sb_2S_3.

Molar mass of $Sb_2S_3 = 2 \times (121.757 \text{ g}) + 3 \times (32.066 \text{ g}) = 339.712$ g/mol

$$1.00 \text{ lb ore} \times \frac{453.59 \text{ g ore}}{1 \text{ lb ore}} \times \frac{10.6 \text{ g Sb}}{100.0 \text{ g ore}} \times \frac{1 \text{ mol Sb}}{121.757 \text{ g Sb}} \times \frac{1 \text{ mol Sb}_2\text{S}_3}{2 \text{ mol Sb}}$$

$$\times \frac{339.712 \text{ g Sb}_2\text{S}_3}{1 \text{ mol Sb}_2\text{S}_3} = 67.1 \text{ g Sb}_2\text{S}_3$$

Check your answer: Sb represents about $\frac{2}{3}$ of the mass of Sb_2S_3. This calculated mass of Sb_2S_3 is about 0.15 pounds, or 15 % of the ore sample mass. Since the %Sb in the ore is 10 %, then approximately $\frac{3}{2}(10 \text{ %})$ of the ore will be Sb_2S_3. This number looks right.

112. *Define the problem:* Given the dimensions of a sample of nickel foil, the density of nickel metal, and the mass of a nickel fluoride compound (Ni_xF_y) produced using that foil as a source of nickel, determine the moles of nickel, and the formula of the nickel fluoride compound.

Develop a plan: Always start with the sample – in this case, the dimensions of the Ni foil sample. Use metric conversion factors and the relationship between length, width, and height and volume to determine the sample's volume. Then use the density to determine the mass of Ni in the sample. Subtract the mass of Ni from the mass of the compound produced to find the mass of F in the sample. Then use the molar masses of Ni and F to find the moles of Ni and moles of F in the sample. Then set up a mole ratio and find the simplest whole number relationship between the atoms.

Execute the plan: The foil is 0.550 mm thick, 1.25 cm long and 1.25 cm wide.

0.550mm thick → 1.25 cm long, 1.25 cm wide

$$V = (\text{thickness}) \times (\text{length}) \times (\text{width})$$

$$V = \left(0.550 \text{ mm} \times \frac{1 \text{ m}}{1000 \text{ mm}} \times \frac{100 \text{ cm}}{1 \text{ m}} \right) \times (1.25 \text{ cm}) \times (1.25 \text{ cm}) = 0.0859 \text{ cm}^3$$

(a) $$0.0859 \text{ cm}^3 \text{ Ni} \times \frac{8.908 \text{ g Ni}}{1 \text{ cm}^3 \text{ Ni}} \times \frac{1 \text{ mol Ni}}{58.69 \text{ g Ni}} = 0.0130 \text{ mol Ni}$$

(b) $$0.0859 \text{ cm}^3 \text{ Ni} \times \frac{8.908 \text{ g Ni}}{1 \text{ cm}^3 \text{ Ni}} = 0.765 \text{ g Ni}$$

Mass of sulfur in the sample = Ni_xS_y compound mass – nickel mass

$$1.261 \text{ g Ni}_xS_y - 0.765 \text{ g Ni} = 0.496 \text{ g F}$$

Find moles of F in sample:

$$0.496 \text{ g F} \times \frac{1 \text{ mol F}}{18.9984 \text{ g F}} = 0.0261 \text{ mol F}$$

Set up mole ratio: 0.0130 mol Ni : 0.0261 mol F

Divide all the numbers in the ratio by the smallest number of moles, 0.0130 mol, and round to whole numbers: 1 Ni : 2 F

Empirical formula: NiF_2

(c) NiF_2 is called nickel(II) fluoride. (It looks like it is composed of nickel(II) ion, Ni^{2+}, and fluoride ion, F^-)

Check your answers: The mole ratio was very close to a whole number ratio, so the empirical formula of NiF_2 makes sense. One of the common ionic forms of nickel is nickel(II) ion, so this also confirms that the result is sensible.

115. (a) Solid lithium nitrate has alternating lattice of Li^+ and NO_3^- ions.

(b) Molten lithium nitrate has ion pairs randomly distributed.

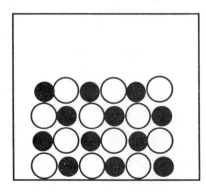

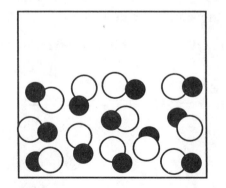

(c)   Molten lithium nitrate when positive and negative electrodes are present will have the anions, NO_3^-, (white circles) crowding around the positive electrode and the cations, Li^+, (the black circles) crowding around the negative electrode:

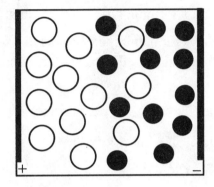

(d) Lithium nitrate solid in the presence of lithium nitrate in water:

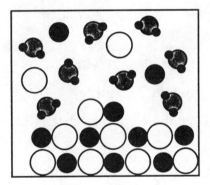

Applying Concepts

117. The student did not properly identify the primary chain. This is a pentane chain (five carbons bonded in a row; see numbers in boxes) with two methyl groups.

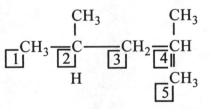

Its name is 2,4-dimethylpentane.

119. Thallium nitrate is $TlNO_3$. Since NO_3^- has a -1 charge, that means that thallium ion has a $+1$ charge, and is represented by Tl^+. The carbonate compound containing thallium will be a combination of Tl^+ and CO_3^{2-}, and the compound's formula will look like this: Tl_2CO_3. The sulfate compound containing thallium will be a combination of Tl^+ and SO_4^{2-}, and the compound's formula will look like this: Tl_2SO_4.

121. (a) Based on the guidelines for naming oxyanions in a series, relate the chloro-oxyanions to the bromo-oxyanions, since chlorine and bromine are in the same group (Group 7A).

    BrO_4^- is perbromate (like ClO_4^- is perchlorate)

    BrO_3^- is bromate (like ClO_3^- is chlorate)

    BrO_2^- is bromite (like ClO_2^- is chlorite)

    BrO^- is hypobromite (like ClO^- is hypochlorite)

    (b) Based on the guidelines for naming oxyanions in a series, relate the sulfur-oxyanions to the selenium-oxyanions.

    SeO_4^{2-} is selenate (like SO_4^{2-} is sulfate)

    SeO_3^{2-} is selenite (like SO_3^{2-} is sulfite)

123. *Define the problem:* Given three samples, determine which has the largest quantity of NH_3.

    Develop a plan: To compare these samples, put them in the same units. Convert all samples into moles.

    Execute the plan:

    (a) 6.022×10^{24} molecules $NH_3 \times \dfrac{1 \text{ mol } NH_3}{6.022 \times 10^{23} \text{ molecules } NH_3} = 10.00$ mol NH_3

    (c) 17.03 g $NH_3 \times \dfrac{1 \text{ mol } NH_3}{17.030 \text{ g } NH_3} = 1.000$ mol NH_3

    Sample (a) has a larger amount of NH_3. (a) 10.00 mol > (c) 1.000 mol > (b) 0.10 mol.

    Check your answer: The first sample is larger than Avogadro's number. The third sample weighs the same as the molar mass. This order makes sense.

Chapter 4: Quantities of Reactants and Products

Stoichiometry

14. *Define the problem:* Given the balanced equation for a reaction, identify the stoichiometric coefficients in this equation, and relate the quantity of products to reactants and vice versa.

Develop a plan: (a) The law of conservation of mass says that the total mass of the reactants is the same as the total mass of the products. (b) The stoichiometric coefficients are the numbers in front of each formula in the equation. (c) The stoichiometric coefficients can be interpreted as the related number of reactants. Use the formula stoichiometry of O_2 to find the number of molecules that reacted. Use equation stoichiometry to find out how many atoms of Mg reacted with that many O_2 molecules.

Execute the plan: First, balance the equation: $2\,Mg(s) + O_2(g) \longrightarrow 2\,MgO(g)$

(a) The total mass of product, MgO, is 1.00 g. The total mass of the reactants (mass of Mg plus mass of O_2) that reacted must also be 1.00 g, due to the conservation of mass.

(b) The stoichiometric coefficients are 2 for Mg, 1 for O_2, and 2 for MgO.

(c) 50 atoms of oxygen make up 25 molecules of O_2, since there are two O atoms in each molecule. Since the stoichiometry is 1:1, that means 50 atoms of Mg were needed to react with this much oxygen.

Check your answers: Though it is an unnecessary calculation, we can check the answer to part (a) by calculating the masses of Mg and O_2 that reacted to make 1.00 gram of MgO,

$$1.00 \text{ g MgO} \times \frac{1 \text{ mol MgO}}{40.30 \text{ g MgO}} \times \frac{1 \text{ mol Mg}}{1 \text{ mol MgO}} \times \frac{24.305 \text{ g Mg}}{1 \text{ mol Mg}} = 0.603 \text{ g Mg}$$

$$1.00 \text{ g MgO} \times \frac{1 \text{ mol MgO}}{40.30 \text{ g MgO}} \times \frac{1 \text{ mol O}_2}{2 \text{ mol MgO}} \times \frac{32.00 \text{ g O}_2}{1 \text{ mol O}_2} = 0.397 \text{ g O}_2$$

then adding them up: $0.603 \text{ g} + 0.397 \text{ g} = 1.000 \text{ g}$. This answer looks right.

16. In the first box, three A_2 molecules are combined with six B atoms. After the reaction occurs, the box contains only six AB molecules. That means symbolically:

$$3 \text{ A}_2 + 6 \text{ B} \longrightarrow 6 \text{ AB}$$

Therefore, the equation given in (b), $\text{A}_2 + 2 \text{ B} \longrightarrow 2 \text{ AB}$, shows the right stoichiometric relationship between reactants and products.

18. Balance Cl atoms (put a "3" in front of Cl_2 and a "2" in front of SbCl_3, so each side of the equation has 6 Cl atoms, $3 \times 2 = 2 \times 3 = 6$). Balance Sb atoms (put a 2 in front of Sb, so each side of the equation has 2 Sb atoms): $2 \text{ Sb} + 3 \text{ Cl}_2 \longrightarrow 2 \text{ SbCl}_3$.

Box (c) represents the correct products.

Classification of Chemical Reactions

19. *Define the problem:* Write the balanced formation equations.

Develop a plan: Write the formula of the compound as the product, and write the reactants in the form of elements. *Note, some elements are in the form of diatomic molecules (e.g., O_2).* Balance the equations.

Execute the plan:

(a) Carbon monoxide = CO(g). $2 \text{ C(s)} + \text{O}_2\text{(g)} \longrightarrow 2 \text{ CO(g)}$

(b) Nickel(II) oxide = NiO(s). $2 \text{ Ni(s)} + \text{O}_2\text{(g)} \longrightarrow 2 \text{ NiO(s)}$

(c) Chromium(III) oxide = Cr_2O_3(s). $4 \text{ Cr(s)} + 3 \text{ O}_2\text{(g)} \longrightarrow 2 \text{ Cr}_2\text{O}_3\text{(s)}$

Check your answers: (a) 2 C & 2 O (b) 2 Ni & 2 O (c) 4 Cr & 6 O

21. *Define the problem:* Write balanced decomposition equations for some carbonate salts.

Develop a plan: Assume that the products of decomposition are oxides and carbon dioxide as described in the text. Write the formula of the compound as the reactant and determine the formula of the metal oxide. Write the products in the form of the metal oxide and carbon dioxide. Balance the equations.

Execute the plan:

(a) $\text{BeCO}_3\text{(s)} \longrightarrow \text{BeO(s)} + \text{CO}_2\text{(g)}$

beryllium oxide, carbon dioxide

(b) $\text{NiCO}_3\text{(s)} \longrightarrow \text{NiO(s)} + \text{CO}_2\text{(g)}$

nickel(II) oxide, carbon dioxide

(c) $\text{Al}_2(\text{CO}_3)_3\text{(s)} \longrightarrow \text{Al}_2\text{O}_3\text{(s)} + 3 \text{ CO}_2\text{(g)}$

aluminum oxide, carbon dioxide

Check your answers: (a) 1 Be, 1 C, & 3 O (b) 1 Ni, 1 C, & 3 O (c) 2 Al, 3 C & 9 O

23. *Define the problem:* Write the balanced combustion equations.

Develop a plan: The products of combustion depend on the elements present in the compound being combusted:

Compound Contains:	C	H
Combustion Product is:	CO_2	H_2O

The compound and O_2 are the reactant. The products are CO_2 and H_2O for these combustion reactions. Balance the equations, starting with C, then H, then O. If the coefficient of O_2 ends up a fraction, multiply everything in the reaction by two, to make all coefficients whole numbers.

Execute the plan:

(a) Unbalanced: $? C_4H_{10}(g) + ? O_2(g) \longrightarrow ? CO_2(g) + ? H_2O(g)$

     Balance C: $C_4H_{10}(g) + ? O_2(g) \longrightarrow 4 CO_2(g) + ? H_2O(g)$ 4 C's

     Balance H: $C_4H_{10}(g) + ? O_2(g) \longrightarrow 4 CO_2(g) + 5 H_2O(g)$ 10 H's

     Balance O: $C_4H_{10}(g) + \dfrac{13}{2} O_2(g) \longrightarrow 4 CO_2(g) + 5 H_2O(g)$ 13 O's

     Multiply by 2: $2 C_4H_{10}(g) + 13 O_2(g) \longrightarrow 8 CO_2(g) + 10 H_2O(g)$

(b) Unbalanced: $? C_6H_{12}O_6(s) + ? O_2(g) \longrightarrow ? CO_2(g) + ? H_2O(g)$

     Balance C: $C_6H_{12}O_6(s) + ? O_2(g) \longrightarrow 6 CO_2(g) + ? H_2O(g)$ 6 C's

     Balance H: $C_6H_{12}O_6(s) + ? O_2(g) \longrightarrow 6 CO_2(g) + 6 H_2O(g)$ 12 H's

     Balance O: $C_6H_{12}O_6(s) + 6 O_2(g) \longrightarrow 6 CO_2(g) + 6 H_2O(g)$ 18 O's

(c) Unbalanced: $? C_4H_8O(\ell) + ? O_2(g) \longrightarrow ? CO_2(g) + ? H_2O(g)$

     Balance C: $C_4H_8O(\ell) + ? O_2(g) \longrightarrow 4 CO_2(g) + ? H_2O(g)$ 4 C's

     Balance H: $C_4H_8O(\ell) + ? O_2(g) \longrightarrow 4 CO_2(g) + 4 H_2O(g)$ 8 H's

     Balance O: $C_4H_8O(\ell) + \dfrac{11}{2} O_2(g) \longrightarrow 4 CO_2(g) + 4 H_2O(g)$ 12 O's

     Multiply by 2: $2 C_4H_8O(\ell) + 11 O_2(g) \longrightarrow 8 CO_2(g) + 8 H_2O(g)$

Check your answers: (a) 8 C, 20 H, & 26 O (b) 6 C, 12 H, & 18 O (c) 8 C, 16 H, & 24 O

24. *Define the problem:* Balance the equation for the reaction between a metal and oxygen.

Develop a plan: The products are metal oxides. Write the product in the form of a neutral metal oxide using the periodic table to predict the metal ion's charge. Balance the equations. Note that the oxide anion has a –2 charge.

Execute the plan:

(a) Magnesium is in Group 2A. Its cation will have +2 charge.

$$2 \, Mg(s) + O_2(g) \longrightarrow 2 \, MgO(s) \qquad \text{magnesium oxide}$$

(b) Calcium is in Group 2A. Its cation will have +2 charge.

$$2 \, Ca(s) + O_2(g) \longrightarrow 2 \, CaO(s) \qquad \text{calcium oxide}$$

(c) Indium is in group 3A. Its cation will have +3 charge.

$$4 \, In(s) + 3 \, O_2(g) \longrightarrow 2 \, In_2O_3(s) \qquad \text{indium(III) oxide}$$

Check your answers: (a) 2 Mg & 2 O (b) 2 Ca & 2 O (c) 4 In & 6 O

26. *Define the problem:* Balance the equation for the reaction between a metal and a halogen.

Develop a plan: Write the product formula in the form of a neutral halide salt. Balance the equations. Note that the halogens form anions with a –1 charge.

Execute the plan:

(a) Potassium is in group 1A. Its cation will have +1 charge.

$$2 \, K(s) + Cl_2(g) \longrightarrow 2 \, KCl(s) \qquad \text{potassium chloride}$$

(b) Magnesium is in group 2A. Its cation will have +2 charge.

$$Mg(s) + Br_2(\ell) \longrightarrow MgBr_2(s) \qquad \text{magnesium bromide}$$

(c) Aluminum is in group 3A. Its cation will have +3 charge.

$$2 \, Al(s) + 3 \, F_2(g) \longrightarrow 2 \, AlF_3(s) \qquad \text{aluminum fluoride}$$

Check your answers: (a) 2 K & 2 Cl (b) 1 Mg & 2 Br (c) 2 Al & 6 F

Balancing Equations

29. *Define the problem:* Balance the given equations.

Develop a plan: When balancing equations, select a specific order in which the elements are balanced. Select first the atoms that are only in one product and one reactant, since they will be easier. Select next those elements that are in more than one reactant or product (such as H or O) and then those which are present in elemental form in either the reactants or products. Be systematic. If any of the coefficients end up fractional, multiply every coefficient by the same constant to eliminate the fraction.

Execute the plan:

(a) $? Fe(s) + ? Cl_2(g) \longrightarrow ? FeCl_3(s)$ Select order: Cl then Fe

   $? Fe(s) + 3 Cl_2(g) \longrightarrow 2 FeCl_3(s)$ 6 Cl

   $2 Fe(s) + 3 Cl_2(g) \longrightarrow 2 FeCl_3(s)$ 2 Fe

(b) $? SiO_2(s) + ? C(s) \longrightarrow ? Si(s) + ? CO(g)$ Select order: O, Si then C

   $SiO_2(s) + ? C(s) \longrightarrow ? Si(s) + 2 CO(g)$ 2 O

   $SiO_2(s) + ? C(s) \longrightarrow Si(s) + 2 CO(g)$ 1 Si

   $SiO_2(s) + 2 C(s) \longrightarrow Si(s) + 2 CO(g)$ 2 C

(c) $? Fe(s) + ? H_2O(g) \longrightarrow ? Fe_3O_4(s) + ? H_2(g)$ Select order: O, Fe then H

   $? Fe(s) + 4 H_2O(g) \longrightarrow Fe_3O_4(s) + ? H_2(g)$ 4 O

   $3 Fe(s) + 4 H_2O(g) \longrightarrow Fe_3O_4(s) + ? H_2(g)$ 3 Fe

   $3 Fe(s) + 4 H_2O(g) \longrightarrow Fe_3O_4(s) + 4 H_2(g)$ 8 H

Check your answers: (a) 2 Fe & 6 Cl (b) 1 Si, 2 O & 2 C (c) 3 Fe, 8 H & 4 O

31. *Define the problem:* Balance the given equations.

Develop a plan: Follow the method described in the answers to Question 29.

Execute the plan:

(a) $? MgO(s) + ? Fe(s) \longrightarrow ? Fe_2O_3(s) + ? Mg(s)$ Select order: O, Mg, Fe

   $3 MgO(s) + ? Fe(s) \longrightarrow Fe_2O_3(s) + ? Mg(s)$ 3 O

   $3 MgO(s) + ? Fe(s) \longrightarrow Fe_2O_3(s) + 3 Mg(s)$ 3 Mg

   $3 MgO(s) + 2 Fe(s) \longrightarrow Fe_2O_3(s) + 3 Mg(s)$ 2 Fe

(b) $? H_3BO_3(s) \longrightarrow ? B_2O_3(s) + ? H_2O(\ell)$ Select order: B, H, O

   $2 H_3BO_3(s) \longrightarrow B_2O_3(s) + ? H_2O(\ell)$ 2 B

   $2 H_3BO_3(s) \longrightarrow B_2O_3(s) + 3 H_2O(\ell)$ 6 H and 6 O

(c) ? $NaNO_3(s)$ + ? $H_2SO_4(aq)$ $\longrightarrow$? $Na_2SO_4(aq)$ + ? $HNO_3(g)$

Select order: Na^+, NO_3^-, H^+, SO_4^{2-}

2 $NaNO_3(s)$ + ? $H_2SO_4(aq)$ $\longrightarrow$ $Na_2SO_4(aq)$ + ? $HNO_3(g)$ 2 Na^+

2 $NaNO_3(s)$ + ? $H_2SO_4(aq)$ $\longrightarrow$ $Na_2SO_4(aq)$ + 2 $HNO_3(g)$ 2 NO_3^-

2 $NaNO_3(s)$ + $H_2SO_4(aq)$ $\longrightarrow$ $Na_2SO_4(aq)$ + 2 $HNO_3(g)$ 2 H^+ and 1 SO_4^{2-}

Check your answers: (a) 3 Mg, 3 O & 2 Fe, (b) 6 H, 2 B & 6 O,

(c) 2 Na, 2 N, 10 O, 2 H & 1 S

33. *Define the problem:* Balance the given equations.

Develop a plan: Follow the method described in the answers to Question 29.

Execute the plan:

(a) ? $CaNCN(s)$ + ? $H_2O(\ell)$ $\longrightarrow$? $CaCO_3(s)$ + ? $NH_3(g)$ Select order: Ca, N, C, O, H

$CaNCN(s)$ + ? $H_2O(\ell)$ $\longrightarrow$ $CaCO_3(s)$ + ? $NH_3(g)$ 1 Ca

$CaNCN(s)$ + ? $H_2O(\ell)$ $\longrightarrow$ $CaCO_3(s)$ + 2 $NH_3(g)$ 2 N and 1 C

$CaNCN(s)$ + 3 $H_2O(\ell)$ $\longrightarrow$ $CaCO_3(s)$ + 2 $NH_3(g)$ 3 O and 6 H

(b) ? $NaBH_4(s)$ + ? $H_2SO_4(aq)$ $\longrightarrow$? $B_2H_6(g)$ + ? $H_2(g)$ + ? $Na_2SO_4(g)$

Select order: B, Na, S, O, H

2 $NaBH_4(s)$ + ? $H_2SO_4(aq)$ $\longrightarrow$ $B_2H_6(g)$ + ? $H_2(g)$ + ? $Na_2SO_4(aq)$ 2 B

2 $NaBH_4(s)$ + ? $H_2SO_4(aq)$ $\longrightarrow$ $B_2H_6(g)$ + ? $H_2(g)$ + $Na_2SO_4(aq)$ 2 Na

2 $NaBH_4(s)$ + $H_2SO_4(aq)$ $\longrightarrow$ $B_2H_6(g)$ + ? $H_2(g)$ + $Na_2SO_4(aq)$ 1 S and 4 O

2 $NaBH_4(s)$ + $H_2SO_4(aq)$ $\longrightarrow$ $B_2H_6(g)$ + 2 $H_2(g)$ + $Na_2SO_4(aq)$ 10 H

(c) ? $H_2S(aq)$ + ? $Cl_2(aq)$ $\longrightarrow$? $S_8(s)$ + ? $HCl(aq)$ Select order: S, H, Cl

8 $H_2S(aq)$ + ? $Cl_2(aq)$ $\longrightarrow$ $S_8(s)$ + ? $HCl(aq)$ 8 S

8 $H_2S(aq)$ + ? $Cl_2(aq)$ $\longrightarrow$ $S_8(s)$ + 16 $HCl(aq)$ 16 H

8 $H_2S(aq)$ + 8 $Cl_2(aq)$ $\longrightarrow$ $S_8(s)$ + 16 $HCl(aq)$ 16 Cl

Check your answers: (a) 1 Ca, 2 N, 6 H & 3 O, (b) 2 Na, 2 B, 10 H, 1 S & 4 O

(c) 16 H, 8 S & 16 Cl

35. *Define the problem:* Balance the given equations.

Develop a plan: Follow the method described in the answers to Question 28.

Execute the plan:

(a) ? Mg + ? HNO_3 $\longrightarrow$? H_2 + ? $Mg(NO_3)_2$ Select order: Mg, N, O, H

     Mg + ? HNO_3 $\longrightarrow$? H_2 + $Mg(NO_3)_2$ 1 Mg

     Mg + 2 HNO_3 $\longrightarrow$? H_2 + $Mg(NO_3)_2$ 2 N and 6 O

     Mg + 2 HNO_3 $\longrightarrow$ H_2 + $Mg(NO_3)_2$ 2 H

(b) ? Al + ? Fe_2O_3 $\longrightarrow$? Al_2O_3 + ? Fe Select order: Fe, Al, O

     ? Al + Fe_2O_3 $\longrightarrow$? Al_2O_3 + 2 Fe 2 Fe

     2 Al + Fe_2O_3 $\longrightarrow$ Al_2O_3 + 2 Fe 2 Al and 3 O

(c) ? S + ? O_2 $\longrightarrow$? SO_3 Select order: O, S

     ? S + 3 O_2 $\longrightarrow$ 2 SO_3 6 O

     2 S + 3 O_2 $\longrightarrow$ 2 SO_3 2 S

(d) ? SO_3 + ? H_2O $\longrightarrow$? H_2SO_4 Select order: S, H, O

     SO_3 + ? H_2O $\longrightarrow$ H_2SO_4 1 S

     SO_3 + H_2O $\longrightarrow$ H_2SO_4 2 H and 4 O

Check your answers: (a) 1 Mg, 2 H, 2 N & 6 O, (b) 2 Al, 2 Fe & 3 O

(c) 2 S & 6 O, (d) 1 S, 4 O & 2 H

The Mole and Chemical Reactions

36. *Define the problem:* Given a balanced chemical equation and the number of moles of a product formed, determine the moles of reactant that was needed.

Develop a plan: Use the stoichiometry of the balanced equation as a conversion factor to convert the moles of product to moles of reactant.

Execute the plan: The balanced equation says: 4 mol HCl are needed to make 1 mol Cl_2.

$$12.5 \text{ mol } Cl_2 \times \frac{4 \text{ mol HCl}}{1 \text{ mol } Cl_2} = 50.0 \text{ mol HCl}$$

Check your answer: More HCl is needed than the Cl_2 is formed. It makes sense that the moles of HCl is greater.

38. *Define the problem:* Given a balanced chemical equation and the number of moles of a reactant, determine the moles and grams of another reactant needed and the grams of product produced.

Develop a plan: Use the stoichiometry of the balanced equation as a conversion factor to convert the moles of the one reactant to moles of the other reactant. Use the molar mass to convert from moles to grams. Use the stoichiometry of the equation to find the moles of product formed, then use the molar mass of the product to find the grams.

Execute the plan: The balanced equation says: 2 mol NO react with 1 mol O_2.

$$2.2 \text{ mol NO} \times \frac{1 \text{ mol } O_2}{2 \text{ mol NO}} = 1.1 \text{ mol } O_2$$

$$1.1 \text{ mol } O_2 \times \frac{32.00 \text{ g } O_2}{1 \text{ mol } O_2} = 35 \text{ g } O_2$$

The balanced equation says: 2 mol NO produces 2 mol NO_2.

$$2.2 \text{ mol NO} \times \frac{2 \text{ mol } NO_2}{2 \text{ mol NO}} \times \frac{46.01 \text{ g } NO_2}{1 \text{ mol } NO_2} = 1.0 \times 10^2 \text{ g } NO_2 \text{ produced}$$

Check your answers: Fewer moles of O_2 are needed than moles of NO, so that answer makes sense. The mass of NO used is 2.2 mol $\times$ (30.01 g/mol) = 66 g. The sum of the reactant masses (66g + 35 g) is equal to the products mass (1.0×10^2 g), within known significant figures. These numbers make sense.

40. *Define the problem:* Given a balanced chemical equation and the number of grams of a reactant, determine the grams of another reactant needed and the moles and grams of product produced.

Develop a plan: Use the molar mass of the first reactant to find the moles of that substance. Then use the stoichiometry of the balanced equation as a conversion factor to convert the moles of the one reactant to moles of the other reactant. Use the molar mass to convert from moles to grams. Use the stoichiometry of the equation to find the moles of product formed, then use the molar mass of the product to find the grams.

Execute the plan: The balanced equation says: 1 mol Fe reacts with 1 mol Cl_2.

$$10.0 \text{ g Fe} \times \frac{1 \text{ mol Fe}}{55.85 \text{ g Fe}} = 0.179 \text{ mol Fe}$$

$$0.179 \text{ mol Fe} \times \frac{1 \text{ mol Cl}_2}{1 \text{ mol Fe}} \times \frac{70.90 \text{ g Cl}_2}{1 \text{ mol Cl}_2} = 12.7 \text{ g Cl}_2$$

The balanced equation says: 1 mol Fe produces 1 mol $FeCl_2$.

$$0.179 \text{ mol Fe} \times \frac{1 \text{ mol FeCl}_2}{1 \text{ mol Fe}} = 0.179 \text{ mol FeCl}_2 \text{ expected}$$

$$0.179 \text{ mol FeCl}_2 \times \frac{126.75 \text{ g FeCl}_2}{1 \text{ mol FeCl}_2} = 22.7 \text{ g FeCl}_2 \text{ expected}$$

Check your answers: The sum of the masses of the reactants must add up to the total mass of the product. 10.0 g + 12.7 g = 22.7 g. This is the same product mass as that calculated above. These numbers make sense.

42. *Define the problem:* Given a balanced chemical equation and the number of grams of a reactant, determine the moles of the reactant and the mass and moles of two products.

Develop a plan: Use the molar mass of the reactant to find the moles of that substance. Then use the stoichiometry of the balanced equation as a conversion factor to convert the moles of reactant to moles of each product formed, then use the molar mass of each product to find the grams.

Execute the plan: Molar mass of $(NH_4)_2PtCl_6 =$

$$2 \times [1 \times (14.01\text{g}) + 4 \times (1.008\text{g})] + (195.08\text{g}) + 6 \times (35.45 \text{ g}) = 443.86 \text{ g/mol}$$

$$12.35 \text{ g } (NH_4)_2PtCl_6 \times \frac{1 \text{ mol } (NH_4)_2PtCl_6}{443.86 \text{ g } (NH_4)_2PtCl_6} = 2.782 \times 10^{-2} \text{ mol } (NH_4)_2PtCl_6$$

The balanced equation says: 3 mol $(NH_4)_2PtCl_6$ react to form 3 mol Pt.

$$2.782 \times 10^{-2} \text{ mol } (NH_4)_2PtCl_6 \times \frac{1 \text{ mol Pt}}{1 \text{ mol } (NH_4)_2PtCl_6} = 2.782 \times 10^{-2} \text{ mol Pt}$$

$$2.782 \times 10^{-2} \text{ mol Pt} \times \frac{195.08 \text{ g Pt}}{1 \text{ mol Pt}} = 5.428 \text{ g Pt}$$

The balanced equation says: 3 mol $(NH_4)_2PtCl_6$ react to form 16 mol HCl.

$$2.782 \times 10^{-2} \text{ mol } (NH_4)_2 PtCl_6 \times \frac{16 \text{ mol HCl}}{3 \text{ mol } (NH_4)_2 PtCl_6} = 0.1484 \text{ mol HCl}$$

$$0.1484 \text{ mol HCl} \times \frac{36.46 \text{ mol HCl}}{1 \text{ mol HCl}} = 5.410 \text{ g HCl}$$

The complete table looks like this:

$(NH_4)_2PtCl_6$	Pt	HCl
12.35 g	5.428 g	5.410 g
0.02782 mol	0.02782 mol	0.1484 mol

Check your answers: The moles are smaller than the grams. This is appropriate. The moles of HCl are larger than the moles of $(NH_4)_2PtCl_6$ and Pt. These numbers make sense.

44. *Define the problem:* Given a balanced chemical equation and the number of grams of a reactant, determine the moles of another reactant that are needed and the masses of the two products expected.

Develop a plan: Use the molar mass of the reactant to find the moles of that substance. Then use the stoichiometry of the balanced equation as a conversion factor to convert the moles of one reactant to moles of the other reactant. Use the stoichiometry to determine the moles of each of the products produced, then use the molar mass of each product to find the grams.

Execute the plan:

(a) The balanced equation says: 1 mol $TiCl_4$ reacts with 2 mol H_2O.

$$14.0 \text{ g } TiCl_4 \times \frac{1 \text{ mol } TiCl_4}{189.68 \text{ g } TiCl_4} = 0.0738 \text{ mol } TiCl_4$$

$$0.0738 \text{ mol } TiCl_4 \times \frac{2 \text{ mol } H_2O}{1 \text{ mol } TiCl_4} = 0.148 \text{ mol } H_2O$$

(b) The balanced equation says: 1 mol $TiCl_4$ produces 1 mol TiO_2.

$$0.0738 \text{ mol } TiCl_4 \times \frac{1 \text{ mol } TiO_2}{1 \text{ mol } TiCl_4} \times \frac{79.88 \text{ g } TiO_2}{1 \text{ mol } TiO_2} = 5.90 \text{ g } TiO_2$$

The balanced equation says: 1 mol $TiCl_4$ produces 4 mol HCl.

$$0.0738 \text{ mol } TiCl_4 \times \frac{4 \text{ mol HCl}}{1 \text{ mol } TiCl_4} \times \frac{36.46 \text{ g HCl}}{1 \text{ mol HCl}} = 10.8 \text{ g HCl}$$

Check your answers: The sum of the masses of the reactants (14.0 g + 2.67 g = 16.7 g) adds up the mass of the product (5.90 g + 10.8 g = 16.7 g). These numbers make sense.

46. *Define the problem:* Given a balanced chemical equation and the number of grams of a reactant, determine the mass of one product formed.

Develop a plan: Convert from kilograms to grams using a standard metric conversion factor. Then use the molar mass of the reactant to find the moles of that substance. Then use the stoichiometry of the balanced equation to determine the moles of the product produced. Then use the molar mass of the product to find the grams.

Execute the plan: The balanced equation says: 1 mol WO_3 produces 1 mol W.

$$1.00 \text{ kg } WO_3 \times \frac{1000 \text{ g } WO_3}{1 \text{ kg } WO_3} \times \frac{1 \text{ mol } WO_3}{231.85 \text{ g } WO_3} \times \frac{1 \text{ mol W}}{1 \text{ mol } WO_3} \times \frac{183.85 \text{ g W}}{1 \text{ mol W}} = 793 \text{ g W}$$

Check your answer: It makes sense that the mass of W is smaller than the mass of WO_3.

48. *Define the problem:* Given a balanced chemical equation and the mass of the reactant, determine the mass of each of the products produced.

Develop a plan: Use the molar mass of the compound to find the moles of that substance. Then use the stoichiometry of the equation to determine the moles of the products produced. Then use the molar mass of each product to find the grams.

Execute the plan: The balanced equation says: 2 mol NH_4NO_3 produces 2 mol N_2.

$$1.0 \text{ kg } NH_4NO_3 \times \frac{1000 \text{ g } NH_4NO_3}{1 \text{ kg } NH_4NO_3} \times \frac{1 \text{ mol } NH_4NO_3}{80.05 \text{ g } NH_4NO_3} \times \frac{2 \text{ mol } N_2}{2 \text{ mol } NH_4NO_3}$$

$$\times \frac{28.02 \text{ g } N_2}{1 \text{ mol } N_2} = 3.5 \times 10^2 \text{ g } N_2$$

The balanced equation says: 2 mol NH_4NO_3 produces 4 mol H_2O.

$$1.0 \text{ kg } NH_4NO_3 \times \frac{1000 \text{ g } NH_4NO_3}{1 \text{ kg } NH_4NO_3} \times \frac{1 \text{ mol } NH_4NO_3}{80.05 \text{ g } NH_4NO_3} \times \frac{4 \text{ mol } H_2O}{2 \text{ mol } NH_4NO_3}$$

$$\times \frac{4 \text{ mol } H_2O}{2 \text{ mol } NH_4NO_3} = 4.5 \times 10^2 \text{ g } H_2O$$

The balanced equation says: 2 mol NH_4NO_3 produces 1 mol O_2.

$$1.0 \text{ kg } NH_4NO_3 \times \frac{1000 \text{ g } NH_4NO_3}{1 \text{ kg } NH_4NO_3} \times \frac{1 \text{ mol } NH_4NO_3}{80.05 \text{ g } NH_4NO_3} \times \frac{1 \text{ mol } O_2}{2 \text{ mol } NH_4NO_3}$$

$$\times \frac{32.00 \text{ g } O_2}{1 \text{ mol } O_2} = 2.0 \times 10^2 \text{ g } O_2$$

Check your answers: The sum of the product masses

$$3.5 \times 10^2 \text{ g} + 4.5 \times 10^2 \text{ g} + 2.0 \times 10^2 \text{ g} = 1.00 \times 10^3 \text{ g}$$

is the same as the total mass of the reactant compound. These numbers make sense.

52. *Define the problem:* Given the reactants and products of a reaction and the mass of a reactant, balance the chemical equation, determine the mass of another reactant, and determine the mass of one product produced.

Develop a plan: Given the formulas of the reactants and products, balance the equation, selecting appropriate order for the systematic balancing of all the atoms. Then use the molar mass of the reactant to find the moles of that substance. Then use the stoichiometry of the equation to determine the moles of the other reactant and product produced. Then use the molar mass of each to find the grams.

Execute the plan:

(a) $? NH_4NO_3 \longrightarrow ? N_2O + ? H_2O$ Select order: N, H, O

$\quad NH_4NO_3 \longrightarrow N_2O + ? H_2O$ 2 N

$\quad NH_4NO_3 \longrightarrow N_2O + 2 H_2O$ 4 H and 3 O

(b) The balanced equation says: 1 mol NH_4NO_3 requires 1 mol N_2O.

$$10.0 \text{ g } NH_4NO_3 \times \frac{1 \text{ mol } NH_4NO_3}{80.05 \text{ g } NH_4NO_3} = 0.125 \text{ mol } NH_4NO_3$$

$$0.125 \text{ mol } NH_4NO_3 \times \frac{1 \text{ mol } N_2O}{1 \text{ mol } NH_4NO_3} \times \frac{44.02 \text{ g } N_2O}{1 \text{ mol } N_2O} = 5.50 \text{ g } N_2O$$

The balanced equation says: 1 mol NH_4NO_3 requires 2 mol H_2O.

$$0.125 \text{ mol } NH_4NO_3 \times \frac{2 \text{ mol } H_2O}{1 \text{ mol } NH_4NO_3} \times \frac{18.02 \text{ g } H_2O}{1 \text{ mol } H_2O} = 4.50 \text{ g } H_2O$$

Check your answers: (a) Check the number of atom of each type in the reactants and products: 2 N, 4 H, 3 O. The equation is properly balanced. (b) The sum of the masses of the products (5.50 g + 4.50 g = 10.00 g) is equal to the mass of the reactant compound (10.0 g). These numbers make sense.

54. *Define the problem:* Given the reactants and products of a reaction and the desired mass of a product, balance the chemical equation and determine the mass of the reactants needed to produce this amount of product.

Develop a plan: Given the formulas of the reactants and products, balance the equation, selecting appropriate order for the systematic balancing of all the atoms. Then use the molar mass of the product to find the moles of that substance. Then use the stoichiometry of the equation to determine the moles of the reactants needed to make that much product. Then use the molar mass of each to find the grams.

Execute the plan:

(a) $? K_2PtCl_4 + ? NH_3 \longrightarrow ? Pt(NH_3)_2Cl_2 + ? KCl$

Select order: N, H, Pt, K, Cl

$? K_2PtCl_4 + 2 NH_3 \longrightarrow Pt(NH_3)_2Cl_2 + ? KCl$ 2 N and 6 H

$K_2PtCl_4 + 2 NH_3 \longrightarrow Pt(NH_3)_2Cl_2 + ? KCl$ 1 Pt

$K_2PtCl_4 + 2 NH_3 \longrightarrow Pt(NH_3)_2Cl_2 + 2 KCl$ 2 K and 4 Cl

(b) The balanced equation says: 1 mol $Pt(NH_3)_2Cl_2$ requires 1 mol K_2PtCl_4.

$$2.50 \text{ g } Pt(NH_3)_2Cl_2 \times \frac{1 \text{ mol } Pt(NH_3)_2Cl_2}{300.05 \text{ g } Pt(NH_3)_2Cl_2} = 0.00833 \text{ mol } Pt(NH_3)_2Cl_2$$

$$0.00833 \text{ mol } Pt(NH_3)_2Cl_2 \times \frac{1 \text{ mol } K_2PtCl_4}{1 \text{ mol } Pt(NH_3)_2Cl_2} \times \frac{415.08 \text{ g } K_2PtCl_4}{1 \text{ mol } K_2PtCl_4}$$

$$= 3.46 \text{ g } K_2PtCl_4$$

The balanced equation says: 1 mol $Pt(NH_3)_2Cl_2$ requires 2 mol NH_3.

$$0.00833 \text{ mol } Pt(NH_3)_2Cl_2 \times \frac{2 \text{ mol } NH_3}{1 \text{ mol } Pt(NH_3)_2Cl_2} \times \frac{17.03 \text{ g } NH_3}{1 \text{ mol } NH_3} = 0.284 \text{ g } NH_3$$

Check your answers: (a) Check the number of atom of each type in the reactants and products: 2 K, 1 Pt, 2 N, 6 H, and 4 Cl. The equation is properly balanced. (b) The mass of K_2PtCl_4 is larger than the mass of $Pt(NH_3)_2Cl_2$, and the mass of NH_3 is less than the mass of $Pt(NH_3)_2Cl_2$. These both make sense, looking at the stoichiometric relationships and the molar masses.

Limiting Reactant

55. *Define the problem:* Given a balanced chemical equation and the masses of both reactants, determine the limiting reactant, the mass of the product produced, and the mass remaining of the excess reactant when the reaction is complete.

Develop a plan: Here, we use a slight variation of what the text calls "the mole method." We will calculate a directly comparable quantity, the moles of product. (a) Use the molar mass of the reactants to find the moles of the reactant substances. Then use the stoichiometry of the equation to determine the moles of the product produced in each case. Identify the limiting reactant from the reactant that produces the least number of products. (b) Use the moles of product produced from the limiting reactant and the molar mass of the product to find the grams. (c) From the limiting reactant quantity, determine the moles of the other reactant needed for complete reaction. Convert that number to grams using the molar mass. Then subtract the quantity used from the initial mass given to get the mass of excess reactant.

Execute the plan:

(a) The balanced equation says: 2 mol Al produces 2 mol $AlCl_3$.

$$2.70 \text{ g Al} \times \frac{1 \text{ mol Al}}{26.98 \text{ g Al}} \times \frac{2 \text{ mol AlCl}_3}{2 \text{ mol Al}} = 0.100 \text{ mol AlCl}_3$$

The balanced equation says: 3 mol Cl_2 produces 2 mol $AlCl_3$.

$$4.05 \text{ g Cl}_2 \times \frac{1 \text{ mol Cl}_2}{70.90 \text{ g Cl}_2} \times \frac{2 \text{ mol AlCl}_3}{3 \text{ mol Cl}_2} = 0.0381 \text{ mol AlCl}_3$$

The number of $AlCl_3$ moles produced from Cl_2 is smaller (0.0381 mol < 0.100 mol), so Cl_2 is the limiting reactant and Al is the excess reactant.

(b) Find the mass of 0.0381 mol $AlCl_3$:

$$0.0381 \text{ mol AlCl}_3 \times \frac{133.33 \text{ g AlCl}_3}{1 \text{ mol AlCl}_3} = 5.08 \text{ g AlCl}_3$$

(c) The balanced equation says: 3 mol Cl_2 react with 2 mol Al.

$$4.05 \text{ g Cl}_2 \times \frac{1 \text{ mol Cl}_2}{70.90 \text{ g Cl}_2} \times \frac{2 \text{ mol Al}}{3 \text{ mol Cl}_2} \times \frac{26.98 \text{ g Al}}{1 \text{ mol Al}} = 1.03 \text{ g Al}$$

2.70 g Al initial – 1.03 g Al used up = 1.67 g Al remains unreacted

Check your answers: There are fewer moles of Cl_2 present than Al, and the equation needs more Cl_2 than Al, so it makes sense that Cl_2 is the limiting reactant. In addition, the calculation in (c) proved that the initial mass of Al was larger than required to react with all of the Cl_2. The sum of the masses of the reactants that reacted (4.05 g + 1.03 g = 5.08 g) equals the mass of the product produced (5.08 g). These answers make sense.

58. *Define the problem:* Given a balanced chemical equation and the masses of both reactants, determine the moles of reactants and products present when the reaction is finished.

Develop a plan: Use the molar mass of the reactants to find the moles of the reactant substances present initially. Then use the stoichiometry of the equation to determine the moles of one of the products. Identify the limiting reactant from the reactant that produces the least number of products. From the limiting reactant quantity, and using the stoichiometry of the balanced equation, determine the moles of the other reactant and products.

Execute the plan: The balanced equation says: 1 mol CH_4 produces 1 mol CO_2.

$$995 \text{ g } CH_4 \times \frac{1 \text{ mol } CH_4}{16.04 \text{ g } CH_4} \times \frac{1 \text{ mol } CO_2}{1 \text{ mol } CH_4} = 62.0 \text{ mol } CO_2$$

The balanced equation says: 2 mol H_2O produces 1 mol CO_2.

$$2510 \text{ g } H_2O \times \frac{1 \text{ mol } H_2O}{18.02 \text{ g } H_2O} \times \frac{1 \text{ mol } CO_2}{2 \text{ mol } H_2O} = 69.6 \text{ mol } CO_2$$

The number of CO_2 moles produced from CH_4 is smaller (62.0 mol < 69.6 mol), so CH_4 is the limiting reactant and H_2O is the excess reactant.

The balanced equation says: 1 mol CO_2 is produced with 4 mol H_2.

$$62.0 \text{ mol } CO_2 \times \frac{4 \text{ mol } H_2}{1 \text{ mol } CO_2} = 248 \text{ mol } H_2$$

The balanced equation says: 1 mol CH_4 reacts with 2 mol H_2O.

$$995 \text{ g } CH_4 \times \frac{1 \text{ mol } CH_4}{16.04 \text{ g } CH_4} \times \frac{2 \text{ mol } H_2O}{1 \text{ mol } CH_4} = 124 \text{ mol } H_2O \text{ reacted}$$

$$2510 \text{ g } H_2O \times \frac{1 \text{ mol } H_2O}{18.02 \text{ g } H_2O} = 139 \text{ mol } H_2O \text{ present initially}$$

$$139 \text{ mol } H_2O \text{ initial} - 124 \text{ mol } H_2O \text{ reacted} = 15 \text{ mol } H_2O \text{ left}$$

Check your answers: The excess moles of H_2O prove that the right limiting reactant was determined, since there are no moles of CH_4 left. The moles of products are stoichiometrically appropriate multiples of 62.0 moles, as they should be. These numbers look right.

60. *Define the problem:* Given a balanced chemical equation and the masses of both reactants, determine the maximum amount of product that can be produced.

Develop a plan: Use a metric conversion and the molar mass of the reactants to find the moles of the reactant substances present initially. Then use the stoichiometry of the equation to determine the moles of one of the products. Identify the limiting reactant from the reactant that produces the least number of products. From the limiting reactant quantity, determine the moles of the product, then use the molar mass of the product to get the grams and a metric conversion to get kilograms.

Execute the plan: The balanced equation says: 1 mol Fe_2O_3 produces 2 mol Fe.

$$2.00 \text{ kg } Fe_2O_3 \times \frac{1000 \text{ g } Fe_2O_3}{1 \text{ kg } Fe_2O_3} \times \frac{1 \text{ mol } Fe_2O_3}{159.69 \text{ g } Fe_2O_3} \times \frac{2 \text{ mol Fe}}{1 \text{ mol } Fe_2O_3} = 25.0 \text{ mol Fe}$$

The balanced equation says: 3 mol CO produces 2 mol Fe.

$$2.00 \text{ kg CO} \times \frac{1000 \text{ g CO}}{1 \text{ kg CO}} \times \frac{1 \text{ mol CO}}{28.01 \text{ g CO}} \times \frac{2 \text{ mol Fe}}{3 \text{ mol CO}} = 47.6 \text{ mol Fe}$$

The number of Fe moles produced from Fe_2O_3 is smaller (25.0 mol < 47.6 mol), so Fe_2O_3 is the limiting reactant and CO is the excess reactant.

Find the mass of the Fe from the limiting reactant:

$$25.0 \text{ mol Fe} \times \frac{55.847 \text{ gFe}}{1 \text{ mol Fe}} \times \frac{1000 \text{ g Fe}}{1 \text{ kg Fe}} = 1.40 \text{ kg Fe}$$

Check your answer: The reactants are present in the same mass quantities, but Fe_2O_3 has a larger molar mass. So, it makes sense that it is the limiting reactant. The mass of iron should be smaller than the mass of iron (III) oxide. These numbers look right.

Percent Yield

62. *Define the problem:* Given the theoretical yield and the actual yield, determine the percent yield.

Develop a plan: Divide the actual yield by the theoretical yield and multiply by 100 % to get percent yield. *(Note that the balanced equation is given too, but you don't need to use it to answer this question.)*

Execute the plan: $\dfrac{100 \text{ g NH}_3 \text{ actual}}{136 \text{ g NH}_3 \text{ theoretical}} \times 100 \text{ \%} = 73.5 \text{ \% yield}$

Check your answer: About three-quarters of the maximum quantity of ammonia was produced, so it makes sense that the percent yield is about 75 %. This number looks right.

64. *Define the problem:* Given the balanced chemical equation, the mass of the limiting reactant and the actual yield, determine the percent yield.

Develop a plan: First, we need to calculate the theoretical yield. We get this by determining the maximum amount of product that could have been made from the given quantities of reactant: Take the mass of the limiting reactant and convert it to moles. Then use stoichiometry to find the moles of product. Then convert to grams using molar mass. Take the given actual yield and divide by the calculated theoretical yield and multiply by 100 % to get percent yield.

Execute the plan: The limiting reactant is $NaBH_4$. From its mass, find the maximum grams of B_2H_6 that could be made. The mole ratio comes from the balanced equation.

$$1.203 \text{ g NaBH}_4 \times \frac{1 \text{ mol NaBH}_4}{37.832 \text{ g NaBH}_4} \times \frac{1 \text{ mol B}_2\text{H}_6}{2 \text{ mol NaBH}_4} \times \frac{27.669 \text{ g B}_2\text{H}_6}{1 \text{ mol B}_2\text{H}_6}$$

$$= 0.4399 \text{ g B}_2\text{H}_6$$

The given mass of B_2H_6 is the actual yield.

$$\frac{0.295 \text{ g B}_2\text{H}_6 \text{ actual}}{0.4399 \text{ g B}_2\text{H}_6 \text{ theoretical}} \times 100 \text{ \%} = 67.1 \text{ \% yield}$$

Check your answer: About two-thirds the maximum quantity of diborane was produced, so it makes sense that the percent yield is about 67 %. This number looks right.

66. *Define the problem:* Given the balanced chemical equation, the desired mass of the product, and the percent yield , determine the amount of limiting reactant that must be used.

Develop a plan: Interpret the percent yield as the relationship between the actual grams and theoretical grams. Use that relationship to find the theoretical yield mass of the product. Then use stoichiometry to find the moles of limiting reactant. Then, using molar mass, convert to actual grams of reactant needed.

Execute the plan: The 51 % percent yield tells us the following:

To make 51 grams of S_2Cl_2, we need to have enough limiting reactant to make 1.19 grams of S_2Cl_2.

$$1.19 \text{ g } S_2Cl_2 \text{ actual} \times \frac{100. \text{ g } S_2Cl_2}{51 \text{ g } S_2Cl_2 \text{ actual}} = 2.3 \text{ g } S_2Cl_2$$

Use this mass of product to determine what mass of limiting reactant to use.

$$2.3 \text{ g } S_2Cl_2 \times \frac{1 \text{ mol } S_2Cl_2}{135.03 \text{ g } S_2Cl_2} \times \frac{3 \text{ mol } SCl_2}{1 \text{ mol } S_2Cl_2} \times \frac{102.97 \text{ g } SCl_2}{1 \text{ mol } S_2Cl_2} = 5.3 \text{ g } SCl_2$$

Check your answer: The yield suggests that we need to try to make about twice as much, so the 2.3 grams makes sense. The stoichiometry is 3:1, so it makes sense that a larger mass of SCl_2 is needed than the mass of S_2Cl_2 formed. These numbers look right.

Empirical Formulas

68. *Define the problem:* Given the mass of a compound, the identity of the elements in the compound, and the identity and masses of all the products produced, determine the empirical formula.

Develop a plan: Use the molar mass of the products to find their moles and use the stoichiometry of their formulas to determine the moles of the elements in the compound. Then use a whole-number mole ratio to determine the empirical formula.

Execute the plan:

The compound is a hydrocarbon and contains only C and H: C_iH_j. Its combustion produced H_2O and CO_2. Use molar mass and formula stoichiometry to determine the moles of C and H.

$$1.481 \text{ g CO}_2 \times \frac{1 \text{ mol CO}_2}{44.010 \text{ g CO}_2} \times \frac{1 \text{ mol C}}{1 \text{ mol CO}_2} = 0.03365 \text{ mol C}$$

$$0.303 \text{ g H}_2\text{O} \times \frac{1 \text{ mol H}_2\text{O}}{18.015 \text{ g H}_2\text{O}} \times \frac{2 \text{ mol H}}{1 \text{ mol H}_2\text{O}} = 0.0336 \text{ mol H}$$

Set up mole ratio and simplify by dividing by the smallest number of moles:

$$0.03365 \text{ mol C} : 0.0336 \text{ mol H}$$

$$1 \text{ C} : 1 \text{ H}$$

Use the whole number ratio for the subscripts in the formula. The empirical formula is CH.

Check your answer: These are not necessary calculations, but we can calculate the masses of each element, C and H.

$$0.03365 \text{ mol C} \times \frac{12.011 \text{ g C}}{1 \text{ mol C}} = 0.4041 \text{ g C} \qquad 0.0336 \text{ mol H} \times \frac{1.008 \text{ g H}}{1 \text{ mol H}} = 0.0339 \text{ g H}$$

The sum of these masses (0.4041 g + 0.0339 g = 0.4380 g) adds up to the mass of the original compound (0.438 g). This answer makes sense.

70. *Define the problem:* Given the mass of a compound, the identity of the elements in the compound, and the identity and masses of all the products produced, determine the empirical formula.

Develop a plan: Combustion uses oxygen. When the compound also contains oxygen, determine the amount of oxygen after the other elements. Use the molar mass of the products to find their moles and use the stoichiometry of their formulas to determine the moles of the elements that are not oxygen in the compound. Find the masses of those elements, and subtract them from the total mass of the compound to get the mass of oxygen in the compound. Then use the molar mass to calculate the moles of oxygen in the compound. Then use a whole-number mole ratio to determine the empirical formula.

Execute the plan: The compound contains C, H, and O: $C_iH_jO_k$. Its combustion produced H_2O and CO_2. Use molar mass and formula stoichiometry to determine the moles of C and H. Use the whole number ratio for the subscripts in the formula.

$$0.421 \text{ g CO}_2 \times \frac{1 \text{ mol CO}_2}{44.010 \text{ g CO}_2} \times \frac{1 \text{ mol C}}{1 \text{ mol CO}_2} = 9.56 \times 10^{-3} \text{ mol C}$$

$$0.172 \text{ g H}_2\text{O} \times \frac{1 \text{ mol H}_2\text{O}}{18.015 \text{ g H}_2\text{O}} \times \frac{2 \text{ mol H}}{1 \text{ mol H}_2\text{O}} = 1.91 \times 10^{-2} \text{ mol H}$$

Calculate the masses of C and H.

$$9.56 \times 10^{-3} \text{ mol C} \times \frac{12.011 \text{ g C}}{1 \text{ mol C}} = 0.115 \text{ g C} \qquad 0.0191 \text{ mol H} \times \frac{1.008 \text{ g H}}{1 \text{ mol H}} = 0.0192 \text{ g H}$$

Calculate the masses of O by subtracting the masses of C and H from the given total compound mass.

$$0.236 \text{ g C}_i\text{H}_j\text{O}_k - 0.115 \text{ g C} - 0.0192 \text{ g H} = 0.102 \text{ g O}$$

Calculate the moles of O.

$$0.102 \text{ g H}_2\text{O} \times \frac{1 \text{ mol O}}{16.00 \text{ g O}} = 6.37 \times 10^{-3} \text{ mol O}$$

Set up mole ratio and simplify by dividing by the smallest number of moles:

$$9.56 \times 10^{-3} \text{ mol C} : 1.91 \times 10^{-2} \text{ mol H} : 6.37 \times 10^{-3} \text{ mol O}$$
$$1.5 \text{ C} : 3 \text{ H} : 1 \text{ O}$$

Multiply by 2, to get a whole number ratio: 3 C : 6 H: 2 O

Use the whole number ratio for the subscripts in the formula. The empirical formula is $C_3H_6O_2$.

Check your answer: The mole ratio came out very close to whole number values. This answer makes sense.

General Questions

72. *Define the problem:* Given the unbalanced chemical equation and the mass of the limiting reactant, determine the mass of a product that can be isolated.

Develop a plan: Balance the equation. Use the molar mass of the reactant to find moles of reactant, then use the stoichiometry of the equation to determine moles of product, then use the molar mass of the product to get the mass of the product.

Execute the plan: Select a balance order: H, O, N, Cu

$$2 \text{ NH}_3 + 3 \text{ CuO} \longrightarrow \text{N}_2 + 3 \text{ Cu} + 3 \text{ H}_2\text{O}$$

Check the balance: 2 N, 6 H, 3 Cu, 3 O

$$26.3 \text{ g NH}_3 \times \frac{1 \text{ mol NH}_3}{17.03 \text{ g NH}_3} \times \frac{1 \text{ mol N}_2}{2 \text{ mol NH}_3} \times \frac{28.02 \text{ g N}_2}{1 \text{ mol N}_2} = 21.6 \text{ g N}_2$$

Check your answer: It makes sense that the initial mass of NH_3 is nearly the same as the mass of N_2 created, since the NH_3 molar mass is about half that of N_2 and the stoichiometry is 2:1.

74. *Define the problem:* Given a balanced chemical equation, the volume of a liquid limiting reactant, and the density of the liquid, determine the maximum theoretical yield of a product.

Develop a plan: Use the density of the reactant to find the grams of the reactant. Then the use molar mass of the reactant to find the moles of reactant. Then use the stoichiometry of the equation to determine moles of product. Then use the molar mass of the product to get the mass of the product.

Execute the plan:

$$25.0 \text{ mL Br}_2 \times \frac{3.1023 \text{ g Br}_2}{1 \text{ mL Br}_2} \times \frac{1 \text{ mol Br}_2}{159.8 \text{ g Br}_2} \times \frac{1 \text{ mol Al}_2\text{Br}_6}{3 \text{ mol Br}_2} \times \frac{533.36 \text{ g Al}_2\text{Br}_6}{1 \text{ mol Al}_2\text{Br}_6}$$

$$= 86.3 \text{ g Al}_2\text{Br}_6$$

Check your answer: The units all cancel, and the numerical multipliers are larger than the dividers. This somewhat larger answer makes sense.

76. *Define the problem:* Given the balanced chemical equation and the mass of one reactant and the moles of the other reactant, determine the maximum theoretical mass of a product.

Develop a plan: Find the limiting reactant by determining moles of a product made from each reactant. Then using the molar mass of the product, determine the mass from the moles produced by the limiting reactant.

Execute the plan:

$$15.5 \text{ g (NH}_4)_2\text{PtCl}_4 \times \frac{1 \text{ mol (NH}_4)_2\text{PtCl}_4}{373.0 \text{ g (NH}_4)_2\text{PtCl}_4} \times \frac{1 \text{ mol Pt(NH}_3)_2\text{Cl}_2}{1 \text{ mol (NH}_4)_2\text{PtCl}_4}$$

$$= 0.0416 \text{ mol Pt(NH}_3)_2\text{Cl}_2$$

$$0.15 \text{ mol NH}_3 \times \frac{1 \text{ mol Pt(NH}_4)_2\text{Cl}_2}{2 \text{ mol NH}_3} = 0.075 \text{ mol Pt(NH}_4)_2\text{Cl}_2$$

$(NH_4)_2PtCl_4$ is the limiting reactant, since fewer moles of $Pt(NH_3)_2Cl_2$ are produced from the $(NH_4)_2PtCl_4$ reactant (0.0416 mol) than produced from the NH_3 reactant (0.075 mol). Therefore, use 0.0416 mol $Pt(NH_3)_2Cl_2$ and the molar mass to determine the grams of $Pt(NH_3)_2Cl_2$ produced:

$$0.0416 \text{ mol Pt(NH}_3)_2\text{Cl}_2 \times \frac{300.05 \text{ g Pt(NH}_3)_2\text{Cl}_2}{1 \text{ mol Pt(NH}_3)_2\text{Cl}_2} = 12.5 \text{ g Pt(NH}_3)_2\text{Cl}_2$$

Check your answer: It is satisfying that $(NH_4)_2PtCl_4$ is the limiting reactant, because that reactant contains the expensive platinum metal and the experimenter would want to make sure all of that got used up. Since the molar mass of the product is somewhat smaller than the molar mass of the limiting reactant and their stoichiometry is 1:1, it makes sense that a slightly smaller mass of product is formed in this reaction.

78. *Define the problem:* Given the mass of a compound, the identity of the elements in the compound, and the identity and masses of all the products produced, determine the empirical formula.

Develop a plan: Use the molar mass of the products to find their moles and use the stoichiometry of their formulas to determine the moles of the elements in the compound. Then use a whole-number mole ratio to determine the empirical formula.

Execute the plan:

The compound is Si_xO_y. Its combustion produced SiO_2 and H_2O. Use molar mass and formula stoichiometry to determine the moles of Si and H.

$$11.64 \text{ g SiO}_2 \times \frac{1 \text{ mol SiO}_2}{60.084 \text{ g SiO}_2} \times \frac{1 \text{ mol Si}}{1 \text{ mol SiO}_2} = 0.1937 \text{ mol Si}$$

$$6.980 \text{ g H}_2\text{O} \times \frac{1 \text{ mol H}_2\text{O}}{18.015 \text{ g H}_2\text{O}} \times \frac{2 \text{ mol H}}{1 \text{ mol H}_2\text{O}} = 0.7749 \text{ mol H}$$

Set up mole ratio and simplify by dividing by the smallest number of moles:

$$0.1937 \text{ mol Si} : 0.7749 \text{ mol H}$$

$$1 \text{ Si} : 4.000 \text{ H}$$

Use the whole number ratio for the subscripts in the formula. The empirical formula is SiH_4.

Check your answer: These are not necessary calculations, but we can calculate the masses of each element, Si and H.

$$0.1937 \text{ mol Si} \times \frac{28.086 \text{ g Si}}{1 \text{ mol Si}} = 5.440 \text{ g Si} \qquad 0.7749 \text{ mol H} \times \frac{1.0079 \text{ g H}}{1 \text{ mol H}} = 0.7810 \text{ g H}$$

The sum of these masses (5.440 g + 0.7810 g = 6.221 g) add up to the mass of the original compound (6.22 g), to the given significant figures. This answer makes sense.

Applying Concepts

82. Two butane molecules react with 13 diatomic oxygen molecules to produce eight carbon dioxide molecules and ten water molecules.

    Two moles of gaseous butane molecules react with 13 moles of gaseous diatomic oxygen molecules to produce eight moles of gaseous carbon dioxide molecules and ten moles of liquid water molecules.

84. Balance the equation for this reaction: $4 A_2 + AB_3 \longrightarrow 3 \underline{\hspace{1.5cm}}$

    Reactant number of A atoms $= 4 \times 2 + 1 = 9$ Product number of A atoms $= 9 = 3 \times \underline{3}$

    Reactant number of B atoms $= 3$ Product number of A atoms $= 3 = 3 \times \underline{1}$

    So, the product molecule has 3 A atoms and 1 B atom. Its formula is A_3B.

$$4 A_2 + AB_3 \longrightarrow 3 A_3B$$

86. *Define the problem:* Given the formula of two binary oxide compounds both containing an unknown element and an unknown amount of oxygen, and given the mass of the both elements in a given sample of each compound, we need to determine the ratio of oxygen atoms in the two formulas. Given the number of oxygen atoms in one formula, determine the identity of the unknown element.

    Develop a plan: We have incomplete information in this problem, so we'll let the ratio help us eliminate the need to know information common between the two substances. Start by converting the mass of O in each sample to moles of O. We then use the mass of A and its molar mass to determine moles of A, then use the formula stoichiometry to determine the moles of AO_x. Dividing the moles of O by the moles of AO_x tells us the value of x. Follow the same procedure to get y. We don't know molar mass of A, but it will cancel in the ratio.

    Execute the plan:

    (a) Find moles of O: $3.2 \text{ g O} \times \dfrac{1 \text{ mol O}}{16.00 \text{ g O}} = 0.20 \text{ mol O}$

    Find moles of AO_x, using M_A for the molar mass of unknown element A:

$$1.2 \text{ g A} \times \dfrac{1 \text{ mol A}}{M_A \text{ g A}} \times \dfrac{1 \text{ mol AO}_x}{1 \text{ mol A}} = \dfrac{1.2}{M_A} \text{ mol AO}_x$$

    Find moles of AO_y in a similar fashion:

$$2.4 \text{ g A} \times \frac{1 \text{ mol A}}{M_A \text{ g A}} \times \frac{1 \text{ mol AO}_y}{1 \text{ mol A}} = \frac{2.4}{M_A} \text{ mol AO}_y$$

Now set up the ratio to get $\dfrac{x}{y}$:
$$\frac{\dfrac{\text{mol O}}{\text{mol AO}_x}}{\dfrac{\text{mol O}}{\text{mol AO}_y}}$$

$$\frac{\dfrac{0.20 \text{ mol O}}{\dfrac{1.2}{M_A} \text{ mol AO}_x}}{\dfrac{0.20 \text{ mol O}}{\dfrac{2.4}{M_A} \text{ mol AO}_y}} = \frac{\dfrac{2 \text{ mol O}}{1 \text{ mol AO}_x}}{\dfrac{1 \text{ mol O}}{1 \text{ mol AO}_y}} \qquad \frac{x}{y} = \frac{2}{1}$$

(b) If $x = 2$, we can use the moles of O to tell use how many moles of A are present in the AO_x sample.

$$0.20 \text{ mol O} \times \frac{1 \text{ mol A}}{2 \text{ mol O}} = 0.10 \text{ mol A}$$

Divide mass of A in the sample by moles of A in the sample to get molar mass of A:

$$\frac{1.2 \text{ g A}}{0.10 \text{ mol A}} = 12 \text{ g A} \qquad \text{The element with this molar mass is carbon, C.}$$

Check your answers: (a) The law of multiple proportions assures us that the ratio of elements in binary compounds is a small whole number, so the ratio looks right. There are probably many other legitimate ways to solve this problem. (b) The molar mass of the unknown element was very close to the molar mass of carbon, and both CO and CO_2 are known compounds, so this result makes sense also.

88. *Define the problem:* Given the number of molecules of all the reactants, determine what molecules are present after a reaction is complete and determine the limiting reactant.

Develop a plan: Use the stoichiometry of the reaction to determine the molecules of the product formed from each reactant. The reactant that produces the least amount of product is the limiting reactant. The excess reactant and the products are present at the end of the reaction. From that information, determine what figure represents the proper diagram of the post-reaction mixture.

Execute the plan:

$$6 \text{ molecules N}_2 \times \frac{2 \text{ molecules NH}_3}{1 \text{ molecules N}_2} = 12 \text{ molecules NH}_3$$

$$12 \text{ molecules } H_2 \times \frac{2 \text{ molecules } NH_3}{3 \text{ molecules } H_2} = 8 \text{ molecules } NH_3$$

Only 8 molecules NH_3 are formed, because the limiting reactant is H_2. The excess reactant is N_2.

Determine the number of molecules of N_2 that actually reacted:

$$12 \text{ molecules } H_2 \times \frac{1 \text{ molecules } N_2}{3 \text{ molecules } H_2} = 4 \text{ molecules } N_2 \text{ reacted}$$

The molecules of N_2 left = 6 molecules of N_2 initial – 4 molecules of N_2 reacted

$$= 2 \text{ molecules of } N_2 \text{ left}$$

These results indicate that Figure 4 is the right figure, and that (b) is the right statement regarding the limiting reactant: "H_2 is the limiting reactant"

Check your answers: There are two molecules of N_2 left after all the molecules of H_2 react, so it makes sense that H_2 is the limiting reactant. The figure shows 8 molecules of NH_3 and 2 molecules of N_2. These answers make sense.

90. Four XY_3 molecules are made from two diatomic X_2 molecules and six diatomic Y_2 molecules. So, symbolically, the reaction is $2 X_2 + 6 Y_2 \longrightarrow 4 XY_3$, and the stoichiometric equation representing that reaction is (b) $X_2 + 3 Y_2 \longrightarrow 2 XY_3$

92. When masses smaller than 1.0 gram of the metal are added, the metal is the limiting reactant, so the mass of the compound produced is directly proportional to the mass of metal present (shown by the straight line rising up to the right on the graph). More metal makes more products when the mass of metal is less than 1.0 gram.

When the mass larger than 1.0 g of metal are added, the bromine is the limiting reactant, so the particular mass of the metal is independent of how much compound is made. Since the amount of bromine is held constant, the mass of compound formed is also a constant (shown by a horizontal line on the graph).

Conclusion: When metal mass is less than 1.0 g the metal is the limiting reactant. When the metal mass is greater than 1.0 g the bromine is the limiting reactant.

Chapter 5: Chemical Reactions

Solubility

11. *Define the problem*: Given a compound's formula, determine whether it is water-soluble.

Develop a plan: Identify the cation and the anion in the salt, then use Table 5.1.

> *Note: Any rule that applies is sufficient to determine the compound's solubility. For example, if Rule 1 applies, there is no need to look for other rules related to the anion.*

> *Note: The Question asks which ions are present, not how many. When answering this Question, numbers that were part of the compound's formula stoichiometry are not included.*

Execute the plan:

(a) $Fe(ClO_4)_2$ contains iron(II) ion and perchlorate ion

Rule six: "All perchlorates are soluble." $Fe(ClO_4)_2$ is soluble.

$Fe(ClO_4)_2$ ionizes to form Fe^{2+} and ClO_4^-.

(b) Na_2SO_4 contains sodium ion (Group 1A) and sulfate ion

Rule one: "All ... sodium ... salts are soluble." Na_2SO_4 is soluble.

Na_2SO_4 ionizes to form Na^+ and SO_4^{2-}.

(c) KBr contains potassium ion (Group 1A) and bromide ion

Rule one: "All ... potassium ... salts are soluble." KBr is soluble.

KBr ionizes to form K^+ and Br^-.

(d) Na_2CO_3 contains sodium ion (Group 1A) and carbonate ion

Rule one: "All ... sodium ... salts are soluble." Na_2CO_3 is soluble.

Na_2CO_3 ionizes to form Na^+ and CO_3^{2-}.

Exchange Reactions

14. *Define the problem*: Given the reactants' formulas, determine if a precipitation reaction occurs. If so, write the balanced equation for the reaction.

Develop a plan: Create appropriate products of the possible exchange reaction, then check Table 5.1 for insoluble products that would form a precipitate, then balance the equation.

Note: Common acids (listed in Table 5.2) are soluble.

Execute the plan:

(a) Possible exchange products are: MnS (Rule 13) Insoluble solid precipitate

NaCl (Rule 1) Soluble

$$MnCl_2(aq) + Na_2S(aq) \longrightarrow MnS(s) + 2\ NaCl(aq)$$

(b) Possible exchange products are: $Cu(NO_3)_2$ (Rule 2) Soluble

H_2SO_4 (Table 5.2) Soluble strong electrolyte

$$HNO_3(aq) + CuSO_4(aq) \longrightarrow \text{"N.R."}\ \text{(see discussion below)*}$$

(c) Possible exchange products are: H_2O Molecular liquid; no significant ionization

$NaClO_4$ (Rule 1) Soluble

$$NaOH(aq) + HClO_4(aq) \longrightarrow \text{"N.R."}\ \text{(see discussion below)*}$$

(d) Possible exchange products are: HgS (Rule 13) Insoluble solid precipitate

$NaNO_3$ (Rule 1) Soluble

$$Hg(NO_3)_2(aq) + Na_2S(aq) \longrightarrow HgS(s) + 2\ NaNO_3(aq)$$

(e) Possible exchange products are: $PbCl_2$ (Rule 3 exception) Insoluble solid precipitate

HNO_3 (Table 5.2) Soluble strong electrolyte

$$Pb(NO_3)_2(aq) + 2\ HCl(aq) \longrightarrow PbCl_2(s) + 2\ HNO_3(aq)$$

(f) Possible exchange products are: $BaSO_4$ (Rule 4 exception) Insoluble solid precipitate

HCl (Table 5.2) Soluble strong electrolyte

$$BaCl_2(aq) + H_2SO_4(aq) \longrightarrow BaSO_4(s) + 2\ HCl(aq)$$

Check your answers: All equations with insoluble products are identified as precipitation reactions. Those equations with all products soluble are identified with "NR." *It is worth noticing in both of those cases that a reaction **does** occur, so it is inappropriate to write "NR," which generally means "no reaction." Both (b) and (c) are acid-base reactions (see Section 5.2). They are not precipitation reactions.

15. *Define the problem*: Given a complete ionic equation, identify the spectator ions, determine the net ionic equation, and describe the reaction type.

Develop a plan: Ions that remain completely unchanged (same physical phase and same ionized form) are spectator ions. Any ions that are the same on the product side as on the reactant side are eliminated to produce the net ionic equation.

Execute the plan:

In this example, the 2 NO_3^- ions are found both in the reactants and the products of the complete ionic equation; therefore, NO_3^-, nitrate ion, is a spectator ion. Everything else changes in one way or another. This is an example of a neutralization reaction that assists in the dissolving of an insoluble solid.

$$2\,H^+(aq) + Mg(OH)_2(s) \longrightarrow 2\,H_2O(\ell) + Mg^{2+}(aq)$$

This is an exchange reaction.

Check your answer: All spectator ions have been identified and eliminated to make the net ionic equation. The atoms are all balanced and the charges are all balanced in the equation.

17. *Define the problem*: Given overall chemical equations, identify the spectator ions and determine the net ionic equations.

Develop a plan: Use the solubility rules to determine the actual physical state of the reactants and products. Ions that remain completely unchanged (same physical phase and same ionized form) are the spectator ions. Any ions that are the same on the product side as on the reactant side are eliminated to produce the net ionic equation.

Note: All soluble ionic compounds and strong acids and bases are ionized in the complete ionic equation. All solids, weak acids and bases, and molecular compounds remain un-ionized.

Execute the plan: All ions have aqueous phase in the equations below.

(a) $CuCl_2$ is soluble (Table 5.1, Rule 3). H_2S is not a common strong acid (Table 5.2), so assume it is weak, CuS is insoluble (Table 5.1, Rule 13), and HCl is a common strong acid (Table 5.2)

$$CuCl_2(aq) + H_2S(aq) \longrightarrow CuS(s) + 2\,HCl(aq)$$

$$Cu^{2+} + 2\,Cl^- + H_2S(aq) \longrightarrow CuS(s) + 2\,H^+ + 2\,Cl^- \quad \text{complete ionic equation}$$

Eliminate spectator ion, Cl^-, chloride ion.

$$Cu^{2+} + H_2S(aq) \longrightarrow CuS(s) + 2\,H^+ \quad \text{net ionic equation}$$

(b) $CaCl_2$ is soluble (Table 5.1, Rule 3). K_2CO_3 is soluble (Table 5.1, Rule 1). $CaCO_3$ is insoluble (Table 5.1, Rule 9). KCl is soluble (Table 5.1, Rule 1).

$$CaCl_2(aq) + K_2CO_3(aq) \longrightarrow CaCO_3(s) + 2\ KCl(aq)$$

$$Ca^{2+} + 2\ Cl^- + 2\ K^+ + CO_3^{2-} \longrightarrow CaCO_3(s) + 2\ K^+ + 2\ Cl^-$$

complete ionic equation

Eliminate spectator ions, K^+ and Cl^-, potassium and chloride ions.

$$Ca^{2+} + CO_3^{2-} \longrightarrow CaCO_3(s)\ \text{net ionic equation}$$

(c) $AgNO_3$ is soluble (Table 5.1, Rule 2). NaI is soluble (Table 5.1, Rule 1). AgI is insoluble (Table 5.1, Rule 3). $NaNO_3$ is soluble (Table 5.1, Rule 1).

$$AgNO_3(aq) + NaI(aq) \longrightarrow AgI(s) + NaNO_3(aq)$$

$$Ag^+ + NO_3^- + Na^+ + I^- \longrightarrow AgI(s) + Na^+ + NO_3^-\ \text{complete ionic equation}$$

Eliminate spectator ions, Na^+ and NO_3^-, sodium and nitrate ions.

$$Ag^+ + I^- \longrightarrow AgI(s)\ \text{net ionic equation}$$

Check your answers: All spectator ions have been identified and eliminated to make the net ionic equation. The atoms are all balanced and the charges are all balanced in each equation.

18. *Define the problem*: Given overall chemical equations, balance them, then write complete and net ionic equations.

Develop a plan: Use standard balancing techniques to balance the overall equation. Use the solubility rules to determine the actual physical state of the reactants and products. Write complete ionic equations using that information. Ions that remain completely unchanged (same physical phase and same ionized form) are the spectator ions. Any ions that are the same on the product side as on the reactant side are eliminated to produce the net ionic equation.

Note: Just like with balancing elements, we can balance ions as units: Select a balancing order and follow it.

Note: All soluble ionic compounds and strong acids and bases are ionized in the complete ionic equation. All solids, weak acids and bases, and molecular compounds remain unionized.

Execute the plan: All ions have aqueous phase in the equations below.

(a) $Zn(s) + 2\ HCl(aq) \longrightarrow H_2(g) + ZnCl_2(aq)$

$Zn(s) + 2\ H^+ + 2\ Cl^- \longrightarrow H_2(g) + Zn^{2+} + 2\ Cl^-$ complete ionic equation

Eliminate spectator ion, Cl^-.

$$Zn(s) + 2\ H^+ \longrightarrow H_2(g) + Zn^{2+}$$ net ionic equation

(b) $Mg(OH)_2(s) + 2\ HCl(aq) \longrightarrow MgCl_2(aq) + 2\ H_2O(\ell)$

$Mg(OH)_2(s) + 2\ H^+ + 2\ Cl^- \longrightarrow Mg^{2+} + 2\ Cl^- + 2\ H_2O(\ell)$

complete ionic equation

Eliminate spectator ion, Cl^-.

$$Mg(OH)_2(s) + 2\ H^+ \longrightarrow Mg^{2+} + 2\ H_2O(\ell)$$ net ionic equation

(c) $2\ HNO_3(aq) + CaCO_3(s) \longrightarrow Ca(NO_3)_2(aq) + H_2O(\ell) + CO_2(g)$

$2\ H^+ + 2\ NO_3^- + CaCO_3(s) \longrightarrow Ca^{2+} + 2\ NO_3^- + H_2O(\ell) + CO_2(g)$

complete ionic equation

Eliminate spectator ion, NO_3^-.

$$2\ H^+ + CaCO_3(s) \longrightarrow Ca^{2+} + H_2O(\ell) + CO_2(g)$$ net ionic equation

(d) $4\ HCl(aq) + MnO_2(s) \longrightarrow MnCl_2(aq) + Cl_2(g) + 2\ H_2O(\ell)$

$4\ H^+ + 4\ Cl^- + MnO_2(s) \longrightarrow Mn^{2+} + 2\ Cl^- + Cl_2(g) + 2\ H_2O(\ell)$

complete ionic equation

Eliminate spectator ions, two of the four Cl^-.

$$4\ H^+ + 2\ Cl^- + MnO_2(s) \longrightarrow Mn^{2+} + Cl_2(g) + 2\ H_2O(\ell)$$ net ionic equation

Check your answers: All of the equations are balanced. The net ionic equations have no spectator ions. The atoms are all balanced and the charges are all balanced in each equation.

20. *Define the problem*: Given overall chemical equations, balance them, then write complete and net ionic equations.

Develop a plan: Use the method described in the answer to Question 18.

Execute the plan: All ions have aqueous phase in the equations below.

(a) $Ca(OH)_2$ is slightly soluble (Table 5.1, Rule 10). HNO_3 is a strong acid (Table 5.2). $Ca(NO_3)_2$ is soluble (Table 5.1, Rule 2). H_2O is a liquid molecular compound.

$$Ca(OH)_2(s) + 2\ HNO_3(aq) \longrightarrow Ca(NO_3)_2(aq) + 2\ H_2O(\ell) \quad \text{balanced}$$

$$Ca(OH)_2(s) + 2\ H^+ + 2\ NO_3^- \longrightarrow Ca^{2+} + 2\ NO_3^- + 2\ H_2O(\ell)$$

complete ionic equation

Eliminate spectator ion, NO_3^-.

$$Ca(OH)_2(s) + 2\ H^+ \longrightarrow Ca^{2+} + 2\ H_2O(\ell) \quad \text{net ionic equation}$$

(b) $BaCl_2$ is soluble (Table 5.1, Rule 3). Na_2CO_3 is soluble (Table 5.1, Rule 1).

$BaCO_3$ is insoluble (Table 5.1, Rule 9), $NaCl$ is soluble (Table 5.1, Rule 1).

$$BaCl_2(aq) + Na_2CO_3(aq) \longrightarrow BaCO_3(s) + 2\ NaCl(aq) \quad \text{balanced}$$

$$Ba^{2+} + 2\ Cl^- + 2\ Na^+ + CO_3^{2-} \longrightarrow BaCO_3(s) + 2\ Na^+ + 2\ Cl^-$$

complete ionic equation

Eliminate spectator ions, Cl^- and Na^+.

$$Ba^{2+} + CO_3^{2-} \longrightarrow BaCO_3(s) \quad \text{net ionic equation}$$

(c) Na_3PO_4 is soluble (Table 5.1, Rule 1). $Ni(NO_3)_2$ is soluble (Table 5.1, Rule2). $Ni_3(PO_4)_2$ is insoluble (Table 5.1, Rule 8). $NaNO_3$ is soluble (Table 5.1, Rule 1).

$$Na_3PO_4(aq) + Ni(NO_3)_2(aq) \longrightarrow Ni_3(PO_4)_2(s) + NaNO_3(aq) \quad \text{unbalanced}$$

Select balancing order: Ni^{2+}, then PO_4^{3-}, then NO_3^-, and Na^+

$$2\ Na_3PO_4(aq) + 3\ Ni(NO_3)_2(aq) \longrightarrow Ni_3(PO_4)_2(s) + 6\ NaNO_3(aq) \quad \text{balanced}$$

$$6\ Na^+ + 2\ PO_4^{3-} + 3\ Ni^{2+} + 6\ NO_3^- \longrightarrow Ni_3(PO_4)_2(s) + 6\ Na^+ + 6\ NO_3^-$$

complete ionic equation

Eliminate spectator ions, NO_3^- and Na^+.

$$2\ PO_4^{3-} + 3\ Ni^{2+} \longrightarrow Ni_3(PO_4)_2(s) \quad \text{net ionic equation}$$

Check your answers: All of the balanced equations are balanced. The net ionic equations have no spectator ions. The atoms are all balanced and the charges are all balanced in each equation.

22. *Define the problem*: Given the names of the reactants and one product of a reaction, balance the equation that occurs.

Develop a plan: Use the solubility rules and acid and base identities (Tables 5.1 and 5.2) to determine solubility of salts and strength of acids and bases to determine the actual physical state of the reactants and exchange products. Use standard balancing techniques to balance the overall equation.

Execute the plan:

Barium hydroxide is $Ba(OH)_2$, a soluble hydroxide compound (Table 5.1, Rule 10). Nitric acid is HNO_3, a strong acid (Table 5.2). Each OH^- ion in the solution reacts with one H^+ ion to make one H_2O molecule. The ionic compound produced during this neutralization is $Ba(NO_3)_2$. It is soluble (Table 5.1, Rule 10).

$$Ba(OH)_2(aq) + 2\ HNO_3(aq) \longrightarrow Ba(NO_3)_2(aq) + 2\ H_2O(\ell) \quad \text{balanced}$$

Check your answer: The equation is balanced. This acid-base neutralization reaction produces liquid water.

24. *Define the problem*: Given an overall chemical equation for a precipitation, balance it, then write complete and net ionic equations.

Develop a plan: Use the method described in the answer to Question 18.

Execute the plan: All ions have aqueous phase in the equations below.

$CdCl_2$ is soluble (Table 5.1, Rule 3). NaOH is soluble (Table 5.1, Rule 1). $Cd(OH)_2$ is insoluble (Table 5.1, Rule 10). NaCl is soluble (Table 5.1, Rule 1)

$$CdCl_2(aq) + 2\ NaOH(aq) \longrightarrow Cd(OH)_2(s) + 2\ NaCl(aq) \quad \text{balanced overall equation}$$

$$Cd^{2+} + 2\ Cl^- + 2\ Na^+ + 2\ OH^- \longrightarrow Cd(OH)_2(s) + 2\ Na^+ + 2\ Cl^-$$

complete ionic equation

Eliminate spectator ions, Cl^- and Na^+.

$$Cd^{2+} + 2\ OH^- \longrightarrow Cd(OH)_2(s) \quad \text{net ionic equation}$$

Check your answer: The atoms and the charges in each equation are balanced. This precipitation reaction produces an insoluble solid.

26. *Define the problem*: Given the reactants of a precipitation reaction, balance the equation.

Develop a plan: Use the method described in the answer to Question 22.

Execute the plan: Lead(II) nitrate is $Pb(NO_3)_2$, a soluble compound (Table 5.1, Rule 2). Potassium chloride is KCl, a soluble compound (Table 5.1, Rule 1). The products of the reaction would be lead(II) chloride, $PbCl_2$, an insoluble compound (Table 5.1, Rule 3) and potassium nitrate, KNO_3, a soluble compound (Table 5.1, Rule 1).

$$Pb(NO_3)_2(aq) + 2\ KCl(aq) \longrightarrow PbCl_2(s) + 2\ KNO_3(aq) \quad \text{balanced overall equation}$$

Check your answer: The equation is balanced. This precipitation reaction produces an insoluble solid.

28. *Define the problem*: Given the reactants of a reaction, balance the equation.

Develop a plan: Use the method described in the answer to Question 22.

Execute the plan: Rhodochrosite is manganese(II) carbonate, $MnCO_3$, an insoluble compound (Table 5.1, Rule 9). Hydrochloric acid, HCl, is a strong acid (Table 5.2). The products of the reaction are manganese(II) chloride, $MnCl_2$, a soluble compound (Table 5.1, Rule 3) and carbonic acid, H_2CO_3, a weak acid (Table 5.2), which decomposes into liquid water and carbon dioxide gas in a gas-forming reaction.

$$MnCO_3(aq) + 2\ HCl(aq) \longrightarrow H_2O(\ell) + CO_2(g) + MnCl_2(aq) \quad \text{balanced overall equation}$$

Check your answer: The equation is balanced. This reaction between a carbonate compound and a strong acid produces H_2CO_3, which decomposes into water and CO_2.

Acids, Bases, and Salts

29. *Define the problem*: Given some chemical formulas, identify if they are acids or bases.

Develop a plan: Acids produce H^+ in aqueous solution. Bases produce OH^- in aqueous solution. Where possible use Table 5.2.

Execute the plan:

(a) KOH is a base, producing K^+ and OH^- ions when dissolved in water.

(b) $Mg(OH)_2$ is a base, producing Mg^{2+} and OH^- ions when dissolved in water.

(c) HClO is an acid, producing H^+ and ClO^- ions when dissolved in water.

(d) HBr is an acid, producing H^+ and Br^- ions when dissolved in water.

(e) LiOH is a base, producing Li^+ and OH^- ions when dissolved in water.

(f) H_2SO_3 is an acid, producing H^+, HSO_3^-, and SO_3^{2-} ions when dissolved in water.

Note: The answers given here do not distinguish the extent of ionization. (See the next Question.)

Check your answers: All of the ions produced are common ions, most of them are found in Figure 3.2 and Table 3.7.

30. *Define the problem*: Given some chemical formulas, identify if they are weak or strong acids and bases.

Develop a plan: These acids and bases were identified in Question 29. In several of these cases, Tables 5.1 or 5.2 can be used to determine whether a large or small amount of ions are produced, by determining if the compound is soluble and/or weak or strong. In some cases, however, it is not possible to look up that information in Chapter 5. However, it is almost always true that, if an acid is **not** one of the common strong acids listed in Table 5.2, it is a **weak** acid. That is the case with the weak acids in this Question.

Execute the plan:

(a) KOH is a strong base. (Given in Table 5.2)

(b) $Mg(OH)_2$ is an insoluble ionic compound (Table 5.1, Rule 10) so few ions are produced in water, though the $Mg(OH)_2$ that dissolves does ionize completely. So, practically, it may be considered weak, because the OH^- ion concentration will never be very large. Technically, it is considered to be strong, because all the dissolved $Mg(OH)_2$ is ionized.

(c) HClO is a weak acid. (It's not listed as a common strong acid in Table 5.2.)

(d) HBr is a strong acid. (Given in Table 5.2)

(e) LiOH is a strong base. (Given in Table 5.2)

(f) H_2SO_3 is a weak acid. (It's not listed as a common strong acid in Table 5.2.)

Check your answers: Most of these are given specifically in Table 5.2. IF you want to look far into your chemistry future, Table 16.2 can also be used to confirm that the two weak acids given here in (c) and (f) are indeed weak.

32. *Define the problem*: Having found the formulas of the acid and the base used to make the given salt (see Question 31), identify the spectator ions, and determine the net ionic equations.

Develop a plan: Set up the complete equation, then use the method described in the answer to Question 18.

Execute the plan: All ions have aqueous phase in the equations below.

(a) HNO_2 is a weak acid. NaOH is a strong base. $NaNO_2$ is a soluble compound.

$$HNO_2(aq) + NaOH(aq) \longrightarrow H_2O(\ell) + NaNO_2(aq) \quad \text{balanced overall equation}$$

$$HNO_2(aq) + Na^+ + OH^- \longrightarrow H_2O(\ell) + Na^+ + NO_2^- \quad \text{complete ionic equation}$$

Eliminate spectator ion, Na^+.

$$HNO_2(aq) + OH^- \longrightarrow H_2O(\ell) + NO_2^- \quad \text{net ionic equation}$$

(b) The first ionization of H_2SO_4 is strong, but HSO_4^- is a weak acid. The reactant $Ca(OH)_2$ is only slightly soluble. The product, $CaSO_4$, is insoluble.

$$H_2SO_4(aq) + Ca(OH)_2(s) \longrightarrow 2\,H_2O(\ell) + CaSO_4(s) \quad \text{balanced overall equation}$$

$$H^+ + HSO_4^- + Ca(OH)_2(s) \longrightarrow 2\,H_2O(\ell) + CaSO_4(s)$$

$$\text{complete ionic and net ionic equations}$$

(c) HI is a strong acid and NaOH is a strong base. The product, NaI, is soluble.

$$HI(aq) + NaOH(aq) \longrightarrow H_2O(\ell) + NaI(aq) \quad \text{balanced overall equation}$$

$$H^+ + I^- + Na^+ + OH^- \longrightarrow H_2O(\ell) + Na^+ + I^- \quad \text{complete ionic equation}$$

Eliminate spectator ions, Na^+ and I^-.

$$H^+ + OH^- \longrightarrow H_2O(\ell) \quad \text{net ionic equation}$$

(d) H_3PO_4 is a weak acid. The reactant $Mg(OH)_2$ is insoluble. The product, $Mg_3(PO_4)_2$, is insoluble.

$$2\,H_3PO_4(aq) + 3\,Mg(OH)_2(s) \longrightarrow 6\,H_2O(\ell) + Mg_3(PO_4)_2(s)$$

$$\text{balanced overall, complete ionic, and net ionic equations}$$

(e) CH_3COOH is a weak acid and NaOH is a strong base. The product, $NaCH_3COO$, is soluble.

$$CH_3COOH(aq) + NaOH(aq) \longrightarrow H_2O(\ell) + NaCH_3COO(aq)$$

$$\text{balanced overall equation}$$

$$CH_3COOH(aq) + Na^+ + OH^- \longrightarrow H_2O(\ell) + Na^+ + CH_3COO^-$$

complete ionic equation

Eliminate spectator ion, Na^+.

$$CH_3COOH(aq) + OH^- \longrightarrow H_2O(\ell) + CH_3COO^- \quad \text{net ionic equation}$$

Check your answers: These acids and bases undergo neutralization to produce the appropriate salt and water. The net ionic equation does not always includes the whole salt, when one or both of the ions of the salt are found to be spectator ions.

34. *Define the problem*: Given the reactants of reactions, classify the reaction that occurs, identify the products, and balance the equations.

Develop a plan: To classify these reactions, determine the exchange products and check their solubility and/or strength using Tables 5.1 and 5.2. Remember that carbonate compounds reacting with acids produce CO_2 gas.

Execute the plan:

(a) When the base and insoluble hydroxide, $Fe(OH)_3$, reacts with the acid HNO_3, we get an **acid-base reaction**. The exchange products are liquid H_2O and $Fe(NO_3)_3$. Checking the solubility of the salt, we find that $Fe(NO_3)_3$ is soluble.

$$Fe(OH)_3(s) + 3\ HNO_3(aq) \longrightarrow 3\ H_2O(\ell) + Fe(NO_3)_3(aq)$$

balanced overall equation

(b) When $FeCO_3$ reacts with H_2SO_4, the exchange products are $FeSO_4$ and H_2CO_3, which decomposes into liquid H_2O and CO_2 gas. That makes this a **gas-forming reaction**. $FeSO_4$ is soluble.

$$FeCO_3(s) + H_2SO_4(aq) \longrightarrow FeSO_4(aq) + H_2O(\ell) + CO_2(g)$$

balanced overall equation

(c) When $FeCl_2$ reacts with $(NH_4)_2S$, the exchange products are NH_4Cl and FeS Checking their solubility, we find that NH_4Cl is soluble, but FeS is insoluble. That makes this a **precipitation reaction**.

$$FeCl_2(aq) + (NH_4)_2S(aq) \longrightarrow 2\ NH_4Cl(aq) + FeS(s) \quad \text{balanced overall equation}$$

(d) When $Fe(NO_3)_2$ reacts with Na_2CO_3, the exchange products are $NaNO_3$ and $FeCO_3$. Checking their solubility, we find that $NaNO_3$ is soluble, but $FeCO_3$ is insoluble. That makes this a **precipitation reaction**.

$$Fe(NO_3)_2(aq) + Na_2CO_3(aq) \longrightarrow 2\,NaCl(aq) + FeCO_3(s) \quad \text{balanced overall equation}$$

Check your answers: Acid-base neutralization reactions form water. Precipitation reactions form insoluble compounds and gas-forming reactions produce gases.

Oxidation-Reduction Reactions

36. *Define the problem*: Given several formulas of compounds, determine the oxidation numbers of the atoms in each of them.

Develop a plan: For this Question, the rules spelled out in Section 5.4 are used extensively. Several elements in compounds have predictable oxidation numbers (Rules 1-3). The oxidation numbers of the remaining atom(s) can be determined using the sum rule (Rule 4).

Execute the plan: The term "oxidation number" is abbreviated below as "Ox. #".

(a) $Fe(OH)_3$ contains a monatomic cation, Fe^{3+}, and a diatomic anion, OH^-. Rule 2 gives us Ox. # Fe = +3. Rule 3 gives us Ox. # O = –2 and Ox. # H = +1.

(b) $HClO_3$ contains oxygen and hydrogen. Rule 3 gives us Ox. # O = –2 and Ox. # H = +1. We use the sum rule to find the Ox. # Cl. This is a molecule, so the sum of the oxidation numbers is zero.

$$0 = 1 \times (+1) + 1 \times (\text{Ox. \# Cl}) + 3 \times (-2)$$

Therefore, Ox. # Cl = +5.

(c) $CuCl_2$ contains a monatomic cation, Cu^{2+}, and a monatomic anion, Cl^-. Rule 2 gives us Ox. # Cu = +2 and Ox. # Cl = –1.

(d) K_2CrO_4 contains a monatomic cation, K^+, and a polyatomic anion, CrO_4^{2-}. Rule 2 gives us Ox. # K = +1. Rule 3 gives us Ox. # O = –2. We use the sum rule to find the Ox. # Cr. For a polyatomic anion, the sum of the oxidation numbers is equal to the ion's charge of 2–.

$$-2 = 1 \times (\text{Ox. \# Cr}) + 4 \times (-2)$$

Therefore, Ox. # Cr = +6.

(e) $Ni(OH)_2$ contains a monatomic cation, Ni^{2+}, and a diatomic anion, OH^-. Rule 2 gives us Ox. # Ni = +2. Rule 3 gives us Ox. # O = –2 and Ox. # H = +1.

(f) N_2H_4 contains hydrogen. Rule 3 gives us Ox. # H = +1. We use the sum rule to find the Ox. # N. This is a molecule, so the sum of the oxidation numbers is zero.

$$0 = 2 \times (\text{Ox. \# N}) + 4 \times (+1)$$

Therefore, $2 \times (\text{Ox. \# N}) = -4$, and Ox. # N = –2.

Check your answer: The non-metal elements farther to the right on the periodic table have consistently more negative oxidation numbers. The metallic elements and nonmetals farther to the left on the periodic table have more positive oxidation numbers.

37. *Define the problem*: Given the several formulas of ions, determine the oxidation numbers of the atoms in each one.

Develop a plan: For this Question, the rules spelled out in Section 5.4 are used extensively. Several elements in compounds have predictable oxidation numbers (Rules 1-3). The oxidation numbers of the remaining atom(s) can be determined using the sum rule (Rule 4).

Execute the plan: The term "oxidation number" is abbreviated below as "Ox. #".

(a) SO_4^{2-}. Rule 3 gives us Ox. # O = –2. We use the sum rule to find the Ox. # S. This is a polyatomic anion, so the sum of the oxidation numbers is equal to the ion's charge of 2–.

$$-2 = 1 \times (\text{Ox. \# S}) + 4 \times (-2)$$

Therefore, Ox. # S = +6.

(b) NO_3^- contains oxygen. Rule 3 gives us Ox. # O = –2. We use the sum rule to find the Ox. # N. This is a polyatomic anion, so the sum of the oxidation numbers is equal to the ion's charge of 1–:

$$-1 = 1 \times (\text{Ox. \# N}) + 3 \times (-2)$$

Therefore, Ox. # N = +5.

(c) MnO_4^- contains oxygen. Rule 3 gives us Ox. # O = –2. We use the sum rule to find the Ox. # Mn. For a polyatomic anion, the sum of the oxidation numbers is equal to the ion's charge of 1–.

$$-1 = 1 \times (\text{Ox. \# Mn}) + 4 \times (-2)$$

Therefore, Ox. # Mn = +7.

(d) $Cr(OH)_4^-$ contains oxygen and hydrogen. Rule 3 gives us Ox. # O = –2 and Ox. # H = +1. We use the sum rule to find the Ox. # Cr. For a polyatomic anion, the sum of the oxidation numbers is equal to the ion's charge of 1–.

$$-1 = 1 \times (\text{Ox. \# Cr}) + 4 \times [1 \times (-2) + 1 \times (+1)]$$

Therefore, Ox. # Cr = +3.

(e) $H_2PO_4^-$ contains oxygen and hydrogen. Rule 3 gives us Ox. # O = –2 and Ox. # H = +1. We use the sum rule to find the Ox. # P. For a polyatomic anion, the sum of the oxidation numbers is equal to the ion's charge of 1–.

$$-1 = 2 \times (+1) + 1 \times (\text{Ox. \# P}) + 4 \times (-2)$$

Therefore, Ox. # P = +5.

(f) *There is a typo in the text. The formula for this ion must have a 2– charge.*

$S_2O_3^{2-}$ contains oxygen. Rule 3 gives us Ox. # O = –2. We use the sum rule to find the Ox. # S. For a polyatomic anion, the sum of the oxidation numbers is equal to the ion's charge of 2–.

$$-2 = 2 \times (\text{Ox. \# S}) + 3 \times (-2)$$

Therefore, $2 \times (\text{Ox. \# S}) = +4$ and Ox. # S = +2.

Check your answers: The non-metal elements farther to the right on the periodic table have consistently more negative oxidation numbers. The metallic elements and nonmetals farther to the left on the periodic table have more positive oxidation numbers.

40. The best reducing agents are elements readily oxidized. Oxidation is losing electrons. The left side of the periodic table makes the best reducing agents, since these elements commonly form stable cations by losing electrons. The best oxidizing agents are elements readily reduced. The elements on the right side of the periodic table (excluding the noble gases) make the best oxidizing agents, since these elements commonly form stable anions by gaining electrons.

42. (a) Ca, (d) Al, and (f) H_2 are common reducing agents. They are all good at reducing other chemicals because they are readily oxidized.

44. *Define the problem*: Given reactants of oxidation-reduction reactions, identify the displacement reactions' products and balance the equation.

Develop a plan: A displacement reaction is when an element reacts with a compound and a new compound is formed along with a different element.

Execute the plan:

(a) $2 K(s) + H_2O(\ell) \longrightarrow 2 KOH(aq) + H_2(g)$

(b) $Mg(s) + 2 HBr(aq) \longrightarrow MgBr_2(aq) + H_2(g)$

(c) $2 NaBr(aq) + Cl_2(aq) \longrightarrow 2 NaCl(aq) + Br_2(aq)$

(d) $WO_3(s) + 3 H_2(g) \longrightarrow 3 H_2O(\ell) + W(s)$

(e) $8 H_2S(aq) + 8 Cl_2(g) \longrightarrow 16 HCl(aq) + S_8(s)$

Check your answers: The atoms in the reactant elements are found in the compound in the products. An element in the reactant compound is found in its elemental form in the products. These are valid displacement reactions.

Activity Series

46. (a) The most reactive metals are in Groups 1A and 2A. (According to Table 5.5, some of the most reactive metals are: Li, K, Ba, Sr, Ca, and Na.) The metals on the right side of the transition elements are the least reactive. (According to Table 5.5, some of the least reactive metals are: Au, Pt, Pd, Ag, Hg, and Cu.)

(b) Table 5.5 says that aluminum is more active than H_2. (H_2 forms when Al is in acid.) H_2 is more active than Ag, which is why it is not formed with Ag in acid, so Al is more active metal than Ag. This is confirmed in Table 5.5. A less reactive metal will not form a more reactive metal, so Ag will not react with Al^{3+} to form Al and Ag^+.

(c) $Pb + 2 H^+ \longrightarrow Pb^{2+} + H_2.$

This happens slowly, so Pb is somewhat more active than H_2.

$$Al + 3 Pb^{2+} \longrightarrow 2 Al^{3+} + Pb.$$

So, Al is more active than Pb.

In (b), we learned that Ag will not react in acid to make H_2. So, Pb is more reactive than Ag. Therefore Pb should react with Ag^+ to form Ag.

$$Pb + 2 Ag^+ \longrightarrow Pb^{2+} + 2 Ag$$

$$Pb(s) + 2 AgNO_3(aq) \longrightarrow Pb(NO_3)_2(aq) + 2 Ag(s)$$

(d) Summarizing the results of (a) to (c) above, the order of decreasing reactivity is:

$$Al(s) > Pb(s) > Ag(s)$$

48. *Define the problem*: Given a set of reactions and using Table 5.5, determine which reaction will occur.

Develop a plan: First check to see if the reaction is a properly balanced oxidation-reduction reaction. It must have one substance oxidizing and one substance reducing. Then, look up the neutral product and the neutral reactant and find which is more active. If the reactant substance is more active than the product substance, then the reaction will occur.

Execute the plan:

(a) No. The equation is not balanced; both metals cannot be oxidized simultaneously.

(b) No. The equation is not balanced; both ions cannot be reduced simultaneously.

(c) No. This reaction will not happen, since Zn (the product) is more active than H_2 (the reactant).

(d) Yes. This reaction will happen, since Mg (the reactant) is more active than Cu (the product).

(e) Yes. This reaction will happen, since Pb (the reactant) is more active than H_2 (the product).

(f) Yes. This reaction will happen, since Cu (the reactant) is more active than Ag (the product).

(g) No. This reaction will not happen, since Al (the product) is more active than Zn (the reactant).

Check your answers: Each time the reactant is more active, the reaction occurs.

Halogens in Redox Reactions

50. *Define the problem*: Given a set of halogen displacement reactions and using Table 5.5, determine which reaction will occur, and in cases where a reaction does occur, predict the products.

Develop a plan: Look up the neutral product and the neutral reactant and find which is more active. If the reactant substance is more active than the product substance, then the reaction will occur.

Execute the plan:

(a) This reaction will not happen, since Br_2 (the product) is more active than I_2 (the reactant). $I_2 + 2\ NaBr \longrightarrow$ N.R.

(b) This reaction will happen, since Br_2 (the product) is more active than I_2 (the reactant). $Br_2 + 2\ NaI \longrightarrow I_2 + 2\ NaBr$

(c) This reaction will happen, since F_2 (the reactant) is more active than Cl_2 (the product). $F_2 + 2\ NaCl \longrightarrow Cl_2 + 2\ NaF$

(d) This reaction will happen, since Cl_2 (the reactant) is more active than Br_2 (the product). $Cl_2 + 2\ NaBr \longrightarrow Br_2 + 2\ NaCl$

(e) This reaction will not happen, since Cl_2 (the reactant) is more active than Br_2 (the product). $Br_2 + 2\ NaCl \longrightarrow N.R.$

(f) This reaction will not happen, since F_2 (the product) is more active than Cl_2 (the reactant). $Cl_2 + 2\ NaF \longrightarrow N.R.$

Check your answers: Each time the reactant is more active, the reaction occurs.

52. *Define the problem*: Take the halogen displacement reactions in Question 50 that do not occur, and rewrite the equation so that the reaction does occur.

Develop a plan: Since these did not occur, it meant that the more active substance was in the products. Reverse the reaction so that the more active substance is a reactant.

Execute the plan:

(a) Br_2 is more active than I_2, so make Br_2 the reactant.

$$Br_2 + 2\ NaI \longrightarrow I_2 + 2\ NaBr$$

(e) Cl_2 is more active than Br_2, so make Cl_2 the reactant.

$$Cl_2 + 2\ NaBr \longrightarrow Br_2 + 2\ NaCl$$

(f) F_2 is more active than Cl_2, so make F_2 the reactant.

$$F_2 + 2\ NaCl \longrightarrow Cl_2 + 2\ NaF$$

Note: (b), (c) and (d) had reactions that occur.

Check your answers: Each time the reactant is more active, the reaction occurs.

Solution Concentrations

53. Formula stoichiometry: 1 mol of $BaCl_2$ contains 1 mol Ba^{2+} ions and 2 mol Cl^- ions.

So, the 0.12 M $BaCl_2$ solutions contains 0.12 M Ba^{2+} ion and $2 \times (0.12\ M)$ Cl^- ion = 0.24 M Cl^- ion.

55. *Define the problem*: Given the mass of the solute and the volume of the solution, find the molarity of the solute, and the concentrations of the ions.

Develop a plan: Use the molar mass to find moles of solute, convert the volume to liters from milliliters, and divide the moles of solute by the volume in liters to get molarity. Use formula stoichiometry to find the concentrations of the ions.

Execute the plan:

(a) Find moles of solute:

$$6.73 \text{ g Na}_2\text{CO}_3 \times \frac{1 \text{ mol Na}_2\text{CO}_3}{105.99 \text{ g Na}_2\text{CO}_3} = 6.35 \times 10^{-2} \text{ mol Na}_2\text{CO}_3$$

$$250. \text{ mL solution} \times \frac{1 \text{ L}}{1000 \text{ mL}} = 0.250 \text{ L solution}$$

$$\text{Molarity} = \frac{6.35 \times 10^{-2} \text{ mol Na}_2\text{CO}_3}{0.250 \text{ L solution}} = 0.254 \text{ M Na}_2\text{CO}_3$$

Notice that the three sequential calculations shown above can be combined into one calculation, as follows:

$$\frac{6.73 \text{ g Na}_2\text{CO}_3}{250. \text{ mL solution}} \times \frac{1 \text{ mol Na}_2\text{CO}_3}{105.99 \text{ g Na}_2\text{CO}_3} \times \frac{1000 \text{ mL}}{1 \text{ L}} = 0.254 \text{ M Na}_2\text{CO}_3$$

This combined version prevents having to write unnecessary intermediate answers and helps cut down on round-off errors in significant figures. Both ways give the right answer, but it is helpful to consolidate your work, when you can.

(b) Na_2CO_3 has a formula stoichiometry that looks like this:

$$1 \text{ mol Na}_2\text{CO}_3 : 2 \text{ mol Na}^+ \text{ ions: } 1 \text{ mol CO}_3^{2-} \text{ ions.}$$

So, the 0.254 M Na_2CO_3 contains $2 \times (0.254 \text{ M}) \text{ Na}^+$ ion = 0.508 M Na^+ ion and 0.254 M CO_3^{2-}.

Check your answers: The number 0.0635 is about one quarter of .250; this value looks right. The concentration of Na^+ is twice than the concentration of CO_3^{2-}. These answers look right.

57. *Define the problem*: Given the volume of the solution and its molarity, find the mass of the solute in the solution.

Develop a plan: Convert the volume of solution to liters from milliliters. Then use the molarity as a conversion factor to determine moles of solute. Then use the molar mass to find grams of solute.

Execute the plan:

$$250. \text{ mL solution} \times \frac{1 \text{ L}}{1000 \text{ mL}} \times \frac{0.0125 \text{ mol KMnO}_4}{1 \text{ L solution}} \times \frac{158.04 \text{ g KMnO}_4}{1 \text{ mol KMnO}_4}$$

$$= 0.493 \text{ g KMnO}_4$$

Check your answer: The units cancel appropriately. The relative size of the answer seems appropriate. This answer looks right.

59. *Define the problem*: Given the mass of the solute and the solution's molarity, find the volume of the solution.

Develop a plan: Use the molar mass to find determine moles of solute. Then use the molarity as a conversion factor to determine volume of the solute in liters. Then convert the volume of solution to milliliters from liters.

Execute the plan:

$$25.0 \text{ g NaOH} \times \frac{1 \text{ mol NaOH}}{40.00 \text{ gNaOH}} \times \frac{1 \text{ L solution}}{0.123 \text{ mol NaOH}} \times \frac{1000 \text{ mL}}{1 \text{ L}} = 5.08 \times 10^3 \text{ mL solution}$$

Check your answer: The units cancel appropriately. The relatively large number of milliliters seems appropriate, since this relatively dilute solution contains a relatively large mass of solute. This answer looks right.

61. *Define the problem*: Given the volume of a concentrated solution, the concentrated solution's molarity, and the final volume of the dilute solution, find the concentration of the dilute solution.

Develop a plan: Moles of solute do not change when water is added. So we can equate the moles of solute in the concentrated solution with the moles of solute in the dilute solution. Get the moles of solute in the concentrated solution by using the molarity of the concentrated solution as a conversion factor to convert the volume into moles. Divide the moles of solute by the new dilute solution's volume to get the dilute solution's concentration. The logical plan outlined here is built into the equation given in Section 5.6 for doing dilution calculations:

$$Molarity(\text{conc}) \times V(\text{conc}) = Molarity(\text{dil}) \times V(\text{dil})$$

Execute the plan:

$$\frac{Molarity(\text{conc}) \times V(\text{conc})}{V(\text{dil})} = \frac{(0.0250 \text{ M CuSO}_4 \text{ conc}) \times (6.00 \text{ mL conc})}{(10.0 \text{ mL dil})}$$

$$= 0.0150 \text{ M CuSO}_4$$

Check your answer: The dilute solution has a smaller concentration than the concentrated solution. That is a sensible result, since water was added to make the new solution.

63. *Define the problem*: Given the desired volume and molarity of a dilute solution, determine which of several dilution processes produces this solution.

Develop a plan: Looking at each choice, it's easy to see that each solution results in a total volume of 1.00 L, which is the desired volume. So, we should focus our attention on which of the choices provides a solution that contains the proper number of moles. Convert milliliters to liters. Then calculate the moles in each choice by multiplying the volume in liters by the concentration of the concentrated solution provided.

Execute the plan: We want 1.00 L of 0.125 M H_2SO_4

$$1.00 \text{ L dil} \times \frac{0.125 \text{ mol } H_2SO_4}{1 \text{ L dil}} = 0.125 \text{ moles of } H_2SO_4$$

So, determine moles H_2SO_4 in each choice and compare it to the 0.125 moles H_2SO_4 desired.

(a) $36.0 \text{ mL conc} \times \dfrac{1 \text{ L}}{1000 \text{ mL}} \times \dfrac{1.50 \text{ mol } H_2SO_4}{1 \text{ L conc}} = 0.00540 \text{ mol } H_2SO_4$ No

(b) $20.8 \text{ mL conc} \times \dfrac{1 \text{ L}}{1000 \text{ mL}} \times \dfrac{6.00 \text{ mol } H_2SO_4}{1 \text{ L conc}} = 0.125 \text{ mol } H_2SO_4$ Yes

(c) $50.0 \text{ mL conc} \times \dfrac{1 \text{ L}}{1000 \text{ mL}} \times \dfrac{3.00 \text{ mol } H_2SO_4}{1 \text{ L conc}} = 0.150 \text{ mol } H_2SO_4$ No

(d) $500. \text{ mL conc} \times \dfrac{1 \text{ L}}{1000 \text{ mL}} \times \dfrac{0.500 \text{ mol } H_2SO_4}{1 \text{ L conc}} = 0.250 \text{ mol } H_2SO_4$ No

Only choice (b) will make the desired solution.

Note: You also could have used the *Molarity*(conc) × *V*(conc) = *Molarity*(dil) × *V*(dil) dilution equation to solve this Question.

Check your answer: Compare moles of solute in the concentrated and dilute solutions:

$$\textit{Molarity}(\text{conc}) \times V(\text{conc}) = (6.00 \text{ M } H_2SO_4 \text{ conc}) \times (20.8 \text{ mL conc}) \times \frac{1 \text{ L}}{1000 \text{ mL}}$$

$$= 0.125 \text{ mol } H_2SO_4$$

Molarity(dil) × *V*(dil) = (0.125 M H_2SO_4 dil) × (1.00 L dil) = 0.125 mol H_2SO_4

Within three significant figures, the moles in the concentrated solution (0.125 mol) are the same as the moles in the dilute solution (0.125 mol), as expected. This answer makes sense.

Calculations for Reactions in Solution

65. *Define the problem*: Given the volume and molarity of a solution containing one reactant and the balanced chemical equation for a reaction, determine the mass of another reactant required for complete reaction.

Develop a plan: The stoichiometry of a balanced chemical equation dictates how the moles of reactants combine, so we will commonly look for ways to calculate moles. Here, the volume and molarity can be used to find the moles of one reactant. Then we will use the equation stoichiometry to find out moles of the other reactant needed. Finally, we will use the molar mass to find the grams. *Note: It is NOT appropriate to use the dilution equation when working with reactions!*

Execute the plan:

We learn from the balanced equation that 2 mol of HNO_3 reacts with 1 mol Na_2CO_3.

$$25.0 \text{ mL } HNO_3 \text{ solution} \times \frac{1 \text{ L}}{1000 \text{ mL}} \times \frac{0.155 \text{ mol } HNO_3}{1 \text{ L } HNO_3 \text{ solution}}$$

$$\times \frac{1 \text{ mol } Na_2CO_3}{2 \text{ mol } HNO_3} \times \frac{105.99 \text{ g } Na_2CO_3}{1 \text{ mol } Na_2CO_3} = 0.205 \text{ g } Na_2CO_3$$

Check your answer: The units cancel appropriately, and the answer is a reasonable size. This result makes sense.

67. *Define the problem*: Given the mass of one reactant, the balanced chemical equation for a reaction, and the molarity of a solution containing the other reactant, determine the volume of the second solution for a complete reaction.

Develop a plan: The stoichiometry of a balanced chemical equation dictates how the moles of reactants combine, so we will commonly look for ways to calculate moles. Here, the mass and molar mass can be used to find the moles of one reactant. Then we will use the equation stoichiometry to find out moles of the other reactant needed. Then we will use the moles and molarity to find volume in liters and convert liters into milliliters.

Note: It is NOT appropriate to use the dilution equation when working with reactions!

Execute the plan:

We learn from the balanced equation that 1 mol of $Ba(OH)_2$ reacts with 2 mol HNO_3.

$$1.30 \text{ g } Ba(OH)_2 \times \frac{1 \text{ mol } Ba(OH)_2}{171.35 \text{ g } Ba(OH)_2} \times \frac{2 \text{ mol } HNO_3}{1 \text{ mol } Ba(OH)_2}$$

$$\times \frac{1 \text{ L } HNO_3 \text{ solution}}{0.125 \text{ mol } HNO_3} \times \frac{1000 \text{ mL}}{1 \text{ L}} = 121 \text{ mL } HNO_3 \text{ solution}$$

Check your answer: The units cancel appropriately, and the answer is a reasonable size. This result makes sense.

71. *Define the problem:* Given the volumes and molarities of separate solutions containing each reactant of a precipitation reaction, determine the maximum mass of product produced and the identity and concentration of the excess reactant.

Develop a plan: First, complete and balance the precipitation equation. Then use the volumes, molarities, and the equation stoichiometry to find the moles of product produced for each reactant. The reactant that produces the smallest number of moles is the limiting reactant. Use the moles of product produced by the limiting reactant to determine the maximum mass of product formed. Determine the number of moles of excess reactant in the solution, using stoichiometry. Determine the excess reactant's final concentration by dividing by the new solution's volume in liters. *Note: It is NOT appropriate to use the dilution equation when working with reactions!*

Execute the plan: The precipitation reaction looks like this:

$$AgNO_3(aq) + NaCl(aq) \longrightarrow AgCl(s) + NaNO_3(aq)$$

We learn from this balanced equation that 1 mol of NaCl produces 1 mol AgCl and that 1 mol of NaCl produces 1 mol AgCl.

$$50.0 \text{ mL AgNO}_3 \text{ solution} \times \frac{1 \text{ L}}{1000 \text{ mL}} \times \frac{0.025 \text{ mol AgNO}_3}{1 \text{ L AgNO}_3 \text{ solution}} \times \frac{1 \text{ mol AgCl}}{1 \text{ mol AgNO}_3}$$

$$= 0.0013 \text{ mol AgCl}$$

$$100.0 \text{ mL NaCl solution} \times \frac{1 \text{ L}}{1000 \text{ mL}} \times \frac{0.025 \text{ mol NaCl}}{1 \text{ L NaCl solution}} \times \frac{1 \text{ mol AgCl}}{1 \text{ mol NaCl}}$$

$$= 0.0025 \text{ mol AgCl}$$

AgCl is the limiting reactant, and NaCl is the excess reactant. Use 0.0013 mol AgCl to determine the maximum grams of AgCl produced:

$$0.0013 \text{ mol AgCl} \times \frac{143.4 \text{ g AgCl}}{1 \text{ mol AgCl}} = 0.18 \text{ g AgCl}$$

If 0.0013 mol AgCl was formed and 0.0025 mol AgCl was expected from NaCl, the difference is how much AgCl was not formed:

$$0.0025 \text{ mol AgCl expected} - 0.0013 \text{ mol AgCl formed} = 0.0012 \text{ mol AgCl not formed}$$

$$0.0012 \text{ mol AgCl not formed} \times \frac{1 \text{ mol NaCl}}{1 \text{ mol AgCl}} = 0.0012 \text{ mol NaCl not reacted}$$

$$\text{Total Volume} = (50.0 \text{ mL} + 100.0 \text{ mL}) \times \frac{1 \text{ L}}{1000 \text{ mL}} = 0.1500 \text{ L}$$

Molarity of NaCl after the AgCl has completely precipitated:

$$\text{Molarity} = \frac{0.013 \text{ mol NaCl}}{0.1500 \text{ L solution}} = 0.0080 \text{ M NaCl}$$

Check your answers: The concentration of the excess reactant after the reaction is over is smaller than it was at the beginning. This make sense, because some of it reacted, and what was left over got diluted. The maximum mass of AgCl produced is a reasonable number. These answers make sense.

73. *Define the problem*: Given the volume and molarity of a solution containing one reactant, the balanced chemical equation, and a series of steps describing the calculation, determine which of the steps is not correct, and correctly determine the mass of the other reactant in a specific volume of its solution.

Develop a plan: Check each step to see if it is right or wrong. If it is wrong, correct it.

Execute the plan:

(a) **Step (i) is correct.**

$$6.42 \text{ mL} \times \frac{1 \text{ L}}{1000 \text{ mL}} \times \frac{9.580 \times 10^{-2} \text{ mol NaOH}}{1 \text{ L}} = 6.15 \times 10^{-4} \text{ mol NaOH}$$

Step (ii) is incorrect.

The equation given in the Question is not balanced. The balanced equation is:

$$C_3H_5O(COOH)_3(aq) + 3 \text{ NaOH}(aq) \longrightarrow Na_3C_3H_5O(COO)_3(aq) + 3 H_2O(\ell)$$

So, the equation stoichiometry gives 1 mol citric acid reacting with 3 mol of NaOH.

$$6.15 \times 10^{-4} \text{ mol NaOH} \times \frac{1 \text{ mol citric acid}}{3 \text{ mol NaOH}} = 2.05 \times 10^{-4} \text{ mol citric acid}$$

Step (ii) has a 3 in front of the moles of citric acid and a 1 in front of the NaOH, resulting in multiplication by 3 instead of division by 3. Thus the moles of citric acid calculated in this step are wrong.

Step (iii) is incorrect, because it uses the erroneous answer from Step (ii); however, the calculation it shows is correct:

$$2.05 \times 10^{-4} \text{ mol citric acid} \times \frac{192.12 \text{ g citric acid}}{1 \text{ mol citric acid}} = 0.0394 \text{ g citric acid}$$

Step (iv) is incorrect, because it uses a different answer than the correct one from Step (iii) and the significant figures on the volume are incorrect; however, the calculation it shows is correct:

$$\frac{0.0394 \text{ g citric acid}}{10.0 \text{ mL}} = 3.94 \times 10^{-3} \text{ g citric acid in 1 mL of soft drink}$$

(b) The right answer is 3.94×10^{-3} g citric acid in 1 mL of soft drink.

Check your answer: The moles of citric acid that react must be less than the moles of NaOH that react. This answer now makes sense.

75. *Define the problem*: Given the mass of one reactant, the balanced chemical equation for a reaction, and the volume of a solution containing the other reactant needed for a complete reaction, determine the molarity of the second solution.

Develop a plan: The mass and molar mass can be used to find the moles of one reactant. Then we will use the equation stoichiometry to find out moles of the other reactant needed. Then we will use the moles and volume in liters to determine the molarity. *Note: It is NOT appropriate to use the dilution equation when working with reactions!*

Execute the plan:

We learn from the balanced equation that 1 mol of Na_2CO_3 reacts with 2 mol HCl.

$$2.050 \text{ g Na}_2\text{CO}_3 \times \frac{1 \text{ mol Na}_2\text{CO}_3}{105.99 \text{ g Na}_2\text{CO}_3} \times \frac{2 \text{ mol HCl}}{1 \text{ mol Na}_2\text{CO}_3} = 3.868 \times 10^{-2} \text{ mol HCl}$$

$$32.45 \text{ mL HCl solution} \times \frac{1 \text{ L}}{1000 \text{ mL}} = 0.03245 \text{ L HCl solution}$$

$$\frac{3.868 \times 10^{-2} \text{ mol HCl}}{0.03245 \text{ L HCl solution}} = 1.192 \text{ M HCl solution}$$

The three separate calculations above can be consolidated into one calculation as follows:

$$\frac{2.050 \text{ g Na}_2\text{CO}_3}{32.45 \text{ mL HCl solution}} \times \frac{1 \text{ mol Na}_2\text{CO}_3}{105.99 \text{ g Na}_2\text{CO}_3} \times \frac{2 \text{ mol HCl}}{1 \text{ mol Na}_2\text{CO}_3} \times \frac{1000 \text{ mL}}{1 \text{ L}}$$
$$= 1.192 \text{ M HCl solution}$$

Both methods will give the right answer, however the consolidated calculation eliminates the need to write down unnecessary intermediate answers and helps eliminate round-off errors.

Check your answer: The units cancel appropriately, and the moles and liters are nearly the same value so it makes sense that the molarity is near 1.

76. *Define the problem*: Given the mass of one reactant, the balanced chemical equation for a reaction, and the volume of a solution containing the other reactant needed for a complete reaction, determine the molarity of the second solution.

Develop a plan: Follow the method described in the answer to Question 75

Execute the plan:

We learn from the balanced equation that 1 mol of $HC_8H_4O_4^-$ reacts with 1 mol OH^-.

We learn from the formulas that 1 mol of $HC_8H_4O_4^-$ is in 1 mol of $KHC_8H_4O_4$ (also called KHP) and 1 mol OH^- is in 1 mol of NaOH.

$$\frac{0.902 \text{ g KHP}}{26.45 \text{ mL NaOH solution}} \times \frac{1 \text{ mol KHP}}{204.223 \text{ g KHP}} \times \frac{1 \text{ mol } HC_8H_4O_4^-}{1 \text{ mol KHP}} \times \frac{1 \text{ mol } OH^-}{1 \text{ mol } HC_8H_4O_4^-}$$

$$\times \frac{1 \text{ mol NaOH}}{1 \text{ mol } OH^-} \times \frac{1000 \text{ mL}}{1 \text{ L}} = 0.167 \text{ M NaOH solution}$$

Check your answer: The units cancel appropriately. The concentration of the NaOH solution is a reasonable size.

77. *Define the problem*: Given the volume and concentration of a solution containing one reactant, the balanced chemical equation for a reaction, and the mass of an impure sample of a salt of the second reactant, determine the percent purity of the impure sample.

Develop a plan: The volume and molarity of the first reactant are used to calculate the moles. Equation and formula stoichiometry can be used to find the moles of the salt. Then we will use the molar mass of the salt to find the mass of the salt. Comparing this mass to the mass of the impure sample we can determine the percent purity. *Note: It is NOT appropriate to use the dilution equation when working with reactions!*

Execute the plan:

We learn from the balanced equation that 2 mol of $S_2O_3^{2-}$ reacts with 1 mol I_2. Formula stoichiometry tells us that 1 mol of $S_2O_3^{2-}$ comes from 1 mol of $Na_2S_2O_3$.

$$40.21 \text{ mL } I_2 \text{ solution} \times \frac{1 \text{ L}}{1000 \text{ mL}} \times \frac{0.246 \text{ mol } I_2}{1 \text{ L } I_2 \text{ solution}} = 9.89 \times 10^{-3} \text{ mol } I_2$$

$$9.89 \times 10^{-3} \text{ mol } I_2 \times \frac{2 \text{ mol } S_2O_3^{2-}}{1 \text{ mol } I_2} \times \frac{1 \text{ mol } Na_2S_2O_3}{1 \text{ mol } S_2O_3^{2-}} \times \frac{158.1 \text{ g } Na_2S_2O_3}{1 \text{ mol } Na_2S_2O_3}$$

$$= 3.13 \text{ g } Na_2S_2O_3$$

The percent purity is calculated by dividing the mass of $Na_2S_2O_3$ by the mass of the impure sample and multiplying by 100 %.

$$\frac{3.13 \text{ g } Na_2S_2O_3}{3.232 \text{ g impure sample}} \times 100\,\% = 96.8\,\% \text{ pure}$$

Check your answer: The units cancel appropriately, and the two masses are very similar, so it makes sense that the percentage is nearly 100 %.

79. *Define the problem*: Given an unknown diprotic acid, the volume and concentration of a solution containing the base that is used to neutralize it, the balanced chemical equation for the neutralization reaction, determine the molar mass of the acid.

Develop a plan: The volume and molarity of the first reactant are used to calculate the moles. Equation stoichiometry can be used to find the moles of the acid. Then we will divide the mass by the moles to get molar mass. *Note: It is NOT appropriate to use the dilution equation when working with reactions!*

Execute the plan:

We learn from the balanced equation that 1 mol of H_2A reacts with 2 mol NaOH.

$$36.04 \text{ mL NaOH solution} \times \frac{1 \text{ L}}{1000 \text{ mL}} \times \frac{0.509 \text{ mol NaOH}}{1 \text{ L NaOH solution}} \times \frac{1 \text{ mol } H_2A}{2 \text{ mol NaOH}}$$

$$= 0.00917 \text{ mol } H_2A$$

$$\frac{0.954 \text{ g } H_2A}{0.00917 \text{ mol } H_2A} = 104 \frac{\text{g } H_2A}{\text{mol } H_2A}$$

The two separate calculations above can be consolidated into one calculation as follows:

$$\frac{0.956 \text{ g } H_2A}{36.04 \text{ mL NaOH solution}} \times \frac{1000 \text{ mL}}{1 \text{ L}} \times \frac{1 \text{ L NaOH solution}}{0.509 \text{ mol NaOH}} \times \frac{2 \text{ mol NaOH}}{1 \text{ mol } H_2A}$$

$$= 104 \frac{\text{g } H_2A}{\text{mol } H_2A}$$

Both methods will give the right answer, however the consolidated calculation eliminates the need to write down unnecessary intermediate answers and helps eliminate round-off errors.

Check your answer: The units cancel appropriately, and the moles are nearly 0.01 times the mass value so it makes sense that the molar mass is near 100.

General Questions

82. (a) $Mg(s) + 4\ HNO_3(q) \longrightarrow Mg(NO_3)_2(aq) + 2\ NO_2(g) + 2\ H_2O(\ell)$

(b) Names of reactants are magnesium and nitric acid. Names of the products are magnesium nitrate, nitrogen dioxide, and water.

(c) $Mg(s) + 4\ H^+(aq) + 2\ NO_3^-(aq) \longrightarrow Mg^{2+}(aq) + 2\ NO_2(g) + 2\ H_2O(\ell)$

(d) This is a redox reaction that is also a gas-forming reaction.

84. (a) Displacement reaction $2\ Li(s) + 2\ H_2O(\ell) \longrightarrow 2\ LiOH(aq) + H_2(g)$

(b) Decomposition reaction: $2\ Ag_2O \xrightarrow{\text{heat}} 4\ Ag(s) + O_2(g)$

(c) Exchange reaction: $Li_2O(s) + H_2O(\ell) \longrightarrow 2\ LiOH(aq)$

(d) Displacement reaction: $I_2(s) + 2\ Cl^-(aq) \longrightarrow NR$

(e) Displacement reaction: $Cu(s) + HCl(aq) \longrightarrow NR$

(f) Decomposition reaction: $BaCO_3(s) \xrightarrow{\text{heat}} BaO(s) + CO_2(g)$

86. The acid used to neutralize this compound is hydrochloric acid, HCl. Two types of reactions occur: neutralization and gas-forming. To balance the equation, it is necessary to add two HCl molecules for the two OH^- ions in the compound, and four HCl molecules for the two CO_3^{2-} ions in the compound. The products will be two water molecules from the neutralization, and two water molecules and two CO_2 gas molecules from the gas-forming reaction. The six Cl^- ions from the acid will become part of aqueous $CuCl_2$ compounds.

$$Cu_3(CO_3)_2(OH)_2(s) + 6\ HCl\ (aq) \xrightarrow{\text{heat}} 3\ CuCl_2(s) + 4\ H_2O(\ell) + 2\ CO_2(g)$$

88. *Define the problem*: Given a set of reactions and using the activity series (Table 5.5), determine which reaction will occur.

Develop a plan: First check to see if the reaction is a properly balanced oxidation-reduction reaction. It must have one substance oxidizing and one substance reducing. Then, look up the neutral product and the neutral reactant and find which is more active. If the reactant substance is more active than the product substance, then the reaction will occur. This is like Question 48.

Execute the plan:

(a) No. This reaction will not occur, since Mg (the product) is more active than Fe (the reactant).

(b) Yes. This reaction will occur, since Ni (the reactant) is more active than Cu (the product).

(c) No. This reaction will not occur, since Cu is not reactive in acid.

(d) Yes. This reaction will occur, since Mg displaces $H_2(g)$ from steam, $H_2O(g)$.

Check your answer: Each time the reactant is more active, the reaction occurs.

90. *Define the problem*: Given a redox reaction, identify the substance oxidized, the substance reduced, the oxidizing agent, the reducing agent, and the change in the oxidation number.

   Develop a plan: The substance reduced is also the oxidizing agent. It will be the reactant whose atoms gain electrons, and end up with a lower (more negative or less positive) oxidation number. The substance oxidized is also the reducing agent. It will be the reactant whose atoms lose electrons, and end up with a higher (more positive or less negative) oxidation number. The difference between the oxidation numbers is the change.

   Execute the plan: The only redox reaction in Question 90 is (c).

   The **substance reduced and the oxidizing agent is the reactant $TiCl_4$**. The oxidation number of Ti goes down from +4 to zero, so the change is $0 - (+4) = -4$. (Note, the Cl atom does not change oxidation state during this reaction.)

   The **substance oxidized and the reducing agent is the reactant Mg**. The oxidation number of Mg goes up from zero to +2, so the change is $+2 - (0) = +2$.

   Check your answer: The two Mg atoms each experience a +2 increase in oxidation number which exactly matches the decrease of –4 experienced by the one Ti atom.

91. (a) $CaF_2(s) + H_2SO_4(aq) \longrightarrow 2\, HF(g) + CaSO_4(s)$

   The reactants are called calcium fluoride and sulfuric acid.

   The products are called hydrogen fluoride and calcium sulfate.

   (b) *Define the problem*: Determine if a reaction is an acid base reaction, an oxidation-reduction reaction, or a precipitation reaction.

   Develop a plan: To decide if a reaction is an acid-base reaction, we need to see if an acid is reacting with a base. To decide if a reaction is an oxidation-reduction reaction, we need to see if any of the elements change oxidation state. To decide if a reaction is a precipitation reaction, look for insoluble ionic compound as a product.

First, check to see if it is an acid-base reaction: The strong reactant acid, H_2SO_4, reacts to form a weak product acid, HF. We don't see a hydroxide compound in the reaction, but in some sense of the word, the ionic fluoride compound is serving as a base. So, there is an acid-base reaction happening here. We'll learn more about these kinds of acid-base reactions in Chapter 16.

Second, check to see if it is an oxidation-reduction reaction: Look at the oxidation numbers for the reactants:

CaF_2 contains a monatomic cation, Ca^{2+}, and a monatomic anion, F^-. Rule 2 gives us Ox. # Ca = +2 and Ox. # F = −1.

H_2SO_4 contains oxygen and hydrogen. Rule 3 gives us Ox. # O = −2 and Ox. # H = +1. We use the sum rule to find the Ox. # S. For a molecule, the sum of the oxidation numbers is equal to zero.

$$0 = 2 \times (+1) + 1 \times (\text{Ox. \# S}) + 4 \times (-2)$$

Therefore, Ox. # P = +6.

Look at the oxidation numbers for the products:

HF contains hydrogen and fluorine. Rule 3 gives us Ox. # H = +1 and Ox. # F = −1.

$CaSO_4$ contains a monatomic cation, Ca^{2+}, and a polyatomic anion, SO_4^{2-}. Rule 2 gives us Ox. # Ca = +2. SO_4^{2-} contains oxygen. Rule 3 gives us Ox. # O = −2. We use the sum rule to find the Ox. # S. For a polyatomic anion, the sum of the oxidation numbers is equal to the ion's charge of 2−.

$$-2 = 1 \times (\text{Ox. \# S}) + 4 \times (-2)$$

Therefore, Ox. # S = +6.

The oxidation numbers don't change, so this is NOT an oxidation-reduction reaction.

Third, check to see if it is a precipitation reaction: An insoluble solid, $CaSO_4(s)$ (Table 5.1, Rule 4), is produced in a solution, so the reaction can also be considered a precipitation reaction.

In conclusion, most people studying Chapter 5 would probably decide this was a precipitation reaction.

Check your answer: The oxidation states determined are typical for these atoms in these compounds. With constant oxidation states, it is clearly not an oxidation-reduction reaction. It is logical to call this either a precipitation reaction or an acid-base reaction, but NOT an oxidation-reduction reaction. Because these kinds of acid-base reactions are described more thoroughly in a later chapter, most students will probably choose the precipitation reaction answer.

91. (continued)

(d) *Define the problem*: Given the identity and the percent mass of each element in a compound, find the empirical formula of the compound.

Develop a plan: This Question might require a review of the "Empirical Formula" calculations introduced in Chapter 3. Choose a convenient mass sample of the chlorofluorocarbon, such as 100.00 g. Using the given mass percents, determine the mass of C, Cl, and F in the sample. Using molar masses of the elements, determine the moles of each element in the sample. Find the whole number mole ratio of the elements C, Cl, and F, to determine the subscripts in the empirical formula.

Execute the plan: A 100.00 gram sample will have 8.74 grams of C, 77.43 grams of Cl, and 13.83 grams of H.

Find moles of C, Cl, and F in the sample:

$$8.74 \text{ g C} \times \frac{1 \text{ mol C}}{12.011 \text{ g C}} = 0.728 \text{ mol C}$$

$$77.43 \text{ g Cl} \times \frac{1 \text{ mol Cl}}{35.453 \text{ g Cl}} = 2.184 \text{ mol Cl}$$

$$13.83 \text{ g F} \times \frac{1 \text{ mol F}}{18.998 \text{ g F}} = 0.7279 \text{ mol F}$$

Mole ratio: 0.728 mol C : 2.184 mol Cl : 0.7279 mol F

Divide each term by the smallest number of moles, 0.7279 mol, to get atom ratio:

Atom ratio: 1 C : 3 Cl : 1 F

The empirical formula is CCl_3F.

Check your answer: This empirical formula makes sense, because four halogens can be bonded to one carbon.

Applying Concepts

92. For the first case: $LiCl(aq)$ and $AgNO_3(aq)$

(a) The two separate solutions of soluble salts would be clear and colorless like water. Once combined, insoluble white $AgCl$ (Table 5.1, Rule 3) would precipitate. We would probably see it eventually sink to the bottom of the beaker, leaving a clear, probably colorless liquid containing aqueous $LiNO_3$ above it.

(b) For proper proportions, these diagrams really need many more water molecules.

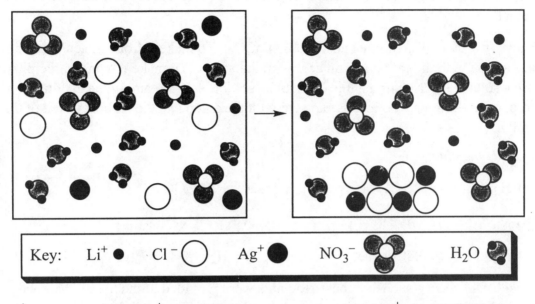

(c) $Li^+(aq) + Cl^-(aq) + Ag^+(aq) + NO_3^-(aq) \longrightarrow AgCl(s) + Li^+(aq) + NO_3^-(aq)$

For the second case: NaOH(aq) and HCl(aq)

(a) The separate acid and base solutions would be clear and colorless like water. Once combined, the solution would still be clear and colorless.

(b) For proper proportions, these diagrams really need many more water molecules.

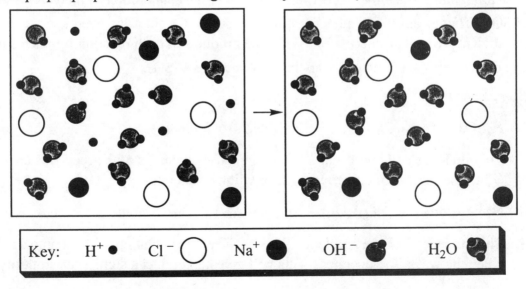

(c) $Na^+(aq) + OH^-(aq) + H^+(aq) + Cl^-(aq) \longrightarrow H_2O(\ell) + Na^+(aq) + Cl^-(aq)$

94. *Define the problem*: Prepare barium sulfate from a given list of chemicals by various means.

Develop a plan: The product needs to be $BaSO_4$. To make it from an acid-base reaction, use a base with the cation and an acid with the anion. To make it from a precipitation, use a soluble salt containing the anion and a soluble salt containing the cation. To make it from a gas-forming reaction, use an acid with the anion and the carbonate salt of the cation.

Execute the plan:

(a) $H_2SO_4(aq) + Ba(OH)_2(aq) \longrightarrow BaSO_4(s) + 2\ H_2O(\ell)$

(b) $Na_2SO_4(aq) + Ba(NO_3)_2(aq) \longrightarrow BaSO_4(s) + 2\ NaNO_3(aq)$

(c) $H_2SO_4(aq) + BaCO_3(aq) \longrightarrow BaSO_4(s) + H_2O(\ell) + CO_2(g)$

Check your answers: All of these reactants were provided in the list of chemicals. The neutralization reaction produces the solid and water. The precipitation reaction produces the solid and a soluble salt. The gas-forming reaction produces the solid and carbon dioxide gas. These make sense.

95. *Define the problem*: Critique the preparation of nickel sulfate from three procedures designed by students.

Develop a plan: The only product remaining after the evaporation of the water should be $NiSO_4$. That means if any other compound is still in the solution, even in aqueous form, the student will not be successful. Write the products of each reaction for each student and determine if anything else is in the water besides $NiSO_4$.

Execute the plan:

(a) $H_2SO_4(aq) + Ni(OH)_2(aq) \longrightarrow NiSO_4(s) + 2\ H_2O(\ell)$

This student should be alright, if stoichiometric quantities of the reactants are mixed. (If he uses excess of either reactant, he will have an impurity.)

(b) $Na_2SO_4(aq) + Ni(NO_3)_2(aq) \longrightarrow NiSO_4(s) + 2\ NaNO_3(aq)$

This student will have problems. The $NaNO_3$ in the solution will become a solid salt when the water is evaporated. This will give his product a significant impurity.

(c) $H_2SO_4(aq) + NiCO_3(aq) \longrightarrow NiSO_4(s) + H_2O(\ell) + CO_2(g)$

This student should be alright, if stoichiometric quantities of the reactants are mixed.

The heating will drive off the last of the CO_2. (If he uses excess of either reactant, he will have an impurity.)

Check your answers: The student who chose a precipitation reaction wasn't thinking about the other soluble product formed. The other two students found reactions where nothing else was left aqueous and water was the only thing left in the liquid phase of the products.

97. *Define the problem*: Determine which of three reagents could be used to distinguish whether a solution contains calcium ions or strontium ions.

Develop a plan: Determine the anion of the added reagent. Determine the products of its reaction with Ca^{2+} and Sr^{2+}. If the reactions are visibly different, then the reagent could be used to distinguish them.

Execute the plan:

First reagent: $NaOH(aq)$ would provide the anion $OH^-(aq)$ for a precipitation reaction.

$Ca(OH)_2$ is slightly soluble, Table 5.1, Rule 10

$$2\ OH^-(aq) + Ca^{2+}(aq) \longrightarrow N.R.$$

$Sr(OH)_2$ is insoluble, Table 5.1, Rule 10

$$2\ OH^-(aq) + Sr^{2+}(aq) \longrightarrow Sr(OH)_2(s)$$

An insoluble precipitate of strontium hydroxide would form if we add a small amount of this reagent to an unknown contained Sr^{2+}, but no visual change (or only slight precipitation) would be seen if the unknown contained only Ca^{2+}. This reagent could be used to distinguish them.

Second reagent: $H_2SO_4(aq)$ would provide the $SO_4^{2-}(aq)$ anion to a precipitation reaction.

$CaSO_4$ is insoluble, Table 5.1, Rule 4

$$SO_4^{2-}(aq) + Ca^{2+}(aq) \longrightarrow CaSO_4(s)$$

$SrSO_4$ is insoluble, Table 5.1, Rule 4

$$SO_4^{2-}(aq) + Sr^{2+}(aq) \longrightarrow SrSO_4(s)$$

Both reactions produce insoluble sulfate precipitate, so this reagent could NOT be used to distinguish them.

Third reagent: $H_2S(aq)$ would provide the $S^{2-}(aq)$ anion to a precipitation reaction.

CaS is sparingly soluble, Table 5.1, Rule 12

$$S^{2-}(aq) + Ca^{2+}(aq) \longrightarrow CaS(s)$$

SrS is insoluble, Table 5.1, Rule 12

$$S^{2-}(aq) + Sr^{2+}(aq) \longrightarrow SrS(s)$$

An insoluble precipitate of strontium sulfide would form if we add a small amount of this reagent to an unknown contained Sr^{2+}, but no visual change (or only slight precipitation) would be seen if the unknown contained only Ca^{2+}. This reagent could be used to distinguish them.

Check your answers: The slight and sparingly soluble nature of the hydroxide and sulfide compounds of calcium ion can be used to distinguish it from strontium. This is not very satisfying, but it makes sense, because these two ions are both in the same group on the periodic table. Their chemical reactions will certainly be similar. So, we must be satisfied with these tests.

99. The products are the results of an oxidizing agents acting on reducing agents. Only the reactants can serve as oxidizing agents and reducing agents.

100. Too much water was added, making the solution too dilute. So, (d) the concentration of the solution is less than 1 M because you added more solvent than necessary.

102. (a) Since the solution is acidic, there are more H^+ ions than OH^- ions in the mixture. So, this statement is TRUE.

(b) This statement is FALSE, since (a) was true.

(c) Only equal quantities of strong acid and strong base make a neutral solution. If either the acid or the base are weak or if the molarities are unequal, then the resulting solution will be basic or acidic, so this statement is FALSE.

(d) Since the resulting solution was acidic, and equal volumes were added, that means the acid's concentration must have been greater than the base's concentration. This statement is TRUE.

(e) While the concentration of H_2SO_4 might have been greater, it is not necessarily true that it MUST have been greater, since only half as many moles of the diprotic acid is needed to neutralize the NaOH base; therefore, it's concentration need only have been anything more than half as large. This statement is FALSE.

104. **(a)** The reaction of magnesium with bromine will produce magnesium bromide, $MgBr_2(s)$. The reaction of calcium with bromine will produce calcium bromide, $CaBr_2(s)$. The reaction of strontium with bromine will produce strontium bromide, $SrBr_2(s)$.

(b)

$$Mg(s) + Br_2(\ell) \longrightarrow MgBr_2(s)$$

$$Ca(s) + Br_2(\ell) \longrightarrow CaBr_2(s)$$

$$Sr(s) + Br_2(\ell) \longrightarrow SrBr_2(s)$$

(c) The reactions here are oxidation-reduction reactions. The oxidation states of the reactants are all zero (Rule 1) and the ionic compounds produced have atoms with non-zero oxidation states. These should definitely not be called gas-forming reactions since no gas is formed. They should also not be called precipitation reactions, since the solid products would be soluble in water if water were present.

(d) *Define the problem:* Use the graph of mass of product vs. mass of metal to confirm the predicted formula of the metal halide produced in these reactions.

Develop a plan: The "crossover" point, where the line changes from linearly rising to horizontal, describes the stoichiometric equivalence of the metal and the product. In other words, that's where we relate the grams of metal reacted to the grams of metal in a known mass of product. Determine the mass of bromine in a specific sample of the product by subtracting the mass of metal (extrapolated on the "Mass of metal" axis at the crossover point) from the sample mass (extrapolated on the "Mass of product" axis at the crossover point). Calculate moles of each and set up a mole ratio to determine empirical formula.

Execute the plan: To find the crossover point, (1) extrapolate the linearly-rising part of the line upward and to the right, (2) extrapolate the horizontal part of the line to the "Mass of product" axis, (3) determine where those two straight lines cross each other, and (4) draw a vertical line down from that crossing point to the "Mass of metal" axis:

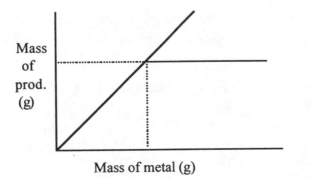

Mass of metal (g)

104. (d) (continued)

The crossover point on the magnesium curve is estimated at 1.6 g Mg and 11.4 g product. That means, in a product sample weighting 11.4 grams, the mass of Br is:

$$11.4 \text{ g sample} - 1.6 \text{ g Mg} = 9.8 \text{ g Br}$$

$$1.6 \text{ g Mg} \times \frac{1 \text{ mol Mg}}{24.305 \text{ g Mg}} = 0.066 \text{ mol Mg}$$

$$9.8 \text{ g Br} \times \frac{1 \text{ mol Br}}{79.90 \text{ g Br}} = 0.12 \text{ mol Br}$$

$$\frac{0.12 \text{ mol Br}}{0.066 \text{ mol Mg}} = 1.9 \approx 2$$

So, the empirical formula of the product is $MgBr_2$, as predicted.

The crossover point on the calcium curve is estimated at 2.5 g Ca and 12.4 g product. That means, in a product sample weighing 12.4 grams, the mass of Br is:

$$12.4 \text{ g sample} - 2.5 \text{ g Ca} = 9.9 \text{ g Br}$$

$$2.5 \text{ g Ca} \times \frac{1 \text{ mol Ca}}{40.078 \text{ g Ca}} = 0.062 \text{ mol Ca}$$

$$9.9 \text{ g Br} \times \frac{1 \text{ mol Br}}{79.90 \text{ g Br}} = 0.12 \text{ mol Br}$$

$$\frac{0.12 \text{ mol Br}}{0.062 \text{ mol Ca}} = 2.0 \approx 2$$

So, the empirical formula of the product is $CaBr_2$, as predicted.

The crossover point on the strontium curve is estimated at 5.5 g Ca and 15.3 g product. That means, in a product sample weighing 15.3 grams, the mass of Br is:

$$15.3 \text{ g sample} - 5.5 \text{ g Sr} = 9.8 \text{ g Br}$$

$$5.5 \text{ g Sr} \times \frac{1 \text{ mol Sr}}{87.62 \text{ g Sr}} = 0.063 \text{ mol Sr}$$

$$9.8 \text{ g Br} \times \frac{1 \text{ mol Br}}{79.90 \text{ g Br}} = 0.12 \text{ mol Br}$$

$$\frac{0.12 \text{ mol Br}}{0.063 \text{ mol Sr}} = 1.9 \approx 2$$

So, the empirical formula of the product is $SrBr_2$, as predicted.

Check your answers: It makes sense that the mass of bromine from all three samples is approximately the same, since all three compounds were determined to contain the same number of bromine atoms. The only common ionic charge found in Group(II) ions is +2 and the only common ionic charge in halide ions is –1, so it makes sense that they would end up combining with one metal ion and two bromide ions. The magnesium and calcium data required obvious rounding to get an integer ratio. The approximations made from the very small graph could have contributed to this degree of imprecision. In a real experiment, the graph would be larger with better-scaled axes so that results extracted from it would be more reliable. The shape of the curve also lends to some approximations. However, even in a real experiment, the data themselves might provide such imprecise results.

106. (a) Na^+ and NO_3^- are spectator ions.

$$Ag^+ + Cl^- \longrightarrow AgCl \text{ (s)}$$
$$Ag^+ + Br^- \longrightarrow AgBr \text{ (s)}$$

(b) The silver halide solid (AgCl) produced by groups C and D is the same. The silver halide solid (AgBr) produced by groups A and B is the same.

(c) Below the mass of 0.75 g, the graph line increases with increasing masses of $AgNO_3$. In this region, Ag^+ is the limiting reactant. (See Question 104 for a more complete explanation.) These reactions both require the same mass of $AgNO_3$ to make their respective silver halide. However, since bromide ion is heavier than chloride ion, the mass of the product will be different and the mass of product where the graph levels out will be different because AgBr is heavier than AgCl. That means the products in groups A and B (AgCl) weigh less than the products in groups C and D (AgBr).

126

Chapter 6: Energy and Chemical Reactions

The Nature of Energy

13. *Define the problem*: Convert a quantity of kilojoules (provided by a piece of cake) into food Calories, and convert a quantity of food Calories into joules.

Develop a plan: Use metric and energy conversion factors to achieve the conversions.

Execute the plan:

(a) $$1670 \text{ kJ} \times \frac{1000 \text{ J}}{1 \text{ kJ}} \times \frac{1 \text{ cal}}{4.184 \text{ J}} \times \frac{1 \text{ kcal}}{1000 \text{ cal}} \times \frac{1 \text{ Cal}}{1 \text{ kcal}} = 399 \text{ Cal}$$

(b) $$\frac{1200 \text{ Cal}}{1 \text{ day}} \times \frac{1 \text{ kcal}}{1 \text{ Cal}} \times \frac{1000 \text{ cal}}{1 \text{ kcal}} \times \frac{4.184 \text{ J}}{1 \text{ cal}} = \frac{5.0 \times 10^6 \text{ J}}{1 \text{ day}}$$

Check your answers: The food Calorie is about four times bigger than a kilojoule. So, the energy quantity in Calories should be about four times smaller than in kilojoules. The sizes of these answers make sense.

15. *Define the problem*: Convert a quantity of kilojoules into joules, calories, and kilocalories.

Develop a plan: Use metric and energy conversion factors to achieve the conversions.

Execute the plan:

(a) $$297 \text{ kJ} \times \frac{1000 \text{ J}}{1 \text{ kJ}} = 2.97 \times 10^5 \text{ J}$$

(b) $$2.97 \times 10^5 \text{ J} \times \frac{1 \text{ cal}}{4.184 \text{ J}} = 7.10 \times 10^4 \text{ cal}$$

(c) $$7.10 \times 10^4 \text{ cal} \times \frac{1 \text{ kcal}}{1000 \text{ cal}} = 71.0 \text{ kcal}$$

Check your answers: A kilocalorie is about four times bigger than a kilojoule. So, the energy quantity in kilocalories should be about four times smaller than in kilojoules. The sizes of these answers make sense.

17. *Define the problem*: Given the wattage of a lightbulb, the time the light is left on, and the cost of electricity, Determine the joules used by the lightbulb, and the cost for that electricity.

Develop a plan: The unit "Watt" is described in Question 16. Use metric and energy conversion factors to calculate the desired quantities

Execute the plan:

$$100 \text{ W} \times 14 \text{ hr} \times \frac{3600 \text{ s}}{1 \text{ hr}} \times \frac{1 \text{ J}}{1 \text{ W} \times 1 \text{ s}} = 5 \times 10^6 \text{ J}$$

$$100 \text{ W} \times 14 \text{ hr} \times \frac{1\text{kW}}{1000 \text{ W}} \times \frac{\$0.09}{1 \text{ kW - hr}} = \$0.13$$

Check your answers: It's rather surprising that it only costs 13 cents to run a light for 14 hours. However, thinking about all the lightbulbs people use all over their homes every day in a month, it adds up to an average electric bill around $50/month just for keeping the lights on. This value seems consistent with the average household electric bill.

Conservation of Energy

19. *Define the problem*: Describe and explain your choice of the system and the surroundings, and describe transfer of energy and materials into and out of the system.

Develop a plan: The system is identified as precisely what we are studying. The surroundings are everything else. The important aspects of the surroundings are usually those things in contact with the system or in close proximity.

Execute the plan:

(a) The System: The plant (stem, leaves, roots, etc.)

The Surroundings: Anything not the plant (air, soil, water, sun, etc.)

(b) To study the plant growing, we must isolate it and see how it interacts with its surroundings.

(c) Light energy and carbon dioxide are absorbed by the leaves and are converted to other molecules storing the energy as chemical energy and using it to increase the size of the plant. Nutrients are absorbed through the soil (minerals and water) to assist in the chemical processes. The plant expels oxygen and other waste materials into the surroundings.

Check your answers: This definition of the system differentiates between the live organism and the materials and energy required for it to stay alive. It seems sensible.

21. *Define the problem*: Describe and explain your choice of the system and the surroundings, describe transfer of energy and materials into and out of the system, and determine if the process is exothermic or endothermic.

Develop a plan: The system is identified as precisely what we are studying. The surroundings are everything else. The important aspects of the surroundings are usually those things in contact with the system or in close proximity. The process is exothermic if the system loses energy. The process is endothermic if the system gains energy.

Execute the plan:

(a) The System: NH_4Cl.

The Surroundings: Anything not NH_4Cl, including the water.

(b) To study the release of energy during the phase change of this chemical, we must isolate it and see how it interacts with the surroundings.

(c) The system's interaction with the surroundings causes heat to be transferred out of the surroundings and into the system. There is no material transfer in this process, but there is a change in the specific interaction between the water and system.

(d) Since changes in the system cause energy to be gained by the system, the process is endothermic.

Check your answers: This definition of the system is restrictive enough to allow us to actually study the reaction excluding the effect on the solvent. Making the water part of the surroundings helps us see that the reaction is consuming energy. It seems sensible.

23. *Define the problem*: Given descriptions and numerical values for work and heat energy changes, determine the ΔE for the system.

Develop a plan: In general, when energy leaves the system, ΔE goes down, and when energy enters the system ΔE is positive. Therefore, if a change in the system causes work to be done or heat to be produced, energy leaves the system and the signs of the values of respective w_{sys} or q_{sys} are negative. And, if a change in the system causes work to done on the system or causes heat to be consumed, energy enters the system and the signs of the values of respective w_{sys} or q_{sys} are positive. *Note: Make sure that the units of energy are the same before adding the energy values.*

Execute the plan: $\Delta E = q_{sys} + w_{sys}$

The system does work, so w_{sys} is negative: $w_{sys} = -75.4$ J.

Heat is transferred into the system, so q_{sys} is positive: $q_{sys} = +25.7$ cal

$$\Delta E = (+25.7 \text{ cal}) \times \frac{4.184 \text{ J}}{1 \text{ cal}} + (-75.4 \text{ J}) = 32 \text{ J}$$

Check your answers: The heat input is more than the work output, so it makes sense that the ΔE is positive.

25. *Define the problem*: Make heat flow diagram for given heat and work energy changes in a system and use that to help determine the ΔE_{sys}.

Develop a plan: In general, when energy leaves the system, ΔE goes down, and when energy enters the system, ΔE is positive. Therefore, if a change in the system causes work to be done or heat to be produced, energy leaves the system and the signs of the values of respective w_{sys} or q_{sys} are negative. And, if a change in the system causes work to done on the system or causes heat to be consumed, energy enters the system and the signs of the values of respective w_{sys} or q_{sys} are positive.

Execute the plan:

Work is done on the surroundings, so w_{sys} is negative: $w_{sys} = -127.6 \text{ kJ}$

Heat is transferred into the system, so q_{sys} is positive: $q_{sys} = 843.2 \text{ kJ}$

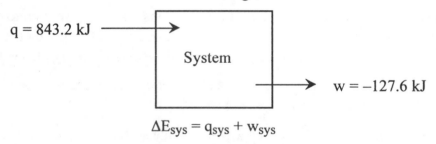

$$\Delta E_{sys} = q_{sys} + w_{sys}$$

$$\Delta E_{sys} = 843.2 \text{ kJ} + (-127.6 \text{ kJ}) = 715.6 \text{ kJ}$$

Check your answer: More heat energy is entering the system than work energy leaving the system, so it makes sense that the ΔE is positive.

Heat Capacity

28. *Define the problem*: Given the mass and initial temperature of two samples made out of two different metals, also given the assumption that they take up heat at the same rate when held in your hand, determine which metal will warm up to body temperature faster.

Develop a plan: First, look up the specific heat capacities of the substances being heated in Table 6.1. The rate of absorption of heat can be related to heat capacity, mass, temperature increase, and elapsed time using Equation 6.2. The elapsed time can then be related to the specific heat, the mass, the change in temperature, and the rate of absorption of heat. The fastest elapsed time will reach body temperature first.

Execute the plan:

The specific heat capacity (c) of Al is $0.902 \text{ J g}^{-1}{}^{\circ}\text{C}^{-1}$ and of Cu is $0.385 \text{ J g}^{-1}{}^{\circ}\text{C}^{-1}$ from Table 6.1.

Dividing Equation 6.2 by Δt, we can write an equation relating the rate of heat absorption to the heat capacity, mass, temperature increase, and elapsed time:

$$\text{rate of heat absorption} = \frac{q}{\Delta t} = c \times m \times \frac{\Delta T}{\Delta t}$$

Solve this equation for the elapsed time, Δt:

$$\Delta t = c \times \frac{m \times \Delta T}{\text{rate}}$$

Comparing both samples, their mass, change in temperature, and the rate of heat absorption are all the same, simplifying the equation above to the following:

$$\Delta t = c \times \text{constant}$$

This equation demonstrates that the elapsed heating time is proportional to the specific heat capacity. That means it takes less time to heat the copper sample than the aluminum sample at a constant rate over a common temperature range.

Check your answer: Specific heat capacity describes the amount of energy needed to increase the temperature of a 1-gram sample by 1 °C. Since these samples have the same mass and temperature change, the object with the smaller specific heat capacity will be faster to heat. This answer makes sense.

30. *Define the problem*: Given the dimensions of a swimming pool and temperature change of the water in it, determine how much thermal energy (in megajoules) is required to heat the water.

Develop a plan: First, look up the specific heat capacity of the substance being heated in Table 6.1. Find the volume of the pool in cubic centimeters and use the density of water (1.00 g/cm^3) to determine the mass of the water sample being heated. Then use Equation 6.2 to calculate the heat required.

Execute the plan:

Table 6.1 tells us that the specific heat capacity (c) of liquid water is $4.184 \text{ J g}^{-1}°\text{C}^{-1}$.

Mass of water:

$$m = (4 \text{ ft}) \times (20 \text{ ft}) \times (75 \text{ ft}) \times \left(\frac{12 \text{ in}}{1 \text{ ft}} \times \frac{2.54 \text{ cm}}{1 \text{ in}} \right)^3 \times \frac{1.00 \text{ g H}_2\text{O}}{1 \text{ cm}^3} = 2 \times 10^8 \text{ g H}_2\text{O}$$

Equation 6.2: $q = c \times m \times \Delta T$

$$q_{Al} = (4.184 \text{ J g}^{-1}°\text{C}^{-1}) \times (2 \times 10^8 \text{ g}) \times (1 \text{ °C}) = 7 \times 10^8 \text{ J}$$

$$7 \times 10^8 \text{ J} \times \frac{1 \text{ MJ}}{10^6 \text{ J}} = 700 \text{ MJ}$$

Check your answer: This is a large amount of water, so it makes sense that the amount of energy is conveniently expressed in units of Megajoules. Using rounded numbers in the intermediate calculations would give an answer of 800 MJ. It is better not to round the intermediate answers during a multi-step process.

31. (a) As described in Question 28, the sample with the lowest specific heat capacity will require less heat to raise the temperature. The specific heat capacity of ethylene glycol is $2.42 \text{ J g}^{-1}°\text{C}^{-1}$ and that of liquid water is $4.184 \text{ J g}^{-1}°\text{C}^{-1}$, according to Table 6.2. That means an equal mass of water requires more energy than ethylene glycol to raise its temperature the same amount.

(b) *Define the problem*: Given the volume of a cooling system, the densities of water and ethylene glycol, and the temperature change for two different samples, compare the thermal energy increase in the two samples.

Develop a plan: First convert the volumes into cubic centimeters, then use the densities to determine the masses of each sample. Then use Equation 6.2 to calculate the heat required in each process and compare them.

Execute the plan:

Volume of the cooling system in cubic centimeters:

$$5.00 \text{ quarts} \times \frac{0.946 \text{ L}}{1 \text{ quart}} \times \frac{1000 \text{ mL}}{1 \text{ L}} \times \frac{1 \text{ cm}^3}{1 \text{ mL}} = 4730 \text{ cm}^3$$

Mass of the two samples:

$$4730 \text{ cm}^3 \times \frac{1.113 \text{ g}}{1 \text{ cm}^3} = 5260 \text{ g ethylene glycol}$$

$$4730 \text{ cm}^3 \times \frac{1.00 \text{ g}}{1 \text{ cm}^3} = 4730 \text{ g water}$$

Equation 6.2: $q = c \times m \times \Delta T$

In both samples, the change in temperature is the same:

$$\Delta T = T_f - T_i = 100.0 \text{ °C} - 25.0 \text{ °C} = 75.0 \text{ °C}$$

$$q_{\text{ethylene glycol}} = (2.42 \text{ J g}^{-1}\text{°C}^{-1}) \times (5260 \text{ g}) \times (75.0 \text{ °C}) = 9.56 \times 10^5 \text{ J}$$

$$q_{\text{water}} = (4.184 \text{ J g}^{-1}\text{°C}^{-1}) \times (4730 \text{ g}) \times (75.0 \text{ °C}) = 1.48 \times 10^6 \text{ J}$$

More energy is absorbed (1.48×10^6 J) by the water sample than is absorbed (9.56×10^5 J) by the ethylene glycol sample.

Check your answers: Water has a much larger specific heat capacity than ethylene glycol, so a somewhat smaller mass of water will still absorb more thermal energy than a larger mass of ethylene glycol. This answer makes sense.

32. *Define the problem*: Given the mass and temperature of a piece of metal and the initial and final temperatures of a cup of coffee, determine how much energy (in kilojoules) was transferred from the metal to the coffee.

Develop a plan: First, look up the specific heat capacity of the metal. Assuming that the temperature stopped dropping once the coffee and the metal reached the same temperature, set the final temperature of the metal to be the same as the final temperature of the coffee. Use Equation 6.2 to calculate the heat gained by the metal from the coffee.

Execute the plan:

Table 6.1 tells us that the specific heat capacity (c) of Al is 0.902 J g^{-1}°C^{-1}.

Equation 6.2: $q = c \times m \times \Delta T$

$$\Delta T = T_f - T_i = 75.0 \text{ °C} - 0.0 \text{ °C} = 75.0 \text{ °C}$$

$$q_{\text{Al}} = (0.902 \text{ J g}^{-1}\text{°C}^{-1}) \times (20.0 \text{ g}) \times (75.0 \text{ °C}) = 1350 \text{ J}$$

$$1350 \text{ J} \times \frac{1 \text{ kJ}}{1000 \text{ J}} = 1.35 \text{ kJ of heat transferred to the aluminum}$$

Check your answer: Because water (the primary component of coffee) has a much larger specific heat capacity than aluminum, a smaller temperature drop is experienced by the water than temperature increase experienced by the aluminum. It is reasonable to see a 1.35 kJ energy transfer when 20.0 grams of aluminum undergoes a temperature increase of 75 °C.

35. *Define the problem:* Given the mass, initial and final temperatures, and energy needed to heat a piece of metal, determine the specific heat capacity and the molar heat capacity.

Develop a plan: Solve Equation 6.2 for c, and plug in the known values. Convert grams to moles with molar mass to get the molar heat capacity

Execute the plan:

(a) Equation 6.2: $q = c \times m \times \Delta T$

$$c = \frac{q}{m \times \Delta T} = \frac{41.0\ \text{J}}{12.3\ \text{g} \times \left(24.7\ ^\circ\text{C} - 17.3\ ^\circ\text{C}\right)} = \frac{41.0\ \text{J}}{12.3\ \text{g} \times 7.4\ ^\circ\text{C}} = 0.45\ \frac{\text{J}}{\text{g}\,^\circ\text{C}}$$

(b) The molar mass of iron is 55.85 g/mol. Use that to calculate the molar heat capacity:

$$0.45\,\frac{\text{J}}{\text{g}\,^\circ\text{C}} \times \frac{55.85\ \text{g}}{1\ \text{mol}} = 25\ \frac{\text{J}}{\text{mol}\,^\circ\text{C}}$$

Check your answers: The specific heat capacity of iron is 0.451 J g^{-1}°C^{-1} according to Table 6.1, so the specific heat capacity calculated here makes sense. The molar mass is larger than 1 gram, so it makes sense that the molar heat capacity is larger than the specific heat capacity.

37. *Define the problem:* Given the mass, initial and final temperatures, and energy needed to heat an unknown element, determine its most probable identity using Table 6.1.

Develop a plan: Adapt the method described in the answer to Question 35.

Execute the plan:

$$c = \frac{q}{m \times \Delta T} = \frac{34.7\ \text{J}}{23.4\ \text{g} \times \left(28.9\ ^\circ\text{C} - 17.3\ ^\circ\text{C}\right)} = 0.128\ \frac{\text{J}}{\text{g}\,^\circ\text{C}}$$

Looking at Table 6.1, the element whose specific heat capacity is closest to this is Au.

Check your answer: The specific heat capacity of Au (0.128 J g^{-1}°C^{-1}) matches the calculated value to three significant figures (0.128 J g^{-1}°C^{-1}). The similarity in the two values gives us confidence in the answer.

39. *Define the problem*: Given the mass of a piece of hot metal, the volume, density, and temperature of a sample of cool water, and the final temperature after the two are combined and heat is transferred, determine the initial temperature of the metal.

Develop a plan: First, look up the specific heat capacities of the metal and the water. Use Equation 6.2 to calculate the heat gained by the water from the metal. The heat gained by the water is the heat lost by the metal. Finally, use Equation 6.2 to calculate the initial temperature of the metal.

Execute the plan:

The specific heat capacity (c) of Al is 0.902 J $g^{-1}°C^{-1}$ and of water is 4.184 J $g^{-1}°C^{-1}$ according to Table 6.1.

$$\text{Equation 6.2:} \qquad q = c \times m \times \Delta T$$

$$\Delta T_{water} = T_f - T_i = 33.6\ °C - 22.0\ °C = 11.6\ °C$$

$$q_{water} = (4.184\ J\ g^{-1}°C^{-1}) \times (500.\ mL) \times \left(\frac{0.98\ g}{1\ ml}\right) \times (11.6\ °C) = 2.4 \times 10^4\ J$$

Heat is gained by the water, so q_{water} is positive. Heat is lost by the aluminum, so $q_{aluminum}$ is negative. The quantity of heat lost by the hot aluminum is absorbed by the cold water. So,

$$q_{aluminum} = -\ q_{water} = -2.4 \times 10^4\ J$$

Using the alternate form of Equation 6.2:

$$\Delta T_{aluminum} = \frac{q_{aluminum}}{c \times m} = \frac{-2.4 \times 10^4\ J}{0.902\ Jg^{-1}°C^{-1} \times 200.\ g} = -130\ °C$$

$$T_i = T_f - \Delta T_{aluminum} - T_f = 33.6\ °C - (-130\ °C) = 160\ °C$$

Check your answer: Because water has a much larger specific heat capacity than aluminum, a smaller temperature rise is experienced by the water than temperature decrease experienced by the aluminum. Quantitatively, we can use $q = c \times m \times \Delta T$ to confirm that the aluminum is losing 2.3×10^4 J at the same time as the water is gaining 2.4×10^4 J. They are the same within the uncertainty limit of $\pm\ 0.1 \times 10^4$ J, so this final temperature makes sense.

41. *Define the problem*: Given the direction of transfer of thermal energy between a system and the surroundings with no work done, determine the algebraic sign of $\Delta T_{surroundings}$ and ΔE_{system}.

Develop a plan: If thermal energy enters the surroundings as heat, then the temperature in the surroundings, $T_{surroundings}$, rises. If thermal energy enters the system as heat, then the temperature in the surroundings, $T_{surroundings}$, drops. If the system gets energy from the surroundings, then the internal energy of the system, E_{system}, rises. If the surroundings gets energy from the system, then the internal energy of the system, E_{system}, lowers.

Execute the plan: Here, the thermal energy enters the surroundings as heat, so the temperature in the surroundings, $T_{surroundings}$, rises:

$$T_{f,surroundings} > T_{i,surroundings}$$

$$\Delta T_{surroundings} = T_{f,surroundings} - T_{i,surroundings} = positive$$

Here, the surroundings gets energy from the system, so the internal energy of the system, E_{system}, lowers:

$$E_{f,system} < E_{i,system}$$

$$\Delta E_{system} = E_{f,system} - E_{i,system} = negative$$

Check your answer: Energy leaves the system, so a negative ΔE_{system} makes sense. The energy must show up in the surroundings, so the fact that these two algebraic signs are opposite signs also makes sense.

Energy and Enthalpy

43. *Define the problem*: Given a tray of ice, the number of cubes in the tray, the mass of each cube, and the thermal energy required to melt a given mass of ice, determine the energy required to melt the whole tray of ice cubes.

    Develop a plan: The tray of ice cubes is the sample. Use the rest of the given information as unit factors to determine the energy.

    Execute the plan:

$$1 \text{ tray} \times \frac{20 \text{ cubes}}{1 \text{ tray}} \times \frac{62.0 \text{ g ice}}{1 \text{ cube}} \times \frac{333 \text{ J}}{1.00 \text{ g ice}} = 4.13 \times 10^5 \text{ J}$$

    Check your answer: The tray has over 1000 grams of ice, so this large number of joules makes sense. Notice, the tray and the number of cubes are countable objects; hence, those numbers are exact with infinite significant figures, so their values do not limit the significant figures of the answer.

45. *Define the problem*: Given the mass and initial temperature of a liquid substance, the boiling point of the liquid, the final temperature of the gaseous substance, the specific heat capacity and the enthalpy of vaporization, determine the energy (in joules) required to complete the transition.

Develop a plan: Using Equation 6.2, determine the heat required to raise the temperature to the boiling point. Then use the enthalpy of vaporization (ΔH_{vap}) to determine the energy required to boil the liquid at the boiling point.

Execute the plan: Define the system as the liquid benzene. To change the temperature of the system, use $q = c \times m \times \Delta T$

$$q_{T\text{-rise}} = c_{benzene} \times m_{benzene} \times \Delta T_{benzene}$$

$$= (1.74 \text{ J g}^{-1}°C^{-1}) \times (1.00 \text{ kg}) \times \frac{1000 \text{ g}}{1 \text{ kg}} \times (80.1 \ °C - 20.0 \ °C) = 1.05 \times 10^5 \text{ J}$$

To change a phase in the system, use $q = m \times \Delta H_{vap}$.

$$q_{boil} = m_{benzene} \times \Delta H_{vap,benzene} = (1.00 \text{ kg}) \times \frac{1000 \text{ g}}{1 \text{ kg}} \times (395 \text{ J/g}) = 3.95 \times 10^5 \text{ J}$$

The sum gives the total heat required.

$$q_{benzene} = q_{T\text{-rise}} + q_{boil} = (1.05 \times 10^5 \text{ J}) + (3.95 \times 10^5 \text{ J}) = 5.00 \times 10^5 \text{ J}$$

Check your answer: To make the final product, the liquid's temperature needed to be raised, and then the liquid needed to be vaporized. Both of these changes require energy to be added to the system. This result makes sense.

47. *Define the problem*: Given the volume and initial temperature of a liquid substance, the freezing point of the liquid, the final temperature of the solid substance, the density of the liquid, the specific heat capacity, and the enthalpy of fusion, determine the thermal energy (in joules) that must be transferred to the surroundings to complete the transition.

Develop a plan: Using Equation 6.2, determine the heat that must be removed to lower the temperature to the freezing point. Then use the enthalpy of fusion (ΔH_{fus}) to determine the thermal energy that must be removed from the liquid to form the solid at the freezing point.

Execute the plan: Define the system as the liquid mercury.

$$m_{mercury} = (1.00 \text{ mL}) \times \frac{1 \text{ cm}^3}{1 \text{ mL}} \times \frac{13.6 \text{ g Hg}}{1 \text{ cm}^3} = 13.6 \text{ g Hg}$$

To change the temperature of the system, use $q = c \times m \times \Delta T$

$q_{\text{T-drop}} = c_{\text{mercury}} \times m_{\text{mercury}} \times \Delta T_{\text{mercury}}$

$$= (0.138 \text{ J g}^{-1}{}^{\circ}\text{C}^{-1}) \times (13.6 \text{ g Hg}) \times (-39 \text{ °C} - 23.0 \text{ °C}) = -116 \text{ J}$$

To change a phase from liquid to solid in the system, use $q_{\text{freeze}} = -q_{\text{fusion}} = -m \times \Delta H_{\text{fus}}$.

$$q_{\text{freeze}} = -m_{\text{mercury}} \times \Delta H_{\text{fus,mercury}} = -(13.6 \text{ g Hg}) \times (11 \text{ J/g}) = -150 \text{ J}$$

The sum gives the total heat required.

$$q_{\text{total}} = q_{\text{T-drop}} + q_{\text{freeze}} = (-116 \text{ J}) + (-1.5 \times 10^2 \text{ J}) = -2.7 \times 10^2 \text{ J}$$

The amount of thermal energy that must be removed is 2.7×10^2 J.

Check your answer: To make the final product, the liquid's temperature needed to be lowered, and then the liquid needed to be frozen. Both of these changes require energy to be removed from the system. This result makes sense.

49. A cooling curve shows how the temperature drops as the heat is removed from the system. In that respect, the lower part of this graph will look like the reverse of Figure 6.10.

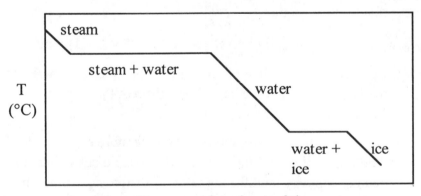

Quantity of heat transferred out of the system

51. The heating graph shown in Question 52 is the one that we are using to answer this question.

   (a) The line on the graph with the steepest vertical slope has the largest temperature change when a fixed amount of heat is added. This substantial temperature increase means the specific heat capacity of the substance in that phase is smaller than in the other phases. Since the slope of the line representing the heating of the solid is steeper than the slopes of the lines representing the heating of the liquid or the gas, X(s) has the smallest specific heat capacity.

(b) The horizontal lines on the graph represent phase transitions. The longest horizontal line on the graph indicates that more energy is required for that phase transition than the other does. Since the line representing the vaporization of the liquid to gas is longer than the line representing the melting of the solid to liquid, the heat of vaporization is larger than the heat of fusion. This is not surprising, since the vaporization of a liquid requires enough energy to overcome the all the attractions between the molecules in the liquid state; whereas the melting process only requires some of the attractive forces be overcome. We always find that the heat of fusion is less than the heat of vaporization.

(c) The algebraic sign of the enthalpy of vaporization is always positive, since heat is required to overcome all of the attractions between the molecules in the liquid state.

Thermochemical Equations

53. *Define the problem*: Given a thermochemical equation, determine if the reaction is exothermic or endothermic.

Develop a plan: Find the sign of ΔH, and use that to determine if the reaction is exothermic or endothermic. When ΔH is negative, the reaction is exothermic. When ΔH is positive, the reaction is endothermic. *Note: If you can't always remembering the Greek origins of these words and find yourself mixing which is positive and which is negative, think of the words "exhale" and "inhale." When you **exhale**, air goes **out** of your lungs (exo = negative); when you **inhale**, air goes **into** your lungs (endo = positive).*

Execute the plan: $\Delta H = +38$ kJ, so this reaction is endothermic.

Check your answer: Endothermic reactions have positive ΔH.

55. In Section 6.7, we learn that breaking bonds is always endothermic. Cutting the giant super-molecule involves breaking carbon-carbon bonds, so this reaction is endothermic.

57. 6.0 kJ of energy is used to convert ice into liquid water. In particular, this energy is measured as heat absorbed at constant standard pressure, which is why we are using the symbol $\Delta H°$.

59. *Define the problem*: Given a thermochemical equation for a phase change, determine the quantity of energy transferred to the surroundings when two different samples undergo that phase change.

Develop a plan: When necessary, convert the sample quantity into moles. Then use the thermochemical equation to create a unit conversion factor relating energy to moles of reactant.

Execute the plan: Reversing the given balanced thermochemical equation to show the freezing reaction, causes the $\Delta H°$ to change sign:

$$H_2O(\ell) \longrightarrow H_2O(s) \qquad \Delta H° = -6.0 \text{ kJ}$$

The thermochemical reaction tells us when 1 mol $H_2O(\ell)$ is frozen, that 6.0 kJ are transferred to the surroundings.

(a) $$34.2 \text{ mol } H_2O(s) \times \frac{-6.0 \text{ kJ}}{1 \text{ mol } H_2O(s)} = -2.1 \times 10^2 \text{ kJ}$$

(b) $$100.0 \text{ g } H_2O(s) \times \frac{1 \text{ mol } H_2O(s)}{18.015 \text{ g } H_2O(s)} \times \frac{-6.0 \text{ kJ}}{1 \text{ mol } H_2O(s)} = -33 \text{ kJ}$$

Check your answers: The freezing reaction is exothermic, since heat must be removed from the reactants to make the products. Also, 34 moles of water weigh more than 100 grams, so it makes sense that the answer in (a) is larger than the answer in (b).

Enthalpy Changes for Chemical Reactions

61. *Define the problem*: Given a thermochemical equation for a reaction, write a thermochemical equation for multiples of the reaction.

Develop a plan: Determine what number the equation must be multiplied by to get the desired number of reactants or products given. To determine the multiplier, divide the stoichiometric coefficient we want to have in the new equation by the stoichiometric coefficient given in the original equation. Multiply every coefficient in the equation and multiply the $\Delta H°$ by that multiplier to set up the new thermochemical equation.

Execute the plan: In each of the samples described, a number of moles is given. Assume these are exact numbers.

(a) The original equation makes 16 mol CO_2. We want to make 4.00 mol CO_2. So,

$$\text{Multiplier} = \frac{4.00 \text{ mol}}{16 \text{ mol}} = 0.250$$

$$0.250 \times [2\, C_8H_{18}(\ell) + 25\, O_2(g) \longrightarrow 16\, CO_2(g) + 18\, H_2O(\ell)]$$

$$\Delta H° = 0.250 \times (-10{,}992 \text{ kJ})$$

$$0.500\, C_8H_{18}(\ell) + 6.25\, O_2(g) \longrightarrow 4.00\, CO_2(g) + 4.50\, H_2O(\ell)$$

$$\Delta H° = -2.75 \times 10^3 \text{ kJ}$$

(b) The original equation burns 2 mol isooctane. We want to burn 100. mol isooctane. So,

$$\text{Multiplier} = \frac{100.\ \text{mol}}{2\ \text{mol}} = 50.0$$

$$50.0 \times [2\ C_8H_{18}(\ell) + 25\ O_2(g) \longrightarrow 16\ CO_2(g) + 18\ H_2O(\ell)]$$

$$\Delta H° = 50.0 \times (-10{,}992\ \text{kJ})$$

$$100.\ C_8H_{18}(\ell) + 1250\ O_2(g) \longrightarrow 800.\ CO_2(g) + 900.\ H_2O(\ell)$$

$$\Delta H° = -5.50 \times 10^5\ \text{kJ}$$

(c) The original equation burns 2 mol isooctane. We want to burn 1.00 mol isooctane. So,

$$\text{Multiplier} = \frac{1.00\ \text{mol}}{2\ \text{mol}} = 0.500$$

$$0.500 \times [2\ C_8H_{18}(\ell) + 25\ O_2(g) \longrightarrow 16\ CO_2(g) + 18\ H_2O(\ell)]$$

$$\Delta H° = 0.500 \times (-10{,}992\ \text{kJ})$$

$$1.00\ C_8H_{18}(\ell) + 12.5\ O_2(g) \longrightarrow 8.00\ CO_2(g) + 9.00\ H_2O(\ell)$$

$$\Delta H° = -5.50 \times 10^3\ \text{kJ}$$

Check your answers: It makes sense that when more moles are involved, the $\Delta H°$ is larger, and when fewer moles are involved, the $\Delta H°$ is smaller.

63. *Define the problem*: Given a thermochemical equation for a reaction, write all the thermostoichiometric factors that can be derived.

Develop a plan: The equation's stoichiometric coefficients are interpreted in units of moles. The thermostoichiometric factors are ratios between the enthalpy of reaction and the moles of all the different reactants and products

Execute the plan: The balanced equation says that 1 mol CaO reacts with 3 mol C to form 1 mol CaC_2 and 1 mole CO with the consumption of 464.8 kJ of thermal energy.

$$\frac{464.8\ \text{kJ}}{1\ \text{mol CaO}} \qquad \frac{464.8\ \text{kJ}}{3\ \text{mol C}} \qquad \frac{464.8\ \text{kJ}}{1\ \text{mol CaC}_2} \qquad \frac{464.8\ \text{kJ}}{1\ \text{mol CO}}$$

The reciprocal of the four factors above are also appropriate factors.

$$\frac{1\ \text{mol CaO}}{464.8\ \text{kJ}} \qquad \frac{3\ \text{mol C}}{464.8\ \text{kJ}} \qquad \frac{1\ \text{mol CaC}_2}{464.8\ \text{kJ}} \qquad \frac{1\ \text{mol CO}}{464.8\ \text{kJ}}$$

Check your answers: In each factor, the enthalpy change is related to the moles of one of the reactants or products. There are four different reactants and products and two ways to set up the ratio, so it makes sense that there are eight factors total.

65. *Define the problem:* Given a thermochemical equation for a reaction and a specific quantity of reactant or product produced, determine the quantity of heat transferred.

Develop a plan: Convert the mass into moles, where necessary. Then use the thermochemical equation to create a unit conversion factor relating energy to moles of reactants or products.

Execute the plan: The balanced thermochemical equation tells us that exactly 1 mol methanol reacts to form exactly 1 mol CO and 2 mol H_2, when 90.7 kJ are consumed by the reaction.

(a) $100 \text{ kg CH}_3\text{OH} \times \dfrac{1000 \text{ g}}{1 \text{ kg}} \times \dfrac{1 \text{ mol CH}_3\text{OH}}{32.04 \text{ g CH}_3\text{OH}} \times \dfrac{90.7 \text{ kJ}}{1 \text{ mol CH}_3\text{OH}}$

$$= 3 \times 10^5 \text{ kJ transferred into the reaction}$$

(b) $400. \text{ mol CO} \times \dfrac{90.7 \text{ kJ}}{1 \text{ mol CO}} = 3.63 \times 10^4 \text{ kJ transferred into the reaction}$

(c) Reversing the balanced thermochemical equation tells us that exactly 1 mol methanol is formed when exactly 1 mol CO and 2 mol H_2 react, and 90.7 kJ are produced by the reaction.

$$43.0 \text{ g CO} \times \dfrac{1 \text{ mol CO}}{28.01 \text{ g CO}} \times \dfrac{90.7 \text{ kJ}}{1 \text{ mol CO}} = 139 \text{ kJ transferred out of the reaction}$$

Check your answers: The larger samples require more energy to be transferred.

67. *Define the problem:* Given a chemical equation for the combustion of a fuel, the mass of fuel burned, and the thermal energy evolved at constant pressure for the reaction, determine the molar enthalpy of combustion of the fuel.

Develop a plan: Molar enthalpy change ($\Delta H°$) is identical to the thermal energy released at constant pressure per mole of substance. So, convert the mass into moles, then divide the thermal energy by the moles to get the molar enthalpy of combustion.

Execute the plan: The balanced thermochemical equation tells us that exactly 1 mol C_2H_5OH burns to produce heat.

$$q = -3.62 \text{ kJ (It is negative since heat is evolved, rather than absorbed.)}$$

$$n = 0.115 \text{ g } C_2H_5OH \times \frac{1 \text{ mol } C_2H_5OH}{46.07 \text{ g } C_2H_5OH} = 0.00250 \text{ mol } C_2H_5OH$$

$$\Delta H^\circ = \frac{q}{n} = \frac{-3.62 \text{ kJ}}{0.00250 \text{ mol}} = -1.45 \times 10^3 \text{ kJ/mol}$$

Check your answer: Ethanol is used as a fuel, so it makes sense that a large amount of heat is produced.

69. *Define the problem:* Given a chemical equation for a reaction, the energy produced per mole of the reactant, and a specific quantity of a reactant, determine the quantity of heat transferred.

Develop a plan: Convert the mass into moles. Then use the thermochemical equation to create a unit conversion factor relating energy to moles of reactant.

Execute the plan: The balanced chemical equation tells us that when exactly 1 mol $(NH_4)_2Cr_2O_7$ reacts, 315 kJ are produced by the reaction.

$$28.4 \text{ g } (NH_4)_2Cr_2O_7 \times \frac{1 \text{ mol } (NH_4)_2Cr_2O_7}{252.08 \text{ g } (NH_4)_2Cr_2O_7} \times \frac{315 \text{ kJ}}{1 \text{ mol } (NH_4)_2Cr_2O_7} = 35.5 \text{ kJ}$$

Check your answer: About one tenth of one mole should produce about one tenth of the molar enthalpy change. This answer looks right.

71. *Define the problem:* Given a thermochemical equation for a reaction and a specific quantity of a reactant, determine the quantity of energy released.

Develop a plan: Convert the mass into moles. Then use the thermochemical equation to create a unit conversion factor relating energy to moles of reactant.

Execute the plan: The balanced chemical equation tells us that when exactly 1 mol CH_2O react, 425 kJ of thermal energy are produced by the reaction.

$$10 \text{ lb } CH_2O \times \frac{454 \text{ g}}{1 \text{ lb}} \times \frac{1 \text{ mol } CH_2O}{30.03 \text{ g } CH_2O} \times \frac{425 \text{ kJ}}{1 \text{ mol } CH_2O} = 6 \times 10^4 \text{ kJ produced}$$

Check your answer: About 150 moles should produce about 150 times the molar enthalpy change. This answer looks right.

Where Does the Energy Come From?

73. The bond with the largest bond enthalpy is the strongest, so H–F (566 kJ/mol) is the strongest of the four hydrogen halide bonds. The others are weaker with smaller bond enthalpies: H–Cl (431 kJ/mol), H–Br (366 kJ/mol), and H–I (299 kJ/mol).

75. *Define the problem*: Given a table of bond enthalpies and the description of two chemical reactions, determine for each reaction (a) the enthalpy change for breaking all the bonds in the reactants, (b) the enthalpy change for forming all the bonds in the products, (c) the enthalpy change for the reaction, and (d) determine which reaction is most exothermic.

Develop a plan: Balance the equations and determine how many moles of bonds are broken and formed. Use each bond's bond enthalpy (ΔH_{bond}) as a conversion factor to determine energy per bond type, then add up all the energies.

Execute the plan: The balanced chemical equations look like this:

$$H_2 + F_2 \longrightarrow 2\ HF$$

$$H_2 + Cl_2 \longrightarrow 2\ HCl$$

(a) In both reactions, one H–H bond and one halogen–halogen bond is broken.

$$\Delta H_{reactants} = 1\ mol \times (\Delta H_{H-H\ bond}) + 1\ mol \times (\Delta H_{halogen-halogen\ bond})$$

For fluorine, $\Delta H_{reactants} = 1\ mol \times (436\ kJ/mol) + 1\ mol \times (158\ kJ/mol) = 594\ kJ$

For chlorine, $\Delta H_{reactants} = 1\ mol \times (436\ kJ/mol) + 1\ mol \times (242\ kJ/mol) = 678\ kJ$

(b) In both reactions, two H–halogen bonds form. The enthalpy of forming a bond is the opposite sign of the enthalpy for breaking a bond, so we add a minus sign in the equation.

$$\Delta H_{products} = -2\ mol \times (\Delta H_{H-halogen\ bond})$$

For fluorine, $\Delta H_{products} = -2\ mol \times (566\ kJ/mol) = -1132\ kJ$

For chlorine, $\Delta H_{products} = -2\ mol \times (431\ kJ/mol) = -862\ kJ$

(c) To get the enthalpy change for the reaction, add the enthalpy change of the reactants to the enthalpy change of the products

$$\Delta H_{total} = \Delta H_{reactants} + \Delta H_{products}$$

For fluorine, $\Delta H_{total} = (594\ kJ) + (-1132\ kJ) = -538\ kJ$

For chlorine, $\Delta H_{total} = (678\ kJ) + (-862\ kJ) = -184\ kJ$

(d) The reaction of fluorine with hydrogen is more exothermic (–538 kJ is more negative) than the reaction of chlorine with hydrogen (–184 kJ is less negative).

Check your answers: We expect the fluorine to be more reactive than chlorine, so more energy would be released.

Measuring Enthalpy Changes: Calorimetry

77. *Define the problem*: Given the mass of a piece of ice, and the volume and temperature of a sample of water, determine the final temperature after the two are combined, the ice is melted and thermal equilibrium is established.

Develop a plan: First, look up the specific heat capacity of water and the enthalpy of fusion of ice. Heat from the drop in temperature of the water is used for two things: to melt the ice and to raise the temperature of the resulting ice water to the final temperature. At the end, all the water reaches the same temperature. Using Equation 6.2, set up one equation showing the heat gained by the water equal to the heat used to melt the ice and to raise its temperature.

Execute the plan:

The specific heat capacity (c) water is $4.184 \ J \ g^{-1}{}^{\circ}C^{-1}$ according to Table 6.1. The enthalpy of fusion of ice (ΔH_{fus}) is 333 kJ/g according to Question 46.

For a temperature change, use Equation 6.2: $q = c \times m \times \Delta T$ and with a phase change use $q = m \times \Delta H_{fus}$. Heat is gained in the conversion of ice to ice water and when the ice water increases in temperature, so q_{ice} is positive. Heat is lost by the original water, so q_{water} is negative. The quantity of heat lost by the water is absorbed by the ice and ice water. So,

$$q_{ice} + q_{ice \ water} = - q_{water}$$

$$m_{ice} \times \Delta H_{fus,ice} + c_{ice \ water} \times m_{ice \ water} \times \Delta T_{ice \ water} = - (c_{water} \times m_{water} \times \Delta T_{water})$$

We will have to assume that the ice is at the freezing point of 0°C, since its temperature was not given.

$$\Delta T_{ice \ water} = T_f - T_{i,ice \ water} = T_f - 0.0 \ ^{\circ}C \qquad \Delta T_{water} = T_f - T_{i,water} = T_f - 25 \ ^{\circ}C$$

$$(15.0 \ g) \times (333 \ kJ/g) + (4.184 \ J \ g^{-1}{}^{\circ}C^{-1}) \times (15.0 \ g) \times (T_f - 0.0 \ ^{\circ}C)$$

$$= - [(4.184 \ J \ g^{-1}{}^{\circ}C^{-1}) \times (200. \ mL) \times \frac{1.00 \ g \ H_2O}{1 \ mL \ H_2O} \times (T_f - 25 \ ^{\circ}C)]$$

$$5.00 \times 10^3 + 62.8 T_f = - 840 T_f + 2.1 \times 10^4$$

$$9 \times 10^2 \times T_f = 1.6 \times 10^4$$

$$T_f = 18 \ ^{\circ}C$$

Check your answer: The final temperature is less than the initial temperature of the water and more than the presumed initial temperature of the ice. That makes sense. Quantitatively, we can use $q = c \times m \times \Delta T$ to confirm that the water is losing 6×10^3 J at the same time as the water is gaining 6.1×10^3 J. They are the same within the uncertainty limit of $\pm 1 \times 10^3$ J, so this final temperature makes sense.

79. *Define the problem*: Given the mass of a soluble ionic solid and the volume and temperature of a sample of water, determine the heat transfer from the system to the surroundings and the enthalpy change ($\Delta H°$) by calculating the energy change per mole of ionic solid.

Develop a plan: Define the system as the NaOH, as described in Question 21. That makes the water part of the surroundings. First, look up the specific heat capacity of water. Using Equation 6.2, find heat gained by the water. Heat from the reaction is used to raise the temperature of the water, so relate the heat gained by the water to that lost in the reaction. Finally, divide the heat by the mass of the ionic compound and convert grams to moles.

Execute the plan:

(a) The specific heat capacity (c) water is 4.184 J g^{-1}°C^{-1} according to Table 6.1.

For a temperature change, use Equation 6.2: $q = c \times m \times \Delta T$. The water represents the part of the surroundings affected by a change in the system.

$$q_{water} = c_{water} \times m_{water} \times \Delta T_{water}$$

$$\Delta T_{water} = T_{f,water} - T_{i,water} = 30.7\ °C - 22.6\ °C = 8.1\ °C$$

$$q_{water} = (4.184\ J\ g^{-1}°C^{-1}) \times (400.0\ mL) \times \frac{1.00\ g\ H_2O}{1\ mL\ H_2O} \times (8.1\ °C)$$

$$= 1.4 \times 10^4\ J\ \text{transferred from the system to the surroundings}$$

(b) Energy is lost by the reaction, so $q_{reaction}$ is negative. Heat is lost by the water, so q_{water} is negative. The quantity of heat lost by the water is absorbed in the reaction. So,

$$q_{reaction} = - q_{water} = -1.4 \times 10^4\ J$$

$$\Delta H° = \frac{-1.4 \times 10^4\ J}{13.0\ g\ NaOH} \times \frac{40.00\ g\ NaOH}{1\ mol\ NaOH} \times \frac{1\ kJ}{1000\ J} = -43\ \frac{kJ}{mol\ NaOH}$$

Check your answer: The ionization of an ionic compound involves separating the cations and anions, then hydrating them. The enthalpy of this change could conceivably be positive or negative, but it is expect that it will be small compared to reactions where more significant bond rearrangement is happening. –43 kJ/mol is smaller (closer to zero) than the rest of the $\Delta H°$ values for other kinds of reactions recently studied. Note that the cooling of the solute in the process has been neglected. This may or may not be appropriate. It would be better if we had the specific heat capacity of the solution, then we could use the entire known mass of the solution, rather than just the water.

81. *Define the problem*: Given the mass of water in a bomb calorimeter, the initial and final temperatures, and the heat capacity of the bomb, determine the energy evolved by a reaction in the bomb.

Develop a plan: First, look up the specific heat capacity of water and use Equation 6.2, to find heat gained by the water. Calculate the heat gained by the bomb using the heat capacity (C_{bomb}). Relate the heat gained by the water and the bomb to that evolved by the reaction.

Execute the plan:

The specific heat capacity (c) water is 4.184 J g^{-1}°C^{-1} according to Table 6.1.

For a temperature change, use Equation 6.2: $q = c \times m \times \Delta T$. For a change in temperature in the bomb with the heat capacity: $q = C_{bomb} \times \Delta T$.

$$q_{total} = c_{water} \times m_{water} \times \Delta T + C_{bomb} \times \Delta T$$

$$\Delta T = T_f - T_i = 22.83 \text{ °C} - 19.50 \text{ °C} = 3.33 \text{ °C}$$

$$q_{total} = (4.184 \text{ J g}^{-1}\text{°C}^{-1}) \times (320. \text{ g}) \times (3.33 \text{ °C}) + (650 \text{ J/°C}) \times (3.33 \text{ °C})$$

$$q_{total} = 4.46 \times 10^3 \text{ J} + 2.2 \times 10^3 \text{ J}$$

$$q_{total} = 6.7 \times 10^3 \text{ J}$$

$$\text{Heat evolved by the reaction} = 6.7 \times 10^3 \text{ J} \times \frac{1 \text{ kJ}}{1000 \text{ kJ}} = 6.7 \text{ kJ}$$

Check your answer: The heat is positive, since the heat is gained by the water and the bomb.

83. *Define the problem*: Given the mass of a reactant of a given reaction in a bomb calorimeter, the mass of water in the calorimeter, the initial and final temperatures, and the heat capacity of the bomb, determine the heat evolved per mole of reactant.

Develop a plan: Adapt the method described in the answer to Question 81.

Execute the plan:

The specific heat capacity (c) water is 4.184 J g^{-1}°C^{-1} according to Table 6.1.

For a temperature change, use Equation 6.2: $q = c \times m \times \Delta T$. For a change in temperature in the bomb with the heat capacity: $q = C_{bomb} \times \Delta T$.

$$q_{total} = c_{water} \times m_{water} \times \Delta T + C_{bomb} \times \Delta T$$

$$\Delta T = T_f - T_i = 27.38\ °C - 25.00\ °C = 2.38\ °C$$

$$q_{total} = (4.184\ J\ g^{-1}°C^{-1}) \times (775\ g) \times (2.38\ °C) + (893\ J°C^{-1}) \times (2.38\ °C)$$

$$q_{total} = 7.72 \times 10^3\ J + 2.13 \times 10^3\ J$$

$$q_{total} = 9.84 \times 10^3\ J$$

$$\text{Heat evolved by the reaction} = 9.84 \times 10^3\ J \times \frac{1\ kJ}{1000\ kJ} = 9.84\ kJ$$

$$\text{Heat per mole} = \frac{-9.84\ kJ}{0.300\ g\ C} \times \frac{12.01\ g\ C}{1\ mol\ C} = -394\ \frac{kJ}{mol\ C}$$

Check your answer: This reaction is the formation reaction for CO_2. Table 6.2 says that its $\Delta H°$ is –393.5 kJ/mol, so the answer looks right.

Hess's Law

85. *Define the problem:* Given three thermochemical equations, determine the enthalpy change of a given reaction.

Develop a plan: Identify unique occurrences of the reactants and/or products in the given reactions to help you determine which and how many of a given reaction to use. If a reactant occurs in the products (or vice versa) then reverse the reaction and change the sign of the reaction. If the substance has a different stoichiometric coefficient than the desired coefficient, determine an appropriate multiplier and multiply the enthalpy by the same multiplier (as was done in Question 61).

Execute the plan: Look at the first reactant: Sr(s). The only given equation that has Sr(s) is the first one, where Sr(s) is also a reactant, so we must include that reaction going forward with its given $\Delta H°$. Look at the second reactant: C(graphite). The only given equation that has C(graphite) is the third one, where C(graphite) is also a reactant, so we must include that reaction going forward with its given $\Delta H°$. The third reactant is O_2. This chemical is in several of the given reactions, so let's skip it. Look at the product: $SrCO_3(s)$. The only given equation that has $SrCO_3(s)$ is the second one, so we must include that reaction going forward with its given $\Delta H°$. We have now planned how we will use all the given reactions, so let's do it:

Add all three reactions, and eliminate reactants that also show up as products: $SrO(s)$, $CO_2(g)$. The remaining reactants and products should form the net reaction, and the sum of all the individual $\Delta H°$ values will give us the $\Delta H°$ for that net reaction:

$$Sr(s) + \tfrac{1}{2} O_2(g) \longrightarrow SrO(s) \qquad \Delta H° = -592 \text{ kJ}$$

$$SrO(s) + CO_2(g) \longrightarrow SrCO_3(s) \qquad \Delta H° = -234 \text{ kJ}$$

$$+ \quad C(graphite) + O_2(g) \longrightarrow CO_2(g) \qquad + \quad \Delta H° = -394 \text{ kJ}$$

$$Sr(s) + C(graphite) + \tfrac{3}{2} O_2(g) \longrightarrow SrCO_3(s) \qquad \Delta H° = -1220. \text{ kJ}$$

Check your answer: The net equation adds up the equation we are looking for. The enthalpy of formation of strontium carbonate is not in the table of formation enthalpies; however, the enthalpy of formation of group-related calcium carbonate is -1206.92 kJ/mol (from Table 6.2). The values are similar, so this seems reasonable.

87. *Define the problem:* Given two thermochemical equations and the mass of a formation reactant sample, determine the enthalpy change of the formation reaction and the heat evolved or absorbed when that sample reacts.

Develop a plan: Write the equation for the desired net reaction. Follow the procedure described in Question 85 and 86. After that, calculate the heat evolved or absorbed by finding the moles and using the thermochemical equation to provide the molar energy change.

Execute the plan: The formation of PbO is written with the standard state elements as reactants and one mole of compound as the product:

$$Pb(s) + \tfrac{1}{2} O_2(g) \longrightarrow PbO(s)$$

Look at the first reactant: Pb(s). The only given equation that has Pb(s) is the first one, but Pb(s) is a product, so the reaction must be reversed. Therefore, we must include that reaction going backwards with the sign of its $\Delta H°$ changed. Look at the second reactant:

$O_2(g)$. The only given equation that has $O_2(g)$ is the second one. It is a reactant, but the stoichiometric coefficient is not what we want it to be.

$$\text{Multiplier} = \frac{\text{coefficient we want}}{\text{coefficient we have}} = \frac{\frac{1}{2}}{1} = \frac{1}{2}$$

Therefore, we must include that reaction multiplied by $\frac{1}{2}$ going forward with $\frac{1}{2} \times \Delta H°$. We have now planned how we will use all the given reactions, so let's do it:

Add all the reactions as planned, eliminate reactants that also show up as products: the CO and the C(graphite). Add all the remaining reactants and products to form the net reaction, and add all the individual $\Delta H°$ values to get the $\Delta H°$ for the net reaction:

$$Pb(s) + CO(g) \longrightarrow PbO(s) + C(\text{graphite}) \qquad \Delta H° = -(106.8 \text{ kJ})$$

$$+ \quad \frac{1}{2} \times [2 \text{ C(graphite)} + O_2(g) \longrightarrow 2 \text{ CO(s)}] \quad + \frac{1}{2} \times [\Delta H° = -221.0 \text{ kJ}]$$

$$Pb(s) + \cancel{CO(g)} \longrightarrow PbO(s) + \cancel{C(\text{graphite})} \qquad \Delta H° = -106.8 \text{ kJ}$$

$$+ \quad \cancel{C(\text{graphite})} + \frac{1}{2} O_2(g) \longrightarrow \cancel{CO(s)} \qquad + \Delta H° = -110.5 \text{ kJ}$$

$$Pb(s) + \frac{1}{2} O_2(g) \longrightarrow PbO(s) \qquad \Delta H° = -217.3 \text{ kJ}$$

The enthalpy change is negative, so energy is evolved during the reaction:

$$250 \text{ g Pb} \times \frac{1 \text{ mol Pb}}{207.2 \text{ g Pb}} \times \frac{217.3 \text{ kJ evolved}}{1 \text{ mol Pb}} = 2.6 \times 10^2 \text{ kJ evolved}$$

Check your answer: The net reaction is the desired reaction. The enthalpy of formation of this lead(II) oxide is given in Appendix J with a $\Delta H° = -217.32$ kJ/mol, so this answer makes sense.

Standard Molar Enthalpies of Formation

89. The formation of AgCl(s) is written with the reactants as standard state elements, Ag(s) and $Cl_2(g)$, and the product as one mole of standard state compound:

$$Ag(s) + \frac{1}{2} Cl_2(g) \longrightarrow AgCl(s) \qquad \Delta H° = 238.7 \text{ kJ}$$

91. (a) The formation of $Al_2O_3(s)$ is written with the reactants as standard state elements, $Al(s)$ and $O_2(g)$, and the product as one mole of standard state compound:

$$2\ Al\ (s) + \frac{3}{2}\ O_2\ (g) \longrightarrow Al_2O_3\ (s)\quad \Delta H° = -1675.7\ kJ$$

(b) The formation of $TiCl_4(\ell)$ is written with the reactants as standard state elements, $Ti(s)$ and $Cl_2(g)$, and the product as one mole of standard state compound:

$$Ti(s) + 2\ Cl_2\ (g) \longrightarrow TiCl_4(\ell)\qquad \Delta H° = -804.2\ kJ$$

(c) The formation of $NH_4NO_3(s)$ is written with the reactants as standard state elements, $N_2(g)$, $H_2(g)$, and $O_2(g)$, and the product as one mole of standard state compound:

$$N_2(g) + 2\ H_2\ (g) + \frac{3}{2}\ O_2\ (g) \longrightarrow NH_4NO_3(s)\qquad \Delta H° = -365.56\ kJ$$

94. *Define the problem:* Given a balanced chemical equation for a reaction and a table of molar enthalpies of formation, determine the enthalpy change of the reaction and whether the reaction is endothermic or exothermic.

Develop a plan: Using molar enthalpies of formation is a common variation of Hess's Law. We will form all the products (using their molar enthalpies as written) and we will destroy – the opposite of formation – the reactants (using their molar enthalpies with opposite sign). That logic is the basis of Equation 6.9:

$$\Delta H° = \Sigma\left[\text{(moles of product)} \times \Delta H°_f(\text{product})\right] -$$
$$\Sigma\left[\text{(moles of reactant)} \times \Delta H°_f(\text{reactant})\right]$$

Use the stoichiometric coefficient of the balanced equation to describe the moles of each of the reactants and products. Set up a specific version of Equation 6.9. Look up the $\Delta H°_f$ for each; remember to check the physical phase and remember that $\Delta H°_f$ for standard state elements is exactly zero. Plug them into the equation and solve for $\Delta H°$. Check the sign of $\Delta H°$ to assess endothermic or exothermic.

Execute the plan:

(a) The products are O_2 and $C_6H_{12}O_6$. The reactants are CO_2 and $H_2O(\ell)$. O_2 is the elemental form for oxygen.

$$\Delta H° = [(1\ mol) \times \Delta H°_f(C_6H_{12}O_6) + (6\ mol) \times \Delta H°_f(O_2)]$$
$$- [(6\ mol) \times \Delta H°_f(CO_2) + (1\ mol) \times \Delta H°_f(H_2O(\ell))]$$

Look up the ΔH_f° values in Table 6.9.

$\Delta H^\circ = [(1 \text{ mol}) \times (-1274.4 \text{ kJ/mol}) + (6 \text{ mol}) \times (0 \text{ kJ/mol})]$

$\qquad - [(6 \text{ mol}) \times (-393.509 \text{ kJ/mol}) + (1 \text{ mol}) \times (-285.830 \text{ kJ/mol})] = 2801.6 \text{ kJ}$

(b) ΔH° is positive, so the reaction is endothermic.

Check your answer: It makes sense that the formation of an organic molecule is endothermic.

96. *Define the problem:* Given a balanced chemical equation for a reaction and a table of molar enthalpies of formation, determine the enthalpy change of the reaction.

Develop a plan: Use the stoichiometric coefficient of the balanced equation to describe the moles of each of the reactants and products. Set up a specific version of Equation 6.9. Look up the ΔH_f° for each. Plug them into the equation and solve for ΔH°.

Execute the plan: The products are $CaCO_3$ and $H_2O(g)$. The reactants are $Ca(OH)_2(s)$ and CO_2.

$\Delta H^\circ = (1 \text{ mol}) \times \Delta H_f^\circ(CaCO_3) + (1 \text{ mol}) \times \Delta H_f^\circ(H_2O(g))$

$\qquad - [(1 \text{ mol}) \times \Delta H_f^\circ(Ca(OH)_2) + (1 \text{ mol}) \times \Delta H_f^\circ(CO_2)]$

Look up the ΔH_f° values in Table 6.9 and Appendix J.

$\Delta H^\circ = (1 \text{ mol}) \times (-1206.92 \text{ kJ/mol}) + (1 \text{ mol}) \times (-241.818 \text{ kJ/mol})$

$\qquad - [(1 \text{ mol}) \times (-986.09 \text{ kJ/mol}) + (1 \text{ mol}) \times (-393.509 \text{ kJ/mol})] = -69.14 \text{ kJ}$

Check your answer: Collectively, the products have slightly more negative formation enthalpies than the reactants, so an exothermic reaction makes sense.

98. *Define the problem:* Given a balanced thermochemical equation for a reaction and some molar enthalpies of formation, determine an unknown molar enthalpy of formation.

Develop a plan: Use the stoichiometric coefficient of the balanced equation to describe the moles of each of the reactants and products. Set up a specific version of Equation 6.9, and solve it for the unknown ΔH_f°. Look up the ΔH_f° for the rest of the reactants and products. Plug them into the equation and solve for ΔH_f°.

Execute the plan: The products are HF and O_2. The reactants are OF_2 and $H_2O(g)$. O_2 is the elemental form for oxygen.

$\Delta H^\circ = [(2 \text{ mol}) \times \Delta H_f^\circ(HF) + (1 \text{ mol}) \times \Delta H_f^\circ(O_2)]$

$\qquad - [(1 \text{ mol}) \times \Delta H_f^\circ(OF_2) + (1 \text{ mol}) \times \Delta H_f^\circ(H_2O(g))]$

$\Delta H° = (2 \text{ mol}) \times \Delta H^{\circ}_f(HF) + (1 \text{ mol}) \times \Delta H^{\circ}_f(O_2)$

$$- (1 \text{ mol}) \times \Delta H^{\circ}_f(OF_2) - (1 \text{ mol}) \times \Delta H^{\circ}_f(H_2O(g))$$

$$\Delta H^{\circ}_f(OF_2) = - \frac{\Delta H°}{(1 \text{ mol})} + 2 \times \Delta H^{\circ}_f(HF) + \Delta H^{\circ}_f(O_2) - \Delta H^{\circ}_f(H_2O(g))$$

Look up the ΔH°_f values for $H_2O(g)$ and HF in Table 6.2.

$\Delta H^{\circ}_f(OF_2) = - (-318 \text{ kJ/mol}) + 2 \times (-271.1 \text{ kJ/mol}) + (0 \text{ kJ/mol})$

$$- (-241.818 \text{ kJ/mol}) = 18 \text{ kJ/mol}$$

Check your answer: The high exothermicity of the reaction can be almost completely accounted for with the destruction of the other reactant and formation of the products. So, it makes sense that the molar enthalpy of formation of OF_2, the reactant, is small.

100. *Define the problem:* Given the mass of a reactant, the description of a reaction, and molar enthalpies of formation, determine the thermal energy transferred out of the system at constant pressure.

Develop a plan: Balanced the equation and use it to describe the moles of each of the reactants and products. Set up a specific version of Equation 6.9. Look up the ΔH°_f for each. Plug them into the equation and solve for the molar enthalpy ($\Delta H°$) – this represents the molar heat produced at constant pressure. Convert mass to moles, and use the molar enthalpy as a conversion factor to get heat produced at constant pressure for the reactant sample.

Execute the plan: The product is Al_2O_3. The reactants are both elements: Al and O_2.

$$2 \text{ Al(s)} + \frac{3}{2} O_2(g) \longrightarrow Al_2O_3(s)$$

$$\Delta H° = (1 \text{ mol}) \times \Delta H^{\circ}_f(Al_2O_3) - [(2 \text{ mol}) \times \Delta H^{\circ}_f(Al) + (\frac{3}{2} \text{ mol}) \times \Delta H^{\circ}_f(O_2)]$$

Look up the ΔH°_f value in Table 6.9.

$$\Delta H° = (1 \text{ mol}) \times (-824.2 \text{ kJ/mol}) - (2 \text{ mol}) \times (0 \text{ kJ/mol}) - (\frac{3}{2} \text{ mol}) \times (0 \text{ kJ/mol})$$

$$= -1675.7 \text{ kJ}$$

That means the thermochemical equation looks like this:

$$2 \text{ Al(s)} + \frac{3}{2} O_2(g) \longrightarrow Al_2O_3(s) \qquad \Delta H° = -1675.7 \text{ kJ}$$

Now, we'll start the calculation with the specific sample give:

$$2.70 \text{ g Al} \times \frac{1 \text{ mol Al}}{26.9815 \text{ g Al}} \times \frac{1675.7 \text{ kJ evolved}}{2 \text{ mol Al}} = 83.8 \text{ kJ evolved}.$$

Check your answer: The sample is a tenth of a mole, and the equation shows that two moles of Al are needed. So, the resulting heat is half of a tenth of the original molar enthalpy. That makes sense.

Chemical Fuels

101. *Define the problem*: Determine the mass of fuel required to generate the heat to melt a mass of ice and raise its temperature to a specified value.

Develop a plan: Use Equation 6.2 and the molar enthalpy of fusion to calculate the heat needed. Then balance the equation describing the combustion of the fuel. Use Equation 6.9, a table of molar enthalpies of formation, and the stoichiometry of the balanced equations to determine the molar enthalpy change for each combustion reaction. Then use that enthalpy and the molar mass as conversion factors to determine the mass of fuel needed.

Execute the plan: To change a phase in the system, use $q = m \times \Delta H_{fus}$; to change the temperature of the water, use $q = c \times m \times \Delta T$.

$$q_{total} = (m_{ice} \times \Delta H_{fus,ice}) + (c_{water} \times m_{water} \times \Delta T_{water})$$

To simplify the calculation, factor out the common mass term $m_{ice} = m_{water} = m$:

$$q_{total} = m \times [(\Delta H_{fus,ice}) + (c_{water} \times \Delta T_{water})]$$

Table 6.1 gives the specific heat capacity of water to be 4.184 J g^{-1}°C^{-1}. Question 46 gives the molar enthalpy of fusion of ice as 333 J/g.

$$q_{total} = 56.0 \text{ g} \times [(333 \text{ J/g}) + (4.184 \text{ J g}^{-1}\text{°C}^{-1}) \times (75.0 \text{ °C} - 0. \text{ °C})] = 3.6 \times 10^4 \text{ J}$$

The reaction describing the combustion of propane is similar to those described in Section 6.11. The fuel reacts with oxygen and making gaseous carbon dioxide and water:

$$C_3H_8(g) + O_2(g) \longrightarrow CO_2(g) + H_2O(g)$$

Balance the chemical equation:

$$C_3H_8(g) + 5 \, O_2(g) \longrightarrow 3 \, CO_2(g) + 4 \, H_2O(g)$$

$$\Delta H° = (3 \text{ mol}) \times \Delta H_f^\circ(CO_2) + (4 \text{ mol}) \times \Delta H_f^\circ(H_2O(g))$$

$$- (1 \text{ mol}) \times \Delta H_f^\circ(C_3H_8) - (5 \text{ mol}) \times \Delta H_f^\circ(O_2)$$

Look up the ΔH_f° value in Table 6.9.

$$\Delta H° = (3 \text{ mol}) \times (-393.509 \text{ kJ/mol}) + (4 \text{ mol}) \times (-241.818 \text{ kJ/mol})$$

$$- (1 \text{ mol}) \times (-103.8 \text{ kJ/mol}) - (5 \text{ mol}) \times (0 \text{ kJ/mol}) = -2044.0 \text{ kJ}$$

Now, we'll figure out the mass needed for the melting and warming of the water sample:

$$3.6 \times 10^4 \text{ J} \times \frac{1 \text{ kJ}}{1000 \text{ J}} \times \frac{1 \text{ mol C}_3\text{H}_8}{2044.0 \text{ kJ}} \times \frac{44.11 \text{ g C}_3\text{H}_8}{1 \text{ mol C}_3\text{H}_8} = 0.78 \text{ g C}_3\text{H}_8$$

Check your answer: The molar enthalpy of C_3H_8 is large and negative because propane is a good fuel. The energy needed for the sample is relatively small, so the mass of fuel is also small.

105. *Define the problem*: Compare the quantity of thermal energy produced per gram (also called the fuel value) when burning four different organic fuels.

Develop a plan: Write the formulas for the four smallest hydrocarbons. Balance their combustion equations. Then use Equation 6.9, a table of molar enthalpies of formation, and the stoichiometry of the balanced equations to determine the molar enthalpy change for each combustion reaction. Then use the molar mass to determine the enthalpy per gram of each fuel and rank them.

Execute the plan: The reaction for the combustion of methane is described in Section 6.11, and those describing the combustion of ethane, propane, and butane are similar to that one. We'll assume that all fuels are initially in the gas state. Look up the ΔH_f° value in Appendix J. *Note, the $\Delta H_f^\circ(C_4H_{10})$ is different between Table 6.2 and Appendix J. The one in Appendix J is correct.*

Methane: $CH_4 + 2\,O_2 \longrightarrow CO_2 + 2\,H_2O$

$$\Delta H°_{methane} = (1 \text{ mol}) \times \Delta H_f^\circ(CO_2) + (2 \text{ mol}) \times \Delta H_f^\circ(H_2O(g))$$

$$- (1 \text{ mol}) \times \Delta H_f^\circ(CH_4) - (2 \text{ mol}) \times \Delta H_f^\circ(O_2)$$

$$= (1 \text{ mol}) \times (-393.509 \text{ kJ/mol}) + (2 \text{ mol}) \times (-241.818 \text{ kJ/mol})$$

$$- (1 \text{ mol}) \times (-74.81 \text{ kJ/mol}) - (2 \text{ mol}) \times (0 \text{ kJ/mol}) = -802.34 \text{ kJ}$$

$$\frac{802.34 \text{ kJ produced}}{1 \text{ mol CH}_4} \times \frac{1 \text{ mol CH}_4}{16.043 \text{ g CH}_4} = 50.013 \frac{\text{kJ produced}}{\text{g CH}_4}$$

105. (continued)

Ethane:
$$C_2H_6 + \frac{7}{2} O_2 \longrightarrow 2\, CO_2 + 3\, H_2O$$

$$\Delta H°_{ethane} = (2\ mol) \times \Delta H_f^°(CO_2) + (3\ mol) \times \Delta H_f^°(H_2O(g))$$

$$- (1\ mol) \times \Delta H_f^°(C_2H_6) - (\frac{7}{2}\ mol) \times \Delta H_f^°(O_2)$$

$$= (2\ mol) \times (-393.509\ kJ/mol) + (3\ mol) \times (-241.818\ kJ/mol)$$

$$- (1\ mol) \times (-84.68\ kJ/mol) - (\frac{7}{2}\ mol) \times (0\ kJ/mol) = -1427.79\ kJ$$

$$\frac{1427.79\ kJ\ produced}{1\ mol\ C_2H_6} \times \frac{1\ mol\ C_2H_6}{30.069\ g\ C_2H_6} = 47.484\ \frac{kJ\ produced}{g\ C_2H_6}$$

Propane:
$$C_3H_8 + 5\, O_2 \longrightarrow 3\, CO_2 + 4\, H_2O$$

Note: We already did the propane calculation in Question 101.

$$\Delta H° = (3\ mol) \times \Delta H_f^°(CO_2) + (4\ mol) \times \Delta H_f^°(H_2O(g))$$

$$- (1\ mol) \times \Delta H_f^°(C_3H_8) - (5\ mol) \times \Delta H_f^°(O_2)$$

$$\Delta H° = (3\ mol) \times (-393.509\ kJ/mol) + (4\ mol) \times (-241.818\ kJ/mol)$$

$$- (1\ mol) \times (-103.8\ kJ/mol) - (5\ mol) \times (0\ kJ/mol) = -2044.0\ kJ$$

$$\frac{2044.0\ kJ\ produced}{1\ mol\ C_3H_8} \times \frac{1\ mol\ C_3H_8}{44.096\ g\ C_3H_8} = 46.353\ \frac{kJ\ produced}{g\ C_3H_8}$$

Butane:
$$C_4H_{10} + \frac{13}{2} O_2 \longrightarrow 4\, CO_2 + 5\, H_2O$$

$$\Delta H°_{butane} = (4\ mol) \times \Delta H_f^°(CO_2) + (5\ mol) \times \Delta H_f^°(H_2O(g))$$

$$- (1\ mol) \times \Delta H_f^°(C_4H_{10}) - (\frac{13}{2}\ mol) \times \Delta H_f^°(O_2)$$

$$= (4\ mol) \times (-393.509\ kJ/mol) + (5\ mol) \times (-241.818\ kJ/mol)$$

$$- (1\ mol) \times (-126.148\ kJ/mol) - (\frac{13}{2}\ mol) \times (0\ kJ/mol) = -2656.978\ kJ$$

$$\frac{2656.978\ kJ\ produced}{1\ mol\ C_4H_{10}} \times \frac{1\ mol\ C_4H_{10}}{58.123\ g\ C_4H_{10}} = 45.713\ \frac{kJ\ produced}{g\ C_4H_{10}}$$

Methane has the highest fuel value, followed by ethane, then propane, and then butane.

Check your answers: While we might not have predicted which of these would be the highest, seeing a consistent trend between these numbers is satisfying.

Food and Energy

107. *Define the problem*: Take the percentages (by mass) of carbohydrate, fat, and protein in some candy and determine the quantity of energy transfer that would occur a given mass of the candy was burned in a bomb calorimeter.

Develop a plan: Start with the sample mass. Using the percentage as a ratio of masses, and the kJ/g for each type of food component given in Section 6.12, calculate the energy related to each of the three components. Then add those three numbers for the total energy.

Execute the plan:

Carbohydrate: $34.5 \times \dfrac{70.\text{ g carbohydrates}}{100 \text{ g M \& M}} \times \dfrac{17 \text{ kJ}}{1 \text{ g carbohydrates}} = 410 \text{ kJ}$

Fat: $34.5 \times \dfrac{21 \text{ g fats}}{100 \text{ g M \& M}} \times \dfrac{38 \text{ kJ}}{1 \text{ g fats}} = 280 \text{ kJ}$

Protein: $34.5 \times \dfrac{4.6 \text{ g proteins}}{100 \text{ g M \& M}} \times \dfrac{17 \text{ kJ}}{1 \text{ g proteins}} = 27 \text{ kJ}$

Total: $410 \text{ kJ} + 280 \text{ kJ} + 27 \text{ kJ} = 720 \text{ kJ}$

Check your answer: We can convert this energy value to food Calories:

$$720 \text{ kJ} \times \dfrac{1 \text{ kcal}}{4.184 \text{ kJ}} \times \dfrac{1 \text{ Cal}}{1 \text{ kcal}} = 170 \text{ Cal}$$

Some vending-machine-size packets of M&M's list the contents as having 400 Calories. So, 170 Cal represents a small handful of M&Ms. This seems a reasonable sample size for this experiment, and therefore a reasonable amount of energy calculated.

109. *Define the problem*: Determine how long you must walk to burn off the Calories of a quarter-pound hamburger.

Develop a plan: Use the caloric value of hamburger (from Table 6.3) determine the Calories in the hamburger. Use the basal metabolic rate (BMR) for a 70 kg person and the multiplier for walking given in Section 6.12, determine the energy burned for that activity per hour. Then use that result to determine how many hours the person would need to walk.

Execute the plan: $0.25 \text{ lb hamburger} \times \dfrac{454 \text{ g}}{1 \text{ lb}} \times \dfrac{3.60 \text{ Cal}}{1 \text{ g hamburger}} = 410 \text{ Cal}$

Section 6.12 gives the following information: A 70-kg person has a BMR of 1750 Cal/day. Walking gives him a multiplier of $2.5 \times$ BMR.

$$2.5 \times \frac{1750 \text{ Cal}}{1 \text{ day}} \times \frac{1 \text{ day}}{24 \text{ hr}} = 182 \frac{\text{Cal}}{\text{hr}}$$

$$410 \text{ Cal} \times \frac{1 \text{ hr}}{182 \text{ Cal}} = 2.2 \text{ hr}$$

Check your answer: Every dieter knows that hamburgers (and most high Calorie junk food) requires significant exercise to work off. This number seems reasonable.

General Questions

111. The metal with the smaller specific heat capacity will increase in temperature faster than a metal with a larger specific heat capacity. From Table 6.1, $c_{Cu} = 0.385 \text{ J g}^{-1}{}^{\circ}\text{C}^{-1}$ and $c_{Au} = 0.128 \text{ J g}^{-1}{}^{\circ}\text{C}^{-1}$, so the Au will reach 100 °C first.

113. *Define the problem*: Given the mass and initial temperature of a sample of ice, the mass and initial temperature of a sample of water, determine how much ice is left after the mixture reaches a common final temperature.

Develop a plan: Using Equation 6.2 and Table 6.1, determine the heat required to lower the temperature of the water. Since the heat lost by the water comes from melting ice, relate those energies. Use the enthalpy of fusion (ΔH_{fus}) from Question 46 to determine amount of ice that melted. Subtract the calculated mass from the initial mass of ice.

Execute the plan: To change the temperature of the water, use $q = c \times m \times \Delta T$:

$$q_{water} = c_{water} \times m_{water} \times \Delta T_{water}$$

Table 6.1 gives the specific heat capacity of water to be $4.184 \text{ J g}^{-1}{}^{\circ}\text{C}^{-1}$.

$$q_{water} = (100.0 \text{ g}) \times (4.184 \text{ J g}^{-1}{}^{\circ}\text{C}^{-1}) \times (0.00 \text{ °C} - 60.0 \text{ °C}) = -2.51 \times 10^4 \text{ J}$$

$$-q_{water} = q_{ice} = 2.51 \times 10^4 \text{ J}$$

To melt the ice, 333 J/g of heat are needed to melt ice:

$$2.51 \times 10^4 \text{ J} \times \frac{1 \text{ g ice melted}}{333 \text{ J}} = 75.4 \text{ g melted}$$

100.0 g ice are present initially, so:

$$100.0 \text{ g ice initial} - 75.4 \text{ g melted} = 24.6 \text{ g ice left}$$

Check your answer: The ice did stop melting before it was all gone, because the thermal energy needed to reach the freezing point was reached in the solution and thermal equilibrium was established.

114. *Define the problem:* Given a balanced thermochemical equation for a reaction and some molar enthalpies of formation, determine an unknown molar enthalpy of formation.

Develop a plan: Use the stoichiometric coefficient of the balanced equation to describe the moles of each of the reactants and products. Set up a specific version of Equation 6.9, and solve it for the unknown ΔH_f°. Look up the ΔH_f° for the rest of the reactants and products. Plug them into the equation and solve for ΔH_f°.

Execute the plan: The products are B_2O_3 and $H_2O(\ell)$. The reactants are B_2H_6 and O_2. O_2 is the elemental form for oxygen.

$$\Delta H^\circ = [(1 \text{ mol}) \times \Delta H_f^\circ(B_2O_3) + (3 \text{ mol}) \times \Delta H_f^\circ(H_2O(\ell))] - (1 \text{ mol}) \times \Delta H_f^\circ(B_2H_6)$$

$$\Delta H_f^\circ(B_2H_6) = -\frac{\Delta H^\circ}{(1 \text{ mol})} + \Delta H_f^\circ(B_2O_3) + 3 \times \Delta H_f^\circ(H_2O(\ell))$$

Look up the ΔH_f° value for $H_2O(\ell)$ in Table 6.2.

$$\Delta H_f^\circ(B_2H_6) = -\frac{(-2166 \text{ kJ})}{(1 \text{ mol})} + (-1273 \text{ kJ/mol}) + 3 \times (-285.830) = 36 \text{ kJ/mol}$$

Check your answer: The high exothermicity of the reaction can be almost completely accounted for by the formation of the products. So, it makes sense that the molar enthalpy of formation of the reactant B_2H_6 is small.

116. *Define the problem:* Given an explanation of a reaction and table of molar enthalpies of formation, determine the thermal energy transferred when a specific reactant mass is used.

Develop a plan: The reaction described is the formation reaction of the P_4O_{10}. Look up the $\Delta H_f^\circ(P_4O_{10})$ value in Appendix J. Convert the sample mass to moles, and use the molar enthalpy of the reaction to determine the thermal energy transferred.

Execute the plan: Making P_4O_{10} from igniting elemental P_4 in air is a formation reaction:

$$P_4(s,\text{white}) + 5\, O_2\,(g) \longrightarrow P_4O_{10}(s)$$

$$\Delta H° = \Delta H^\circ_f(P_4O_{10}) = -2984 \text{ kJ}$$

$$9.08 \times 10^5 \text{ g } P_4 \times \frac{1 \text{ mol } P_4}{123.895 \text{ g } P_4} \times \frac{2984 \text{ kJ evolved}}{1 \text{ mol } P_4} = 2.19 \times 10^7 \text{ kJ consumed}$$

Check your answer: In Question 68, we were given 85.3 kJ evolved when 3.56 g P_4 burned. Using this information we get the same result.

$$9.08 \times 10^5 \text{ g } P_4 \times \frac{85.3 \text{ kJ evolved}}{3.56 \text{ g } P_4} = 2.19 \times 10^7 \text{ kJ consumed}$$

So, these results make sense.

118. *Define the problem:* Given a balanced thermochemical equation for a reaction and some molar enthalpies of formation, determine an unknown molar enthalpy of formation.

    Develop a plan: Use the stoichiometric coefficient of the balanced equation to describe the moles of each of the reactants and products. Set up a specific version of Equation 6.9, and solve it for the unknown ΔH°_f. Look up the ΔH°_f for the rest of the reactants and products. Plug them into the equation and solve for ΔH°_f.

    Execute the plan: The products are N_2 and $H_2O(g)$. The reactants are N_2H_4 and O_2. N_2 and O_2 are elemental forms.

    $$\Delta H° = (2 \text{ mol}) \times \Delta H^\circ_f(H_2O(g)) - (1 \text{ mol}) \times \Delta H^\circ_f(N_2H_4)$$

    $$\Delta H^\circ_f(N_2H_4) = -\frac{\Delta H°}{(1 \text{ mol})} + 2 \times \Delta H^\circ_f(H_2O(g))$$

    Look up the ΔH°_f value for $H_2O(\ell)$ in Table 6.2.

    $$\Delta H^\circ_f(N_2H_4) = -\frac{(-534 \text{ kJ})}{(1 \text{ mol})} + 2 \times (-241.818 \text{ kJ/mol}) = 50. \text{ kJ/mol}$$

    Check your answer: The high exothermicity of the reaction can be almost completely accounted for due to the formation of the product $H_2O(g)$. So, it makes sense that the molar enthalpy of formation of the reactant N_2H_4 is small. You can also compare this to the answer you got on Question 104.

120. *Define the problem:* Given a balanced chemical equation for two reactions and a table of molar enthalpies of formation, determine the enthalpy change of the first reaction and the thermal energy evolved by the second.

Develop a plan: Use the stoichiometric coefficient of the balanced equation to describe the moles of each of the reactants and products. Set up a specific version of Equation 6.9. Look up the ΔH_f° for each. Plug them into the equation and solve for ΔH°.

Execute the plan:

(a) The products of the first reaction are N_2O and $H_2O(g)$. The reactant is $NH_4NO_3(s)$.

$$\Delta H^{\circ} = (1 \text{ mol}) \times \Delta H_f^{\circ}(N_2O) + (2 \text{ mol}) \times \Delta H_f^{\circ}(H_2O(g))$$

$$- (1 \text{ mol}) \times \Delta H_f^{\circ}(NH_4NO_3(s))$$

Look up the ΔH_f° values in Table 6.9 and Appendix J.

$$\Delta H^{\circ} = (1 \text{ mol}) \times (82.05 \text{ kJ/mol}) + (2 \text{ mol}) \times (-241.818 \text{ kJ/mol})$$

$$- (1 \text{ mol}) \times (-365.56 \text{ kJ/mol}) = -36.03 \text{ kJ}$$

(b) The products of the second reaction are N_2, $H_2O(g)$, and O_2. The reactant is $NH_4NO_3(s)$. N_2 and O_2 are elemental forms, with ΔH_f° exactly zero.

$$\Delta H^{\circ} = (4 \text{ mol}) \times \Delta H_f^{\circ}(H_2O(g)) - (2 \text{ mol}) \times \Delta H_f^{\circ}(NH_4NO_3(s))$$

Look up the ΔH_f° values in Table 6.9 and Appendix J.

$$\Delta H^{\circ} = (4 \text{ mol}) \times (-241.818 \text{ kJ/mol}) - (2 \text{ mol}) \times (-365.56 \text{ kJ/mol}) -236.15 \text{ kJ}$$

The balanced chemical equation says that 2 mol NH_4NO_3 react with the evolution of 236.15 kJ of thermal energy.

$$8.00 \text{ kg NH}_4\text{NO}_3 \times \frac{1000 \text{ g}}{1 \text{ kg}} \times \frac{1 \text{ mol NH}_4\text{NO}_3}{80.05 \text{ g NH}_4\text{NO}_3} \times \frac{236.15 \text{ kJ}}{2 \text{ mol NH}_4\text{NO}_3} = 1.18 \times 10^4 \text{ kJ}$$

Check your answer: These explosive reactions produce large amounts of heat. The sign and size of these answers make sense.

122. *Define the problem:* Given a balanced chemical equation for two reactions and a table of molar enthalpies of formation, determine the enthalpy change of the first reaction and the thermal energy evolved by the second.

Develop a plan: Use the stoichiometric coefficient of the balanced equations to describe the moles of each of the reactants and products. Set up specific versions of Equation 6.9. Look up the ΔH_f° for each species. Plug them into the equations and solve for the three ΔH°. Then add (as described in Question 85) the three chemical reactions together to make the overall reaction and determine the overall ΔH°.

Execute the plan:

Step 1: The products of the reaction are $H_2SO_4(\ell)$ and $HBr(g)$. The reactants are SO_2, $H_2O(g)$, and $Br_2(g)$. $Br_2(g)$ is NOT elemental bromine.

$$\Delta H^\circ = (1 \text{ mol}) \times \Delta H_f^\circ(H_2SO_4(\ell)) + (2 \text{ mol}) \times \Delta H_f^\circ(HBr(g))$$

$$- [(1 \text{ mol}) \times \Delta H_f^\circ(SO_2) + (2 \text{ mol}) \times \Delta H_f^\circ(H_2O(g)) + (1 \text{ mol}) \times \Delta H_f^\circ(Br_2(g))]$$

Look up the ΔH_f° values in Table 6.9 and Appendix J.

$$\Delta H^\circ = (1 \text{ mol}) \times (-813.989 \text{ kJ/mol}) + (2 \text{ mol}) \times (-36.40 \text{ kJ/mol})$$

$$- (1 \text{ mol}) \times (-296.830 \text{kJ/mol}) - (2 \text{ mol}) \times (-241.818 \text{kJ/mol})$$

$$- (1 \text{ mol}) \times (30.907 \text{kJ/mol}] = -137.23 \text{ kJ}$$

Step 2: The products of the reaction are $H_2O(g)$, SO_2, and O_2. The reactant is $H_2SO_4(\ell)$. $O_2(g)$ is in elemental form.

$$\Delta H^\circ = (1 \text{ mol}) \times \Delta H_f^\circ(H_2O(g)) + (1 \text{ mol}) \times \Delta H_f^\circ(SO_2) - (1 \text{ mol}) \times$$

$$\Delta H_f^\circ(H_2SO_4(\ell))$$

Look up the ΔH_f° values in Table 6.9 and Appendix J.

$$\Delta H^\circ = (1 \text{ mol}) \times (-241.818 \text{ kJ/mol}) + (1 \text{ mol}) \times (-296.830 \text{ kJ/mol}) -$$

$$(1 \text{ mol}) \times (-813.989 \text{ kJ/mol}) = 275.341 \text{ kJ}$$

. Step 3: The products of the reaction are $H_2(g)$ and $Br_2(g)$. The reactant is $HBr(g)$. $H_2(g)$ is in elemental form, but $Br_2(g)$ is not.

$$\Delta H^\circ = (1 \text{ mol}) \times \Delta H_f^\circ(Br_2(g)) - (2 \text{ mol}) \times \Delta H_f^\circ(HBr(g))$$

Look up the ΔH_f° values in Table 6.9 and Appendix J.

$$\Delta H^\circ = (1 \text{ mol}) \times (30.907 \text{ kJ/mol}) - (2 \text{ mol}) \times (-36.40 \text{ kJ/mol}) = 103.71 \text{ kJ}$$

Add the three thermochemical equations to get the net equation:

$$SO_2(g) + 2 H_2O(g) + Br_2(g) \longrightarrow H_2SO_4(\ell) + 2 HBr(g) \qquad \Delta H^\circ = -137.23 \text{ kJ}$$

$$H_2SO_4(\ell) \longrightarrow H_2O(g) + SO_2(g) + \frac{1}{2}O_2(g) \qquad \Delta H^\circ = 275.341 \text{ kJ}$$

$$+ \quad 2 HBr(g) \longrightarrow H_2(g) + Br_2(g) \qquad\qquad + \Delta H^\circ = 103.71 \text{ kJ}$$

$$H_2O(g) \longrightarrow H_2(g) + \frac{1}{2}O_2(g) \qquad\qquad \Delta H^\circ = 241.82 \text{ kJ}$$

The net equation is endothermic.

Check your answers: The net reaction is the reverse of the $\Delta H_f^\circ(H_2O(g))$, which has a value of -241.818 kJ/mol. The positive value makes sense because the reaction is reversed, and the size of these two numbers is very similar. This answer makes sense.

Applying Concepts

124. Ice in a system melts in warm conditions and the molecules are moving around more quickly in the liquid state than in the solid state, so energy must be absorbed by the surroundings. Hence, melting ice is an endothermic process. Freezing water in a system happens only in cold conditions and the molecules are moving more slowly in the solid state than in the liquid, so energy must be lost by the water into the surroundings to freeze it. Hence, freezing water is an exothermic process.

126. Look at Equation 6.2: $q = m \times c \times \Delta T$. This equation says that energy transferred is proportional to ΔT for constant mass samples, using the specific heat capacity, c. So, the slower the temperature rises with the transfer of energy, the larger the specific heat capacity. On the graph, the shallowest line (the line with the least steep slope) has the highest specific heat capacity. Here, that is Substance A.

128. Thermal energy content is greater in Beaker 1 than in Beaker 2. A larger mass of water will contain larger amount of heat at a given temperature.

130. Each of these terms describes the thermal energy changes at constant pressure. However, they each represent different chemical reactions.

132. The given reaction produces 2 mol SO_3. Formation enthalpy from Table 6.2 is for the production of 1 mol SO_3.

133. (a) We predict the temperature of experiment to be 26.6 °C, because, above $C_6H_8O_6$ masses of 8.81g, the other reactant NaOH appears to be the limiting reactant.

(b) $C_6H_8O_6$ limits in Experiments 1, 2, and 3, which is why the final temperature increases proportionally to the mass of $C_6H_8O_6$ from experiment to experiment in those three. NaOH limits in Experiments 3,4, and 5, as seen by the constant temperature increase in the available data for those three experiments, even though larger masses of $C_6H_8O_6$ are used.

(c) Experiment 3 seems to be the stoichiometric equivalence point, where both reactants run out at the same time. Before that experiment $C_6H_8O_6$ limits, and after that experiment, NaOH limits. Use the mass of $C_6H_8O_6$ and the volume and molarity of the NaOH to determine the moles of each present in that experiment.

$$100. \text{ mL NaOH} \times \frac{1 \text{ L}}{1000 \text{ mL}} \times \frac{0.500 \text{ mol NaOH}}{1 \text{ L NaOH}} = 0.0500 \text{ mol NaOH}$$

$$8.81 \text{ g } C_6H_8O_6 \times \frac{1 \text{ mol } C_6H_8O_6}{176.12 \text{ g } C_6H_8O_6} = 0.0500 \text{ mol } C_6H_8O_6$$

Because equal quantities of each reactant are present at the stoichiometric equivalence point, ascorbic acid, $C_6H_8O_6$, must have just one hydrogen ion.

135. If the balloon is fully inflated, the volume will not change and PV work cannot be done; hence, w = 0 J. The heating of the contents of the balloon increases its thermal energy, so q = 310 J. Because $\Delta E = q + w$, $\Delta E = 310 \text{ J} + 0 \text{ J} = 310 \text{ J}$.

Chapter 7: Electron Configurations and the Periodic Table

Electromagnetic Radiation

14. When an atom absorbs a photon in the visible region, the energy of the electrons changes, but the particles in the nucleus (the protons and neutrons) are not affected.

16. Electromagnetic radiation that is high in energy consists of waves that have short wavelengths and high frequencies. The high-energy end of the spectrum includes γ rays and x-rays as seen in Figure 7.1.

18. Use Figure 7.1

(a) Radio waves have lower frequency, thus less energy, than infrared light.

(b) Microwaves have higher frequency than radio waves.

20. *Define the problem*: Given the frequency of electromagnetic radiation, determine the wavelength in meters, the energy of one photon, and the energy of one mole of photons.

Develop a plan: Use equations described in Section 7.1 and 7.2.

Execute the plan: $v = 1.00 \times 10^{11} \text{ s}^{-1}$

(a)
$$\lambda = \frac{c}{v} = \frac{2.998 \times 10^8 \text{ m/s}}{1.00 \times 10^{11} \text{ s}^{-1}} = 3.00 \times 10^{-3} \text{ m}$$

(b) $E = hv = (6.626 \times 10^{-34} \text{ J·s}) \times (1.00 \times 10^{11} \text{ s}^{-1}) = 6.63 \times 10^{-23} \text{ J}$ for one photon

(c)
$$\frac{6.63 \times 10^{-23} \text{ J}}{1 \text{ photon}} \times \frac{6.022 \times 10^{23} \text{ photons}}{1 \text{ mol photons}} = 39.9 \text{ J/mol}$$

Check your answers: High frequency has short wavelength. The energy for one photon is a tiny number, but a mole of photons has a sizable energy.

22. Use Figure 7.1. The energy of a photon is directly proportional to the frequency (v):

Lowest energy   (d) radiowaves < (c) microwaves

< (a) green light < (b) X-rays   Highest energy

24. *Define the problem*: Given the wavelength of light, determine the frequency.

Develop a plan: Use appropriate length conversions and the equation in Section 7.1. 1 Hz is defined as 1 s^{-1}.

Execute the plan: $\lambda = 495$ nm

$$\nu = \frac{c}{\lambda} = \frac{2.998 \times 10^8 \, m/s \times \dfrac{1 \, Hz}{1 \, s^{-1}}}{495 \, nm \times \dfrac{1 \times 10^{-9} \, m}{1 \, nm}} = 6.06 \times 10^{14} \, Hz$$

Check your answer: Nanometer wavelengths are fairly small, so it makes sense that the frequency is high. Figure 7.1 also shows visible light in the 10^{14} - 10^{16} Hz frequency range.

26. *Define the problem*: Given the wavelength of light, determine the energy of one photon.

Develop a plan: Use appropriate length conversions and the equation in Section 7.2.

Execute the plan: $\lambda = 450$ nm

$$E_{photon} = \frac{hc}{\lambda} = \frac{6.626 \times 10^{-34} \, J \cdot s \times 2.998 \times 10^8 \, m/s}{450 \, nm \times \dfrac{1 \times 10^{-9} \, m}{1 \, nm}} = 4.4 \times 10^{-19} \, J$$

Check your answer: The energy of this blue photon is similar to that calculated in Section 7.2 for another blue photon (with a similar wavelength).

28. This question requires some interpretation and some assumptions. Certain visible light interacts with the cells of our eyes. Ultraviolet light causes sunburns, which is probably in an interaction with skin cells. Infrared light feels like heat to us, that might also be a cellular response of the skin cells. Microwave energy is used to cook food, which is made from cells, too; therefore it appears to interact with cells also. X-rays can trigger cellular damage. In conclusion, many types of electromagnetic radiation interact with molecules at the cellular level. There is a discussion of effects of electromagnetic radiation on matter at the end of Section 7.3.

30. *Define the problem*: Given the wavelength of electromagnetic radiation, determine the frequency and energy of this radiation.

Develop a plan: Use equations in Sections 7.1 and 7.2, and appropriate metric conversion factors. The energy calculated is the energy of one photon of this light.
Execute the plan: $\lambda = 270$ nm

$$\nu = \frac{c}{\lambda} = \frac{2.998 \times 10^8 \, \text{m/s}}{270 \, \text{nm} \times \dfrac{1 \times 10^{-9} \, \text{m}}{1 \, \text{nm}}} = 1.1 \times 10^{15} \, \text{s}^{-1}$$

$$E_{\text{photon}} = h\nu = \left(6.626 \times 10^{-34} \, \text{J} \cdot \text{s}\right) \times \left(1.1 \times 10^{15} \, \text{s}^{-1}\right) = 7.4 \times 10^{-19} \, \text{J}$$

Check your answers: The wavelength calculated coincides with that of ultraviolet light, according to Figure 7.1. This ultraviolet photon's energy is also more than those of visible light calculated in Section 7.2.

32. *Define the problem*: Given the wavelength of electromagnetic radiation, determine the energy of one photon of this radiation and compare it to the given energy of another kind of photon.

Develop a plan: Use appropriate length conversions and the equation in Section 7.2.

Execute the plan: $\lambda = 2.36 \, \text{nm}$

$$E_{\text{photon}} = \frac{hc}{\lambda} = \frac{6.626 \times 10^{-34} \, \text{J} \cdot \text{s} \times 2.998 \times 10^8 \, \text{m/s}}{2.36 \, \text{nm} \times \dfrac{1 \times 10^{-9} \, \text{m}}{1 \, \text{nm}}} = 8.42 \times 10^{-17} \, \text{J}$$

$$8.42 \times 10^{-17} \, \text{J} > 3.18 \times 10^{-19} \, \text{J}$$

Check your answers: The energy of an x-ray photon is more than 250 times more than that of a visible photon. That make sense, since the wavelength of the x-ray is about 250 times shorter than the wavelengths of visible light (400 nm - 700 nm).

34. *Define the problem*: Given the wavelength of electromagnetic radiation, determine the energy possessed by one mole of photons of this radiation.

Develop a plan: Use appropriate length conversions and the equation in Section 7.2 to find the energy of the photon.

Execute the plan: $\lambda = 1.00 \times 10^9 \, \text{m}$. According to Figure 9.1, this is a typical wavelength for an x-ray.

$$E_{\text{photon}} = \frac{hc}{\lambda} = \frac{6.626 \times 10^{-34} \, \text{J} \cdot \text{s} \times 2.998 \times 10^8 \, \text{m/s}}{1.00 \times 10^{-9} \, \text{m}} = 1.99 \times 10^{-16} \, \text{J}$$

$$\frac{1.99 \times 10^{-16} \, \text{J} \times}{1 \, \text{photon}} \times \frac{6.022 \times 10^{23} \, \text{photons}}{1 \, \text{mol photons}} = 1.20 \times 10^8 \, \frac{\text{J}}{\text{mol}}$$

Check your answer: X-rays have very high energy, so this large amount of energy per mole makes sense.

Photoelectric Effect

36. Photons of light with long wavelength are low in energy. It is clear from the description that the energy of these photons is insufficient to cause electrons to be ejected. Increasing the intensity only increases the number of photons, not their energy.

38. *Define the problem*: Given the wavelength of light and the minimum energy for ejecting electrons from a metal surface, determine if the light has sufficient energy to eject electrons from that metal's surface.

    Develop a plan: Use appropriate length conversions and the equation in Section 7.2 to find the energy of the photon. Compare the photon energy to the minimum energy. If the photon energy is larger, then the photon can eject electrons; if the energy is smaller, it cannot.

    Execute the plan: $\lambda = 600.$ nm

$$E_{photon} = \frac{hc}{\lambda} = \frac{6.626 \times 10^{-34} \text{ J} \cdot \text{s} \times 2.998 \times 10^8 \text{ m/s}}{600. \text{ nm} \times \frac{1 \times 10^{-9} \text{ m}}{1 \text{ nm}}} = 3.31 \times 10^{-19} \text{ J}$$

$$E_{minimum} = 3.69 \times 10^{-19} \text{ J} > E_{photon}$$

   Therefore it has insufficient energy to eject electrons.

    Check your answer: The minimum energy is similar to ones calculated for visible light photons in Section 7.2. These two energies are close, so we probably would not have been able to predict this result before doing the calculation. This result does make sense.

Atomic Spectra and the Bohr Atom

40. Line emission spectra are mostly dark, with discrete bands of light. Sunlight is a continuous rainbow of color. This is described in detail in Section 7.3.

42. Energy is emitted from an atom when an electron moves from the **higher-energy** (excited) state to a **lower-energy** state (maybe the ground state, maybe another less-

excited state). The energy of the emitted radiation corresponds to the **difference** between the two energy states.

44. *Define the problem*: Given the values of n for the initial and final states of an electron in hydrogen, determine whether energy is absorbed or emitted.

Develop a plan: The size of n indicates the relative energy of the state. If the final state has a larger n value than the initial state, then energy must be absorbed. If the final state has a smaller n value than the initial state, then energy must be emitted.

Execute the plan:

(a) n = 1 to n = 3. Energy absorbed.

(b) n = 5 to n = 1. Energy emitted.

(c) n = 2 to n = 4. Energy absorbed.

(d) n = 5 to n = 4. Energy emitted.

Check your answers: Energy is used to get the electron to a higher n value state. Energy is lost when moving the electron to a lower n value state. These answers make sense.

46. The difference between the energies of the two levels is the energy emitted. The smaller the energy, the longer the wavelength. When we look at Figure 7.8, we need to find which of the given levels is closer together than the energy levels represented by n = 1 and n = 4. The energies levels represented by (a) n = 2 and n = 4, (b) n = 1 and n = 3, and (d) n = 3 and n = 5, are closer together than n = 1 and n = 4. So, (a), (b) and (d) will require radiation with longer wavelength.

48. (a) *Define the problem*: Given the wavelength of the emission line, determine the energy difference for this transition.

Develop a plan: Using metric conversions and the equation given in Section 7.1, determine the energy of the photon. Use that to determine the energy of the transition.

Execute the plan: $\lambda = 555.4$ nm

$$E_{photon} = \frac{hc}{\lambda} = \frac{6.626 \times 10^{-34} \text{ J} \cdot \text{s} \times 2.998 \times 10^8 \text{ m/s}}{555.4 \text{ nm} \times \frac{1 \times 10^{-9} \text{ m}}{1 \text{ nm}}} = 3.577 \times 10^{-19} \text{ J}$$

The energy of the photon emitted comes from the **loss** of energy by the electron. That means ΔE_{e-} will be negative and we must change its sign to get the positive energy of the photon.

$$\Delta E_{e^-} = -E_{photon} = -3.577 \times 10^{-19} \text{ J}$$

Check your answer: The energy of this photon is similar to those calculated for visible light in Section 7.2.

(b) *The value of n_i cannot be found without an equation describing the energy levels of the electrons in BARIUM. We only have equations for the energy levels of HYDROGEN.*

50. *Define the problem*: Given the wavelength of the emission line, determine the energy difference for this transition. (Note: *The value of n_i cannot be found without an equation describing the energy levels of the electrons in SODIUM. We only have equations for the energy levels of HYDROGEN.*)

Develop a plan: Using metric conversions and the equation given in Section 7.1, determine the energy of the photon. Use that to determine the energy of the transition.

Execute the plan: $\lambda = 589.6$ nm

$$E_{photon} = \frac{hc}{\lambda} = \frac{6.626 \times 10^{-34} \text{ J} \cdot \text{s} \times 2.998 \times 10^8 \text{ m/s}}{589.6 \text{ nm} \times \dfrac{1 \times 10^{-9} \text{ m}}{1 \text{ nm}}} = 5.085 \times 10^{-19} \text{ J}$$

The energy of the photon emitted comes from the **loss** of energy by the electron. That means ΔE_{e^-} will be negative and we must change its sign to get the positive energy of the photon.

$$\Delta E_{e^-} = -E_{photon} = -5.085 \times 10^{-19} \text{ J}$$

Check your answer: The energy of this photon is similar to those calculated for visible light in Section 7.2.

52. *Define the problem*: Given the values of n for the initial and final states of an electron in hydrogen, determine the energy and wavelength of the emission.

Develop a plan: Using equations given in Section 7.3, calculate the energy of the photon emitted during the transition. Use the equation given in Section 7.1 and appropriate metric conversions to calculate the wavelength in nanometers.

Execute the plan: Here, $n_i = 2$ and $n_f = 5$:

$$\Delta E_{e^-} = \left(2.179 \times 10^{-18} \text{ J}\right)\left(\frac{1}{n_i^2} - \frac{1}{n_f^2}\right) = \left(2.179 \times 10^{-18} \text{ J}\right)\left(\frac{1}{2^2} - \frac{1}{5^2}\right) = 4.576 \times 10^{-19} \text{ J}$$

The energy of the photon absorbed is equal to the energy gained by the electron as it moves to a higher-energy state. $E_{photon} = \Delta E_{e^-}$. The wavelength of the photon absorbed relates to the energy:

$$\lambda = \frac{hc}{E_{photon}} = \frac{\left(6.626 \times 10^{-34} \text{J} \cdot \text{s}\right)\left(2.998 \times 10^{8} \text{m/s}\right)}{4.576 \times 10^{-19} \text{J}} \times \frac{1 \text{ nm}}{10^{-9} \text{ m}} = 434.0 \text{ nm}$$

Check your answers: The wavelength of radiation produced from the reverse of this transition is reported in Figure 7.8.

de Broglie Wavelength

54. *Define the problem*: Given the mass and speed of a particle, determine the de Broglie wavelength.

Develop a plan: Use the equation in Section 7.4, the method described in Problem-Solving Example 7.4, and appropriate unit conversions.

Execute the plan: Problem-Solving Example 7.4 shows how $h = 6.626 \times 10^{-34}$ kg·m^{2}s^{-1}

Find speed in meters per second:

$$v = (0.05) \times (2.998 \times 10^{8} \text{ m/s}) = 1 \times 10^{7} \text{ m/s}$$

Now, find the de Broglie wavelength:

$$\lambda_{DeBroglie} = \frac{h}{mv} = \frac{6.626 \times 10^{-34} \text{ kg} \cdot \text{m}^2\text{s}^{-1}}{\left(9.11 \times 10^{-31} \text{ kg}\right)\left(1 \times 10^{7} \frac{\text{m}}{\text{s}}\right)} = 5 \times 10^{-11} \text{ m}$$

Check your answer: The electron is a sub-nanoscale particle, so it makes sense that its de Broglie wavelength is similar (50 pm) to its size dimensions.

Quantum Numbers

56. *Define the problem*: Give the four quantum numbers for various electrons.

Develop a plan: The set of three quantum numbers representing each orbital in an atom must be different (Pauli Exclusion Principle). ℓ values must be less than n, but not negative. m_ℓ values must be between $-\ell$ and $+\ell$. m_s values must be $+\frac{1}{2}$ or $-\frac{1}{2}$. Electrons with the same n, ℓ, and m_ℓ must have opposite signs for m_s. Electrons with the same n

and ℓ, but different m_ℓ must have the same m_s value (Hund's rule) until all values of m_ℓ

have been used once. Be systematic.

Execute the plan:

(a) There are five electrons in boron:

First electron	$n = 1$	$\ell = 0$	$m_\ell = 0$	$m_s = +\dfrac{1}{2}$
Second electron	$n = 1$	$\ell = 0$	$m_\ell = 0$	$m_s = -\dfrac{1}{2}$
Third electron	$n = 2$	$\ell = 0$	$m_\ell = 0$	$m_s = +\dfrac{1}{2}$
Fourth electron	$n = 2$	$\ell = 0$	$m_\ell = 0$	$m_s = -\dfrac{1}{2}$
Fifth electron	$n = 2$	$\ell = 1$	$m_\ell = 1$	$m_s = +\dfrac{1}{2}$

(The fifth electron could have different m_ℓ and m_s values also, as long as they are

within the appropriate ranges.)

(b) Valence electrons are defined in Section 7.6 as the highest n value electrons.
Magnesium has two valence electrons:

First valence electron of Mg	$n = 3$	$\ell = 0$	$m_\ell = 0$	$m_s = +\dfrac{1}{2}$
Second valence electron of Mg	$n = 3$	$\ell = 0$	$m_\ell = 0$	$m_s = -\dfrac{1}{2}$

(c) A 3d electron in iron atom:

A 3d electron in Fe	$n = 3$	$\ell = 2$	$m_\ell = 2$	$m_s = +\dfrac{1}{2}$

(The values of m_ℓ and m_s can be different here, as long as they are within the appropriate

ranges.)

58. (a) $n = 2$, $\ell = 1$, $m_\ell = 2$, $m_s = +\dfrac{1}{2}$ could not occur because m_ℓ must be between $-\ell$ and

$+\ell$.

(b) $n = 3$, $\ell = 2$, $m_\ell = 0$, $m_s = -\dfrac{1}{2}$ can occur.

(c) $n = 1$, $\ell = 0$, $m_\ell = 0$, $m_s = 1$ could not occur because m_s must be $+\dfrac{1}{2}$ or $-\dfrac{1}{2}$.

(d) $n = 3$, $\ell = 3$, $m_\ell = 2$, $m_s = -\dfrac{1}{2}$ could not occur because ℓ must be less than n.

(e) $n = 2$, $\ell = 0$, $m_\ell = 0$, $m_s = +\dfrac{1}{2}$ can occur.

60. (a) The up-spin electron in the 4s orbital has the following four quantum numbers:

$n = 4$	$\ell = 0$	$m_\ell = 0$	$m_s = +\dfrac{1}{2}$

(b) The down-spin electron in the first of three 3p orbitals has the following four quantum numbers:

$n = 3$	$\ell = 1$	$m_\ell = 1$	$m_s = -\dfrac{1}{2}$

Note: Two other values of m_ℓ (0 and –1) would also be valid.

(c) The up-spin electron in the third of three 3d orbitals has the following four quantum numbers:

$n = 3$	$\ell = 2$	$m_\ell = 0$	$m_s = +\dfrac{1}{2}$

Note: Four other values of m_ℓ (+2, +1, –1, and –2) would also be valid.

Quantum Mechanics

62. There are n subshells in the nth energy level; they have $\ell = n - 1, \ldots, 0$. That means four different subshells (designated $\ell = 3, 2, 1,$ and 0) are found in the 4th energy level.

64. The wave mechanical model of the atom tells us that electrons do not follow simple paths as do planets. Rather, they occupy regions of space having certain shapes and varying distances around the nucleus. Hence there are subshells and orbitals that were not part of the Bohr model.

66. Orbits have predetermined paths – position and momentum are both exactly known at all times. Heisenberg's uncertainty principle says that we cannot know both position and momentum simultaneously.

68. The n = 3 shell has orbitals with ℓ = 2, 1, and 0. That means it has d, p, and s orbitals. nine orbitals total. There are n^2 orbitals in the nth shell. That means there are nine orbitals in the n = 3 shell.

Electron Configurations

70. Find the atomic number and make sure you use that number of electrons:

(a) $_{13}$Al electron configuration: $1s^2 2s^2 2p^6 3s^2 3p^1$

(b) $_{16}$S electron configuration: $1s^2 2s^2 2p^6 3s^2 3p^4$

72. $_{32}$Ge electron configuration: $1s^2 2s^2 2p^6 3s^2 3p^6 3d^{10} 4s^2 4p^2$

74. $_8$O, oxygen, is an element in Group 6A. The valance electron configuration of this element is $2s^2 2p^4$. All of the elements in this group have six valence electrons, so their electron configurations are: $ns^2 np^4$

76. (a) In the ground state of the iron atom, the lower-energy 4s orbital must be full.

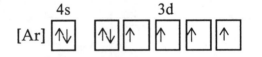

(b) The orbital labels must be 3, not 2. The electrons in 3p subshell should be in separate orbitals with parallel spin (Hund's Rule).

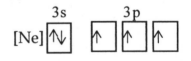

(c) Electrons should be removed from 5p orbitals to make the cation, not from 4d orbitals.

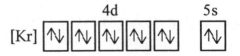

78. (a) V   (b) V^{2+}

(c) V^{4+}

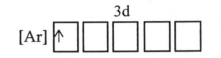

80. 18 elements are in the fourth period of the periodic table. It is not possible for there to be another element in this period because all possible orbital electron combinations are already used.

82. When transition metals form cations, the electrons lost first are those from their valence ns orbitals.

$_{25}$Mn electron configuration: $1s^2 2s^2 2p^6 3s^2 3p^6 3d^5 4s^2$. Mn has 5 unpaired electrons:

$_{25}$Mn^{2+} electron configuration: $1s^2 2s^2 2p^6 3s^2 3p^6 3d^5$. Mn^{2+} has 5 unpaired electrons:

$_{25}$Mn^{3+} electron configuration: $1s^2 2s^2 2p^6 3s^2 3p^6 3d^4$. Mn^{3+} has 4 unpaired electrons:

84. $_{23}$V electron configuration: $1s^2 2s^2 2p^6 3s^2 3p^6 3d^3 4s^2$ or $[Ar]3d^3 4s^2$

86. (a) $_{63}$Eu electron configuration: $[Xe]4f^7 6s^2$ or $[Xe]4f^6 5d^1 6s^2$

   (b) $_{70}$Yb electron configuration: $[Xe]4f^{14}6s^2$ or $[Xe]4f^{13}5d^1 6s^2$

Valence Electrons

88. (a) ·Sr·   (b) :B̈r·   (c) ·Ġa·   (d) ·S̈b·

90. (a) $_{20}$Ca^{2+} electron configuration: $1s^2 2s^2 2p^6 3s^2 3p^6$ or $[Ar]$

(b) $_{19}K^+$ electron configuration: $1s^2 2s^2 2p^6 3s^2 3p^6$ or [Ar]

(c) $_8O^{2-}$ electron configuration: $1s^2 2s^2 2p^6$ or [Ne]

$_{20}Ca^{2+}$ and $_{19}K^+$ are isoelectronic.   They have the same number of electrons.

92. $_{50}Sn$ electron configuration: $[Kr]4d^{10}5s^2 5p^2$

$_{50}Sn^{2+}$ electron configuration: $[Kr]4d^{10}5s^2$

$_{50}Sn^{4+}$ electron configuration: $[Kr]4d^{10}$

Paramagnetism and Unpaired Electrons

94. Ferromagnetism is a property of permanent magnets. It occurs when the spins of unpaired electrons in a cluster of atoms (called a domain) in the solid are all aligned in the same direction. Only metals in the Fe, Co, Ni subgroup (Group 8B) exhibit this property. This is described at the end of Section 7.7.

96. In both paramagnetic and ferromagnetic substances, atoms have unpaired spins and so are attracted to magnets. Ferromagnetic substances retain their aligned spins after an external magnetic field has been removed, so they can function as magnets. Paramagnetic substances lose their aligned spins after a time and therefore cannot be used as permanent magnets.

Tools of Chemistry: NMR and MRI

98. In magnetic resonance imaging (MRI), the intensity of the emitted signal is related to the **number** of hydrogen nuclei and the relaxation time is related to the type of tissues being examined.

Periodic Trends

100. *Define the problem*: Without using the table that gives numerical sizes, list elements in order of increasing size.

Develop a plan: Periodic Trends in atom size are related to the radius of the atoms:

- Comparing atoms across the period (row) of the periodic table, the atom sizes increase from right to left.

- Comparing atoms down a group (column) of the periodic table, the atom sizes increase from top to bottom.

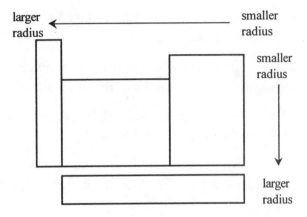

Execute the plan: Looking at a Periodic Table, the smallest of the ions given is P, since its the closest to the top right corner. Next is Ge, diagonally down and to the left of P. Then comes Ca, in the same period as Ge, but further to the left. Then comes Sr, in Group 2A, below Ca. Then comes Rb, in the same period as Sr, but further to the left. So the order is: P < Ge < Ca < Sr < Rb.

Check your answers: Using Figures 7.15 and 7.16, we can confirm these predictions.

102. *Define the problem*: Without using the table that gives numerical sizes, determine which element of a pair of atoms and ions has smaller radius.

Develop a plan: Use the periodic table, atom size trends (as described in the answers to Questions 99 and 100), and the following periodic trends.

Periodic trends comparing atoms to ions:

- The cation of an element will always be smaller than the atom of the same element (since fewer electrons are attracted more closely to the nucleus).

- The anion of an element will always be larger than the atom of the same element (since more electrons will be less attracted to the nucleus).

Periodic trends comparing ions to ions:

- When comparing isoelectronic ions, the element with the larger atomic number is smaller (since it has more protons to attract the electrons).

- When comparing ions of the same element, the ion with the larger positive or less negative charge is smaller (since fewer electrons are attracted more closely to the nucleus).

Execute the plan:

(a) Using the Periodic Table, Rb has a smaller radius than Cs atom, because they are both in the same group and Cs is below Rb.

(b) The O atom has a smaller radius than O^{2-} ion, since anions are larger than the neutral atom. (8 protons attract 8 electrons better than they can attract 10 electrons.)

(c) Using the Periodic Table, Br has a smaller radius than As atom, because they are both in the same period and Br is further to the right.

(d) The Ba^{2+} ion has a smaller radius than Ba atom, since cations are smaller than the neutral atom. (56 protons attract 54 electrons better than they can attract 56 electrons.)

(e) The Cl^- ion has a larger radius than Ca^{2+} atom, since they are isoelectronic and the atomic number of Cl (17) is smaller than the atomic number of Ca (20). (20 protons can attract 18 electrons better than 17 protons can.)

Check your answers: Using Figures 7.15 - 7.17, we can confirm these predictions.

104. *Define the problem*: List elements in order of increasing first ionization energy.

Develop a plan: Sometimes we will not have a chart with numbers like Figure 7.18, so let's get an idea of what periodic trends are exhibited in first ionization energies:

- Comparing atoms in the period (row) of the periodic table, the first ionization energy increases from left to right.

- Comparing atoms in a group (column) of the periodic table, the first ionization energy increases from bottom to top.

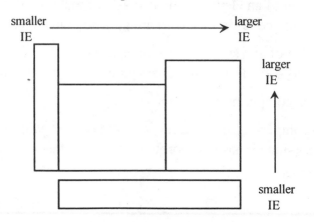

- Elements in group 2A, 2B, and 5A seem to be exceptions to this general trend, having higher IE than elements to their immediate right.

Execute the plan: Looking at a Periodic Table, the element the furthest to the left in period 3 is Mg; however, it is in one of the exception groups (Group 2A). So, we'll predict that Al has the lowest first ionization energy, then Mg in the same period. Then comes P, to the right of Al in the same period. Then comes F, to the right and in the period above the others. So the order is: Al < Mg < P < F

Check your answer: Using Figure 7.18, we can confirm these predictions.

106. *Define the problem*: List elements in order of increasing first ionization energy.

Develop a plan: Sometimes we will not have a chart with numbers like Figure 7.18, so follow the periodic trends described in the answers to Question 104.

Execute the plan: Looking at a Periodic Table, the element the furthest to the left is Li, so, we'll predict that it has the lowest first ionization energy. Then comes Si, to the right of the metal. Then comes C, above Si in the same group. Then comes Ne, to the right of C in the same period. So the correct order is: Li < Si < C < Ne. That is choice (c).

Check your answer: Using Figure 7.18, we can confirm these predictions.

108. Looking at Table 7.9, it is clear that the second ionization energies are all higher than the first ionization energies and once core electrons are being removed, the ionization energy rises dramatically. To have a large gap between the first and second ionization energies, the element must have one valence electron. The only element given that has one valence electron is Na, from Group 1A.

110. *Define the problem*: Given a list of elements, determine which is most metallic, determine which has largest radius, and place the list in order of increasing of ionization energy.

Develop a plan: Metallic character follows these general trends:

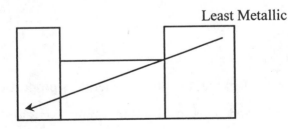

Least Metallic

Most Metallic

Use the periodic table for atom size trends and ionization energy trends (as described in the answers to Questions 100 and 104).

Execute the plan: Locate the elements: B, Al, C, and Si on a Periodic Table.

(a) We find Al closest to the bottom left corner. That means Al is most metallic.

(b) We find Al closest to the bottom left corner. That means Al is the largest of these four atoms.

(c) We find Al closest to the bottom left corner. That means Al has the smallest first ionization energy. B is next, because it is above Al in the same group. Last comes C, to the right of B in the same period. So the order is: Al < B < C

Check your answers: Using Figures 7.15, 7.16, and 7.18, we can confirm these predictions.

112. *Define the problem*: Given pairs of ions and/or atoms, determine which member of the pair has the larger radius.

Develop a plan: Use the periodic table and atom and ion size trends (as described in the answers to Questions 100 - 102).

Execute the plan:

(a) H^- is larger than H^+. The anion has two electrons and the cation has none!

(b) N^{3-} is larger than N, since anions are larger than the neutral atom. (7 protons attract 7 electrons better than they can attract 10 electrons.)

(c) The F^- ion has a larger radius than an F atom, since anions are larger than the neutral atom. (9 protons attract 9 electrons better than they can attract 10 electrons.) Using the Periodic Table, an F atom has a larger radius than an Ne atom, because they are both in the same period and Ne is further to the right. Hence, an F^- ion also has a larger radius than an Ne atom.

Check your answers: Using Figures 7.15 and 7.16, we can confirm these predictions.

General Questions

114. Using Figure 7.1, and the fact that the photon energy is proportional to frequency:

Lowest-energy photons: red < yellow < green < blue   :Highest-energy photons

116. Selenium has the following electron configuration: $1s^22s^22p^63s^23p^63d^{10}4s^24p^4$. That means the first two p sublevels are full, and the third one has 4 electrons. Look at the orbital diagrams of the p sublevels to find out the number of pairs:

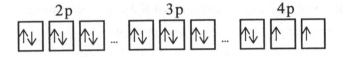

From this, it is very easy to tell that there are seven pairs of electrons in p orbitals.

118. (a) For a ground state element to be in Group 8A but have no p electrons, it must have fewer than five electrons, because the fifth electron goes into a 2p orbital (i.e., $1s^22s^22p^1$). This Group 8 element is He ($1s^2$).

(b) For a ground state element to have a single electron in the 3d subshell, it must have this electron configuration: $1s^22s^22p^63s^23p^63d^1$. This element is Sc.

(c) For a ground state element's cation to have a +1 charge and an electron configuration of $1s^22s^22p^6$, the element's electron configuration would have one more electron: $1s^22s^22p^63s^1$. This element is Na.

120. *Define the problem*: Given lists of elements, place one list in order of increasing atomic radius and determine which member of another list has the largest ionization energy.

Develop a plan: Use the periodic table, atom size trends, and ionization energy trends (as described in the answers to Questions 100 and 104).

Execute the plan:

(a) Locate the elements: O, S, and F on a Periodic Table. We find F closest to the top right corner. That means F is the smallest of these three atoms. Then comes O, to the left of the F in the same period. Then comes S, below O in the same group. So the order is: F < O < S

(b) Locate the elements: P, Si, S, and Se on a Periodic Table. We find S closest to the top right corner. That means S has the largest first ionization energy.

Check your answers: Using Figures 7.15, 7.16, and 7.18, we can confirm these predictions.

122. (a) The ground state element whose atoms have an electron configuration of $1s^22s^22p^63s^23p^4$ has 16 electrons and 16 protons. This is sulfur, S, with the atomic number of 16.

(b) The largest *known* element in the alkaline earth group (Group 2A) is the one at the bottom of the group on the periodic table. This is radium, Ra.

(c) The largest first ionization energy of elements in the group is the group member at the top of the group on the periodic table. In Group 5A, this is nitrogen, N.

(d) For a ground state element's cation to have a +2 charge and an electron configuration of $[Kr]4d^6$, the element's electron configuration would have two more electrons: $[Kr]4d^65s^2$. This element is ruthenium, Ru.

(e) The ground state element whose neutral atoms have an electron configuration of $[Ar]3d^{10}4s^1$ has 29 electrons and 29 protons. This is copper, Cu, with the atomic number of 29.

124. Considering $X = [Ar]3d^84s^2 = Ni$ and $Z = [Ar]3d^{10}4s^24p^5 = Br$

(a) Using the periodic trends for sizes of first ionization energies (as described in the answers to Questions 104 and 105), we find that Br has a larger first ionization energy than Ni, because they are both in period 4, and Br is found further to the right. That means Z is the answer.

(b) Using the periodic trends for atom sizes (as described in the answers to Questions 99 and 100), we find that Br has a smaller radius than the Ni, because they are both in period 4, and Br is found further to the right. That means Z is the answer.

126. Cations likely to form are ones that have lost only valence electrons. Any ion that has a more positive charge than its number of valence electrons is not likely to form. Anions likely to form are ones that add enough electrons to the atom to increase the number of valence electrons to eight. Any ion that has a more negative charge than the difference between eight and the number of valence electrons is not likely to form. Free ions with charges larger than ±3 are rare, because making the ion requires us to remove electrons from ions that are already very positive or add electrons to ions that are already very negative (i.e., Notice how large the high successive ionization energies get in Table 7.9).

Cs has one valence electrons; Cs^+ **is** likely to form.

In has three valence electrons; In^{4+} is **not** likely to form because a core electron must be removed.

Fe has two valence electrons and six d electrons, so core electrons are being removed; however, Fe^{5+} is **not** likely to form, because the positive charge is too large.

Te has six valence electrons. It needs $8 - 6 = 2$ electrons. Te^{2-} **is** likely to form.

Sn has four valence electrons; Sn^{5+} is **not** likely to form because a core electron must be removed and the positive charge is too large.

I has seven valence electrons. It needs $8 - 7 = 1$ electron. I^- **is** likely to form.

128. Use Figure 7.1 and the relationships between wavelength, frequency, and photon energies described in Sections 7.1 and 7.2

(a) "The wavelength of green light is longer than that of red light." This statement is false. We can fix it in several ways, such as with this related true statement: "The wavelength of green light is **shorter** than that of red light." Red light has the longest wavelength of all visible light.

(b) "Photons of green light have greater energy than those of red light." This statement is true. Red photons are the lowest energy visible photons.

(c) "The frequency of green light is greater than that of red light." This statement is true. Red light has the lowest frequency of all visible light.

(d) "In the electromagnetic spectrum, frequency and wavelength are directly related." This statement is false. We can fix it in several ways, such as with this related true statement: "In the electromagnetic spectrum, frequency and wavelength are inversely related." The word "directly" implies directly proportional, and these two quantities are inversely proportional.

130. (a) This statement: "The energy of a photon is inversely related to its frequency." is false. Actually, the energy of a photon is **directly** related to its frequency by the equation $E = h\nu$.

(b) This statement: "The energy of the hydrogen electron is inversely proportional to its principle quantum number n." is false. Actually, the energy of the hydrogen electron is inversely proportional to the **square of** its principle quantum number n.

$$E = - \frac{2.179 \times 10^{-18} \text{ J}}{n^2}$$

(c) This statement: "Electrons start to fill the fourth energy level as soon as the third level is full." is false. Several different slight changes can make it into a true statement, for example: "Electrons start to fill the fourth energy level **before** the third level is full" This can be seen in the electron configuration for potassium and calcium: $1s^2 2s^2 2p^6 3s^2 3p^6 3d^4 4s^1$ and $1s^2 2s^2 2p^6 3s^2 3p^6 3d^4 4s^2$ where the 4s sublevel fills before the 3d sublevel starts to fill.

(d) This statement: "Light emitted by an n = 4 to n = 2 transition will have a longer frequency than that from an n = 5 to n = 2 transition." is false. Several different slight changes can make it into a true statement, for example: "Light emitted by an n = 4 to n = 2 transition will have a longer **wavelength** than that from an n = 5 to n = 2 transition." The energy difference is smaller in the first transition than the second, so the frequency of the emitted photon is smaller, and its wavelength is longer.

132. *Define the problem*: Given a list of types of radiation, determine which is needed to ionize hydrogen.

Develop a plan: Use the ionization energy of hydrogen, from Figure 7.18 in kJ/mol, and Avogadro's number to determine the energy required to ionize one hydrogen atom, then determine the wavelength of light with that energy using the equation from Section 7.2.

Execute the plan:

$$\frac{1312 \text{ kJ}}{1 \text{ mol}} \times \frac{1000 \text{ J}}{1 \text{ kJ}} \times \frac{1 \text{ mol}}{6.022 \times 10^{23} \text{ atoms}} \times \frac{1 \text{ atom}}{1 \text{ photon}} = 2.179 \times 10^{-18} \text{ J/photon}$$

$$\lambda = \frac{hc}{E_{photon}} = \frac{6.626 \times 10^{-34} \text{ J·s} \times 2.998 \times 10^8 \text{ m/s}}{2.179 \times 10^{-18} \text{ J}} \times \frac{1 \text{ nm}}{1 \times 10^{-9} \text{ m}} = 91.18 \text{ nm}$$

To ionize hydrogen, we need light with a wavelength of 91.18 nm or shorter. This is the ultraviolet region of the electromagnetic spectrum.

Check your answer: It is satisfying to get 2.179×101^{-18} J/photon, since that is the number used in the Bohr model calculations for electrons changing quantum states in hydrogen:

$$\Delta E_{e^-} = \left(2.179 \times 10^{-18} \text{ J}\right)\left(\frac{1}{n_i^2} - \frac{1}{n_f^2}\right)$$

To ionize the electron, we must move it from $n = 1$ (the ground state) to above $n = \infty$ (the last state in the atom). This should be the minimum photon energy needed for ionization:

$$\Delta E_{e^-} = \left(2.179 \times 10^{-18} \text{ J}\right)\left(\frac{1}{1^2} - \frac{1}{\infty^2}\right) = 2.179 \times 10^{-18} \text{ J} = E_{ionization \ photon}$$

134. This question asks us to use the predictive power of the periodic table. Suppose an element, Et, has atomic number 113. That means atoms of the neutral element have 113 protons and 113 electrons.

(a) The ground state electron configuration has 113 electrons: $[Rn]5f^{14}6d^{10}7s^27p^1$
   The Rn noble gas core has 87 electrons, and we add $14 + 10 + 2 + 1$ outer electrons, to make a total of 113 electrons.

(b) The valence electrons are the $n = 7$ electrons ($7s^2 7p^1$) of which there are three. That means we will predict this element to be a member of group 3A. There are several elements in this group, including boron.

(c) Group 3A metals will combine with nonmetals forming ionic compounds. The metallic elements in that group can be predicted to form cations with a 3+ charge. They will combine with the anions of oxygen and chlorine (oxide, O^{2-}, and chloride, Cl^-) to form Et_2O_3 and $EtCl_3$, respectively.

Low on the periodic table, the valence s electrons have much lower energy than the valence p electrons, so stable ions in which only the valence p electrons have been lost might also form. This is true of thallium, Tl (atomic number 81), which is known to make both Tl^+ and Tl^{3+} ions. If our Et element follows this pattern, we might also find that compounds Et_2O and $EtCl$ also form. However, the question asks us to predict the resulting compounds from the periodic trends we have learned in this chapter. This second pair of answers uses more information about the chemistry of Group 3A than you might currently have known.

Applying Concepts

136. The smallest halogen is the element at the top of Group 7A, fluorine. The neutral atom loses an electron during the process of ionization:

$$F\ (1s^2 2s^2 2p^5) \longrightarrow F^+\ (1s^2 2s^2 2p^4) + e^- \qquad \text{First ionization}$$

So the electron configuration is: $1s^2 2s^2 2p^4$

138. The element, X, with electron configuration: $1s^2 2s^2 2p^6 3s^1$ has one valence electron ($3s^1$), placing it in Group 1A. Group 1A metals will combine with nonmetals forming ionic compounds. The Group 1A elements form cations with a 1+ charge. They will combine with the anion of chlorine (chloride, Cl^-) to form XCl.

140. (a) $1s^2 2s^1$ is a ground state electron configuration. The 1s sublevel has the lowest energy it is full. The 2s sublevel has the second lowest energy and the remaining electron is there. This is the lowest energy electron configuration, so it would be called the ground state.

(b) $1s^2 2s^2 2p^3$ could be ground state or excited state. If each of the p electrons is in a separate orbital with the same spin, that is the ground state. However, if any of the spins are reversed or if any of the electrons are paired, this is a different and higher energy state, hence an excited state. We would need to see an orbital energy diagram of this state to determine an unambiguous answer:

Ground state 2p orbital energy diagram:

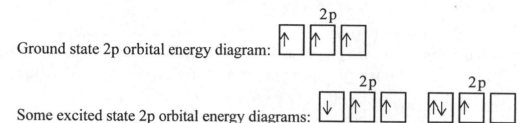

Some excited state 2p orbital energy diagrams:

(c) [Ne]$3s^2 3p^3 4s^2$ is an excited state. The lower-energy 3p sublevel needs to be full ($3p^6$) before the 4s sublevel gets any electrons; therefore, this possible electron configuration is not the lowest energy state, and would be called an excited state. The ground state electron configuration would be [Ne]$3s^2 3p^5$.

(d) [Ne]$3s^2 3p^6 4s^3 3d^2$ is impossible. The 4s sublevel has only one orbital, so the maximum number of electrons is 2. This electron configuration has three electrons in that orbital.

(e) [Ne]$3s^2 3p^6 4f^4$ is an excited state. Several lower-energy sublevels need to be full before the 4f sublevel gets any electrons; therefore, this possible electron configuration is not the lowest energy state, and would be called an excited state. The ground state electron configuration would be [Ne]$3s^2 3p^6 3d^2 4s^2$.

(f) $1s^2 2s^2 2p^4 3s^2$ is an excited state. The lower-energy 2p sublevel needs to be full ($2p^6$) before the 3s sublevel gets any electrons; therefore, this possible electron configuration is not the lowest energy state, and would be called an excited state. The ground state electron configuration would be $1s^2 2s^2 2p^6$.

142. This question asks us to use the predictive power of the periodic table. Table 7.4 and Figure 7.13 are very helpful here. Since the atomic number of our undiscovered element is 164, the number of protons and electrons in a neutral atom is also 164. We need to extend Figure 7.13 using Table 7.4, to predict what sublevels the new electrons fill. After 7p, the 8s orbitals fill. Then we fill the first g orbitals, 5g. Then we fill 6f, 7d, 8p, 9s, and so on. Table 7.4 tells us the maximum capacity of the g sublevel is 18. Constructing the electron configuration in the order in which the electrons fill looks like this:

$$[Rn]7s^2 5f^{14} 6d^{10} 7p^6 8s^2 5g^{18} 6f^{14} 7d^{10} 8p^2$$

$(86 + 2 + 14 + 10 + 6 + 2 + 18 + 14 + 10 + 2 =)$ 164 electrons. We have added enough electrons now. To conform to those shown in Figure 7.6, we then rearrange the sublevels into shell order:

$$[Rn]5f^{14} 5g^{18} 6d^{10} 6f^{14} 7s^2 7p^6 7d^{10} 8s^2 8p^2$$

There are four valence electrons in this element ($8s^2 8p^2$), so we will predict it is part of Group 4A.

144. (a) Based in the graphical data, ionization energies increase left to right and decrease top to bottom on the periodic table.

(b) The largest first ionization energy is helium (atomic number = 2).

(c) The peaks of the fourth ionization energies will occur at the atomic number of boron (5) and at the atomic number of aluminum (13).

(d) He has only two electrons, so after two ionizations, there are no electrons left.

(e) First electron in lithium is a valence electron, but the second electron is a core electron.

(f) The atomic number of 12 belongs to magnesium, $_{12}Mg$.

$$Mg^{2+}(aq) \longrightarrow Mg^{3+}(aq) + e^-$$

Chapter 8: Covalent Bonding

Lewis Structures

16. *Define the problem*: Write Lewis structures for a list of ions and molecules.

Develop a plan: Following a systematic plan will give you reliable results every time. Trial and error often works for small molecules, but as molecules get more complex, it's better to follow a procedure than to try to guess where electrons will end up. There are a number of methods. The one described here is the same as that described in the text and it works all the time.

(A) Count the total number of valence electrons. If there is a nonzero charge, adjust the electron count appropriately. Add electrons for negative charges and subtract electrons for positive charges.

(B) Determine which atom is the central atom. This is usually the first element in the formula, and it will have a smaller electronegativity (see Section 8.7) than the rest of the atoms in the formula. There is an exception: H will **never** be the central atom.

Put the rest of the atoms (called the outer electrons or lone pair electrons) around the central atom and connect them with single bonds.

(C) Subtract the electrons used for single bonds from the total. Each bond is composed of two electrons.

(D) Complete the octets of all of the outer electrons by adding pairs of dots (called lone pairs) so that each of them ends up with a total of eight electrons, including the two shared with the central atom. There is an exception: H only need two electrons, so **never** put dots on H.

(E) Subtract the electrons used for lone pairs from the total.

(F) If you still have unused electrons, put all the remaining electrons on the central atom. (This may occasionally result in the central atom having more than eight electrons.)

(G) Check the octet of the central atom.

(i) If it has eight or more electrons, then the structure is complete.

(ii) If it has less than eight electrons, move a lone pair of electrons from one of the outer atoms to make a new shared pair, forming a multiple bond to the central atom. Repeat this procedure until the central atom has an octet.

(H)   If the structure is an ion, put it in brackets and designate the net charge in the upper right corner.

Execute the plan:

(a) (A)   The atoms Cl and F are both in Group 7A: $7 + 7 = 14$ e$^-$

(B)   The Cl atom is the central atom with the F atom bonded to it.

$$Cl\text{---}F$$

(C)   Connecting the F atom uses two electrons: 14 e$^-$ $- 2$ e$^-$ $= 12$ e$^-$

(D)   Complete the octet of the F atom.

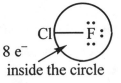

8 e$^-$

inside the circle

(E)   Completing the octet of the F atom uses 6 electrons:

12 e$^-$ $- 6$ e$^-$ $= 6$ e$^-$

(F)   Put the last six electrons on the Cl.

$$:\overset{..}{\underset{..}{Cl}}\text{---}\overset{..}{\underset{..}{F}}:$$

(G)   Cl has eight electrons, so the structure is complete.

(b) (A)   The H atoms are in Group 1A and the Se atom is in Group 6A:

$2 \times (1) + 6 = 8$ e$^-$

(B)   The Se atom is the central atom with the H atoms around it.

$$H\text{---}Se\text{---}H$$

(C)   Connecting the H and Se atoms uses four electrons: 8 e$^-$ $- 4$ e$^-$ $= 4$ e$^-$

(D)   H atoms need no more electrons.

$$H\text{---}Se\text{---}H$$

(E)   No electrons used in this step. So we still have 4 e$^-$.

(F)   Put the last four electrons on the Se.

$$H\text{---}\overset{..}{\underset{..}{Se}}\text{---}H$$

16. (continued)

    (G)   Se has eight electrons, so the structure is complete.

(c) (A)   The B atom is in Group 3A and the F atoms are in Group 7A. The – charge **adds** one extra electron: $3 + 4 \times (7) + 1 = 32\ e^-$

    (B)   The B atom is the central atom with the F atoms around it.

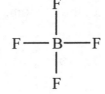

    (C)   Connecting four F atoms uses eight electrons: $32\ e^- - 8\ e^- = 24\ e^-$

    (D)   Complete the octets of all the F atoms.

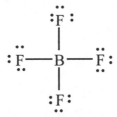

    (E)   Completing the octets of the four F atoms uses 24 electrons:

        $24\ e^- - 24\ e^- = 0\ e^-$

    (F)   No more electrons are available.

    (G)   B has eight electrons, so the structure is complete.

    (H)   The structure is an ion, so put it in brackets and add the – charge.

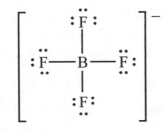

(d) (A)   The P atom is in Group 5A and the O atoms are in Group 6A. The 3– charge **adds** three extra electrons: $5 + 4 \times (6) + 3 = 32\ e^-$

    (B)   The P atom is the central atom with the O atoms around it.

(C)   Connecting four O atoms uses eight electrons: $32 \ e^- - 8 \ e^- = 24 \ e^-$

(D)   Complete the octets of all the O atoms.

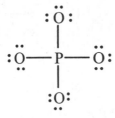

(E)   Completing the octets of the four O atoms uses 24 electrons:

$24 \ e^- - 24 \ e^- = 0 \ e^-$

(F)   No more electrons are available.

(G)   P has eight electrons, so the structure is complete.

(H)   The structure is an ion, so put it in brackets and add the 3– charge.

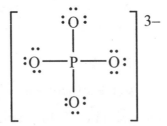

Check your answers: All the Period 2 and 3 elements have an octet of electrons. The H atoms have two electrons. All valance electrons are accounted for in the structures, either as members of shared pairs or of lone pairs. These answers look right.

18. *Define the problem*: Write Lewis structures for a list of ions and molecules.

Develop a plan: Follow the systematic plan given in the answers to Question 16. Use what you learned about organic molecules in Chapter 3 to determine how atoms are bonded in the organic structures.

Execute the plan:

(a) The C atom is in Group 4A, the H atom is in Group 1A, and the Cl atom is in Group 7A: $4 + 3 \times (1) + 7 = 14$ e⁻. The C atom is the central atom with the other atoms around it. Connecting four atoms uses eight electrons: 14 e⁻ $- 8$ e⁻ $= 6$ e⁻. H needs no more electrons. Complete the octet of the Cl atom using six electrons: 6 e⁻ $- 6$ e⁻ $= 0$ e⁻. No more electrons are available. The central atom, C, has eight electrons in shared pairs, so it needs no more. We get the following structure:

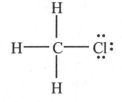

(b) The Si atom is in Group 4A and the O atoms are in Group 6A. The 4– charge **adds** four extra electrons: $4 + 4 \times (6) + 4 = 32$ e⁻. The Si atom is the central atom with the O atoms around it. Connecting four O atoms uses eight electrons: 32 e⁻ $- 8$ e⁻ $= 24$ e⁻. Complete the octets of all the O atoms using 24 electrons. 24 e⁻ $- 24$ e⁻ $= 0$ e⁻. No more electrons are available. The central atom, Si, has eight electrons in shared pairs, so it needs no more. The structure is an ion, so put it in brackets and add the 4– charge:

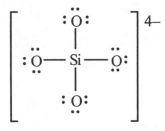

(c) The atoms Cl and F are both in Group 7A. The + charge **removes** one electron: $7 + 4 \times (7) - 1 = 34$ e⁻. The Cl atom is the central atom with the F atoms around it. Connecting four F atoms uses eight electrons: 34 e⁻ $- 8$ e⁻ $= 26$ e⁻. Complete the octets of all the F atoms, using 24 electrons: 26 e⁻ $- 24$ e⁻ $= 2$ e⁻. Put the last two electrons on the Cl. The central atom, Cl, has ten electrons already, so it needs no more. The structure is an ion, so put it in brackets and add the + charge:

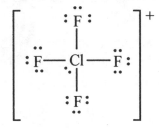

(d) The C atom is in Group 4A and the H atoms are in Group 1A: $2 \times (4) + 6 \times (1) = 14$ e⁻. As described in Section 3.3, the ethane molecule has three H atoms bonded to each C, and the two carbons bonded to each other. Connecting eight atoms uses 14 electrons: 14 e⁻ $- 14$ e⁻ $= 0$ e⁻ No more electrons are available. Each C has eight electrons in shared pairs, so they need no more. We get the following structure:

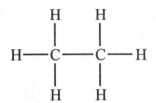

Check your answers: All the Period 2 elements have an octet of electrons. The H atoms have two electrons. All the Period 3 elements have eight or more electrons. All valance electrons are accounted for in the structures, either as members of shared pairs or of lone pairs. These answers look right.

20. *Define the problem*: Write Lewis structures for a list of ions and molecules.

Develop a plan: Follow the systematic plan given in the answers to Question 16. Use what you learned about organic molecules in Chapter 3 to determine how atoms are bonded in the organic structures.

Execute the plan:

(a) $2 \times (4) + 4 \times (7) = 36$ e⁻. This organic molecule looks like ethene with all the H atoms changed to F atoms. The two carbon atoms are bonded to each other, and each has two F atoms bonded to it. Any other arrangement would make F–F bonds. F has the highest electronegativity (see Section 8.7) and won't bond with another F atom if any other element is present. Connecting six atoms with single bonds uses 10 electrons: 36 e⁻ $- 10$ e⁻ $= 26$ e⁻. For the purposes of this method, let's call the first C the central atom. Complete the octets of the F atoms and the second C atom, using 26 electrons: 26 e⁻ $- 26$ e⁻ $= 0$ e⁻. At this point, the structure looks like this:

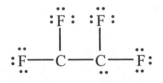

The first C atom has only six electrons in shared pairs, so it needs two more, but there are no more electrons available for lone pairs. Therefore, we must move one lone pair of electrons from the C atom to make a new shared pair, forming a double bond to the C atom:

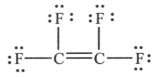

The first C atom now has eight electrons in shared pairs and this is the proper Lewis Structure.

(b) $3 \times (4) + 3 \times (1) + 5 = 20 \ e^-$. As described in Section 3.1, the structural formula given here tells us that this organic compound has two H atoms and a C atom bonded to the first C atom, an H atom and a C atom bonded to the second C atom, and an N atom bonded to the third C atom. Connecting seven atoms uses 12 electrons: $20 \ e^- - 12 \ e^- = 8 \ e^-$. For the purposes of this method, let's call the second and third C atoms "central atoms" so we can start from the ends and work toward the middle. H needs no more electrons. Complete the octet of the N atom and the octet of the first carbon using eight electrons: $8 \ e^- - 8 \ e^- = 0 \ e^-$. At this point, we have the following structure:

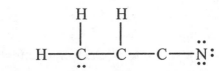

The second C atom has only six electrons in shared pairs and the third C atom has only four electrons in shared pairs, but there are no more electrons available. Therefore, we must move one lone pair of electrons from the first C atom to make a new shared pair between the first and second C atoms, making a double bond. We must also move two lone pairs of electrons from the N atom to the third C atom to make two new shared pairs, forming a triple bond between the N atom and C atom:

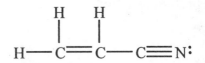

The second and third C atoms now both have eight electrons in shared pairs and this is the proper Lewis Structure. *NOTE: Acrylonitrile's structure is incorrectly given in Question 71. Remove the H atom from the last C atom.*

Check your answers: All the Period 2 elements have an octet of electrons. The H atoms have two electrons. All valence electrons are accounted for in the structures, either as members of shared pairs or of lone pairs. These answers look right.

22. *Define the problem*: Determine if given Lewis structures are correct and explain what is wrong with the incorrect ones.

Develop a plan: A Lewis structure is correct if all valance electrons are accounted for in the structures, either as members of shared pairs or of lone pairs; if all the Period 2 elements have an octet of electrons; if the H atoms have two electrons; and if all the Period 3 and higher elements have eight or more electrons. So, first count the electrons in the structure and compare that with the total number of valence electrons. If the count is correct, then check the octets of all the elements and check H atoms for two electrons. Only elements in period 3 and higher are allowed to exceed eight electrons.

Execute the plan:

(a) OF_2

Check electron count: $6 + 2 \times (7) = 20$ e$^-$. The given structure has 8 electrons (two lone pairs and two single bonds), so the total electron count is wrong.

This structure is incorrect.

(b) O_2

Check electron count: $2 \times (6) = 12$ e$^-$. The given structure has 10 electrons (two lone pairs and a triple bond), so the total electron count is wrong.

This structure is incorrect.

(c) CCl_2O

Check electron count: $4 + 2 \times (7) + 6 = 24$ e$^-$. The given structure also has 24 electrons (eight lone pairs, two single bonds, and one double bond), so the total electron count is okay.

Check octets, etc.: The O atom, the C atom, and both Cl atoms each have eight electrons, so the octet rule is satisfied.

This structure is correct.

(d) CH_3Cl

Check electron count: $4 + 3 \times (1) + 7 = 14$ e$^-$. The given structure has 16 electrons (four lone pairs, and four single bonds), so the total electron count is wrong.

This structure is incorrect.

(e) NO_2^-

Check electron count: $5 + 2 \times (6) + 1 = 18$ e$^-$. The given structure has 16 electrons (five lone pairs, one single bond, and one double bond), so the total electron count is wrong.

This structure is incorrect.

The incorrect structures are (a) OF_2, (b) O_2, (d) CH_3Cl, and (e) NO_2^-.

Check your answers: We used what we know about Lewis structures to determine the incorrect structures. Let's draw correct Lewis structures for those that were incorrect.

(a) The correct structure for OF_2 needs 12 more electrons in the form of three more lone pairs on each F atom:

$$:\ddot{F}:\ddot{O}:\ddot{F}:$$

(b) The correct structure for O_2 needs 2 more electrons, two more lone pairs, and one less shared pair:

$$:\ddot{O}=\ddot{O}:$$

(d) The correct structure has two fewer electrons and the carbon bonded to the chlorine without an H atom between them. Remember, H atom only shares two electrons:

$$\begin{matrix} & H & & \\ H : & \ddot{C} : & \ddot{Cl}^: & \\ & H & & \end{matrix}$$

(e) The correct structure has two more electrons on N (and brackets with a charge):

$$\left[:\ddot{O}-\ddot{N}=\ddot{O}\right]^-$$

The corrected structures are different from the incorrect ones, so these answers look right.

Bonding in Hydrocarbons

24. *Define the problem*: Write structural formulas for all the branched-chain compounds with a given formula.

Develop a plan: Be systematic. Start with a long chain that has only one methyl branch. Move the methyl around (but don't put it on the end carbon and don't put it on a carbon past the first half of the chain). Then make two methyl branches, and move them around similarly. Continue this process until the main chain is too short for methyl branches. Then make an ethyl branch and move it around (but don't put it on the end carbon or the carbon next to the end carbon and don't put it on a carbon past the first half of the chain). Follow a similar pattern with ethyls as with methyls.

Execute the plan: The straight chain isomer of C_6H_{14} has six carbons, so start with a five-carbon chain.

One methyl branch can go on the five-carbon chain in two different ways. The methyl branch can go on the second carbon or on the third carbon.

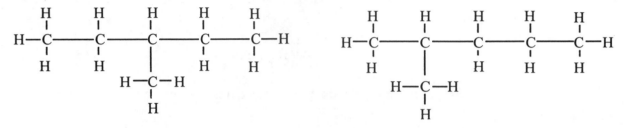

Two methyl branches can go on the four-carbon chain in two ways. They can both go on the second carbon or one can go on the second carbon and one can go on the third carbon.

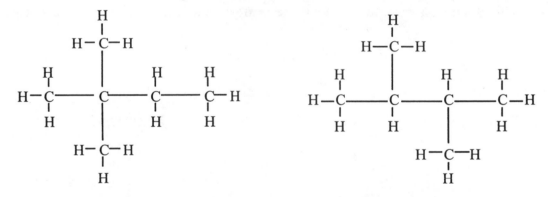

Three methyl branches cannot all three be attached to the one middle carbon in a three-carbon chain, so we're done with methyl branches.

We could try to make a four carbon chain with an ethyl branch. However, ethyl branches can't be attached to chain-carbons within two carbons of the end of the chain. (If you do that, the molecules "longest" chain will actually include your branch!) A four-carbon chain isn't long enough and therefore we can't use ethyl branches. So we have found all the branched isomers of the hydrocarbon with the formula C_6H_{14}.

Check your answers: The molecules have one or more branches off the three-carbon chain. They also all have six C atoms and 14 H atoms. These answers makes sense.

26. *Define the problem*: Given formulas of hydrocarbons determine if they are alkane, alkene, or alkyne.

Develop a plan: If we assume that each of these straight chain molecules has at most one multiple bond, we can use the formula pattern to determine the class. The formulas of alkanes are C_nH_{2n+2}, for $n = 1$ and higher. Each multiple bond removes two electrons. The formulas of alkenes are C_nH_{2n}, for $n = 2$ and higher. The formulas of alkynes are C_nH_{2n-2}, for $n = 2$ and higher. Using the number of C atoms, set n. Then figure out $2n+2$, $2n$, or $2n-2$. Compare these numbers to the number of H atoms. Form a conclusion based on the that comparison.

Execute the plan:

(a) C_5H_8 $n = 5$, with this n, $2n+2 = 12$, $2n = 10$, $2n-2 = 8$. This hydrocarbon is an alkyne. Note, that it could also be an alkene, if there are two double bonds in the molecule.

(b) $C_{24}H_{50}$ $n = 24$, with this n, $2n+2 = 50$, $2n = 48$, $2n-2 = 46$. This hydrocarbon is an alkane.

(c) C_7H_{14} $n = 7$, with this n, $2n+2 = 16$, $2n = 14$, $2n-2 = 12$. This hydrocarbon is an alkene.

Check your answers: One and only one of the three calculations using n gave a matching number. These answers makes sense.

28. *Define the problem*: Given the name of a hydrocarbon, write *cis-* and *trans-*isomers for it.

Develop a plan: Use the prefix to determine number of carbons, and the numeral provided in the name to determine where the double bond starts along the chain. Add H atoms to complete the octets. Look at the fragments attached at the double bond put the two larger ones on the same side for the *cis-*isomer, and on opposite sides for the *trans-*isomer.

Execute the plan: 2-pentene has a five-carbon chain with the double bond starting at the second carbon.

cis-2-pentene (methyl and   trans-2-pentene (methyl and

ethyl on the same side)   ethyl on the opposite sides)

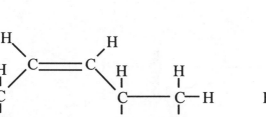

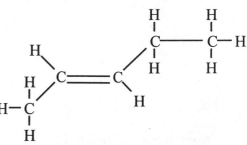

Check your answers: These are the same isomers discovered when answering Question 25. The structure for *cis*-2-pentene given in Question 30 is consistent with this structure. These answers makes sense.

30. *Define the problem*: Given some isomers' structures or structural formulas, determine the other *cis*-or *trans*-isomers.

Develop a plan: Switching the positions of the fragments on the left carbon, gives the other isomer.

Execute the plan:

(a) Switch the positions of the H and the Cl on the left carbon to make *cis*-1,2-dichloropropene. (It's *cis*- because the two Cl atoms are on the same side.)

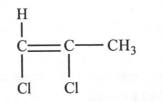

(b) Switch the positions of the H and the CH₃ on the left carbon to make *trans*-2-pentene. (It's *trans*- because the ethyl and methyl fragments are opposite sides.)

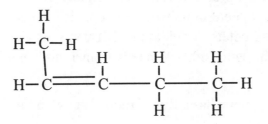

(c) Switch the positions of the H and the CH₃ on the left carbon to make *trans*-3-hexene. (It's *trans*- because the two ethyl fragments are on opposite sides.)

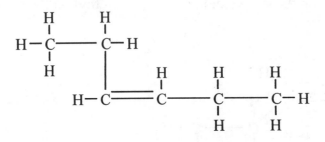

(d) Switch the positions of the H and the CH₃ on the left carbon to make *cis*-2-hexene. (It's *cis*- because the methyl and propyl fragments are on the same side.)

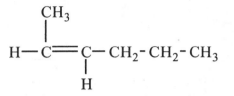

Check your answers: We've seen some of these isomers in previous examples. The particular isomers formed can be correlated to the name change. These answers makes sense.

Bond Properties

32. *Define the problem*: Given a series of pairs of bonds, predict which of the bonds will be shorter.

    Develop a plan: Use periodic trends in atomic radii to identify the smaller atoms. The bond with the smaller atoms, will have a shorter bond. In cases where bonds between the same atoms are compared, triple bonds are shorter than double bonds, which are shorter than single bonds.

    Execute the plan:

    (a) Both bonds have Cl, so compare the sizes of B and Ga. B is smaller than Ga. (It is higher in the same group of the periodic table.) So, B–Cl is shorter than Ga–Cl.

    (b) Both bonds have O, so compare the sizes of C and Sn. C is smaller than Sn. (It is higher in the same group of the periodic table.) So, C–O is shorter than Sn–O.

    (c) Both bonds have P, so compare the sizes of O and S. O is smaller than S. (It is higher in the same group of the periodic table.) So, P–O is shorter than P–S.

    (d) Both bonds have C, so compare the sizes of C and O. O is smaller than C. (It is further to the right in the same period of the periodic table.) So, C=O is shorter than C=C.

    Check your answers: Several of these predictions are confirmed in Table 8.1.

34. *Define the problem*: Given some bonds and only a periodic table, predict which of the bonds will be strongest.

    Develop a plan: Ionic bonds are stronger than polar covalent bonds. Polar covalent bonds are stronger than purely covalent bonds. Use the periodic trend for electronegativity (EN):

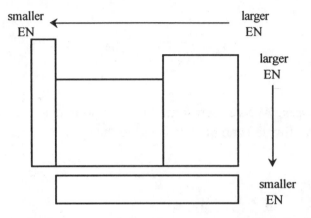

The larger the difference in electronegativity, the stronger the bond will be.

Execute the plan: (a) Si–F, (b) P–S, (c) P–O

Looking up these atoms on the periodic table, we find that Si has lower electronegativity than P. (It is further to the left in the same period of the periodic table.) F has higher electronegativity than O. (It is further to the left in the same period of the periodic table.) So the Si–F bond has the largest electronegativity difference; hence, it is the strongest.

Check your answers: These predictions are confirmed in Figure 8.6.

36. *Define the problem*: Given two chemical formulas, predict which has the shorter carbon-oxygen bond.

Develop a plan: First, write Lewis structures for the two molecules. In cases where bonds between the same two atoms are compared, triple bonds are shorter and stronger than double bonds, which are shorter and stronger than single bonds.

Execute the plan: There is only one plausible Lewis structure for formaldehyde, H_2CO, with 12 valence electrons.

$$\text{H}$$
$$|$$
$$\text{H}\!-\!\text{C}\!=\!\ddot{\ddot{\text{O}}}$$

In formaldehyde, the carbon-oxygen bond is a double bond.

There is only one plausible Lewis structure for carbon monoxide, CO, with 10 valence electrons.

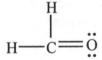

In carbon monoxide, the carbon-oxygen bond is a triple bond.

That means the shorter bond is the bond in CO.

Check your answers: The Lewis structures obey the octet rule and have the right number of valance electrons. These predictions are upheld in general with values given in Table 8.1.

38. *Define the problem*: Given two chemical formulas, predict which has the shorter carbon-oxygen bond.

Develop a plan: First, write resonance structures for the two molecules. In cases where bonds between the same two atoms are compared, triple bonds are shorter and stronger than double bonds, which are shorter and stronger than single bonds. If more than one resonance structure is possible, average their contribution.

Execute the plan: Consider formate ion first. Two equivalent plausible Lewis structures exist for HCO_2^- with 18 valence electrons:

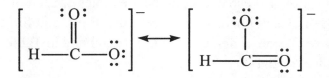

In the formate ion, one resonance structure has a single carbon-oxygen bond and one has a double carbon-oxygen bond. We predict that the carbon-oxygen bond will be halfway between a single and double bond.

There are three plausible equivalent Lewis structures for carbonate ion, CO_3^{2-}, with 24 valence electrons:

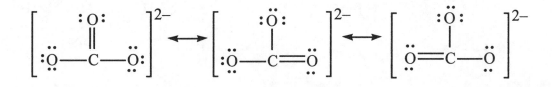

In the carbonate, one resonance structure has a double carbon-oxygen bond and two have single carbon-oxygen bond. We predict that the carbon-oxygen bond will be closer to a single bond than a double bond.

That means the shortest bond is the bond in HCO_2^-.

Check your answers: The Lewis structures obey the octet rule and have the right number of valance electrons. These predictions are upheld in general with values given in Table 8.1.

Bond Energies and Enthalpy Changes

40. *Define the problem:* Given a description of a chemical equation for a reaction and a table of bond energies, estimate the standard enthalpy change of the reaction and determine whether the reaction is exothermic or endothermic.

Develop a plan: First, balance the chemical equation. Then use another variation of Hess's Law to estimate $\Delta H°$. We will break all the bonds in the reactants (by putting their bond energies into the system) and then we will form all the bonds – the opposite of breaking – in the products (by removing their bond energy from the system). That is the logic behind the equation given in Section 8.6.

$$\Delta H° = \sum\left[(\text{moles of bond}) \times D(\text{bond broken})\right] - \sum\left[(\text{moles of bond}) \times D(\text{bond formed})\right]$$

Get the balanced equation and use it to describe the moles of each of the reactants and products. Count the moles of bonds of each type that break and form. Set up a specific version of the above equation. Look up the D for each bond in Table 8.2, plug them into the equation and solve for $\Delta H°$.

Execute the plan: The reactants are nitrogen, N_2, and hydrogen, H_2. The product is ammonia, NH_3. The balanced equation is

$$N_2 \quad + \quad 3\,H_2 \quad \longrightarrow \quad 2\,NH_3$$

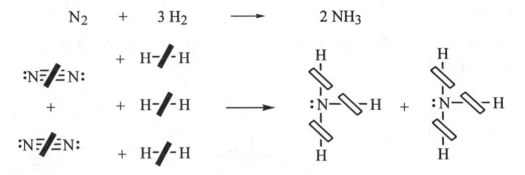

That means we break one mole of N≡N bond and three moles of H–H bonds We form six moles of N–H bonds:

$$\Delta H° = (1 \text{ mol of N≡N}) \times D_{N≡N} + (3 \text{ mol H–H}) \times D_{H–H} - (6 \text{ mol N–H}) \times D_{N–H}$$

Look up the D values in Table 8.2.

$$\Delta H° = (1 \text{ mol}) \times (946 \text{ kJ/mol}) + (3 \text{ mol}) \times (436 \text{ kJ/mol}) - (6 \text{ mol}) \times (391 \text{ kJ/mol})$$

$$\Delta H° = -92 \text{ kJ}$$

This reaction is exothermic.

Check your answer: The reaction is twice the formation reaction for NH_3. Appendix J tells us that NH_3 has $\Delta H_f^\circ = -46.11$ kJ/mol. Twice this value produces a $\Delta H^\circ = -92.22$ kJ/mol. This is very close to the estimate calculated here.

42. *Define the problem:* Given a list of molecules, a description of their chemical reactions and a table of bond energies, identify which has the strongest bond and estimate the standard enthalpy change of the reaction and which is the most exothermic.

Develop a plan: The strongest chemical bond has the largest bond energy. Balance the chemical equations for the reactions. Use the equations to describe the moles of each of the reactants and products. Count the moles of bonds of each type that break and form. Set up an equation as described in the answer to Questions 40. Look up the D for each bond in Table 8.2, plug them into the equation, and solve for ΔH°.

Execute the plan: $D_{H-F} = 566$ kJ/mol, $D_{H-Cl} = 431$ kJ/mol, $D_{H-Br} = 363$ kJ/mol, and $D_{H-I} = 299$ kJ/mol. The largest D is for the H–F bond so the strongest of the four bonds is the one in HF.

Using X to represent the halogen, the chemical reaction looks like this:

$$H_2 + X_2 \longrightarrow 2\ HX$$

That means we break one mole of H–H bonds and one mole of X–X bonds. We form two moles H–X bonds:

$$\Delta H^\circ = (1\ \text{mol of H-H}) \times D_{H-H} + (1\ \text{mol X-X}) \times D_{X-X} - (2\ \text{mol H-X}) \times D_{H-X}$$

Look up the D values in Table 8.2.

When X = F

$$\Delta H^\circ = (1\ \text{mol}) \times (436\ \text{kJ/mol}) + (1\ \text{mol}) \times (158\ \text{kJ/mol}) - (2\ \text{mol}) \times (566\ \text{kJ/mol})$$

$$= -538\ \text{kJ}$$

When X = Cl

$$\Delta H^\circ = (1\ \text{mol}) \times (436\ \text{kJ/mol}) + (1\ \text{mol}) \times (242\ \text{kJ/mol}) - (2\ \text{mol}) \times (431\ \text{kJ/mol})$$

$$= -184\ \text{kJ}$$

When X = Br

$$\Delta H^\circ = (1\ \text{mol}) \times (436\ \text{kJ/mol}) + (1\ \text{mol}) \times (193\ \text{kJ/mol}) - (2\ \text{mol}) \times (363\ \text{kJ/mol})$$

$$= -97\ \text{kJ}$$

When X =I

$\Delta H° = (1 \text{ mol}) \times (436 \text{ kJ/mol}) + (1 \text{ mol}) \times (151 \text{ kJ/mol}) - (2 \text{ mol}) \times (299 \text{ kJ/mol})$

$= -11 \text{ kJ}$

The $H_2 + F_2$ reaction is the most exothermic.

Check your answer: The reaction is twice the formation reaction for HX. Appendix J tells us that HF has $\Delta H°_f = -271.1$ kJ/mol, HCl has $\Delta H°_f = -92.307$ kJ/mol, and HBr has $\Delta H°_f = -36.40$ kJ/mol. Twice these values produce $\Delta H° = -542.2$ kJ/mol, $\Delta H° = -184.614$ kJ/mol, and $\Delta H° = -72.80$ kJ/mol. These are reasonably close to the estimates calculated here. The bromine value is farther away because Appendix J uses standard state liquid bromine, not gas-phase as assumed in the bond energy numbers. HI would be far off since the standard state of I_2 is solid.

Electronegativity and Bond Polarity

44. (a) Look up electronegativity values (in Figure 8.6) for the atoms in these bonds.

$EN_C = 2.5, EN_N = 3.0, EN_H = 2.1, EN_{Br} = 2.8, EN_S = 2.5, N_O = 3.5$

N is more electronegative in C–N,

C is more electronegative in C–H,

Br is more electronegative in C–Br, and

O is more electronegative in S–O.

(b) Bonds are more polar when the electronegativity difference is larger. Find ΔEN to determine which is most polar.

$\Delta EN_{C-N} = EN_N - EN_C = 3.0 - 2.5 = 0.5$

$\Delta EN_{C-H} = EN_C - EN_H = 2.4 - 2.1 = 0.4$

$\Delta EN_{C-Br} = EN_{Br} - EN_C = 2.8 - 2.5 = 0.3$

$\Delta EN_{S-O} = EN_O - EN_S = 3.5 - 2.5 = 1.0$ most polar

46. (a) Bonds are polar if the atoms' electronegativities are different.

$EN_H = 2.1, EN_N = 3.0, EN_C = 2.5, EN_O = 3.5$

Since all the electronegativities are different, the bonds are all somewhat polar. None of the bonds are nonpolar.

(b) Bonds are more polar when the electronegativity difference is larger.

$$\Delta EN_{N-H} = EN_N - EN_H = 3.0 - 2.1 = 0.9$$

$$\Delta EN_{C-N} = EN_N - EN_C = 3.0 - 2.5 = 0.5$$

$$\Delta EN_{C-O} = EN_O - EN_C = 3.5 - 2.5 = 1.0$$

The largest electronegativity difference is for the C=O bond, so it is the most polar.

$$C \underset{\delta^+}{\makebox[1.5em]{}} \underset{\delta^-}{O}$$

The O atom is the partial negative end of this bond, because it has the larger electronegativity.

Formal Charge

48. The total of the formal charges in a molecule must be zero, because molecules have no net charge. The total of the formal charges in an ion must be its ionic charge.

50. *Define the problem*: Given the formulas of molecules or ions, write the correct Lewis structure and assign formal charges to each atom.

Develop a plan: Write the Lewis structures. Then determine the number of lone pair electrons and bonding electrons around each atom. Use the method described in Section 8.9 to determine the formal charges on each atom.

Formal charge = (valence electrons in the atom) –

$$[(lone\ pair\ electrons) + (\tfrac{1}{2}\ number\ of\ bonding\ electrons)]$$

Execute the plan:

(a) Lewis structure for SO_3 molecule: 24 electrons total

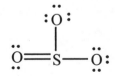

The formal charges are calculated for each atom using the number of valence electrons, the number of lone pair electrons and the number of bonding electrons. Let's set up a chart for these values:

Don't like those charts

refer to do 4/h →

	S	=O	−O
Valence electrons	6	6	6
Lone pair electrons	0	4	6
Bonding electrons	8	4	2
Formal charge	$6 - (0 + 4) = +2$	$6 - (4 + 2) = 0$	$6 - (6 + 1) = -1$

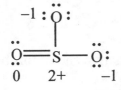

There are other resonance structures that could be written for SO_3 with the double bond moved to each of the other two O atoms and with more than one double bond to the sulfur atom. (A more involved description of writing and judging the feasibility of resonance structures is found in the answers to Questions 54 - 57.)

(b) Lewis structure for NCCN molecule: 18 electrons total.

$$:N\equiv C\text{---}C\equiv N:$$

The formal charges are calculated for each atom using the number of valence electrons, the number of lone pair electrons and the number of bonding electrons. Let's set up a chart for these values:

	C	N
Valence electrons	4	5
Lone pair electrons	0	2
Bonding electrons	8	6
Formal charge	$4 - (0 + 4) = 0$	$5 - (2 + 3) = 0$

$$:N\equiv C\text{---}C\equiv N:$$
$$0\quad\;\;0\quad\;\;\;0\quad\;\;0$$

(c) Lewis structure for NO_2^- ion: 18 electrons total

$$\left[\ddot{O} = \ddot{N} - \ddot{O} \colon \right]^-$$

The formal charges are calculated for each atom using the number of valence electrons, the number of lone pair electrons and the number of bonding electrons. Let's set up a chart for these values:

	N	=O	–O
Valence electrons	5	6	6
Lone pair electrons	2	4	6
Bonding electrons	6	4	2
Formal charge	$5 - (2 + 3) = 0$	$6 - (4 + 2) = 0$	$6 - (6 + 1) = -1$

$$\left[\ddot{O} = \ddot{N} - \ddot{O} \colon \right]^-$$
$$00-1$$

One other resonance structure could be written for NO_2^- with the double bond moved to the other O atom. (A more involved description of writing resonance structures is found in the answers to Questions 54 and 55.)

Check your answers: The sum of the formal charges is zero for the neutral molecules and the ionic charge for the charged ion. The atoms with more bonds have more positive formal charges than those with fewer bonds and more lone pairs. These answers make sense.

52. *Define the problem*: Given the formulas of molecules or ions, write the correct Lewis structure and assign formal charges to each atom.

Develop a plan: Write the Lewis structures. Then determine the number of lone pair electrons and bonding electrons around each atom. Use the method described in Section 8.9 and the answers to Question 50 to determine the formal charges on each atom.

Execute the plan:

(a) The Lewis structure for CH_3CHO molecule: 18 electrons total.

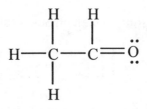

Set up the chart:

	C–	C=	H	O
Valence electrons	4	4	1	6
Lone pair electrons	0	0	0	4
Bonding electrons	8	8	2	4
Formal charge	$4 - (0 + 4) = 0$	$4 - (0 + 4) = 0$	$1 - (0 + 1) = 0$	$6 - (4 + 2) = 0$

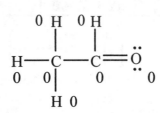

(b) There are three possible Lewis structures for N_3^- ion: 16 electrons total.

First structure:

$$\left[\ddot{N}\!=\!N\!=\!\ddot{N} \right]^{-}$$

Set up the chart:

	N=	=N=
Valence electrons	5	5
Lone pair electrons	4	0
Bonding electrons	4	8
Formal charge	$5 - (4 + 2) = -1$	$5 - (0 + 4) = +1$

$$\left[\ddot{N}\!=\!N\!=\!\ddot{N} \right]^{-}$$
$$\quad -1 \qquad +1 \qquad -1$$

52. (b) (continued)

Second structure:

$$\left[\ :N \equiv N \text{—} \ddot{\underset{..}{N}}: \ \right]^{-}$$

Set up the chart:

	Left-most N	Middle N	Right-most N
Valence electrons	5	5	5
Lone pair electrons	2	0	6
Bonding electrons	6	8	2
Formal charge	$5 - (2 + 3) = 0$	$5 - (0 + 4) = +1$	$5 - (6 + 1) = -2$

$$\left[\ :N \equiv N \text{—} \ddot{\underset{..}{N}}: \ \right]^{-}$$
$$\quad 0 \qquad +1 \qquad -2$$

Third structure:

$$\left[\ :\ddot{\underset{..}{N}} \text{—} N \equiv N: \ \right]^{-}$$

Set up the chart:

	Left-most N	Middle N	Right-most N
Valence electrons	5	5	5
Lone pair electrons	6	0	2
Bonding electrons	2	8	6
Formal charge	$5 - (6 + 1) = -2$	$5 - (0 + 4) = +1$	$5 - (2 + 3) = 0$

$$\left[\ :\ddot{\underset{..}{N}} \text{—} N \equiv N: \ \right]^{-}$$
$$\quad -2 \qquad +1 \qquad 0$$

(c) Lewis structure for CH_3CN molecule: 16 electrons total

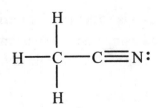

Set up the chart:

	C	H	N
Valence electrons	4	1	5
Lone pair electrons	0	0	2
Bonding electrons	8	2	6
Formal charge	$4-(0+4)=0$	$1-(0+1)=0$	$5-(2+3)=0$

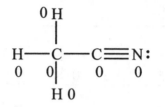

Check your answers: The sum of the formal charges is zero for the neutral molecules and the ionic charge for the charged ion. The atoms with more bonds have more positive formal charges than those with fewer bonds and more lone pairs. These answers make sense.

Resonance

54. *Define the problem*: Given the formulas of molecules or ions, write all the resonance structures.

Develop a plan: Write the Lewis structure. Each resonance structure differs only by where the electrons for a multiple bond come from. When two or more atoms with lone pairs are bonded to an atom that needs more electrons, any one of them can supply the electron pair for a multiple bond. To write all resonance structures, systematically and sequentially supply the central atom with needed electrons from each of the possible sources. Separate these different structures with a double-headed arrow to show that they are resonance structures.

Execute the plan:

(a) There are three plausible Lewis structures for nitric acid, HNO_3, with 24 valence electrons. They are formed using one lone pair from a different one of the outer O atoms to make the second bond in the double bond to complete the octet of the N atom.

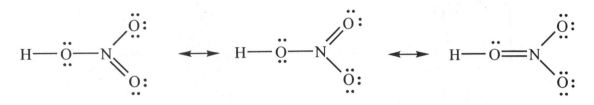

These structures are not equally plausible. (See Question 57 for details.)

(b) There are three plausible Lewis structures for nitrate ion, NO_3^-, with 24 valence electrons. They are formed using one lone pair from a different one of the outer O atoms to make the second bond in the double bond to complete the octet of the N atom.

Check your answers: The structures drawn all follow the octet rule and have the right number of valence electrons. They differ by which outer atom is double bonded to the N atom. These answers make sense.

56. *Define the problem*: Write resonance structures using all single bonds, one, two and three double bonds, and use formal charges to predict the most plausible one.

Develop a plan: Write the Lewis structure. To write all resonance structures, systematically and sequentially supply the central atom with needed electrons from each of the possible sources. Keep in mind that Period 2 elements must have an octet, but Period 3 elements can have 8 or more electrons. Separate these different structures with a double-headed arrow to show that they are resonance structures. To determine the relative plausibility of the structures, determine the formal charges (as described in the answers to Question 50) then use the rules described in Section 8.8:

- Smaller formal charges are more favorable than larger ones.

- Negative formal charges should reside on the more electronegative atoms. Conversely, positive formal charges should reside on the least electronegative atoms.

- Like charges should not be on adjacent atoms.

Execute the plan: BrO_4^- has 32 electrons. Formal charges for single bonded O atoms are always −1. Formal charges for double bonded O are always 0. Each time electrons are moved from lone pairs into bonding pairs the positive formal charge on Br goes down: The Lewis structure that follows the octet rule for all the atoms is the first one:

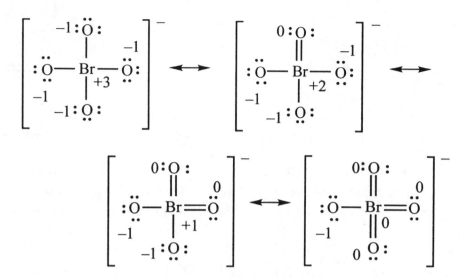

Since smaller formal charges are more favorable than larger ones, we predict that this fourth resonance structure, with three double bonds and the most zero formal charges, is the most plausible.

Check your answers: The Period 2 O atoms in all these structures follow the octet rule. The Period 3 Br atom has eight or more electrons. All structures have the right number of valence electrons. They differ by which outer atom forms the multiple bonds to the central atom. While it might feel strange to select a resonance structure that does not follow the octet rule as being the most plausible, it is clear from the formal charges that this structure is preferred over the one that does follow the octet rule. These answers make sense.

Exceptions to the Octet Rule

59. Follow the systematic plan given in the answers to Question 16.

(a) BrF₅ has 42 valance electrons. Use ten of the electrons to connect the six F atoms to the central Br atom. Use 30 more of them to fill the octets of the F atoms. Put the last two on Br.

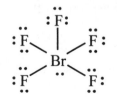

(b) IF_5 has 42 valance electrons. Use ten of the electrons to connect the five F atoms to the central I atom. Use 30 more of them to fill the octets of the F atoms. Put the last two on I.

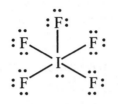

(b) IBr_2^- has 22 valance electrons. Use four of the electrons to connect the Br atoms to the central I atom. Use 12 more of them to fill the octets of the Br atoms. Put the last six on I.

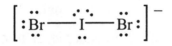

Check your answers: All the Period 2 elements have an octet of electrons. All the Period 3 elements have eight or more electrons. All valance electrons are accounted for in the structures, either as members of shared pairs or of lone pairs. These answers look right.

61. Elements in Periods 3 or higher can form compounds with five or six pairs of valence electrons surrounding their atoms. The Period 2 elements cannot. Using a periodic table, we find that (b) P, (e) Cl, (g) Se, and (h) Sn can, and (a) C, (c) O, (d) F and (f) B cannot.

Aromatic Compounds

63. "All carbon-to-carbon bond lengths are the same" argues against the presence of C=C bonds in benzene, since the C=C bonds would be shorter than the C–C bonds.

65. Adapting the structural notation given in Section 8.11, anthracene looks like this:

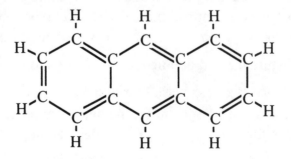

That means it has a formula of $C_{14}H_{10}$.

General Questions

67. Use the periodic table and trends in electronegativities to determine which pair is farthest apart. That will mean that their electronegativities are most different and the bond will be most polar.

Here, all the choices have F atom in common so find out which of the other elements is the farthest from F. Si is farthest from F on the periodic table, so (c) Si–F is more polar than these other choices (a) C–F, (b) S–F, (d) O–F

69. Yes, it is a good generalization, because elements close together in the periodic table have similar electronegativities and the bonds would be formed by shared electrons (polar covalent). If they are far apart on the periodic table, their electronegativities will likely be very different and the bond more likely to be ionic.

A few exceptions exist, of course. For example, H atom is fairly far away from most of the nonmetals on the periodic table, yet it is also a nonmetal and forms polar covalent bonds. We will learn later that metal-metal bonds are somewhat different from the traditional polar covalent bonds studied in this chapter.

71. *Warning: The structure given in the problem is incorrect. There should be no H atom bonded to the C atom with the triple bond. This error does not affect the answers here.*

The resonance structure given in the problem is the only resonance structure with zero formal changes, so it is the dominant form for the molecule.

(a) There are two carbon-carbon bonds, a C=C and a C–C. Double bonds are shorter than single bonds, so C=C is the shortest.

(b) Double bonds are stronger than single bonds, so C=C is the strongest.

(c) The most polar bond is the bond with atoms that have the largest electronegativity difference. The difference between the electronegativities of C (EN = 2.5) and H (EN = 2.1) is 0.4. The difference between the electronegativities of C and N (EN = 3.0) is 0.5. So, the carbon-nitrogen bond is slightly more polar. Nitrogen has the higher electronegativity, so it is the partial negative end of the bond.

73. Use the Lewis Structure method outlined in the answers to Question 16.

(a) SCl_2 has 20 valence electrons. Use four of them to attach the two Cl atoms to the central S atom. Use the 12 of them for the lone pairs on the Cl atoms. Use the remaining four of them for the lone pairs on the S atom.

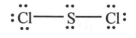

(b) The Cl_3^+ ion has 20 valence electrons, just as in (a). One of the Cl atoms is the central atom, while the other two Cl atoms are single bonded to the central Cl atom:

$$\left[:\ddot{C}l-\ddot{C}l-\ddot{C}l:\right]^+$$

(c) $ClOClO_3$ has 38 valence electrons. We are told there is a Cl–O–Cl bond, so we will put the last three O atoms on the second Cl. Use ten electrons to attach the atoms together using single bonds. Use the remaining 28 of them to complete the octets.

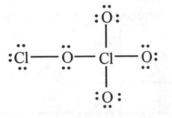

The second Cl atom has a +3 formal charge in this Lewis structure, so we can make a lower formal charge resonance structure for this molecule by expanding the octet of the Period 3 Cl atom, as we did for Br atom in Question 56.

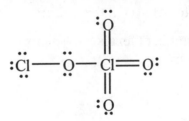

Now, all the formal charges are zero.

(d) $SOCl_2$ has 26 valence electrons. Use six of them to attach the O atoms to the central Cl atom. Use the remaining 18 of them complete the octets of the O atoms. Use the remaining 2 to complete the octet on S.

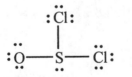

The S atom has a +1 formal charge in this Lewis structure, so we can make a lower formal charge resonance structure for this molecule by expanding the octet of the Period 3 S atom and making one double bond to the S atom, similarly to what we did in (c) for Cl.

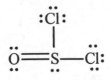

75. The strongest chemical bond has the largest bond energy. Look up the bond energies for the bonds in Table 8.2:

$$D_{O-H} = 467 \text{ kJ/mol}, D_{O=O} = 498 \text{ kJ/mol}, D_{O=C} = 695 \text{ to } 803 \text{ kJ/mol},$$

$$D_{O-O} = 146 \text{ kJ/mol}, \text{ and } D_{O-Cl} = 205 \text{ kJ/mol}.$$

The order of increasing strength: O–O < Cl–O < O–H < O=O < O=C

77. Given formulas of hydrocarbons, determine if they are alkane, aromatic, or neither. We can try what was done in the answers to Question 26.

	n	2n+2 (alkane)	2n (alkene)	2n–2 (alkyne)
$C_6H_?$	6	? = 14	? = 12	? = 10
$C_8H_?$	8	? = 18	? = 16	? = 14
$C_{10}H_?$	10	? = 22	? = 20	? = 18

We find the answers to (c) - (f) in this chart:

(c) C_6H_{12} has the formula of an alkene, so this is neither an alkane nor aromatic, though it could be a cyclic alkane.

(d) C_6H_{14} is an alkane.

(e) C_6H_{18} is an alkane.

(f) C_6H_{10} is an alkyne, so this is neither an alkane nor aromatic.

Let's draw some structures that fit the other two formulas as aromatic compounds with the help of Section 8.11 :

(a) Xylene is given in Section 8.11, and its formula is C_8H_{10} (*para*-xylene shown here).

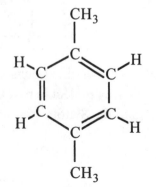

(b) Naphthalene is given in Section 8.11, and its formula is $C_{10}H_8$.

Applying Concepts

79. The number of valence electrons in SF_5^+ is $6 + 5 \times (7) - 1 = 42$ e⁻. It looks like the student forgot to subtract one electron for the positive charge, since the given structure has $6 + 5 \times (7) = 41$ e⁻.

81. Resonance structures must only be different by where the electrons are. The bonding arrangement of the atoms (what atom is bonded to what atom) must be the same from structure to structure. In the first structure, the C atom has only one S atom bonded to it. In the second structure the C atom has an N atom and an S atom bonded to it. These bonding arrangements differ; therefore these structures represent different molecules, not resonance structures of one molecule.

83. In this fictional universe, a nonet is nine electrons. We will assume that the number of electrons in the atom called "O" in the other universe is still six, the number of electrons in the atom called "H" is still one, and the number of electrons in the atom called "F" is still seven.

The 6 electrons in the O atom would need three more electrons to make a total of nine. That means it would combine with three H atoms to make this molecule:

$$H : O : H$$
$$\overset{\cdot\cdot\cdot}{}$$
$$H$$

The 7 electrons in the F atom would need two more electrons to make a total of nine. That means it would combine with two H atoms to make this molecule:

$$H : F :$$
$$\overset{\cdot\cdot\cdot}{}$$
$$H$$

85.   Using the periodic trend for electronegativity (EN)

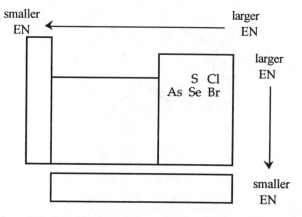

We certainly know that Cl's EN is the largest, so we'll assign it the value of 3.0. We certainly know that As's EN is the lowest, so we'll assign it value of 2.1 The trend shows that EN's of S and Br are both larger than that of Se, so we'll assign them both the value of 2.5. That leaves the EN value of 2.4 for Se. With the values of 2.5 and 2.4 so close together, the assignment of those values are uncertain.

87. All of these compounds require the central atom to have more than eight electrons around it in order to have bonds to five atoms. Three of the four of the central atoms are in Periods 3 or higher, making it possible to expand their octets. The Period 2 N atom cannot have more than eight electrons, so NF_5 is least likely to exist.

Chapter 9: Molecular Structures

Molecular Shape

16. *Define the problem*: Write Lewis Structures for a list of formulas and identify their shape.

Develop a plan: Follow the systematic plan for Lewis structures given in the answers to Question 8.16, then determine the number of bonded atoms and lone pairs on the central atom, determine the designated type (AX_nE_m) and use Table 9.1. *Note: It is not important to expand the octet of a central atom solely for the purposes of lowering its formal charge, because the shape of the molecule will not change. Hence, we will write Lewis structures that follow the octet rule unless the atom needs more than eight electrons.*

Execute the plan:

(a) BeH_2 (4 e⁻)   H——Be——H   The type is AX_2E_0, so it is linear.

(b) CH_2Cl_2 (20 e⁻)   The type is AX_4E_0, so it is tetrahedral.

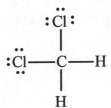   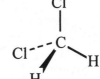

(c) BH_3 (6 e⁻)   The type is AX_3E_0, so it is triangular planar.

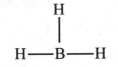   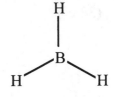

(d) SCl_6 (48 e⁻)   The type is AX_6E_0, so it is octahedral.

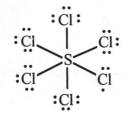   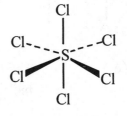

(e) PF_5 (40 e$^-$)   The type is AX_5E_0, so it is triangular bipyramidal.

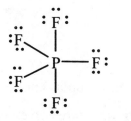   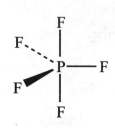

18. Adapt the method given in the answers to Question 16.

(a) ClF_2^+ (20 e$^-$)   The type is AX_2E_2, so the electron-pair geometry is tetrahedral and the molecular geometry is angular (109.5°).

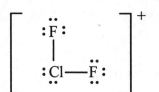

(b) $SnCl_3^-$ (26 e$^-$)   The type is AX_3E_1, so the electron-pair geometry is tetrahedral and the molecular geometry is triangular pyramidal.

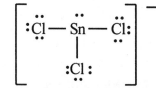   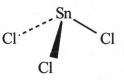

(c) PO_4^{3-} (32 e$^-$)   The type is AX_4E_0, so the electron-pair geometry and the molecular geometry are both tetrahedral.

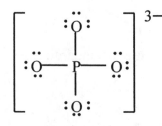   

(d) CS_2 (16 e$^-$)   The type is AX_2E_0, so the electron-pair geometry and the molecular geometry are both linear.

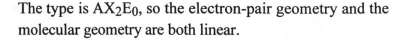

20. Adapt the method given in the answers to Question 9.16.

(a) BO_3^{3-} (24 e⁻)   The type is AX_3E_0, so both the electron-pair geometry and the molecular geometry are triangular planar.

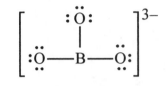   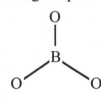

(b) CO_3^{2-} (24 e⁻) AX_3E_0, so both the electron-pair geometry and the molecular geometry are triangular planar.

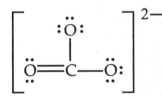   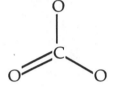

(c) SO_3^{2-} (26 e⁻) AX_3E_1, so the electron-pair geometry is tetrahedral and the molecular geometry is triangular pyramidal.

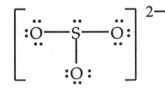   

(d) ClO_3^- (26 e⁻) AX_3E_1, so the electron-pair geometry is tetrahedral and the molecular geometry is triangular pyramidal.

All of these ions and molecules have one central atom with three O atoms bonded to it. The number and type of bonded atoms is constant. The geometries vary depending on how many lone pairs are on the central atom. The structures with the same number of valance electrons all have the same geometry.

22. Adapt the method given in the answers to Question 9.16.

(a) SiF_6^{2-} (48 e⁻)   The type is AX_6E_0, so both the electron-pair geometry and the molecular geometry are octahedral.

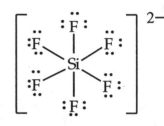   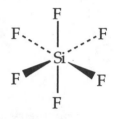

(b) SF_4 (34 e⁻)   The type is AX_4E_1, so the electron-pair geometry is triangular bipyramid and the molecular geometry is seesaw.

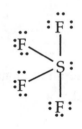   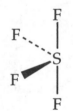

(c) PF_5 (40 e⁻)   The type is AX_5E_0, so the electron-pair geometry is triangular bipyramidal and the molecular geometry is triangular bipyramidal.

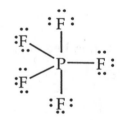   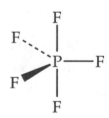

(d) XeF_4 (36 e⁻)   The type is AX_4E_2, so the electron-pair geometry is octahedral and the molecular geometry is square pyramidal.

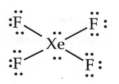   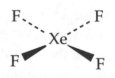

24. Adapt the method given in the answers to Question 16 to get the electron-pair geometry of the second atom in the bond. Use this to predict the approximate bond angle.

(a) SO_2 (18 e⁻)

The type is AX_2E_1, so the electron-pair geometry is triangular planar and the approximate O–S–O angle is 120°.

$$\overset{\cdot\cdot}{\underset{\cdot\cdot}{O}}=\overset{\cdot\cdot}{S}-\overset{\cdot\cdot}{\underset{\cdot\cdot}{O}}:$$

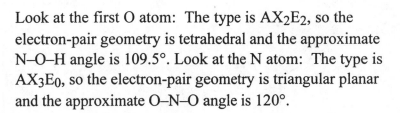

(b) BF_3 (24 e⁻)

The type is AX_3E_0, so the electron-pair geometry is triangular planar and the approximate F–B–F angle is 120°.

$$:\overset{\cdot\cdot}{F}:$$
$$|$$
$$:\overset{\cdot\cdot}{F}-B-\overset{\cdot\cdot}{F}:$$

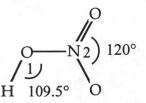

(c) HNO_3 (24 e⁻)

Look at the first O atom: The type is AX_2E_2, so the electron-pair geometry is tetrahedral and the approximate N–O–H angle is 109.5°. Look at the N atom: The type is AX_3E_0, so the electron-pair geometry is triangular planar and the approximate O–N–O angle is 120°.

$$H-\overset{\cdot\cdot}{\underset{\cdot\cdot}{O}}-N$$

(c) CH_2CHCN (20 e⁻)

Look at the first C atom: The type is AX_3E_0, so the electron-pair geometry is triangular planar and the approximate H–C–H angle is 120°. Look at the third C atom: The type is AX_2E_0, so the electron-pair geometry is linear and the approximate C–C–N angle is 180°.

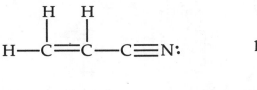

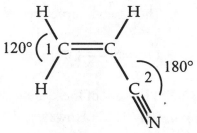

26. Adapt the method given in the answers to Question 24.

(a) SeF_4 (34 e⁻)

The type is AX_4E_1, so the electron-pair geometry is trigonal bipyramid. The $F_{equitorial}$–Se–$F_{equitorial}$ angle is 120°, the $F_{equitorial}$–Se–F_{axial} angles are 90°, and the F_{axial}–Se–F_{axial} angle is 180°.

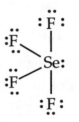

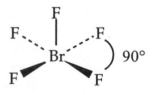

(b) SOF_4 (40 e⁻)

The type is AX_5E_0, so the electron-pair geometry is triangular bipyramidal, equatorial-F–S–O angles are 120° and the axial-F–S–O angles are 90°.

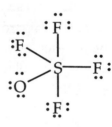

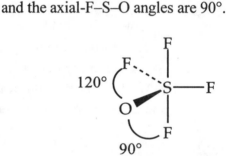

(c) BrF_5 (42 e⁻)

The type is AX_5E_1, so the electron-pair geometry is octahedral and all the F–Br–F angles are 90°.

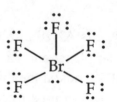

28. NO_2 molecule and NO_2^- ion differ by only one electron. NO_2 has 17 electrons and NO_2^- has 18 electrons. Their Lewis structures are quite similar:

According to VESPR, both of these are triangular planar (AX_2E_1), where the O–N–O angle is approximately 120°. However, the single unpaired electron in NO_2 molecule is not as good at repelling other electrons as a full pair would be. When a nonbonded pair of electrons is repelling the bonding pairs, it pushes them closer together making a narrower angle. Hence, we will predict that the O–N–O angle in NO_2^- ion is slightly smaller than the angle in NO_2 molecule.

Hybridization and Multiple Bonds

30. Write a Lewis structure for $HCCl_3$ and use VSEPR to determine the molecular geometry as described in the answer to Question 16. Use Table 9.2 to determine the hybridization of the central atom using the electron-pair geometry. The molecule is AX_4E_0 type, so its electron-pair geometry and its molecular geometry are tetrahedral.

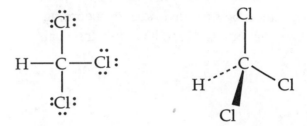

To make four equal bonds with an electron-pair geometry of tetrahedral, the C atom must be sp^3 hybridized, according to Table 9.2. The H atom and Cl atoms are not hybridized.

32. Draw the Lewis structure and use VSEPR to determine the geometry of the central atom as described in the answer to Question 16. Use Table 9.2 or Table 9.3 to determine the hybridization of the central atom using the electron-pair geometry.

(a) GeF_4 (32 e⁻)   The type is AX_4E_0, so both the electron-pair geometry and the molecular geometry are tetrahedral. The Ge atom must be sp^3 hybridized, according to Table 9.2, to make these four bonds.

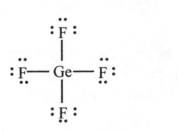
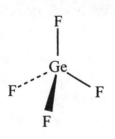

(b) SeF$_4$ (34 e$^-$)   The type is AX$_4$E$_1$, so the electron-pair geometry is triangular bipyramidal and the molecular geometry is seesaw. The Se atom must be sp^3d hybridized, according to Table 9.3, to make these four bonds and to have one lone pair.

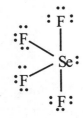 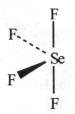

(c) XeF$_4$ (36 e$^-$)   The type is AX$_4$E$_2$, so the electron-pair geometry is octahedral and the molecular geometry is square pyramidal. The Xe atom must be sp^3d^2 hybridized, according to Table 9.3, to make these four bonds and to have two lone pairs.

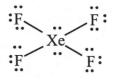 

34. (a) One s and three p orbitals combine to make sp^3 hybrid orbitals.

    (b) One s, three p, and two d orbitals combine to make sp^3d^2 hybrid orbitals.

    (c) One s and two p orbitals combine to make sp^2 hybrid orbitals.

36. The bond angles are associated with the electron-pair geometry, so use Figure 9.4.

    (a) Tetrahedral sp^3 hybrid orbitals are generally associated with bond angles of 109.5°.

    (b) Octahedral sp^3d^2 hybrid orbitals are generally associated with bond angles of 90°.

    (c) Triangular planar sp^2 hybrid orbitals are generally associated with bond angles of 120°.

38. Draw the Lewis structure and use VSEPR to determine the electron-pair geometry of each of the C atoms. Use Table 9.2 or Table 9.3 to determine the hybridization of the central atom using the electron-pair geometry. The bond angles are associated with the electron-pair geometry, also, so use Figure 9.4.

(a) Look at the first C atom: The type is AX_4E_0, so the electron-pair geometry is tetrahedral and this C atom's hybridization is sp^3. Look at the second C atom: The type is AX_3E_0, so the electron-pair geometry is triangular planar and this C atom's hybridization is sp^2.

(b) The sp^3-hybridized C atom has tetrahedral H–C–H and H–C–C bond angles of approximately 109.5°. The sp^2-hybridized C atom has triangular planar C–C–O and O–C–O bond angles of approximately 120°.

40. Use the Lewis structure of alanine and the VSEPR model to determine the electron-pair geometry of each of the atoms. Use Table 9.2 to determine the hybridization of the central atom using the electron-pair geometry. The bond angles are associated with the electron-pair geometry also, so use Figure 9.4.

Look at the first C atom (the one farthest to the left): The type is AX_4E_0, so the electron-pair geometry is tetrahedral and this C atom's hybridization is sp^3. The sp^3-hybridized C atom has tetrahedral bond angles of approximately 109.5°.

Look at the second C atom (the center one): The type is AX_4E_0, so the electron-pair geometry is tetrahedral and this C atom's hybridization is sp^3. The sp^3-hybridized C atom has tetrahedral bond angles of approximately 109.5°.

Look at the third C atom (the one farthest to the right): The type is AX_3E_0, so the electron-pair geometry is triangular planar and this C atom's hybridization is sp^2. The sp^2-hybridized C atom has triangular planar bond angles of approximately 120°.

Look at the N atom: The type is AX_3E_1, so the electron-pair geometry is tetrahedral and this N atom's hybridization is sp^3. The sp^3-hybridized N atom has tetrahedral bond angles of approximately 109.5°.

Look at the O atom on the right: The type is AX_2E_2, so the electron-pair geometry is tetrahedral and this O atom's hybridization is sp^3. The sp^3-hybridized O atom has tetrahedral bond angles of approximately 109.5°.

The top O atom has only one atom bonded to it. It is not hybridized and has no bond angles.

42. (a) Write the Lewis structure. Use the VSEPR model to determine the electron-pair geometry of each of the atoms. Use Table 9.2 to determine the hybridization of the central atom using the electron-pair geometry. The bond angles are associated with the electron-pair geometry also, so use Figure 9.4.

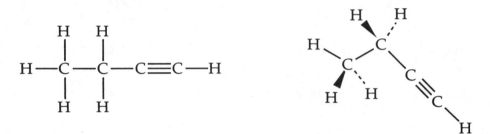

Look at the first two C atoms (the two farthest to the left): They are both of type AX_4E_0, so the electron-pair geometry is tetrahedral and these C atoms have sp^3 hybridization. The sp^3-hybridized C atoms have tetrahedral bond angles of approximately 109.5°.

Look at the second two C atoms (the two farthest to the right): They are both of type AX_2E_0, so the electron-pair geometry is linear and these C atoms have sp hybridization. The sp-hybridized C atoms have linear bond angles of approximately 180°.

(b) The shortest carbon-carbon bond is the triple bond. That can be confirmed by looking up the bond lengths in Table 8.1.

(c) The strongest carbon-carbon bond is the triple bond. That can be confirmed by looking up the bond energies in Table 8.2.

44. The first bond between two atoms must always be a σ bond. When more than one pair of electrons are shared between atoms, they are always part of π bonds. So the second bond in a double bond and the second and third bonds in a triple bond are always π bonds.

(a)

$$:\overset{..}{O}\underset{\sigma}{\overset{\pi}{=\!\!=}}C\underset{\sigma}{\overset{\pi}{=\!\!=}}\overset{..}{S}:$$

(b)

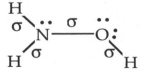

(c)

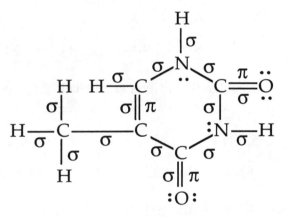

(d)

46. Adapt the method described in the answers to Questions 30 and 44.

(a) Counting the labeled lines representing pairs of electrons above, we find that there are fifteen σ bonds.

(b) Counting the labeled lines representing pairs of electrons above, we find that there are three π bonds.

(c) Each of the carbon atoms in the ring has three bonded atoms and no lone pairs, that makes them each AX_3E_0 type, so the electron-pair geometry is triangular planar and the hybridization is sp^2.

(d) Each of the nitrogen atoms in the ring has three bonded atoms and one lone pair, that makes them each AX_3E_1 type, and we would predict the electron-pair geometry is tetrahedral and the hybridization is sp^3. However, to create an extended π system using the lone pairs on nitrogen, they would actually need to be in a unhybridized p orbitals, so we find electron-pair geometry is actually triangular planar and the hybridization is sp^2.

(e) The carbon atom not in the ring has four bonded atoms and no lone pairs, that makes it AX_4E_0 type, so the electron-pair geometry is tetrahedral and the hybridization is sp^3.

Molecular Polarity

48. We need the Lewis structures and molecular shapes of H_2O, NH_3, CS_2, ClF, and CCl_4.

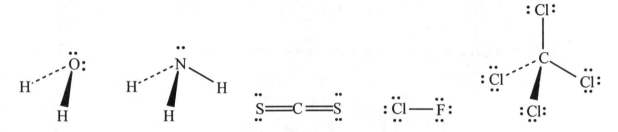

(a) The bond polarity is related to the difference in electronegativity. Use Figure 8.6 to get electronegativity (EN) values:

$$EN_O = 3.5, \ EN_H = 2.1, \ EN_N = 3.0, \ EN_C = 2.5, \ EN_{Cl} = 3.0$$

$$\Delta EN_{O-H} = EN_O - EN_H = 3.5 - 2.1 = 1.4 \quad \text{O–H is most polar.}$$

$$\Delta EN_{N-H} = EN_N - EN_H = 3.0 - 2.1 = 0.9$$

$$\Delta EN_{C-O} = EN_O - EN_C = 3.5 - 2.5 = 1.0$$

$$\Delta EN_{Cl-F} = EN_F - EN_{Cl} = 3.5 - 3.0 = 0.5$$

$$\Delta EN_{C-Cl} = EN_{Cl} - EN_C = 3.0 - 2.5 = 0.5$$

The most polar bonds are in H_2O.

(b) Use the description given in Section 9.4 to determine if the molecule is polar:

H_2O is polar, since the terminal atoms are not symmetrically arranged around the O atom.

NH_3 is polar, since the terminal atoms are not symmetrically arranged around the N atom.

CO_2 is not polar, since the two terminal atoms are all the same, they are symmetrically arranged (in a l shape) around the C atom, and they have the same partial charge.

$$C \text{—} O$$
$$\delta^+ \quad \delta^-$$

CCl_4 is not polar, since the four terminal atoms are all the same, they are symmetrically arranged (in a tetrahedral shape) around the C atom, and they all have the same partial charge.

$$C \text{—} Cl$$
$$\delta^+ \quad \delta^-$$

So, CO_2 and CCl_4 are the molecules in the list that are not polar.

(c) The more negatively charged atom of a pair of bonded atoms is the atom with the largest electronegativity. In ClF, the F atom is more negatively charged.

50. We need the Lewis structures and molecular shapes of the molecules:

(a)   (b)   (c)   (d)

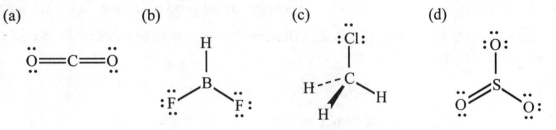

The molecules that are not polar have asymmetrical atom arrangement, such as (b) HBF_2 and (b) CH_3Cl. The others have symmetrical arrangements of identical atoms with the same partial charge. (Note: Three equivalent resonance structures following the octet rule can be written for SO_3, making each S–O bond the same length and strength.)

The bond polarities related to the difference in electronegativity. Use Figure 8.6 to get electronegativity (EN) values:

$$EN_B = 2.0, \ EN_H = 2.1, \ EN_F = 4.0, \ EN_C = 2.5, \ EN_{Cl} = 3.0$$

$$\Delta EN_{B-H} = EN_H - EN_B = 2.1 - 2.0 = 0.1$$

$$\Delta EN_{B-F} = EN_F - EN_B = 4.0 - 2.0 = 2.0$$

$$\Delta EN_{C-H} = EN_C - EN_H = 2.5 - 2.1 = 0.4$$

$$\Delta EN_{C-Cl} = EN_{Cl} - EN_C = 3.0 - 2.5 = 0.5$$

The bond pole arrows' lengths are related to their ΔEN, and points toward the atom with the more negative EN. So, the B–F arrows are much longer than the HB arrow, and a net dipole points toward the F atoms' side of the molecule, making the F atoms' side of the molecule the partial negative end and the H atom's side of the molecule the partial positive end.

The C–Cl arrow points toward Cl, and the C–H points toward the C, so all the arrows point toward the C. (Left right, and back forward cancel due to the symmetry of the triangular orientation of the H atoms). That means a net dipole points toward the Cl atom's side of the molecule, making the Cl atom's side of the molecule the partial negative end and the H atoms' side of the molecule the partial positive end.

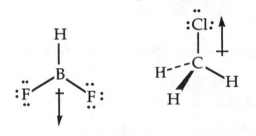

52. The dipole moment relates both to the strength of the bond poles and their directionality.

The bond polarities are related to the difference in electronegativity. Use Figure 8.6 to get electronegativity (EN) values:

$$EN_{Br} = 2.8, \ EN_F = 4.0, \ EN_{Cl} = 3.0$$

$$\Delta EN_{Br-F} = EN_F - EN_{Br} = 4.0 - 2.8 = 1.2$$

$$\Delta EN_{Cl-F} = EN_F - EN_{Cl} = 4.0 - 3.5 = 0.5$$

The bond polarity of the bond in BrF is much greater than that in ClF, which explains the significant dipole moment difference.

$$EN_H = 2.1, \ EN_O = 3.5, \ EN_S = 2.5$$

$$\Delta EN_{O-H} = EN_O - EN_H = 3.5 - 2.1 = 1.4$$

$$\Delta EN_{S-H} = EN_S - EN_H = 2.5 - 2.1 = 0.4$$

The bond polarity of the O–H bonds is greater than that of the O–S bonds, which explains the dipole moment difference.

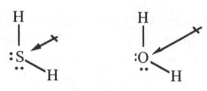

Noncovalent Interactions

53. Water and ethanol can both form hydrogen bonds, hence they have similar strong interactive hydrogen bonding forces. These two substances would easily interact, hence they would mix readily (i.e., they are miscible).

    Cyclohexane, a six-carbon-ring hydrocarbon, has no capability for forming hydrogen bonds. Because of its size, it interacts primarily with London forces. The symmetry of the molecule and the only slightly polar bonds give it very little capacity for dipole-dipole interactions. Water molecules would have to give up hydrogen bonds between other water molecules to interact with cyclohexane using only much weaker London forces. These two substances would not interact very well, hence they would not mix readily (i.e., they are not miscible).

55. Boiling points increase as the strength of the interactive forces increases. The strongest interactive force is hydrogen bonding, then dipole-dipole, then London forces. London forces increase with increasing number of electrons (found in larger atoms) and increased molecular complexity.

    (a) Group IV hydrides all interact by London forces. The larger the Group IV atom, the stronger the London forces and the higher the boiling point, because it has more electrons.

    (b) NH_3 molecules interact via hydrogen bonding so its boiling point is quite high compared to the other Group V hydrides. The other Group V hydrides interact mostly by London forces and dipole-dipole forces. Dipole forces are not very different between one and the next; however, the larger the Group V atom, and the stronger the London forces the higher the boiling point.

    (c) H_2O molecules interact via hydrogen bonding, so its boiling point is quite high compared to the other Group VI hydrides. The other Group VI hydrides interact mostly by London forces and dipole-dipole forces. Dipole forces are not very different between one and the next; however, the larger the Group VI atom, and the stronger the London forces, the higher the boiling point.

(d) HF molecules interact via hydrogen bonding, so its boiling point is quite high compared to the other Group VII hydrides. The other Group VII hydrides interact mostly by dipole-dipole and London forces. Dipole forces are not very different between one and the next; however, the larger the Group VII atom, and the stronger the London forces, the higher the boiling point.

57. Hydrogen bonds form between very electronegative atoms in one molecule to H atoms bonded to a very electronegative atom ($EN \geq 3.0$) in another molecule. If H atoms are present in a molecule but they are bonded to lower electronegativity atoms, such as C atoms, the molecule cannot use those H atoms for hydrogen bonding.

(a) No, the H atoms are bonded to C atoms and the highest electronegativity atom in the molecule is Br ($EN = 2.8$), so this molecule cannot form hydrogen bonds.

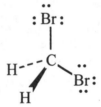

(b) No, the H atoms are bonded to C atoms, so this molecule cannot form hydrogen bonds with other molecules of the same compound. It does have a high electronegativity O atom ($EN = 3.5$), so this molecule could interact with other molecules, such as H_2O, which can provide the H atoms for hydrogen bonding to the O atom on this molecule.

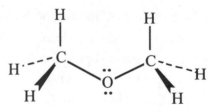

(c) Yes, three H atoms are bonded to high electronegativity atoms ($EN_O = 3.5$ and $EN_N = 3.0$). Those H atoms (circled in the structure below) can form hydrogen bonds with the N and O atoms in neighboring molecules.

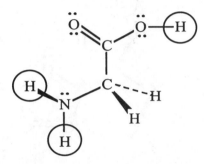

(d) Yes, two H atoms are bonded to high electronegativity O atoms (EN = 3.5). Those H atoms (circled in the structure below) can form hydrogen bonds with the O atoms in neighboring molecules.

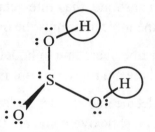

(e) Yes, one H atom is bonded to a high electronegativity O atom (EN = 3.5). That H atom (circled in the structure below) can form hydrogen bonds with the O atoms in neighboring molecules.

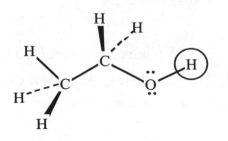

59. Vitamin C is capable of forming hydrogen bonds with water. Four H atoms are bonded to high electronegativity O atoms (EN = 3.5). Those H atoms (circled in the structure below) can form hydrogen bonds with H_2O molecules.

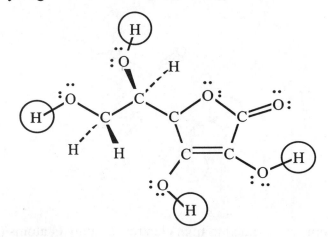

This molecule is quite polar and would not interact well with fats, because they interact primarily using weaker London forces.

61. (a) Hydrocarbons have no capability of forming hydrogen bonds. Because of the size of octane molecules used in gasoline, they interact primarily with London forces. The London forces between the molecules must be overcome to evaporate gasoline.

(b) No interactive forces must be overcome to liquefying a gas, since the molecules start in the gas phase, where there are no significant interactions between molecules.

(c) Decomposing a large molecule into smaller molecules requires breaking covalent bonds, so the forces that must be overcome are the intramolecular (covalent) forces.

(d) The double strands of DNA are shown in Figure 9.25. The interaction between the two strands are hydrogen bonds. These hydrogen bonding forces must be overcome to "unzip" the DNA double helix.

Chirality in Organic Compounds

64. Chiral centers are C atoms with four bonds, where each fragment bonded is different.

(a) The second C atom has four different fragments: CH_3, HO, CH_2OH, and H. The first and third C atoms have three and two H atoms bonded, respectively, so they are not chiral centers.

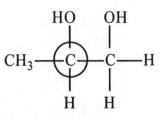

(b) This molecule has no chiral centers. The first two C atoms do not have four bonds. The third C atom has two H atoms bonded.

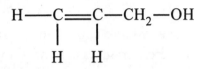

(c) The third C atom has four different fragments: CH_3CCl_2, F, Cl, and H. The first C atom has two H atoms bonded. The second C atom has two Cl atoms bonded, so they are not chiral centers.

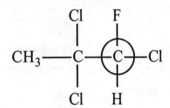

66. Chiral centers are C atoms with four bonds, where each fragment bonded is different.

(a) The second C atom has four different fragments: CH_3, Br, CH_3CH, and Cl. The other C atoms have more than one H atoms bonded, so they are not chiral centers.

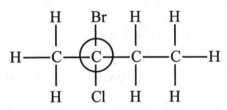

(b) This molecule has no chiral centers. The end C atoms have three H atoms bonded, and the third C atom has three CH_3 fragments.

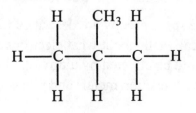

(c) The fourth C atom has four different fragments: $CH_3CH=CH$, H, CH_3, and Br. The end C atoms have three H atoms bonded, and the second and third C atoms have only three bonds, so they are not chiral centers.

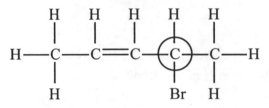

68. A molecule is chiral if it has one or more chiral centers, so look at the structural formula and find an atom with four different fragments bonded to it. For example:

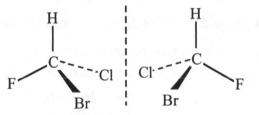

The C in this formula (circled) has four different fragments: F, H, Cl, and Br. This chiral center makes the molecule chiral and allows for the creation of two nonsuperimposeable mirror images of the same molecule:

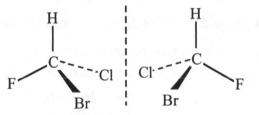

Molecular Structure Determination by Spectroscopy

70. The "Tools of Chemistry" box in Section 9.2 describes infrared spectroscopy. The "Tools of Chemistry" box in Section 9.4 describes ultraviolet-visible spectroscopy.

(a) Ultraviolet and visible spectroscopy are described together and some of the given examples that specifically talk about what electrons are involved refer to visible light. The electrons that would be affected by ultraviolet light (or high-energy visible light) are d electrons in some transition metals and pi electrons in the double bonds of organic compounds.

(b) Infrared spectroscopy tells us about the frequencies of specific vibrational motions in molecules.

Biomolecules

72.   G-C interacts with three hydrogen bonds; whereas, A-T interacts with only two hydrogen bonds. Stronger attractive forces require higher melting temperature.

General Questions

74. NO_2Cl has three resonance structures.   The first two are equivalent and the third is not as good because the formal charges are higher.

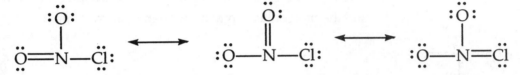

The central N atom has three atoms bonded to it and no lone pairs, so the type is AX_3E_0. The electron-pair geometry and the molecular geometry are both triangular planar. The triangular planar electron geometry indicates approximate O–N–O and O–N–Cl angles of 120°.

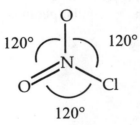

78. (a) BF_3 (24 e⁻)   The type is AX_3E_0, so the electron-pair geometry and the molecular geometry are both triangular planar.

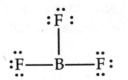

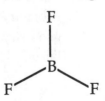

(b) CF$_4$ (32 e$^-$)   The type is AX$_4$E$_0$, so both the electron-pair geometry and the molecular geometry are tetrahedral.

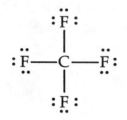

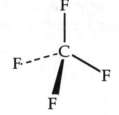

(c) NF$_3$ (26 e$^-$)   The type is AX$_3$E$_1$, so the electron-pair geometry is tetrahedral and the molecular geometry is triangular pyramidal.

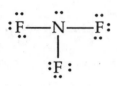

(d) OF$_2$ (20 e$^-$)   The type is AX$_2$E$_2$, so the electron-pair geometry is tetrahedral and the molecular geometry is angular (109.5°).

(e) HF (8 e$^-$)   There are two atoms in the molecule so the electron-pair geometry is not defined by VSEPR. The electrons have the geometry of the atomic orbitals. The molecular geometry is linear.

H—F:

The similarities are that all of the molecules have F atoms in terminal positions (except HF),. The different geometries result in the differences between the central atoms and the number of F atoms.

80. (a) CH_4 (8 e$^-$)   The type is AX_4E_0, so the electron-pair geometry is tetrahedral and the H–C–H angles are predicted to be 109.5°.

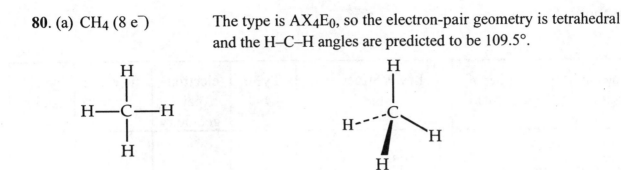

(b) CH_3^- (26 e$^-$) The type is AX_3E_1, so the electron-pair geometry is tetrahedral and the molecular geometry is triangular pyramidal and the H–C–H angles are predicted to be slightly less than 109.5° because of the lone pair effect.

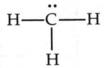   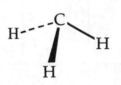

(c) CO_3^{2-} (24 e$^-$) AX_3E_0, so both the electron-pair geometry and the molecular geometry are triangular planar and the O–C–O angles are predicted to be 120°.

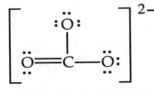   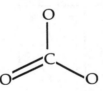

(d) CO_2 (16 e$^-$) AX_2E_2, so both the electron-pair geometry and the molecular geometry are linear and the O–C–O angles are predicted to be 180°.

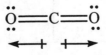

84. A molecule with polar bonds can have a dipole moment of zero if the bond poles are equal and point in opposite directions. CO_2 is a good example of a nonpolar molecule with polar bonds.

Applying Concepts

85.

Molecule or Ion	Number of electrons	Lewis Structure	Type AX_nE_m	electron-pair geometry	molecular geometry	hybrid-ization
ICl_2^+	$7 + 2(7) - 1$ $= 20$		AX_2E_2	tetrahedral	angular	sp^3
I_3^-	$3(7) + 1$ $= 22$		AX_2E_3	triangular bipyramidal	linear	sp^3d
ICl_3	$7 + 3(7)$ $= 28$		AX_3E_2	triangular bipyramidal	T-shaped	sp^3d
ICl_4^-	$7 + 4(7) + 1$ $= 36$		AX_4E_2	octahedral	square planar	sp^3d^2
IO_4^-	$7 + 4(6) + 1$ $= 32$		AX_4E_0	tetrahedral	tetrahedral	sp^3
IF_4^+	$7 + 4(7) - 1$ $= 34$		AX_4E_1	triangular bipyramidal	seesaw	sp^3d

85. continued

Molecule or Ion	Number of electrons	Lewis Structure	Type AX_nE_m	electron-pair geometry	molecular geometry	hybrid-ization
IF_5	$7 + 5(7)$ $= 42$	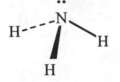	AX_4E_1	octahedral	square pyramid	sp^3d^2
IF_6^+	$7 + 6(7) - 1$ $= 48$	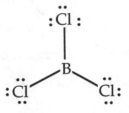	AX_6E_0	octahedral	octahedral	sp^3d^2

87. (a) XH_3 would have a central atom with one lone pair of electrons, if X = N.

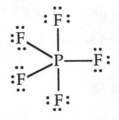

(b) XCl_3 would have no lone pairs on the central atom, if X = B.

(c) XF_5 would have no lone pairs on the central atom, if X = P.

(d) XCl_3 would have two lone pairs on the central atom, if X = I.

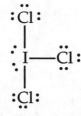

89. Five water molecules could hydrogen bond to an acetic acid molecule: one to each of the four lone pairs on O atoms in acetic acid and one to the H atom bonded to an O atom in acetic acid.

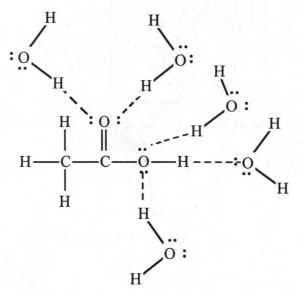

91. (a) This is incorrect, because the H atoms are not bonded to highly electronegative atoms.

(b) This is correct. The H atom is bonded to an O atom (EN = 3.5) and is hydrogen bonding to another O atom.

(c) This is incorrect. The H atom is bonded F. This is a covalent bond, not an example of the noncovalent interactive force called a hydrogen bond.

(d) This is incorrect. The H atoms are bonded to O atom (EN = 3.5), but they will not form a hydrogen bond to each other!

The only answer that is correct is (b).

Chapter 10: Gases and the Atmosphere

Properties of Gases

10. *Define the problem*: Convert a series of pressure quantities into other pressure units.

Develop a plan: Use Table 10.2 to design conversion factors to achieve the conversions.

Execute the plan:

(a)
$$720 \text{ mmHg} \times \frac{1 \text{ atm}}{760 \text{ mmHg}} = 0.95 \text{ atm}$$

(b)
$$1.25 \text{ atm} \times \frac{760 \text{ mmHg}}{1 \text{ atm}} = 950. \text{ mmHg}$$

(c)
$$542 \text{ mmHg} \times \frac{760 \text{ torr}}{760 \text{ mmHg}} = 542 \text{ torr}$$

(d)
$$740 \text{ mmHg} \times \frac{1 \text{ atm}}{760 \text{ mmHg}} \times \frac{101.325 \text{ kPa}}{1 \text{ atm}} = 99 \text{ kPa}$$

(e)
$$700 \text{ kPa} \times \frac{1 \text{ atm}}{101.325 \text{ kPa}} = 7 \text{ atm}$$

Check your answers: The unit of atm represents a lot more pressure than the units of kPa, torr, and mmHg, so it makes sense that the numbers of atm are always much smaller than the pressure expressed in these other units. The unit of kPa represents more pressure than the units of torr and mmHg, so it makes sense that the numbers of kPa are always smaller than the pressure expressed in the other units of torr and mmHg. Torr and mmHg are the same size, so their quantities should be identical. These answers make sense.

12. *Define the problem*: Given the density of mercury, the density of an oil used to construct a barometer, and the atmospheric pressure, determine the height in meters of the oil column in the oil barometer.

Develop a plan: Use Table 10.2 to convert the pressure into mmHg. Pressure on the liquid pushes the liquid up the barometer until its mass exerts the same force per unit area as the air pressure: $P_{liquid} = \text{Force/Area} = mg/\text{Area}$. Since g is a constant, the force per unit area of the liquid counteracting the air pressure is proportional to just the mass per unit area. Relate the mass per unit area of mercury to the mass per unit area of the oil for 1 atm pressure. Relate the mass of each liquid to its respective density and volume. Relate the volume of each liquid to the dimensions of its respective barometers, including the height. Relate the height of mercury in the mercury barometer to the height of the oil in the oil barometer.

Execute the plan: According to Table 10.2, at 1.0 atm the height of a column of mercury in a mercury barometer is 760 mm. Let m = mass of the liquid, A = cylindrical area of the barometer's column, d = density of the liquid, V = volume of the liquid, and h = the height of the liquid in the barometer. Now, equate the mass per unit area of each barometer and derives an equation relating their heights:

$$m_{Hg}/A_{Hg} = m_{oil}/A_{oil}$$

$$d_{Hg}V_{Hg}/A_{Hg} = d_{oil}V_{oil}/A_{oil}$$

$$V = (Area) \times (height) = Ah$$

$$d_{Hg}A_{Hg}h_{Hg}/A_{Hg} = d_{oil}A_{oil}h_{oil}/A_{oil}$$

$$d_{Hg}h_{Hg} = d_{oil}h_{oil}$$

$$h_{oil} = \frac{d_{Hg}h_{Hg}}{d_{oil}} = \frac{\left(13.596 \text{ g/cm}^3\right)\left(760 \text{ mm}\right)}{\left(0.75 \text{ g/cm}^3\right)} \times \frac{1 \text{ m}}{1000 \text{ mm}} = 14 \text{ m}$$

Check your answer: A larger mass of oil will be needed to counterbalance the atmospheric pressure because it is less dense than mercury. A higher column of oil makes sense.

14. With a perfect vacuum at the top of the well, this system would resemble a water barometer. Using the equation derived in the answer to Question 12:

$$h_{water} = \frac{d_{Hg}h_{Hg}}{d_{water}} = \frac{\left(13.596 \text{ g/cm}^3\right)\left(760 \text{ mm}\right)}{\left(1.00 \text{ g/cm}^3\right)} \times \frac{1 \text{ m}}{1000 \text{ mm}} \times \frac{3.281 \text{ feet}}{1 \text{ m}} = 34 \text{ feet}$$

That means atmospheric pressure can only push water up to about 34 feet. So, the well cannot be deeper than that, not even using a high quality vacuum pump.

The Atmosphere

16. Nitrogen serves to moderate the reactiveness of oxygen by diluting it. Oxygen sustains animal life as a reactant in the conversion of food to energy. Oxygen is produced by plants in the process of photosynthesis.

18. *Define the problem*: Convert all the numbers in Table 10.3 to parts per million (ppm) and parts per billion (ppb)

Develop a plan: To accomplish these conversions, we need to describe a relationship between percent and ppm and ppb:

$$\text{Percent gas in air} = \frac{\text{L of gas}}{100 \text{ L of air}} \qquad \text{Parts per million gas in air} = \frac{\text{L of gas}}{1,000,000 \text{ L of air}}$$

So, 1% = 10,000 ppm. Multiply the number in units of percent by 10,000 to get ppm.

$$\text{Parts per billion gas in air} = \frac{\text{L of gas}}{1,000,000,000 \text{ L of air}}$$

So, 1ppm = 1,000 ppb. Multiply the number in units of ppm by 1,000 to get ppm.

Execute the plan:

Molecule	ppm	ppb	
N_2	780,840	780,840,000	↑
O_2	209,480	209,480,000	
Ar	9,340	9,340,000	
CO_2	330	330,000	
Ne	18.2	18,200	> 1 ppm
H_2	10.	10,000	
He	5.2	5,200	
CH_4	2	2,000	↓ ↑
Kr	1	1,000	
CO	0.1	100	between
Xe	0.08	80	1ppm
O_3	0.02	20	and
NH_3	0.01	10	1ppb
NO_2	0.001	1	↓
SO_2	0.0002	0.2	< 1ppb

Check your answer: The numbers are different only by the appropriate factor of 10. These answers make sense.

20. *Define the problem*: Given the mass of a sample of coal, the percentage of sulfur in the coal, the weight fraction of SO_2 in the atmosphere, and (from Question 19) the mass of the atmosphere, determine the mass of SO_2 added to the atmosphere, and the total amount of SO_2 in the atmosphere.

Develop a plan: To answer the first question, use metric conversions and the mass fraction as a conversion factor to determine the mass of sulfur. Use mole and molar mass conversion factors to determine the mass of SO_2, assuming that all the sulfur in the coal is converted to SO_2 and released into the atmosphere. Convert the mass back to metric tons. To answer the second question, use the mass of the atmosphere and the mass fraction of SO_2 in terms of metric tons as a conversion factor to determine the total mass of SO_2.

Execute the plan: First find the mass of S:

$$3.1 \times 10^9 \text{ metric tons coal} \times \frac{1000 \text{ kg coal}}{1 \text{ metric ton coal}} \times \frac{2.5 \text{ kg S}}{100 \text{ kg coal}} \times \frac{1000 \text{ g S}}{1 \text{ kg S}} = 7.8 \times 10^{13} \text{ g S}$$

Then, find the mass of SO_2:

$$7.8 \times 10^{13} \text{ g S} \times \frac{1 \text{ mol S}}{32.066 \text{ g S}} \times \frac{1 \text{ mol } SO_2}{1 \text{ mol S}} \times \frac{64.065 \text{ g } SO_2}{1 \text{ mol } SO_2} = 1.5 \times 10^{14} \text{ g } SO_2$$

Then, convert the mass of SO_2 back to metric tons:

$$1.5 \times 10^{14} \text{ g } SO_2 \times \frac{1 \text{ kg } SO_2}{1000 \text{ g } SO_2} \times \frac{1 \text{ metric ton } SO_2}{1000 \text{ kg } SO_2}$$

$$= 1.5 \times 10^8 \text{ metric tons } SO_2 \text{ was added to the atmosphere in 1980}$$

Get the total mass of SO_2 currently in the atmosphere.

$$5.3 \times 10^{15} \text{ metric tons air} \times \frac{0.4 \text{ metric tons } SO_2}{1,000,000,000 \text{ metric tons air}} = 2 \times 10^6 \text{ metric tons } SO_2$$

Check your answers: It is clear that some of the SO_2 presumably released into the atmosphere in 1980 is no longer there, since the total mass of SO_2 is less than what was introduced that year. SO_2 is a reactive gas, getting oxidized to SO_3 in the presence of air and then producing sulfuric acid when reacting with rainwater. This removes the sulfur from the air. See Section 10.12 for more details on acid rain and SO_2 as a primary pollutant.

Kinetic-Molecular Theory

22. *Define the problem*: Given equal volumes of two gases in separate flasks, their molecular identity, their temperatures, and their pressures, compare (a) their average kinetic energy per molecule, (b) their average molecular velocity, and (c) the number of molecules.

Develop a plan: (a) Kinetic energy is proportional to temperature, so compare their temperatures to relate their kinetic energy. (b) Kinetic energy is related to mass and velocity. Since we determine the relative kinetic energy in (a), use that and their molar masses to determine their relative velocity. (c) Use the ideal gas law and the relative P, T, and V to determine which has a larger number (in moles) of molecules.

Execute the plan:

(a) The two samples have the same temperature, so the average kinetic energy per molecule in each sample is the same.

(b) $$E_{kin} = \frac{1}{2}mv^2 \qquad \text{So, } v^2 = \frac{2E}{m}$$

Because the average kinetic energy of the two samples is the same, only the mass of the molecules affects the velocity. The molecules with the smaller mass have the faster velocity. Here, H_2 (molar mass = 2.0 g/mol) is lighter than CO_2 (molar mass = 44.0 g/mol), so the molecules in the H_2 sample are moving with a higher velocity than those in the CO_2 sample.

(c) Because both the volume and temperature are the same, the only thing affecting the number of moles is the pressure. The ideal gas law ($PV = nRT$) tells is that the pressure is directly proportional to the number of moles, if the temperature and volume are fixed. Here, the CO_2 sample's pressure (2 atm) is twice that of H_2 (1 atm), so the CO_2 sample has twice as many molecules as the H_2 sample.

Check your answer: It makes sense that lightweight things go faster. The postulates of kinetic-molecular theory help us see why these comparisons are sensible.

24. *Define the problem*: Given the formulas of molecules, put their gases in order of increasing average molecular speed.

Develop a plan: Kinetic energy is proportional to temperature. Assuming all of the samples are at the same temperature, their average kinetic energies are the same. Kinetic energy is related to mass and velocity. $E_{kin} = \frac{1}{2}mv^2$. Velocity is related to kinetic energy and mass: $v^2 = \frac{2E}{m}$ Therefore, molecules with smaller mass have the faster molecular speed. To rank the molecules with increasing speed, rank them from the largest molar mass to the smallest.

Execute the plan: Find the molar masses:

SO_2 molar mass = 64 g/mol, Cl_2 molar mass = 71 g/mol,

$SOCl_2$ molar mass = 119 g/mol, Cl_2O molar mass = 87 g/mol.

slowest speed: $SOCl_2 < Cl_2O < Cl_2 < SO_2$:fastest speed

Check your answer: It makes sense that lightweight things go faster.

Gas Behavior and the Ideal Gas Law

26. *Define the problem*: Given the volume of a sample of air at STP and the volume fraction of CO in ppm, determine the moles of CO.

Develop a plan: Use the volume fraction in terms of liters as a conversion factor to determine the liters of CO. Use the molar volume of a gas at STP as a conversion factor to determine the moles of CO.

Execute the plan:

$$1.0 \text{ L air} \times \frac{950 \text{ L CO}}{1,000,000 \text{ L air}} \times \frac{1 \text{ mol CO}}{22.414 \text{ L CO}} = 4.2 \times 10^{-5} \text{ mol CO}$$

Check your answer: Air has a very small proportion of CO. This small sample of air has a very small amount of CO. This number makes sense.

28. *Define the problem*: Given the volume and pressure of a sample of gas in one flask and the volume of a flask it is transferred to at the same temperature, determine the new pressure.

Develop a plan: Use Boyle's law to relate volume to pressure.

$$P_1 V_1 = P_2 V_2 \qquad \text{(unchanging T and n)}$$

Execute the plan:

$$P_2 = \frac{P_1 V_1}{V_2} = \frac{(75.0 \text{ mmHg}) \times (256 \text{ mL})}{(125 \text{ mL})} = 154 \text{ mmHg}$$

Check your answer: The smaller volume should have a larger pressure. This answer makes sense.

30. Use Boyle's law to relate volume to pressure, as described in the answer to Question 27.

$$P_2 = \frac{P_1 V_1}{V_2} = \frac{(735 \text{ mmHg}) \times (3.50 \text{ L})}{(15.0 \text{ L})} = 172 \text{ mmHg}$$

32. *Define the problem*: Given the original volume and temperature of a sample of gas in a syringe (presumably at atmospheric pressure) and the new temperature of the sample, determine the new volume (presumably still at atmospheric pressure).

Develop a plan: Convert the temperatures to Kelvin. Use Charles' law to relate volume to absolute temperature.

$$\frac{V_1}{T_1} = \frac{V_2}{T_2} \qquad \text{(P and n constant)}$$

Execute the plan: *(Assume the T_1 reading has the same precision as the reading for T_2.)*

$$T_1 = 20. \,°C + 273 = 293 \text{ K}$$

$$T_2 = 37 \,°C + 273 = 310. \text{ K}$$

$$V_2 = V_1 \times \frac{T_2}{T_1} = (25.0 \text{ mL}) \times \frac{(310. \text{ K})}{(293 \text{ K})} = 26.5 \text{ mL}$$

Check your answer: Gas at higher temperature should have a larger volume. This answer makes sense.

34. Use Charles' law to relate volume to absolute temperature, as described in the answer to Question 32.

(Assume the T_1 reading is $\pm 1 \,°C$.) $T_1 = 80. \,°C + 273 = 353 \text{ K}$

$$T_2 = T_1 \times \frac{V_2}{V_1} = (353 \text{ K}) \times \frac{(1.25 \text{ L})}{(2.50 \text{ L})} = 176 \text{ K}$$

$$176 \text{ K} - 273 = -97 \,°C$$

36. *Define the problem*: Given the original pressure and temperature of gas in a tire, the assumption that volume is unchanged, and the new temperature, determine the new pressure exerted by the gas in the tire.

Develop a plan: Convert the temperature to Kelvin. Use the combined gas law to relate pressure to absolute temperature.

$$\frac{P_1 V_1}{T_1} = \frac{P_2 V_2}{T_2} \qquad \text{(n constant)}$$

At constant volume $\qquad \dfrac{P_1}{T_1} = \dfrac{P_2}{T_2} \qquad$ (V and n constant)

Execute the plan: $\qquad T_1 = 15\ ^\circ\text{C} + 273 = 288\ \text{K}$

$$T_2 = 35\ ^\circ\text{C} + 273 = 308\ \text{K}$$

$$P_2 = P_1 \times \frac{T_2}{T_1} = (3.74\ \text{atm}) \times \frac{(308\ \text{K})}{(288\ \text{K})} = 4.00\ \text{atm}$$

Check your answer: A gas with a higher temperature should exert a higher pressure. This answer makes sense.

38. *Define the problem*: Given the original volume, pressure, and temperature of a gas sample, the new temperature, and the new pressure, determine the new volume of the sample.

Develop a plan: Convert the temperature to Kelvin. Use the combined gas law to relate pressure to absolute temperature.

$$\frac{P_1 V_1}{T_1} = \frac{P_2 V_2}{T_2} \qquad \text{(n constant)}$$

Execute the plan: $\qquad T_1 = 22\ ^\circ\text{C} + 273 = 295\ \text{K}$

$$T_2 = 42\ ^\circ\text{C} + 273 = 313\ \text{K}$$

$$V_2 = V_1 \times \frac{T_2}{T_1} \times \frac{P_1}{P_2} = (754\ \text{mL}) \times \frac{(313\ \text{K})}{(295\ \text{K})} \times \frac{(165\ \text{mmHg})}{(265\ \text{mmHg})} = 501\ \text{mL}$$

Check your answer: The temperature fraction: $\dfrac{(313\ \text{K})}{(295\ \text{K})}$ is larger than one, consistent with increasing the volume due to the increased temperature. The pressure fraction: $\dfrac{(165\ \text{mmHg})}{(265\ \text{mmHg})}$ is smaller than one, consistent with decreasing the volume due to an increased pressure. Clearly these two effects counteract each other, but this pressure change affects the volume more than the temperature change does, so the answer makes sense.

40. *Define the problem*: Given the mass, identity, temperature, and volume of a gas sample, determine the pressure of the sample.

Develop a plan: Use the ideal gas law:

$$PV = nRT \qquad\qquad R = 0.08206 \; \frac{L \cdot atm}{mol \cdot K}$$

The units of R remind us to determine the moles of gas (using mass and molar mass), to convert the temperature to Kelvin, and to convert the volume to liters.

Execute the plan: *(Assume the T reading is ± 1 °C.)*

$$1.55 \text{ g Xe} \times \frac{1 \text{ mol Xe}}{131.29 \text{ g Xe}} = 0.0118 \text{ mol Xe}$$

$$T = 20. \,°C + 273 = 293 \text{ K}$$

$$V = 560 \text{ mL} \times \frac{1 \text{ L}}{1000 \text{ mL}} = 0.56 \text{ L}$$

$$P = \frac{nRT}{V} = \frac{(0.0118 \text{ mol}) \times \left(0.08206 \dfrac{L \cdot atm}{mol \cdot K} \right) \times (293 \text{ K})}{(0.56 \text{ L})} = 0.51 \text{ atm}$$

Check your answer: The small fraction of a mole makes sense with the small mass. The units in the pressure calculation cancel properly to give atm. This answer make sense.

42. *Define the problem*: Given a set of gas samples, determine which has the largest number of molecules and which has the smallest number of molecules.

Develop a plan: Some of the samples are at STP, so use the molar volume of a gas at STP to determine the number of molecules (in units of moles). In the other cases, use the ideal gas law as described in the answer to Question 40. Once all the moles are calculated, identify which sample has the most moles and which sample has the least moles.

Execute the plan: (a) and (b) are at STP. 1 mol of **any** gas occupies 22.414 L.

$$1.0 \text{ L} \times \frac{1 \text{ mol gas}}{22.414 \text{ L}} = 0.045 \text{ mol gas}$$

(a) 0.045 mol H_2

(b) 0.045 mol O_2

(c) $T = 27\ ^{\circ}C + 273 = 300.\ K$

$$P = 760\ \text{mmHg} \times \frac{1\ \text{atm}}{760\ \text{mmHg}} = 1.0\ \text{atm}$$

$$n = \frac{PV}{RT} = \frac{(1.0\ \text{atm}) \times (1.0\ \text{L})}{\left(0.08206 \dfrac{\text{L} \cdot \text{atm}}{\text{mol} \cdot \text{K}}\right) \times (300.\ \text{K})} = 0.041\ \text{mol}$$

(d) *(Assume P has as many sig. figs as in (c))*

$$T = 0\ ^{\circ}C + 273 = 273\ K$$

$$P = 800\ \text{mmHg} \times \frac{1\ \text{atm}}{760\ \text{mmHg}} = 1.1\ \text{atm}$$

$$n = \frac{PV}{RT} = \frac{(1.0\ \text{atm}) \times (1.0\ \text{L})}{\left(0.08206 \dfrac{\text{L} \cdot \text{atm}}{\text{mol} \cdot \text{K}}\right) \times (273\ \text{K})} = 0.047\ \text{mol}$$

Of these samples, (d) has the most molecules (0.047 moles CO_2) and (c) has the least molecules (0.041 moles H_2).

Check your answers: To keep a 1.0-L gas sample at standard temperature and still have larger than standard pressure suggests that there must be more molecules hitting the walls than a sample at STP. To have a 1.0-L gas sample at higher than standard temperature and still stay at standard pressure suggests that there must be fewer molecules hitting the walls harder, than a sample at STP. These answers make sense.

Quantities of Gases in Chemical Reactions

44. *Define the problem*: Given the mass of sucrose, the formula, and the product of a reaction, determine the maximum volume of CO_2 produced at STP. Compare that volume with the typical volume of two loaves of French bread.

Develop a plan: Balance the equation. Determine the moles of sucrose from the molar mass, then use the stoichiometric relationships given in the balanced equation to determine the moles of CO_2 produced. Last, use the molar volume of a gas at STP to determine the volume of CO_2. Estimate the total volume of two loaves of French bread assuming they are cylinders. Compare the two volumes.

Execute the plan: Balance the equation:

$$C_{12}H_{22}O_{11}(s) + 12\ O_2(g) \longrightarrow 12\ CO_2\ (g) + 11\ H_2O\ (\ell)$$

$$2.4\ g\ C_{12}H_{22}O_{11} \times \frac{1\ mol\ C_{12}H_{22}O_{11}}{342\ g\ C_{12}H_{22}O_{11}} \times \frac{12\ mol\ CO_2}{1\ mol\ C_{12}H_{22}O_{11}} \times \frac{22.414\ L\ CO_2\ at\ STP}{1\ mol\ CO_2}$$

$$= 1.9\ L\ CO_2$$

Assume one loaf of French bread is a cylinder, 3.0 inches in diameter and 18 inches long.

$$r = 1.5\ in$$

$$A = \pi r^2 = \pi(1.5\ in)^2 = 7.1\ in^2$$

$$V = A\ell = (7.1\ in^2) \times (18\ in) = 130\ in^3$$

$$130\ in^3 \times \left(\frac{2.54\ cm}{1\ in}\right)^3 \times \frac{1\ mL}{1\ cm^3} \times \frac{1\ L}{1000\ mL} = 2.1\ L$$

Two loaves would have twice this volume: $2 \times (2.1\ L) = 4.2\ L$. The CO_2 bubbles produced in the bread are nearly half of its volume.

Check your answers: Slicing open French bread we see that it has a vast "honeycomb" of bubble-shaped spaces in it. It makes sense that approximately half the loaf's volume can be associated with the CO_2 bubbles formed by the yeast when the bread was rising. These answers make sense.

46. *Define the problem*: Given the balanced chemical equation for a chemical reaction and the volume of one reactant at a specified pressure and temperature, determine the volume of the other reactant at a specified pressure and temperature that will cause complete reaction, and the volume of one of the products at a specified pressure and temperature that will be produced.

Develop a plan: Avogadro's Law allows us to interpret a balanced equation with gas reactants and products in terms of gas volumes, as long as their temperatures and pressures are the same. Use the stoichiometry to relate liters of $SiH_4(g)$ that react with liters of $O_2(g)$ and liters of $H_2O(g)$.

Execute the plan: The balanced equation tells us that one volume of $SiH_4(g)$ reacts with two volumes of $O_2(g)$ to make two volumes of $H_2O(g)$, since all the volumes are measured at the same temperature and pressure.

$$5.2 \text{ L } H_2(g) \times \frac{2 \text{ L } O_2(g)}{1 \text{ L } SiH_4(g)} = 10.4 \text{ L } O_2(g)$$

$$5.2 \text{ L } H_2(g) \times \frac{2 \text{ L } H_2O(g)}{1 \text{ L } SiH_4(g)} = 10.4 \text{ L } H_2O(g)$$

Note: Multiplying a number by an exact whole number, n, is like adding that number to itself n times. $2 \times (5.2) = 5.2 L + 5.2 L = 10.4 L$, that is why we use the addition rule for assessing the significant figures of the results here.

Check your answers: Twice as many O_2 molecules are needed compared to the number of SiH_4 molecules, forming twice as many H_2O molecules. So, it makes sense that both the volume of O_2 and the volume of H_2O are twice the volume of SiH_4.

48. *Define the problem*: Given the balanced chemical equation for a chemical reaction, the mass of one reactant, and excess other reactant, determine the pressure of the product produced at a specified volume and temperature.

Develop a plan: Convert from grams to moles. Use the stoichiometric relationship from the balanced equation to determine moles of product. Use the ideal gas law to determine the pressure of the product.

Execute the plan:

$$0.050 \text{ g } B_4H_{10} \times \frac{1 \text{ mol } B_4H_{10}}{53.32 \text{ g } B_4H_{10}} \times \frac{10 \text{ mol } H_2O}{2 \text{ mol } B_4H_{10}} = 0.0047 \text{ mol } H_2O$$

$$T = 30. \text{ °C} + 273 = 303 \text{ K}$$

$$P = \frac{n_{H_2O}RT}{V} = \frac{(0.0047 \text{ mol } H_2O) \times \left(0.08206 \frac{\text{L} \cdot \text{atm}}{\text{mol} \cdot \text{K}}\right) \times (303 \text{ K})}{(4.25 \text{ L})} = 0.027 \text{ atm}$$

$$0.027 \text{ atm} \times \frac{760 \text{ mmHg}}{1 \text{ atm}} = 21 \text{ mmHg}$$

Check your answer: The relative quantities of B_4H_{10} and H_2O seem sensible. All units cancel properly in the calculation of atmosphere. This is a relatively low pressure for water but the sample is also small.

50. *Define the problem*: Given the balanced chemical equation for a chemical reaction, the mass of the solid ionic reactant, the formula of the ionic compound and the periodic group that its metal is found in, the pressure of a gas-phase product with a specified volume and temperature, determine the molar mass of the reactant.

Develop a plan: The formula of the ionic compound, $M_x(CO_3)_y$, can be used to determine the values of x and y, since M is from Group 2A and the anion carbonate has a known charge. The ideal gas law can be used to find the moles of CO_2 produced. The stoichiometry is then used to find the moles more solid that reacted. Divide the mass by the moles for the molar mass.

Execute the plan: Group 2A metals have 2+ charges and carbonate ion has a 2– charge:

$$x\ M^{2+} + y\ CO_3^{2-} \longrightarrow M_x(CO_3)_y$$

From this equation, we can see that $x = y = 1$, and the heating of the metal oxide reaction can be simplified to the following:

$$MCO_3(s) \longrightarrow MO(s) + CO_2(g)$$

$$T = 25\ °C + 273 = 298\ K$$

$$69.8\ mmHg \times \frac{1\ atm}{760\ mmHg} = 0.0918\ atm$$

$$285\ mL \times \frac{1\ L}{1000\ mL} = 0.285\ L$$

$$n_{CO_2} = \frac{PV}{RT} = \frac{(0.0918\ atm) \times (0.285\ L)}{\left(0.08206\dfrac{L \cdot atm}{mol \cdot K}\right) \times (298\ K)} = 1.07 \times 10^{-3}\ mol\ CO_2$$

$$1.07 \times 10^{-3}\ mol\ CO_2 \times \frac{1\ mol\ MCO_3}{1\ mol\ CO_2} = 1.07 \times 10^{-3}\ mol\ MCO_3$$

$$Molar\ Mass = \frac{0.158\ g\ MCO_3}{1.07 \times 10^{-3}\ mol\ MCO_3} = 148\ \frac{g}{mol}$$

Check your answer: If this metal M is really a Group 2A element, it's molar mass should be similar to one of them. We can subtract the molar mass of one C atom and three O atoms from the molar mass of the solid and get the molar mass of M:

$$148\ g/mol - 12\ g/mol - 3 \times (16\ g/mol) = 88\ g/mol\ M$$

This is close to that of Sr (molar mass = 87.62 g/mol), a group 2A metal. This answer makes sense.

52. (a) This equation is given in Problem-Solving Exercise 10.11:

$$2 C_8H_{18}(\ell) + 25 O_2(g) \longrightarrow 16 CO_2(g) + 18 H_2O(g)$$

(b) *Define the problem*: Given the length in miles of a trip, the fuel efficiency of a car, the density of the liquid fuel, and the temperature and pressure, determine the volume of a gas–phase product produced during the trip.

Develop a plan: Use the miles, the fuel efficiency, volume conversions, the density, and the stoichiometry to find the moles of the product. Use the ideal gas law to determine the volume of the product.

Execute the plan:

$$10.\ \text{miles} \times \frac{1\ \text{gal gasoline}}{32\ \text{miles}} \times \frac{3.785\ \text{L gasoline}}{1\ \text{gal gasoline}} \times \frac{1000\ \text{mL}}{1\ \text{L}} \times \frac{1\ \text{cm}^3}{1\ \text{mL}}$$

$$= 1.2 \times 10^3\ \text{cm}^3\ \text{gasoline}$$

$$1.2 \times 10^3\ \text{cm}^3\ \text{gasoline} \times \frac{0.692\ \text{g gasoline}}{1\ \text{cm}^3\ \text{gasoline}} \times \frac{1\ \text{mol gasoline}}{114.22\ \text{g gasoline}} \times \frac{16\ \text{mol } CO_2}{2\ \text{mol gasoline}}$$

$$= 57\ \text{mol } CO_2$$

$$T = 25\ °C + 273 = 298\ K$$

$$V = \frac{n_{CO_2}RT}{P} = \frac{(57\ \text{mol } CO_2) \times \left(0.08206 \dfrac{L \cdot atm}{mol \cdot K}\right) \times (298\ K)}{(1.0\ atm)} = 1.4 \times 10^3\ L$$

Check your answer: This is a large volume of CO_2! However, it is not an unreasonable quantity considering how many gallons of gasoline are used and how much CO_2 is generated from each octane molecule. Comparing the results to the answer in Question 53, less CO_2 is generated by methanol as a fuel than octane.

Gas Density and Molar Mass

54. *Define the problem*: Given the identity of a gaseous compound, the mass of the compound, the volume, and the pressure, determine the temperature of the gas.

Develop a plan: Use the formula and the molar mass to get moles, then use the ideal gas law to get volume.

Execute the plan:
$$4.25 \text{ g SiH}_4 \times \frac{1 \text{ mol SiH}_4}{32.12 \text{ g SiH}_4} = 0.132 \text{ mol SiH}_4$$

$$580 \text{ mL} \times \frac{1 \text{ L}}{1000 \text{ mL}} = 0.58 \text{ L}$$

$$T = \frac{PV}{nR} = \frac{(1.2 \text{ atm}) \times (0.58 \text{ L})}{(0.132 \text{ mol SiH}_4) \times \left(0.08206 \dfrac{\text{L} \cdot \text{atm}}{\text{mol} \cdot \text{K}}\right)} = 64 \text{ K}$$

$$64 \text{ K} - 273 = -209 \text{ °C}$$

Check your answer: The units properly canceling, and the reasonable size of the numbers makes us think that this answer looks perfectly good; however, we have made one assumption that turns out to be false. We assumed that SiH_4 remains completely in the gas phase at this temperature. According to data given in the last chapter (Table 9.6 in Section 9.5), SiH_4 condenses (the reverse of boiling) into the liquid state at -112 °C (also the boiling point). We learned in Chapter 6 that as soon as the sample reaches the normal condensation/boiling point of -112 °C, its temperature will stop changing while the phase change occurs. Since our gas sample still has a gas pressure of 1.2 atm, that means the liquid and gas are both together. Therefore, the final temperature of the mixture of gas and liquid forms of SiH_4 will really be -112 °C, or:

$$T_{condensation} = -112 \text{ °C} + 273.15 = 161 \text{ K}$$

Solids have an even smaller volume than gases, so the mixture would not have to be as cold as if the substance were all in the gas state. This answer makes sense, *but it is not the answer one would predict while focusing on how to use the gas laws.*

56. *Define the problem*: Given the identity of a gas and its volume, pressure, and temperature, determine the mass of the gas.

Develop a plan: Use the ideal gas law to determine the moles of the gas. Then, convert from moles to grams.

Execute the plan:
$$T = 25 \text{ °C} + 273 = 298 \text{ K}$$

$$n_{He} = \frac{PV}{RT} = \frac{(1.1 \text{ atm}) \times (5.0 \text{ L})}{\left(0.08206 \dfrac{\text{L} \cdot \text{atm}}{\text{mol} \cdot \text{K}}\right) \times (298 \text{ K})} = 0.22 \text{ mol He}$$

$$0.22 \text{ mol He} \times \frac{4.0026 \text{ g He}}{1 \text{ mol He}} = 0.90 \text{ g He}$$

Note: It is perfectly acceptable to use the equations derived in Section 10.7 to streamline your approach to answering these questions.

$$m_{He} = \frac{PVM}{RT} = \frac{(1.1\ atm) \times (5.0\ L) \times \left(\dfrac{4.0026\ g\ He}{1\ mol\ He}\right)}{\left(0.08206\dfrac{L \cdot atm}{mol \cdot K}\right) \times (298\ K)} = 0.90\ g\ He$$

This can save time, but don't lose sight of what you're doing. If you find yourself looking through the book for an equation with all the right variables, that is a sign that you are losing sight of what you're doing.

Check your answer: The units all cancel and the size of the number make sense.

58. *Define the problem*: Given the molar mass of a gas and its pressure and temperature, determine the density of the gas.

Develop a plan: We have insufficient information to take a straightforward approach to this task, so let's look at what we know: Density is mass per unit volume. If we can calculate the molar volume (the volume per mole) and divide that into the molar mass (M, the grams per mole), we can get the density. Use the ideal gas law to determine the molar volume of the gas.

$$molar\ volume = \frac{V}{n} = \frac{RT}{P}$$

$$d = \frac{m}{V} = \frac{M}{\left(\dfrac{RT}{P}\right)} = \frac{MP}{RT}$$

Execute the plan: $T = -23\ °C + 273 = 250.\ K$

$$0.20\ mmHg \times \frac{1\ atm}{760\ mmHg} = 2.6 \times 10^{-4}\ atm$$

$$d = \frac{MP}{RT} = \frac{\left(29.0\dfrac{g}{mol}\right) \times (2.6 \times 10^{-4}\ atm)}{\left(0.08206\dfrac{L \cdot atm}{mol \cdot K}\right) \times (250.\ K)} = 3.7 \times 10^{-4}\ \frac{g}{L}$$

Check your answer: The units all cancel. The low density makes sense at the very low pressure. These numbers make sense.

Partial Pressures of Gases

60. *Define the problem*: Given the partial pressure of several gases in a sample of the atmosphere, and the total pressure of the sample, determine the partial pressure of O_2, the mole fraction of each gas, and the percent by volume. Compare the percentages to Table 10.3.

Develop a plan: Dalton's law and its applications are described in Section 10.8. Dalton's law of partial pressures states that the total pressure (P) exerted by a mixture of gases is the sum of their partial pressures (p_1, p_2, p_3, etc.), if the volume (V) and temperature (T) are constant.

$$P_{tot} = p_1 + p_2 + p_3 + ... \qquad \text{(V and T constant)}$$

Because $p_i = X_i P_{tot}$, we can use the total pressure and the partial pressure of a component to determine its mole fraction:

$$X_i = \frac{p_i}{P_{tot}}$$

According to Avogadro's law, moles and gas volumes are proportional, so the mole fraction is equal to the volume fraction. To get percent, multiply the volume fraction by 100 %.

Execute the plan:

(a) $P_{tot} = p_{N_2} + p_{O_2} + p_{Ar} + p_{CO_2} + p_{H_2O}$

$p_{O_2} = P_{tot} - p_{N_2} - p_{Ar} - p_{CO_2} - p_{H_2O}$

$= (740.\text{ mmHg}) - (575\text{ mmHg}) - (6.9\text{ mmHg}) - (0.2\text{ mmHg}) - (4.0\text{ mmHg})$

$= 154\text{ mmHg}$

(b) $X_{N_2} = \dfrac{p_{N_2}}{P_{tot}} = \dfrac{575\text{ mmHg}}{740.\text{ mmHg}} = 0.777$ $\qquad X_{O_2} = \dfrac{p_{O_2}}{P_{tot}} = \dfrac{154\text{ mmHg}}{740.\text{ mmHg}} = 0.208$

$X_{Ar} = \dfrac{p_{Ar}}{P_{tot}} = \dfrac{6.9\text{ mmHg}}{740.\text{ mmHg}} = 0.0093$ $\qquad X_{CO_2} = \dfrac{p_{CO_2}}{P_{tot}} = \dfrac{0.2\text{ mmHg}}{740.\text{ mmHg}} = 0.0003$

$X_{H_2O} = \dfrac{p_{H_2O}}{P_{tot}} = \dfrac{4.0\text{ mmHg}}{740.\text{ mmHg}} = 0.0054$

(c)
$$\% \ N_2 = X_{N_2} \times 100 \ \% = 0.777 \times 100 \ \% = 77.7 \ \%$$

$$\% \ O_2 = X_{O_2} \times 100 \ \% = 0.208 \times 100 \ \% = 20.8 \ \%$$

$$\% \ Ar = X_{Ar} \times 100 \ \% = 0.0093 \times 100 \ \% = 0.93 \ \%$$

$$\% \ CO_2 = X_{CO_2} \times 100 \ \% = 0.0003 \times 100 \ \% = 0.03 \ \%$$

$$\% \ H_2O = X_{H_2O} \times 100 \ \% = 0.0003 \times 100 \ \% = 0.54 \ \%$$

The Table 10.3 figures are slightly different. This sample is wet, whereas the proportions given in Table 10.3 are for dry air.

Check your answers: The percentages are very close to those provided in the table, and the variations are explainable. The sum of the mole fractions is 1, and the sum of the percentages is 100 %. These answers makes sense.

62. *Define the problem*: Given the density of a gas and its pressure and temperature, determine the molar mass of the gas.

Develop a plan: Density (d) is mass (m) per unit volume (V), so dV = m. Therefore, if we calculate the molar volume (the volume per mole) and multiply it by the density, we can get the molar mass (M = grams per mole). Use the ideal gas law to determine the molar volume of the gas.

$$d \times \frac{V}{n} = \frac{m}{n} = M$$

$$\text{molar volume} = \frac{V}{n} = \frac{RT}{P}$$

$$d \times \frac{RT}{P} = M$$

The sum of the molar masses of each component weighted by its mole fraction must be equal to the average molar mass. The sum of the mole fractions must be equal to 1.

Execute the plan:

(a)
$$T = -63 \ °C + 273 = 210. \ K$$

$$42 \ mmHg \times \frac{1 \ atm}{760 \ mmHg} = 0.055 \ atm$$

$$92 \, \frac{g}{m^3} \times \left(\frac{1 \, m}{100 \, cm} \right)^3 \times \frac{1 \, cm^3}{1 \, mL} \times \frac{1000 \, mL}{1 \, L} = 0.092 \, \frac{g}{L}$$

$$M = \frac{dRT}{P} = \frac{\left(0.092 \, \frac{g}{L} \right) \times \left(0.08206 \frac{L \cdot atm}{mol \cdot K} \right) \times (210. \, K)}{(0.055 \, atm)} = 29 \, \frac{g}{mol}$$

(b) Molar mass of N_2 is 28 g/mol. Molar mass of O_2 is 32 g/mol. Assuming that air is only N_2 and O_2, setting the mole fraction of N_2 equal to X, means that the mole fraction of O_2 is $1 - X$.

$$X(28 \, g/mol) + (1 - X)(32 \, g/mol) = 29 \, g/mol$$

$$- 4X + 32 = 29$$

$$- 4X = 3$$

$$X = 0.8 = X_{N_2} \qquad\qquad 1 - X = 0.2 = X_{O_2}$$

Check your answers: The small number of significant figures makes it difficult to compare these numbers in detail. They are approximately the same as described in Table 10.3. These answers make sense.

64. *Define the problem*: Given the partial pressure of a gas in a sample of the atmosphere, determine its partial pressure.

Develop a plan: Use $p_i = X_i P_{tot}$, to determine its the mole fraction.

Execute the plan: Assume that the $P_{tot} = 760$ mmHg, standard pressure.

$$X_{H_2O} = \frac{P_{H_2O}}{P_{tot}} = \frac{25 \, mmHg \, H_2O}{760 \, mmHg \, air} = 0.033$$

The mole fraction of water is not given in Table 10.3, because the air described in that table is dry (i.e., $X_{H_2O} = 0$, exactly). This air is humid air. It has significantly more water in it.

Check your answer: The partial pressure is small compared to the total pressure, so the small mole fraction makes sense.

66. *Define the problem*: Given the balanced equation for a reaction, the mass of a reactant, the volume, temperature and pressure of a product collected over water, and the vapor pressure of water at that same temperature, determine the percent yield of the reaction.

Develop a plan: Calculate the theoretical yield from the mass of reactant. Use Dalton's law to calculate the pressure of the product gas. Use the idea gas law to determine the moles of product. Convert the moles to grams to get actual yield. Divide the actual yield by the theoretical yield and multiply by 100 % to get percent yield.

Execute the plan: 1 mole CaC_2 produces 1 mol C_2H_2.

$$2.62 \text{ g CaH}_2 \times \frac{1 \text{ mol CaH}_2}{64.10 \text{ g CaH}_2} \times \frac{1 \text{ mol C}_2\text{H}_2}{1 \text{ mol CaH}_2} \times \frac{26.04 \text{ g C}_2\text{H}_2}{1 \text{ mol C}_2\text{H}_2} = 1.08 \text{ g C}_2\text{H}_2$$

$$P_{tot} = P_{C_2H_2} + P_{H_2O}$$

$$P_{C_2H_2} = P_{tot} - P_{H_2O} = (735.2 \text{ mmHg}) - (23.8 \text{ mmHg}) = 711.4 \text{ mmHg}$$

$$711.4 \text{ mmHg} \times \frac{1 \text{ atm}}{760 \text{ mmHg}} = 0.9361 \text{ atm}$$

$$T = 25.0 \text{ °C} + 273.15 = 298.2 \text{ K}$$

$$795 \text{ mL} \times \frac{1 \text{ L}}{1000 \text{ mL}} = 0.795 \text{ L}$$

$$n_{C_2H_2} = \frac{PV}{RT} = \frac{(0.9361 \text{ atm}) \times (0.795 \text{ L})}{\left(0.08206 \dfrac{\text{L} \cdot \text{atm}}{\text{mol} \cdot \text{K}}\right) \times (298.2 \text{ K})} = 0.0304 \text{ mol C}_2\text{H}_2$$

$$0.0304 \text{ mol C}_2\text{H}_2 \times \frac{26.04 \text{ g C}_2\text{H}_2}{1 \text{ mol C}_2\text{H}_2} = 0.792 \text{ g C}_2\text{H}_2$$

$$\frac{0.792 \text{ g C}_2\text{H}_2 \text{ actual}}{1.08 \text{ g C}_2\text{H}_2 \text{ theoretical}} \times 100 \text{ %} = 73.6 \text{ %}$$

Check your answer: The percent yield is a realistic size for collection of a gas.

The Behavior of Real Gases

68. This is a standard conversion factor problem:

$$1 \text{ mol H}_2\text{O} \times \frac{18.02 \text{ g H}_2\text{O}}{1 \text{ mol H}_2\text{O}} \times \frac{1 \text{ mL H}_2\text{O}}{1.0 \text{ g H}_2\text{O}} = 18 \text{ mL H}_2\text{O liquid}$$

1 mol H_2O at STP occupies 22.4 L.

Table 10.5 gives the vapor pressure of water at 0 °C to be 4.6 mmHg. At pressures higher than this, water would liquefy. We cannot achieve 1 atm pressure of water vapor at this low temperature, so we cannot achieve the standard state condition for water vapor.

70. The behavior of real gases is discussed in Section 10.9. At low temperatures, the molecules are moving relatively slowly; however, when the pressure is very low, they are still quite far apart. As external pressures increases, the gas volume decreases, the slow molecules are squeezed closer together, and the attractions among the molecules get stronger. Figure 10.16 shows that a gas molecule strikes the walls of the container with less force due to the attractive forces between it and its neighbors. This makes the mathematical product PV smaller than the mathematical product nRT.

Chemical Reactions in the Atmosphere

75. Introduced in Section 10.10, a free radical is an atom or group of atoms with one or more unpaired electrons; as a result, it is highly reactive. Methyl radicals are $\cdot CH_3$. When two of them react with each other, the unpaired electron on each forms a bond pair:

$$\cdot CH_3 + \cdot CH_3 \longrightarrow CH_3-CH_3$$

77. These reactions can be found in Section 10.10.

(a) $\qquad O + O_2 \longrightarrow O_3$

(b) $\qquad O_3 \xrightarrow{h\nu} O + O_2$

(c) $\qquad O + SO_2 \longrightarrow SO_3$

(d) $\qquad H_2O \longrightarrow \cdot H + \cdot OH$

(e) $\qquad NO_2 + NO_2 \longrightarrow N_2O_4$

79. (a) An example of a reaction that creates a free radical is the decomposition of ozone into oxygen and a free radical oxygen atom:

$$O_3 \longrightarrow O_2 + \cdot O \cdot$$

(b) An example of a reaction that creates a hydroxyl radical is a free radical oxygen atom reacting with a water molecule:

$$\cdot O \cdot + H_2O \longrightarrow 2 \cdot OH$$

Ozone and Ozone Depletion

81. CF_4 has no C–Cl bonds, which in CCl_4 are readily broken when exposed to UV light. Looking at the bond energies, C–Cl (327 kJ/mol) is much weaker than C–F (486 kJ/mol). In fact, the bond energy of C–F is very close to the bond energy of O=O (498 kJ/mol)!

83. CFCs must have at least one F and one Cl atom. They must also have four halogens (either F or C) attached to the C atom. The total set of possible variations are:

$$FCCl_3, \quad F_2CCl_2, \quad F_3CCl$$

85. New CFCs cannot catalyze the destruction of ozone at night, because sunlight is required to initiate the first step and create the ·Cl radical. However, additional reactions recreate the ·Cl, which can continue to destroy ozone even at night.

Chemistry and Pollution in the Troposphere

87. Primary pollutants are substances that are introduced into the air directly from their source.

- Particle pollutants: pollutants made out of particles:
    - aerosols: particles incorporated into water droplets.
    - particulates: larger solid particles
- Sulfur dioxide: pollutant produced when sulfur or sulfur compounds are burned in air.
- Nitrogen oxides: pollutant produced when nitrogen and oxygen react at high temperatures.
- Hydrocarbons: pollutants produced from many organic sources; their identity is small hydrocarbons from CH_4 to ones with six or seven carbons.

Secondary pollutants are substances produced from reactions of a primary pollutants.

- Ozone: ozone in the troposphere is produced from a reaction of O_2 with O_2 in the presence of intense energy (e.g., spark, lightening, etc.)
- Sulfur trioxide: SO_3 is produced from the reaction of SO_2.
- PAN (peroxyacetylnitrate): produced by the reaction of various free radicals in urban air.

Look at Section 10.12 for the specific ways these pollutants are harmful.

89. Adsorption is the process of firmly attaching to a surface. Absorption is the process of drawing a substance into the bulk of a solid or liquid.

91. *NOTE: The first sentence was left off of this question: "Approximately 65 million metric tons of SO_2 enter the atmosphere every year from the burning of coal."*

Define the problem: Given the above information and the percentage of sulfur in the coal, determine the mass (in metric tons) of coal burned and determine the number of hours this quantity of coal will burn.

Develop a plan: Determine the moles of sulfur in the product, then determine the mass of coal that contains this amount of sulfur. Use the metric tons of coal per hour to determine the hours one plant would need to burn this quantity.

Execute the plan:

$$65 \times 10^6 \text{ metric tons } SO_2 \times \frac{1000 \text{ kg } SO_2}{1 \text{ metric tons } SO_2} \times \frac{1000 \text{ g } SO_2}{1 \text{ kg } SO_2} = 6.5 \times 10^{13} \text{ g } SO_2$$

$$6.5 \times 10^{13} \text{ g } SO_2 \times \frac{1 \text{ mol } SO_2}{64.07 \text{ g } SO_2} \times \frac{1 \text{ mol S}}{1 \text{ mol } SO_2} \times \frac{32.07 \text{ g S}}{1 \text{ mol S}} = 3.3 \times 10^{13} \text{ g S}$$

$$3.3 \times 10^{13} \text{ g S} \times \frac{100 \text{ g coal}}{2 \text{ g S}} \times \frac{1 \text{ kg coal}}{1000 \text{ g coal}} \times \frac{1 \text{ metric tons coal}}{1000 \text{ kg coal}} = 1.6 \times 10^9 \text{ metric tons coal}$$

$$1.6 \times 10^9 \text{ metric tons coal} \times \frac{1 \text{ hr}}{700 \text{ metric tons coal}} = 2 \times 10^6 \text{ hr}$$

That's about 30 decades.

Check your answer: That seems like a long time. It is clear that there are many power plants adding SO_2 to the atmosphere for this amount to be produced in one year.

Urban Air Pollution

93. Photochemical reactions require the absorption of a photon of light. An example of a photochemical reaction is given in this equation:

$$O_2 \xrightarrow{h\nu} O + O$$

Not all photons of light have sufficient energy to cause photochemical reactions. Looking at the example above, according to the discussion of Section 10.11, we find that only photons with wavelengths of less than 242 nm are able to break this bond.

Therefore, this reaction does not occur if the light is visible light, or even low-energy ultraviolet (with wavelength 242 - 300 nm).

95. The reducing nature of industrial (London) smog is due to a sulfur oxide, SO_2. The burning of coal and oil as fuels produce this oxide. Further oxidation happens in air according to this equation:

$$2\ SO_2 + O_2 \longrightarrow 2\ SO_3$$

97. The atmospheric reaction that favors the formation of nitrogen monoxide, NO, is given in this chemical equation:

$$N_2 + O_2 \xrightarrow{\text{heat}} 2\ NO$$

The formation of NO in a combustion chamber is similar to the formation of NH_3 in a reactor designed to manufacture ammonia, because in both cases a reaction takes elemental nitrogen and makes a compound of nitrogen.

98. Stratospheric ozone is beneficial as a sunscreen. It screens out harmful ultraviolet light that can cause increased incidences of skin cancer in humans. Tropospheric ozone causes diminished respiratory capacity in humans. As a result, it poses a health hazard to people, especially those individuals who are susceptible to lung ailments.

General Questions

101. *NOTE: A sentence is left out of this problem. It should say "Worldwide, about 100 million metric toms of H_2S are produced annually from sources that include the oceans, bogs, swamps, and tidal flats."*

Define the problem: Given the information above, the description of a reaction, and the mass of a gaseous reactant, determine the mass of product produced.

Determine a plan: Use the masses, moles, and the stoichiometry of the balanced equations to determine the masses of the product.

Execute the plan:
$$H_2S + O_3 \longrightarrow SO_2 + H_2O$$

$$2\ SO_2 + O_2 \longrightarrow 2\ SO_3$$

$$SO_3 + H_2O \longrightarrow H_2SO_4$$

$$100\times10^6 \text{ metric ton } H_2S \times \frac{1000 \text{ kg } H_2S}{1 \text{ metric ton } H_2S} \times \frac{1000 \text{ g } H_2S}{1 \text{ kg } H_2S} = 1\times10^{14} \text{ g } H_2S$$

$$1 \times 10^{14} \text{ g H}_2\text{S} \times \frac{1 \text{ mol H}_2\text{S}}{34 \text{ g H}_2\text{S}} \times \frac{1 \text{ mol SO}_2}{1 \text{ mol H}_2\text{S}} \times \frac{2 \text{ mol SO}_3}{2 \text{ mol SO}_2} \times \frac{1 \text{ mol H}_2\text{SO}_4}{1 \text{ mol SO}_3}$$

$$= 3 \times 10^{12} \text{ mol H}_2\text{SO}_4$$

$$3 \times 10^{12} \text{ mol H}_2\text{SO}_4 \times \frac{98 \text{ g H}_2\text{SO}_4}{1 \text{ mol H}_2\text{SO}_4} \times \frac{1 \text{ kg H}_2\text{SO}_4}{1000 \text{ g H}_2\text{SO}_4} \times \frac{1 \text{ metric ton H}_2\text{SO}_4}{1000 \text{ kg H}_2\text{SO}_4}$$

$$= 3 \times 10^8 \text{ metric tons H}_2\text{SO}_4$$

Check your answer: 300 million metric tons of H_2SO_4 makes sense, because the molar mass of H_2SO_4 is about three times the molar mass of H_2S and the stoichiometry is 1:1, overall.

Applying Concepts

103. The molar masses of C_2H_4 and N_2 are 26 g/mol and 28 g/mol. The total pressure of the gases is 750 mmHg, and the partial pressure of N_2 is 500 mmHg. Dalton's law tells us that the sum of the partial pressures is the total pressure, and so the partial pressure of C_2H_4 is 250 mmHg, or half that of the N_2 molecules. The partial pressure is directly proportional to the mole fraction of the molecules in the container, so there should be half as many C_2H_4 molecules than N_2 molecules. That fact eliminates choices (b) and (c). So, graph (a) with more N_2 molecules than C_2H_4 molecules (indicated by smaller vertical rise in the curve) and the lower average speed for the heavier N_2 (indicated by the graph curve peaking at a slightly smaller speed value) is the best representation for this mixture.

105. Boyle's law: $V \propto \dfrac{1}{P}$ (unchanging T and n)

$V = b\left(\dfrac{1}{P}\right)$ where b is a proportionality constant.

Boyle's Law

$V = b\left(\dfrac{1}{P}\right)$

V

slope = b

$\dfrac{1}{P}$

Charles' law: $V \propto T$ (unchanging P and n)

$V = cT$ where c is a proportionality constant.

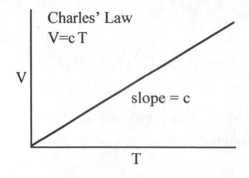

Avogadro's law: $V \propto n$ (unchanging T and P)

$V = an$ where a is a proportionality constant.

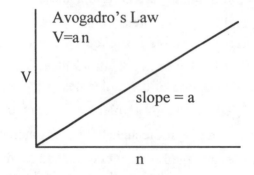

107. The initial volume is 40 cm^3 and the final volume is 60 cm^3. This 2:3 ratio in the gas volumes means for every two molecules of gas reactants there must be three molecules of gas products. The reaction that exhibits this ratio is (c):

$$2\,AB_2(g) \longrightarrow A_2(g) + 2\,B_2(g)$$

109. Use methods described in the answer to Questions 26, 3.77, 8.26, and 8.77.

(a) The density times the molar volume gives the molar mass:

$$\frac{1.87\ \text{g}}{1\ \text{L}} \times \frac{22.414\ \text{L}}{1\ \text{mol}} = 41.9\ \frac{\text{g}}{\text{mol}}$$

(b)

$$85.7\ \text{g C} \times \frac{1\ \text{mol C}}{12.011\ \text{g C}} = 7.1\ \text{mol C}$$

$$14.3\ \text{g H} \times \frac{1\ \text{mol H}}{1.0079\ \text{g H}} = 14.2\ \text{mol H}$$

Mole ratio: 7.1 mol C : 14.2 mol H

Simplified: 1C : 2 H

Empirical formula: CH_2

Empirical formula molar mass: 14.0 g/mol

$$(CH_2)_n \qquad n = \frac{41.9 \text{ g/mol}}{14.0 \text{ g/mol}} = 3$$

Molecular formula: C_3H_6

(c) C_3H_6 must have a ring or a double bond.

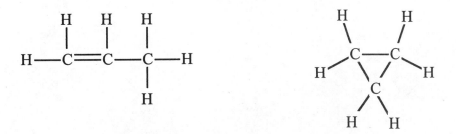

111. No, it is wise not to agree with this assumption immediately, since the gas volume might have dropped due to low temperature, not necessarily because the balloon is defective.

274

Chapter 11: Liquids, Solids, and Materials

The Liquid State

12. Surface tension is based on the ability of a liquid to interact with other particles in the liquid. At higher temperatures, the molecules move around more. The increased random motion disrupts the intermolecular interactions responsible for surface tension.

14. A liquid can be converted to a vapor without changing the temperature by reducing the pressure above it. This can be accomplished by putting the liquid in a container whose pressure can be altered using a vacuum pump. Pumping the gases out of the container reduces the atmospheric pressure acting on the liquid allowing it to reach the boiling point at a much lower temperature. Incidentally, a liquid in an open container will eventually evaporate at temperatures much lower than its boiling point because it absorbs energy from the surroundings.

16. The molecules of water in your sweat have a wide distribution of molecular speeds. The fastest of these molecules are more likely to escape the liquid state into the gas phase. The low-speed molecules left behind will have a lower average speed, and therefore a lower average kinetic energy, that, according to kinetic-molecular theory, implies that the temperature lowers.

18. *Define the problem*: Given the molar heat of vaporization of a compound, determine the heat required to vaporize a given mass of a compound.

    Develop a plan: Convert the mass to moles, and use the molar heat of vaporization as a conversion factor to get total heat.

    Execute the plan:

$$1.0 \text{ metric ton NH}_3 \times \frac{10^3 \text{ kg}}{1 \text{ metric ton}} \times \frac{1000 \text{ g}}{1 \text{ kg}} \times \frac{1 \text{ mol NH}_3}{17.03 \text{ g NH}_3} \times \frac{25.1 \text{ kJ}}{1 \text{ mol NH}_3} = 1.5 \times 10^6 \text{ kJ}$$

    Check your answer: Units cancel appropriately, and it makes sense that a large amount of ammonia needs a large amount of heat.

20. *Define the problem*: Given the molar heat of vaporization of a compound and the density of the liquid, determine the heat required to vaporize a given volume of the liquid.

    Develop a plan: Convert the volume to moles, then use the molar heat of vaporization to get total heat.

Execute the plan:

$$250. \text{ mL } CH_3OH \times \frac{0.787 \text{ g } CH_3OH}{1 \text{ mL } CH_3OH} \times \frac{1 \text{ mol } CH_3OH}{32.04 \text{ g } CH_3OH} \times \frac{38.0 \text{ kJ}}{1 \text{ mol } CH_3OH} = 233 \text{ kJ}$$

Check your answer: Units cancel appropriately, and it makes sense that a quantity of about six moles needs about six times the molar enthalpy.

22. *Define the problem*: Given the heat of vaporization of two compounds and the density of their liquids, determine the heat required to vaporize a given volume of each liquid and compare them.

Develop a plan: Convert to grams and use the heat of vaporization for the mercury. Convert the volume of water to moles, then use the molar heat of vaporization to get total heat for the water.

Execute the plan:

$$0.500 \text{ mL Hg} \times \frac{13.6 \text{ g Hg}}{1 \text{ mL Hg}} \times \frac{294 \text{ J}}{1 \text{ g Hg}} \times \frac{1 \text{ kJ}}{1000 \text{ J}} = 2.00 \text{ kJ to vaporize Hg sample}$$

$$0.500 \text{ mL } H_2O \times \frac{1.00 \text{ g } H_2O}{1 \text{ mL } H_2O} \times \frac{1 \text{ mol } H_2O}{18.02 \text{ g } H_2O} \times \frac{40.7 \text{ kJ}}{1 \text{ mol } H_2O}$$

$$= 1.13 \text{ kJ to vaporize the same volume of water}$$

Check your answers: Units cancel appropriately. Equal volumes of water and Hg have very different masses, due to the large difference in their densities. So, while water has a larger heat of vaporization (2.26×10^3 J/g), the number of grams of water in the sample volume is quite a bit smaller. These answers make sense.

24. NH_3 has a relatively large boiling point because the molecules interact using relatively strong hydrogen bonding intermolecular forces. The increase in the boiling points of the series PH_3, AsH_3, and SbH_3 is related to the increasing London dispersion intermolecular forces experienced due to the larger central atom in the molecule. (size: P < As < Sb)

Vapor Pressure

26. Methanol molecules are capable of hydrogen bonding, whereas formaldehyde molecules use dipole-dipole forces to interact. Molecules experiencing stronger intermolecular forces (such as methanol here) will have higher boiling points and lower vapor pressures compared to molecules experiencing weaker intermolecular forces (such as formaldehyde here).

28. *Define the problem*: Given the altitude on a mountain above sea level, a simple relationship between altitude and pressure changes, and Figure 11.5, determine the atmospheric pressure on the mountain and the boiling point of water at that altitude.

Develop a plan: Use the altitude/pressure relationship to get the atmospheric pressure on the mountain, then look up the boiling temperature on the figure.

Execute the plan:

$$22834 \text{ ft} \times \frac{3.5 \text{ mbar decrease}}{100 \text{ ft}} \times \frac{1 \text{ bar}}{1000 \text{ mbar}} \times \frac{10^5 \text{ Pa}}{1 \text{ bar}} \times \frac{1 \text{ kPa}}{1000 \text{ Pa}} \times \frac{1 \text{ atm}}{101.325 \text{ kPa}}$$

$$= 0.79 \text{ atm decrease}$$

$$1.00 \text{ atm} - 0.79 \text{ atm} = 0.21 \text{ atm}$$

$$0.21 \text{ atm} \times \frac{760 \text{ mm Hg}}{1 \text{ atm}} = 160 \text{ mm Hg}$$

Looking at Figure 10.5, this corresponds to a boiling temperature between 40 °C and 60 °C, around 57 °C.

Check your answer: It makes sense that a lower pressure causes a liquid to have a lower boiling temperature.

Phase Changes: Solids, Liquids, and Gases

30. A high melting point and a high heat of fusion tell us that a large amount of energy is required to melt a solid. That is the case when the interparticle interactions between the particles in the solid are very strong, such as in solids composed of ions.

32. A higher heat of fusion occurs when the intermolecular forces are stronger. The intermolecular forces between the molecules of H_2O in the solid (hydrogen bonding) are stronger than the intermolecular forces between the molecules of H_2S (dipole-dipole).

34. *Define the problem*: Given the molar heats of fusion and vaporization of a compound, determine the heat required to raise a given number of moles of solid to the melting point, melt it, raise the liquid to the boiling point and boil it.

Develop a plan: Find the mass from the moles. Use the molar heats to get total heat for both of the phase transitions. Use Equation 6.2 and Table 6.1 to get the heat for changing the temperature of the solid and liquid water (as described in Chapter 6, Section 6.3). Add together the heat for each stage to get the total quantity of heat.

Execute the plan:

$$0.50 \text{ mol H}_2\text{O} \times \frac{18.02 \text{ g H}_2\text{O}}{1 \text{ mol H}_2\text{O}} = 9.0 \text{ g H}_2\text{O}$$

Total heat is the sum of the heat to warm the ice to the melting point (0 °C), the heat required to melt the ice, the heat to warm the water to the boiling point (100 °C), and the heat required to vaporize the water.

$$q_{tot} = (c_{ice} \times m \times \Delta T) + (n\Delta H_{fus}) + (c_{liquid} \times m \times \Delta T) + (n\Delta H_{vap})$$

$$q_{tot} = \left(\frac{2.06 \text{ J}}{g \,°C}\right) \times \left(\frac{1 \text{ kJ}}{1000 \text{ J}}\right) \times (9.0 \text{ g}) \times [0 \,°C - (-5 \,°C)] + (0.50 \text{ mol}) \times \left(\frac{6.020 \text{ kJ}}{mol}\right)$$

$$+ \left(\frac{4.184 \text{ J}}{g \,°C}\right) \times \left(\frac{1 \text{ kJ}}{1000 \text{ J}}\right) \times (9.0 \text{ g}) \times [100 \,°C - (0 \,°C)] + (0.50 \text{ mol}) \times \left(\frac{40.07 \text{ kJ}}{mol}\right)$$

$$= 0.02 \text{ kJ} + 3.0 \text{ kJ} + 3.8 \text{ kJ} + 20. \text{ kJ} = 27 \text{ kJ}$$

Check your answer: The relative size of the four terms seems sensible, comparing the sizes of the heat capacities and the heats of the phase transitions.

36. A higher melting point is a result of stronger interparticle forces. Coulomb's Law describes the attraction between charged particles; the ion-ion coulombic interaction is stronger with smaller distance between the charges because the charges are more localized and closer together. According to the periodic trends described in Section 7.9, Li^+ and F^- are smaller than Cs^+ and I^-. Therefore, LiF experiences higher coulombic interactions than those in solid CsI, and LiF has a higher melting point.

38. The highest melting point is a result of strongest interparticle forces. Covalent interactions are stronger forces in solids than intermolecular forces between separate molecules. SiC solid is described as a nonoxide ceramic in Section 11.11. It is held together by an extended covalent network. Rb is a metal from Group 1A, and alkali metals have relatively low melting points. The other two interact by London forces, so the highest boiling point is (a) SiC.

The lowest melting point is a result of the weakest intermolecular forces. We need to compare the intermolecular forces in the two molecules, I_2 (molar mass 254 g/mol) and $CH_3CH_2CH_2CH_3$ (molar mass 54 g/mol). The larger I_2 molecules can experience more significant London forces than the smaller $CH_3CH_2CH_2CH_3$ molecules, so (d) $CH_3CH_2CH_2CH_3$ has the lowest melting point.

40. The freezer compartment of a frost-free refrigerator keeps the air so cold and dry that ice inside the freezer compartment undergoes sublimation at normal pressures, the direct conversion of solid to gaseous form. The hailstones would eventually disappear.

42. *Define the problem*: Calculate the density of an ideal gas from the conditions of the critical point for a compound and calculate the actual density of a real gas from the given relationship between moles and volume.

Develop a plan: Use methods described in Chapter 10 to find the density of an ideal gas at a given temperature and pressure. Convert that number to g/cm^3. Use the moles and volume of the real gas to find the density.

Execute the plan: Figure 11.16 shows the phase diagram for CO_2 with a triple point at 73 atm and 31 °C.

$$31 \text{ °C} + 273 = 304 \text{ K}$$

$$d = \frac{PM}{RT} = \frac{(73 \text{ atm}) \times \left(44.0 \frac{g}{mol}\right)}{\left(0.08206 \frac{L \cdot atm}{mol \cdot K}\right) \times (304 \text{ K})} = 130 \frac{g}{L}$$

$$\frac{130 \text{ g}}{1 \text{ L}} \times \frac{1 \text{ L}}{1000 \text{ mL}} \times \frac{1 \text{ mL}}{1 \text{ cm}^3} = 0.13 \frac{g}{cm^3} = \text{density of an ideal gas}$$

$$\frac{1 \text{ mol } CO_2}{94 \text{ cm}^3} \times \frac{44.0 \text{ g } CO_2}{1 \text{ mol } CO_2} = 0.47 \frac{g}{cm^3} = \text{density of } CO_2$$

Check your answer: The triple point conditions are not ideal gas conditions. Especially the extreme pressure makes it easy to believe that CO_2 under these conditions will not be ideal, so the much larger real density compared to the ideal gas density makes sense.

Types of Solids

44. (a) P_4O_{10} is a molecule, so it forms a molecular solid.

(b) Brass is composed of various metals, so it is a metallic solid.

(c) Graphite is an extended array of covalently bonded carbon, so it is a network solid.

(d) $(NH_4)_3PO_4$ is composed of common ions, NH_4^+ and PO_4^{3-}, so it is an ionic solid.

46. Pure substances have fixed melting temperatures, whereas mixtures and amorphous solids have ill-defined melting points. Network and ionic solids have much higher melting points, because of stronger interparticle bonds. Solids that conduct are made from metals. Liquids that conduct could be metals or ions.

(a) A soft, slippery solid that has no definite melting point is probably an amorphous solid.

(b) Violet crystals with moderate (not high) melting point that do not conduct in either phase are probably a molecular solid.

(c) Hard colorless crystal with a high melting point and liquid that conducts is probably an ionic solid.

(d) A hard solid that melts at a high temperature and conducts in both solid and liquid states is a metallic solid.

48. Use the same characterization conclusions as described in the answer to Question 46:

(a) A solid that melts below 100 °C and is insoluble in water is probably a nonpolar molecular solid.

(b) An ionic solid will conduct electricity only when melted.

(c) A solid that is insoluble in water and conducts electricity is probably a metallic solid, though it might be a network solid like graphite.

(d) A noncrystalline solid that has a wide melting point range is an amorphous solid.

Crystalline Solids

50. See Figure 11.19 and its description.

52. *Define the problem*: Given the length of the edge of a face-centered cubic unit cell of xenon, determine the radius of the xenon atoms.

Develop a plan: Use the figure and method similar to those described in Problem-Solving Example 11.5. The face diagonal is distance is the hypotenuse of an equilateral triangle. Diagonal distance = $\sqrt{2} \times$ (edge). We can see from the figure that the diagonal distance represents four times the radius of the atom.

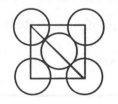

Use these two relationships to find the radius of the atom.

Execute the plan:

$$\text{Diagonal distance} = \sqrt{2} \times (\text{edge length}) = \sqrt{2} \times (620 \text{ pm}) = 880 \text{ pm}$$

$$\text{Radius} = \frac{\text{Diagonal distance}}{4} = \frac{880 \text{ pm}}{4} = 220 \text{ pm}$$

Check your answer: The geometric relationships are logical. We expect the radius to be less than the edge length. The only troubling thing about this result is that Figure 7.15 gives the atomic radius for Xenon to be 131 pm. These two data are inconsistent.

54. Looking at Figure 11.21, we see one Cs^+ ion in the middle of the cell and eight corners of Cl^- ions. These corners representing one eighth of a Cl^- ion, so that means the unit cell has one Cs^+ ion and one Cl^- ion.

56. *Define the problem*: Given the radii of cesium and chloride ions, determine the length of the body diagonal and the length of the side of the unit cell.

Develop a plan: Use Figure 11.21 showing the unit cell. The body diagonal runs through one whole Cs^+ ion and from the center to the edge of two Cl^- ions. The body diagonal is the hypotenuse of a triangle with sides represented by the face diagonal and the edge, so:

$$\left(\text{Body diagonal length}\right)^2 = (\text{edge length})^2 + \left(\text{face diagonal length}\right)^2$$

The face diagonal is the hypotenuse of an equilateral triangle.

$$\text{Face diagonal length} = \sqrt{2} \times (\text{edge length})$$

Use these three relationships to find the length of the body diagonal and the length of the side of the unit cell.

Execute the plan: Diagonal distance $= 2 \times \text{Radius } Cs^+ + 2 \times \text{Radius } Cs^+$

$$= 2 \times 169 \text{ pm} + 2 \times 181 \text{ pm} = 700. \text{ pm}$$

$$\left(\text{Body diagonal length}\right)^2 = (\text{edge length})^2 + \left[\sqrt{2} \times (\text{edge length})\right]^2 = 3 \times (\text{edge length})^2$$

$$(\text{edge length})^2 = \frac{\left(\text{Body diagonal length}\right)^2}{3} = \frac{\left(700. \text{ pm}\right)^2}{3} = 1.63 \times 10^3 \text{ pm}^2$$

$$\text{edge length} = \sqrt{1.63 \times 10^3 \text{ pm}^3} = 404 \text{ pm}$$

Check your answer: The geometric relationships are logical. The relative sizes of the atom diameters, the body diagonal length and the edge length are all reasonable. This answer makes sense.

58. The ratio of ions in the unit cell must reflect the empirical formula of the compound. Here, the compound $CaCl_2$ has a 1:2 ratio between Ca^{2+} and Cl^-. The compound NaCl has a 1:1 ratio between the Na^+ and Cl^- ions, so it is not possible to have the same structure.

60. *Define the problem*: Given the lengths of the sides of the unit cells for solid lithium metal and solid lithium iodide, the type of structure, and some assumptions of which atoms are touching, determine the radius of Li atom and the radii of Li^+ and I^-.

Develop a plan: Use the geometrical relationships described in the answer to Question 56.

Execute the plan:

(a) Looking at Figure 11.19 at the body-centered cubic, and assuming that the atoms touch along the body diagonal, we get:

$$\text{Body diagonal length} = 4r_{Li}$$

Given the edge length of the Li unit cell, 351 pm, calculate the body diagonal length:

$$\text{Body diagonal length} = \sqrt{3 \times (\text{edge length})^2} = \sqrt{3} \times 351 \text{ pm} = 608 \text{ pm}$$

Use body diagonal length to get radius of Li atom:

$$r_{Li} = \frac{\text{Body diagonal length}}{4} = \frac{608 \text{ pm}}{4} = 152 \text{ pm}$$

(b) Looking at Figure 11.23, and assuming that the I^- ions touch each other along the face diagonal and the I^- ions touch the Na^+ ion along the edge:

$$\text{Face diagonal length} = 4 \, r_{I^-}$$

$$\text{Edge length} = 2 \, r_{Li^+} + 2 \, r_{I^-}$$

Given the edge length of the LiI unit cell, 600. pm, calculate the face diagonal length:

(For sig. figs, assume that this length is as precise as the length in (a), ± 1 pm.)

$$\text{Face diagonal length} = \sqrt{2} \times (\text{edge length}) = \sqrt{2} \times 600. \text{ pm} = 849 \text{ pm}$$

Use face diagonal length to get radius of I^- ion:

$$r_{I-} = \frac{\text{Body diagonal length}}{4} = \frac{849 \text{ pm}}{4} = 212 \text{ pm}$$

Use the edge length of the LiI unit cell radius of I^- ion to get the radius of Li^+ ion:

$$2 \, r_{Li+} = \text{Edge length} - 2 \, r_{I-} = 600. \text{ pm} - 2 \times 212 \text{ pm} = 176 \text{ pm}$$

$$r_{Li+} = 88.0 \text{ pm}$$

(c) It is reasonable that the Li atom is larger than the Li^+ cation. Figure 7.17 gives exactly the same value for the radius of the Li atom (152 pm). It gives a slightly smaller value for Li^+ ion (76 pm) and slightly larger value for I^- (220 pm). The assumption that I^- anions touch each other seems unreasonable, there would be some repulsion and probably a small gap or distortion. In addition, the assumption that the very tiny Li^+ ion spans the entire gap between the two I^- ions in the unit cell might not be a good one.

Check your answers: The geometric relationships are logical. The relative sizes of the atom and ion radii, the body diagonal length, and the edge length are all reasonable. The sum of the Figure 7.17 radii to back-calculate the edge length (2×76 pm $+ 2 \times 220$ pm) gives 592 pm, a little less than the actual measured length, suggesting that the ions do not touch. Similarly, the face diagonal length (4×220 pm) gives 880 pm, a little more than the actual measured length, suggesting that the ions may be distorted to make that length shorter than the sum of the radii. These answers make sense.

Tools of Chemistry: X-Ray Crystallography

62. *Define the problem*: Using the length of the edge of a unit cell as the wavelength of light, determine the frequency of light, the energy per photon, and the energy per mole.

Develop a plan: Use methods described in Chapter 7 (Sections 7.1 and 7.2).

Execute the plan: Problem-Solving Example 11.6 gives the length of a unit cell of NaCl to be 566 pm. Use that as the wavelength for the light, and calculate the frequency.

$$v = \frac{c}{\lambda} = \frac{2.998 \times 10^8 \, \frac{m}{s}}{566 \text{ pm} \times \frac{10^{-12} \text{ m}}{1 \text{ pm}}} = 5.30 \times 10^{17} \text{ s}^{-1}$$

(a) $E = h\nu = (6.626 \times 10^{-34} \text{ J·s}) \times (5.30 \times 10^{17} \text{ s}^{-1}) = 3.51 \times 10^{-16}$ J for one photon

(b) $$\frac{3.51 \times 10^{-16} \text{ J}}{1 \text{ photon}} \times \frac{6.022 \times 10^{23} \text{ photons}}{1 \text{ mol photon}} = 2.11 \times 10^{8} \frac{\text{J}}{\text{mol}}$$

This photon is in the x-ray region of the electromagnetic spectrum.

Check your answers: It makes sense that the light is an x-ray, since x-ray crystallography uses x-rays to examine crystal structures.

Metals, Semiconductors, and Insulators

64. In a conductor, the valence band is only partially filled, whereas, in an insulator, the valence band is completely full, the conduction band is empty, and there is a wide energy gap between the two. In a semiconductor, the gap between the valence band and the conduction band is very small so that electrons are easily excited into the conduction band.

66. (c) Ag has the greatest electrical conductivity because it is a metal. (d) P_4 has the smallest electrical conductivity because it is a nonmetal. (The other two are metalloids.)

68. A superconductor is a substance that is able to conduct electricity with no resistance. Two examples are found near the end of Section 11.8: $YBa_2Cu_3O_7$ and $HgBa_2Ca_2Cu_4O_8$.

Silicon and the Chip

70. These reactions are given and described in Section 11.9:

$SiO_2(s) + 2 C(s) \longrightarrow Si(s) + 2 CO(g)$; Si is being reduced and C is being oxidized.

$SiCl_2(\ell) + 2 Mg(s) \longrightarrow Si(s) + 2 MgCl_2(s)$; Si is being reduced and Mg is being oxidized.

72. Doping is described in Section 11.9. It is the intentional addition of small amounts of specific impurities into very pure silicon. Group III elements are used because they have one less electron per atom than the group IV silicon. Group V elements are used because they have one more electron per atom.

Network Solids

74. Carbon atoms in diamond are sp^3 hybridized and are tetrahedrally bonded to four other carbon atoms. Carbon atoms in pure graphite are sp^2 hybridized and bonded with a trianglar planar shape to other carbon atoms. These bonds are partially double bonded so they are shorter than the single bonds in diamond. However, the planar sheets of sp^2 hybridized carbon atoms are only weakly attracted by intermolecular forces to adjacent layers, so these interplanar distances in graphite are much longer than the C–C single bonds in the diamond. The net result is that graphite is less dense than diamond.

76. Diamond is an electrical insulator because all the electrons are in single bonds, which are shared between two specific atoms and cannot move around. However, graphite is a good conductor of electricity because its electrons are delocalized in conjugated double bonds, which allow the electrons to move easily through the graphite sheets.

Cements, Ceramics, and Glasses

78. Amorphous solids are compared to crystalline solids in Section 11.5 and glasses are discussed in more detail in Section 11.11. The amorphous solids known as glasses are different from NaCl, because they lack symmetry or long range order, whereas ionic solids such as NaCl are extremely symmetrical. NaCl must be heated to melting temperatures, then cooled very slowly, to make a glass.

80. Ceramics are described in Section 11.11. Two examples of oxide ceramics: Al_2O_3 and MgO. Two examples of nonoxide ceramics: Si_3N_2 and SiC.

General Questions

82. *Define the problem*: Given the normal boiling point, molar heat of vaporization, and the specific heat capacities of the gas and liquid states of a compound, determine the heat evolved when a given mass of the substance is cooled from an initial to a final temperature.

Develop a plan: Use techniques similar to those described in the answer to Question 34.

Execute the plan: $10. \text{ kg NH}_3 \times \dfrac{1000 \text{ g NH}_3}{1 \text{ kg NH}_3} = 1.0 \times 10^4 \text{ g NH}_3$

$$10. \text{ kg NH}_3 \times \dfrac{1000 \text{ g NH}_3}{1 \text{ kg NH}_3} \times \dfrac{1 \text{ mol NH}_3}{17.03 \text{ g NH}_3} = 5.9 \times 10^2 \text{ mol NH}_3$$

A change in temperature in Kelvin degrees is the same as the change in temperature in Celsius degrees. So, use $c_{gas} = 2.2$ J/g °C and $c_{liqiud} = 4.7$ J/g °C

Find the heat absorbed when heating the liquid to the boiling point (– 33.4 °C)

$$q_{heating\ liquid} = c_{liquid} \times m \times \Delta T$$

$$q_{heating\ liquid} = \left(\frac{4.7\ J}{g\ °C}\right) \times (1.0 \times 10^4\ g\ NH_3) \times [-33.4\ °C - (-50.0\ °C)] = 7.8 \times 10^5\ J$$

$$q_{heating\ liquid} = 7.8 \times 10^5\ J \times \left(\frac{1\ kJ}{1000\ J}\right) = 780\ kJ$$

The total heat absorbed to get to 0.0 °C is the sum of the heat absorbed heating the liquid, the heat absorbed when vaporizing the liquid to a gas, and the heat absorbed when heating the gas to the final temperature (0.0 °C).

$$q_{tot} = q_{heating\ liquid} + (n \times \Delta H_{vap}) + (c_{gas} \times m \times \Delta T)$$

$$q_{tot} = 780\ kJ + 5.9 \times 10^2\ mol \times \left(\frac{23.5\ kJ}{mol}\right)$$

$$+ \left(\frac{2.2\ J}{g\ °C}\right) \times (1.0 \times 10^4\ g\ NH_3) \times [0.0\ °C - (-33.4\ °C)] \times \left(\frac{1\ kJ}{1000\ J}\right)$$

$$= 780\ kJ + (14000\ kJ) + (730\ kJ) = 16000\ kJ\ (rounded\ to\ thousands\ place)$$

$$q_{tot} = 1.6 \times 10^4\ kJ$$

Check your answer: The relative size of the three terms seems sensible, comparing the sizes of the heat capacities and the heat of vaporization.

84. (a) The shape of the SO_2 molecule is determined in the answer to Question 9.19(c):

     SO_2 (18 e⁻)   The type is AX_2E_1, so the electron-pair geometry is triangular planar and the molecular geometry is angular (120°).

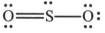 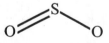

The asymmetric shape of the SO_2 molecule means that it is polar. Therefore, the strongest intermolecular forces between molecules in solid and liquid SO_2 are dipole-

dipole forces. All molecules experience London forces, so that force is part of the attractions in the liquid and solid states of SO_2, also.

(b) The normal boiling point is smaller when the intermolecular attractions are weaker. The normal boiling point is larger when the intermolecular attractions are stronger.

CH_4 (−161.5 °C) < NH_3 (− 33.4 °C) < SO_2 (−10 °C) < H_2O (100 °C)

Ordering the molecules by boiling points indicates the surprising result that the intermolecular attractions in SO_2 are stronger than the attractions in NH_3. This is probably a result of variable size. The molar mass of SO_2 is 64 g/mol, but the molar mass of NH_3 is 17 g/mol. Therefore, SO_2 molecules have much more significant London forces.

Applying Concepts

86. (a) To find the equilibrium vapor pressure for ethyl alcohol at room temperature, look at Figure 11.5 and find the pressure where the ethyl alcohol curve passes a temperature of 25 °C. This is about 80 mm Hg.

(b) To find the temperature when the equilibrium vapor pressure for diethyl ether is 400 mmHg, look at Figure 11.5 and find the temperature where the diethyl ether curve passes a pressure of 400 mmHg. This is about 18 °C.

(c) Find the equilibrium vapor pressure for water at 95 °. To do this, look at Figure 11.5 and find the pressure where the water curve passes a temperature of 95 °C. This is about 740 mm Hg.

(d) To find out which substances will be gases under specific condition of 200 mmHg and 60 °C, look at Figure 11.5 and find the pressure where each curve passes a temperature of 60 °C. If their vapor pressure is above 200 mmHg, they will be gases. Here, both diethyl ether and ethanol will evaporate readily. Water will evaporate only slowly.

(e) To find out which substance will evaporate in your hand, look at Figure 11.5 and find the pressure where each curve passes body temperature of 37 °C. If their pressure is close to 760 mmHg they will readily evaporate. Here, diethyl ether will evaporate readily.   The ethanol and water would evaporate more slowly.

(f) Water has the strongest intermolecular attractions as empirically indicated by having a lower vapor pressure than the other two liquids at any common temperature.

88. The butane in the lighter is under great enough pressure that the vapor pressure of butane at room temperature is less than the pressure inside the lighter. Hence, it exists as a liquid.

90. Refer to Section 11.3 if you need to be reminded about the different parts of a phase diagram.

Nanoscale diagram 1 looks like the substance atoms are all in the gas phase. That is region C on the phase diagram.

Nanoscale diagram 2 looks like some of the substance atoms are in the solid phase (atoms piled up in a regular array) and some of the substance atoms are in the gas phase. That is a point on the line described by E on the phase diagram.

Nanoscale diagram 3 looks like the substance atoms are all in the liquid phase. That is region B on the phase diagram.

Nanoscale diagram 4 looks like some of the substance atoms are in the solid phase and some of the substance atoms are in the liquid phase. That is a point on the line described by F on the phase diagram.

Nanoscale diagram 5 looks like some of the substance atoms are in the liquid phase and some of the substance atoms are in the gas phase. That is a point on the line described by G on the phase diagram.

Nanoscale diagram 6 looks like some of the substance atoms are in the liquid phase, some of the substance atoms are in the liquid phase, and some of the substance atoms are in the gas phase. That is the point described by H on the phase diagram.

Nanoscale diagram 7 looks like the substances is in a very strange phase, perhaps this is the critical point, though there is no reason to describe the atoms as larger or different colors just because they are at or above the critical point. By identifying diagrams for all of the other points, we can attribute this diagram to region D on the phase diagram.

Nanoscale diagram 8 looks like the substance atoms are all in the solid phase. That is region A on the phase diagram.

92. Each has the same fraction of filled space. The fraction of spaces filled by closest packed equal-sized spheres is the same, no matter what the size of the spheres.

Chapter 12: Fuels, Organic Chemicals, and Polymers

Fuel Sources and Products

24. (a) The gasoline fraction, with the hydrocarbons that will provide fuel for most people's cars, has a temperature range of 20 - 200 °C.

    (b) The octane rating of the straight-run fraction is 55.

    (c) Do not use the straight-run fraction as a motor fuel. The octane rating is far lower than regular gasoline we buy at the pump (87 - 92), that means it would cause far more preignition than we expect from the gasoline. It would need to be reformulated to make it an acceptable motor fuel.

26. Gasolines contain molecules in the liquid phase that, at ambient temperatures, can easily overcome their intermolecular forces and escape into the vapor phase. That means all gasolines evaporate easily.

28. Ethanol has the following structural formula: CH_3-CH_2-OH

30. Greenhouse effect is the trapping of heat by atmospheric gases. Global warming is the increase of the average global temperature. Global warming is related to an increase in the amount of greenhouse gases in the atmosphere.

32. CO_2 gets into the atmosphere by animal respiration, by burning fossil fuels and other plant materials, and by the decomposition of organic matter. CO_2 gets removed from the atmosphere by plants during photosynthesis, when it is dissolved in rain water, and when it is incorporated into carbonate and bicarbonate compounds in the oceans. Currently, atmospheric CO_2 production exceeds the CO_2 removal processes.

Alcohols

34. (a) A primary alcohol is one that has the –OH group attached to a C atom that is only bonded to one other C atom. An example of a primary alcohol is 1-propanol:

$$CH_3-CH_2-CH_2-OH$$

(b) A secondary alcohol is one that has the –OH group attached to a C atom that is bonded to two other C atoms. An example of a secondary alcohol is 2-propanol:

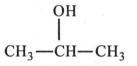

(c) A tertiary alcohol is one that has the –OH group attached to a C atom that is bonded to three other C atoms. An example of a tertiary alcohol is 2-methyl-1-propanol:

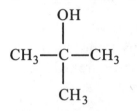

36. (a) 2-methyl-2-pentanol has a five-carbon main chain with a methyl branch on the second C atom of the main chain and an OH group on the second carbon of the main chain:

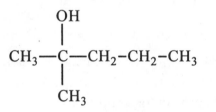

(b) 2,3-dimethyl-1-butanol has a four-carbon main chain with two methyl branches, one on the second C atom and one on the third C atom of the main chain. It also has an OH group on the first C atom of the main chain:

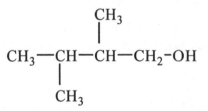

(c) 4-methyl-2-pentanol has a five-carbon main chain with one methyl fragment on the fourth C atom and an OH group on the second C atom of the main chain:

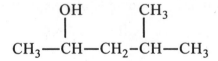

(d) 2-methyl-3-pentanol has a five-carbon main chain with one methyl fragment on the second C atom and an OH group on the third C atom of the main chain:

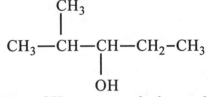

(e) *tertiary*-butyl alcohol has an OH group attached to a *t*-butyl alkyl group (refer to Table 3.5):

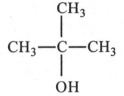

(f) isopropyl alcohol has an OH group attached to a isopropyl alkyl group (refer to Table 3.5):

$$CH_3—CH—CH_3$$
(with OH above the central CH)

38. Oxidation of a primary alcohol gives an aldehyde. Oxidation of an aldehyde gives a carboxylic acid.

(a)

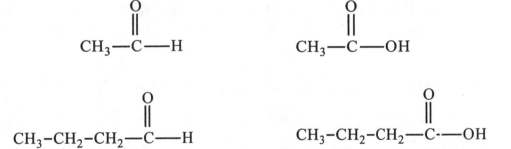

(b)

40. Oxidation of a secondary alcohol gives a ketone. Oxidation of a primary alcohol gives an aldehyde. Oxidation of an aldehyde gives a carboxylic acid.

(a) This aldehyde is produced from the oxidation of a primary alcohol:

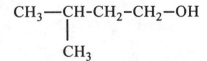

(b) This ketone is produced from the oxidation of a secondary alcohol:

$$
\begin{array}{c}
\text{OH} \\
| \\
\text{CH}_3\text{–CH}_2\text{—CH—CH}_2\text{–CH}_3
\end{array}
$$

(c) This carboxylic acid is produced from the oxidation of a primary alcohol and subsequent oxidation of the resulting aldehyde. The original alcohol is this one:

$$
\begin{array}{c}
\text{CH}_3\text{–CH}_2\text{—CH—CH}_2\text{-OH} \\
| \\
\text{CH}_3
\end{array}
$$

42. Wood alcohol (methanol) is made by heating hardwoods such as beach, hickory, maple, or birch. Grain alcohol (ethanol) is made from the fermentation of plant materials, such as grains.

44. –OH groups are a common site of hydrogen bonding intermolecular forces. Their presence would increase the solubility of the biological molecule in water and create specific interactions with other biological molecules.

Carboxylic Acids and Esters

46. The carboxylic acid group loses OH, and the alcohol group loses H (in the formation of water). Connect the remaining fragments to get the product of the condensation reaction produces the resulting ester.

(a) $CH_3COO\underline{H} + \underline{H}OCH_2CH_3 \longrightarrow CH_3COOCH_2CH_3 + H_2O$

So the product ester is: **$CH_3COOCH_2CH_3$**

(b) $CH_3CH_2COO\underline{H} + \underline{H}OCH_2CH_2CH_3 \longrightarrow CH_3CH_2COOCH_2CH_2CH_3 + H_2O$

So the product ester is: **$CH_3CH_2COOCH_2CH_2CH_3$**

(c) $CH_3CH_2COO\underline{H} + \underline{H}OCH_3 \longrightarrow CH_3CH_2COOCH_3 + H_2O$

So the product ester is: **$CH_3CH_2COOCH_3$**

48. Break the ester product of the condensation reaction between the two O atoms. Add OH to the C=O to make the carboxylic acid, and add H to the other fragment to make the alcohol group.

(a) $CH_3CH_2CO\underline{OH} + \underline{H}OCH_3 \longrightarrow CH_3CH_2COOCH_3 + H_2O$

So the reactants are: **CH_3CH_2COOH and CH_3OH**

(b) $HCOO\underline{H} + \underline{H}OCH_2CH_3 \longrightarrow HCOOCH_2CH_3 + H_2O$

So the reactants are: **$HCOOH$ and CH_3CH_2OH**

(c) $CH_3CO\underline{OH} + \underline{H}OCH_2CH_3 \longrightarrow CH_3COOCH_2CH_3 + H_2O$

So the reactants are: **CH_3COOH and CH_3CH_2OH**

Organic Polymers

50. Examples of thermoplastics are milk jugs (polyethylene), cheap sun glasses and toys (polystyrene) and CD audio discs (polycarbonates). Thermoplastics soften and flow when heated.

52. (a) 1-butene has a double bond between the first and second C atoms in the four-atom chain, $CH_2=CHCH_2CH_3$, where the addition polymerization occurs The saturated two-carbon end of the butane will put an ethyl branch on every other carbon in the long polymer chain

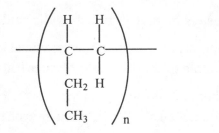

(b) 1,1-dichloroethylene has a double bond between the two C atoms where the addition polymerization occurs The two Cl atoms on the first C atom in the monomer mean that there will be two Cl atoms on every other carbon in the long polymer chain:

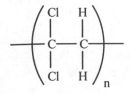

(c) Vinyl acetate, $CH_3COOCH=CH_2$, has a double bond between the two vinyl C atoms where the addition polymerization occurs The acetate group in the monomer means that every other carbon in the long polymer chain will have an acetate group attached to it:

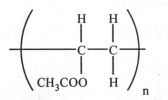

54. The monomer, methyl methacrylate, $CH_2=C(CH_3)COOCH_3$, has a double bond between the first two C atoms where the addition polymerization occurs. The methyl branch and the $-COOCH_3$ group on the first carbon in the monomer mean that every other carbon in the long polymer chain will have a methyl and a $-COOCH_3$ group attached to it:

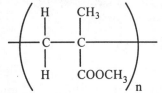

56. The monomer of natural rubber is isoprene, 2-methyl-1,3-butadiene. The isomer present in natural rubber is the *cis*-isomer.

58. The formation of polyesters involves the carboxylic acid and alcohol functional groups.

60. The formation of polyamides involves the carboxylic acid and amine functional groups. The most common example of this class of polymer is nylon.

62. One major difference between proteins and most other polyamides is that the protein polymer's monomers are not all alike. Different side chains on the amino acids change the properties of the protein.

66. The carboxylic acid group loses OH, and the alcohol group loses H (in the formation of water). Connect the resulting fragments to get the product of the condensation reaction for each ester linkage.

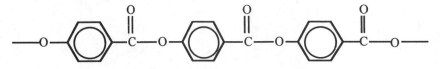

68. *This subject is not discussed in the textbook.* If you think about it, plastics are composed of large organic molecules, some contain benzene rings and other functional groups. The products of incomplete combustion could be carbon monoxide, hydrocarbons, aromatic hydrocarbons, hydrogen cyanide, and other toxic materials.

Proteins and Polysaccharides

70. The biological molecules that have monomer units that are not all alike are proteins, DNA, and RNA.

72. The monomers alanine, glycine, and phenylalanine are given in Table 12.8. We will link them using peptide linkages.

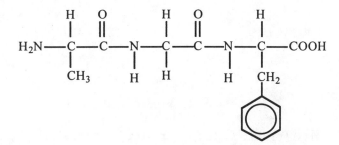

74. (a) A monosaccharide is a molecule composed of one simple sugar molecule, while disaccharides are molecules composed of two simple sugar molecules.

(b) Disaccharides have only two simple sugar molecules; whereas, polysaccharides have many.

76. Starch and cellulose are polysaccarides that yield only D-glucose upon complete hydrolysis.

78. (a) Glycogen contains glucose linked together with the glycosidic linkages in "*cis*-positions" and cellulose contains glucose with the glycosidic linkages in "*trans*-positions." Humans do not have the enzyme required to break the *trans*-linkage in cellulose.

(b) Cows do have the enzymes for breaking the *trans*-linkage of cellulose.

General Questions

80. $CH_3CH_2CH_2CH_2CH_2CH_2CH_2CH_2CH_2CH_2OH$ is a larger molecule than CH_3CH_2OH. The polar end of the alcohol group will interact well with the water; however, the nonpolar end of the molecule will not. The longer nonpolar end of the decanol will not be miscible in water, lowering the solubility compared to smaller, more polar ethanol.

82. Vulcanized rubber has short chains of sulfur atoms that bond together (crosslink) the polymer chains of natural rubber.

84. The hydrolysis of a triglyceride breaks up the ester linkage by adding water and forming the acid and the tri-alcohol glycerol.

86. (a) This equation is given in Problem-Solving Exercise 10.11:

$$2\ C_8H_{18}(\ell) + 25\ O_2(g) \longrightarrow 16\ CO_2(g) + 18\ H_2O(g)$$

(b) *Define the problem*: Given the length in miles of a trip, the fuel efficiency of a car, the density of the liquid fuel, the temperature and pressure, determine the volume of a gas-phase product produced during the trip.

Develop a plan: Use the miles, the fuel efficiency, volume conversions, the density, and the stoichiometry to find the moles of the product. Use the ideal gas law (from Chapter 10) to determine the volume of the product.

Execute the plan:

$$10.\ \text{miles} \times \frac{1\ \text{gal gasoline}}{32\ \text{miles}} \times \frac{3.785\ \text{L gasoline}}{1\ \text{gal gasoline}} \times \frac{1000\ \text{mL}}{1\ \text{L}} \times \frac{1\ \text{cm}^3}{1\ \text{mL}}$$

$$= 1.2 \times 10^3\ \text{cm}^3\ \text{gasoline}$$

$$1.2 \times 10^3\ \text{cm}^3\ \text{gasoline} \times \frac{0.692\ \text{g gasoline}}{1\ \text{cm}^3\ \text{gasoline}} \times \frac{1\ \text{mol gasoline}}{114.22\ \text{g gasoline}} \times \frac{16\ \text{mol } CO_2}{2\ \text{mol gasoline}}$$

$$= 57\ \text{mol } CO_2$$

$$T = 25\ °C + 273 = 298\ K$$

$$V = \frac{n_{CO_2}RT}{P} = \frac{(57 \text{ mol } CO_2) \times \left(0.08206 \frac{L \cdot atm}{mol \cdot K}\right) \times (298 \text{ K})}{(1.0 \text{ atm})} = 1.4 \times 10^3 \text{ L}$$

Check your answer: This is a large volume of CO_2! However, it is not an unreasonable quantity considering how many gallons of gas are used and how much CO_2 is generated from each octane molecule. Comparing the results to the answer in Question 87, less CO_2 is generated by methanol as a fuel than octane.

88. Glycogen has the glycosidic linkages in "*cis*-positions," whereas cellulose has glycosidic linkages in "*trans*-positions."

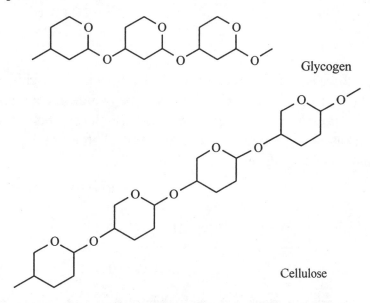

Glycogen

Cellulose

As a result, the glycogen molecules are more dense and can curl into granules. The cellulose is stretched out and more rigid and forms sheets.

Applying Concepts

90. The hydrogen bonding between molecules occurs when a hydrogen-bonding H atom (one that is bonded to a highly electronegative element) is attracted to another highly electronegative element in a neighboring molecule.

In propanoic acid, there is one hydrogen-bonding H atom and two highly electronegative O atoms in the molecule. That means there will be four different kinds of hydrogen-bonding interactions between the given molecule (in the box below) and its neighbors: two ways for the H atom to be attracted to neighboring O atoms and two ways for O atoms to be attracted to neighboring H atoms:

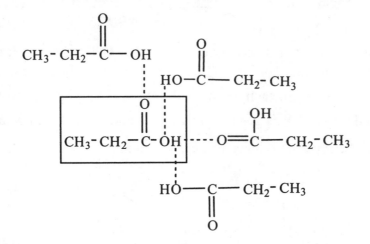

In 1-butanol, there is one hydrogen-bonding H atom and one highly electronegative O atom in the molecule. That means there will be two different kinds of hydrogen-bonding interactions between the given molecule (in the box below) and its neighbors: one way for the H atom to be attracted to neighboring O atoms and one way for O atoms to be attracted to neighboring H atoms:

$$HO-CH_2-CH_2-CH_2-CH_3$$

$$\boxed{CH_3-CH_2-CH_2-CH_2-OH}$$

$$HO-CH_2-CH_2-CH_2-CH_3$$

The more extensive hydrogen bonding in the propanoic acid suggests that it will have a higher boiling point, since the larger the intermolecular attractions, the higher the boiling point.

92. Find the repeating unit, then put an extra bond between the two C atoms at that point in the chain. The repeating unit looks like this.

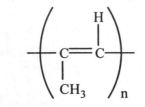

That means we need $CH_3-C\equiv C-H$ for the monomer.

94. The ester linkages are formed between the POH bonds on the phosphate (playing the role of the carboxylic acid) and the alcohol sites on the sugar. The phosphate ester linkages are the bonds formed where the condensation occurred.

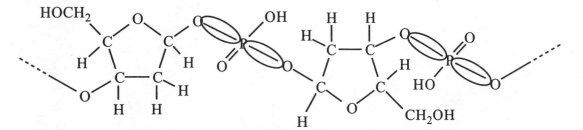

96. Some data that would be needed are:

* Sources and amounts of CO_2 generated over time to determine additional CO_2.

* Photosynthesis rate of depletion per tree per year.

* Average number of trees per acre.

* The number of acres of land in Australia that could support trees.

* The allowable tree density.

Other information would also be necessary. Try to think of other things you would need to know to determine the validity of the assertion.

Chapter 13: Chemical Kinetics: Rates of Reactions

Reaction Rate

14. (a) *Define the problem*: Given the side length of a cube involved in a chemical reaction and the assumption that the reaction rate is proportional to the surface area, determine by what factor the reaction rate will change if the cube is cut in half.

Develop a plan: Find the area of the cube in contact with the other reactant before and after the cut, then find a ratio to determine the increase in the reaction rate.

Execute the plan: $A_{face} = (side)^2 = (1.0 \text{ cm})^2 = 1.0 \text{ cm}^2$

A cube has six sides:

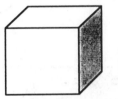

So the area in contact with the other reactant before the cut is

$$6 \times (1.0 \text{ cm}^2) = 6.0 \text{ cm}^2$$

Once the cube is cut in half, the newly exposed surfaces represent two new faces.

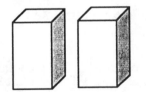

So the area in contact with the other reactant after the cut is

$$6.0 \text{ cm}^2 + 2 \times (1.0 \text{ cm}^2) = 8.0 \text{ cm}^2$$

Because the rate is proportional to the surface area, the ratio of the surface area is equal to the rate ratio.

$$\text{factor} = \frac{\text{rate}_{cut}}{\text{rate}_{uncut}} = \frac{\text{surface}_{cut}}{\text{surface}_{uncut}} = \frac{8.0 \text{ cm}^3}{6.0 \text{ cm}^3} = 1.3 \text{ times faster}$$

Check your answer: We would expect the fractional increase in the surface area to translate into a slightly faster reaction. In every day life we see reactions go faster when the particles are small, such as the dissolving of sugar or salt in water. This answer makes sense.

(b) To speed the reaction up as fast as possible without changing the temperature, we can grind the aluminum metal into dust.

15. (a) A graph of concentration verses time look like this

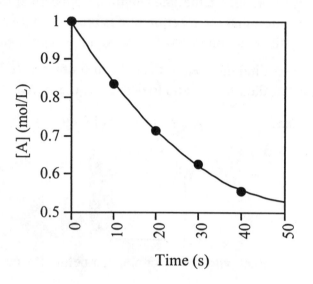

Define the problem: Given data of concentrations at various times during the course of a chemical reaction, find the rate of change for each time interval.

Develop a plan: To calculate the average rate for each time interval, use the method described in Section 13.1:

$$\text{Rate} = \frac{-\Delta[\text{reactant}]}{\Delta t} = \frac{-\left([\text{reactant}]_{\text{final}} - [\text{reactant}]_{\text{initial}}\right)}{t_{\text{final}} - t_{\text{initial}}}$$

$$= \frac{[\text{reactant}]_{\text{initial}} - [\text{reactant}]_{\text{final}}}{t_{\text{final}} - t_{\text{initial}}}$$

Execute the plan:

From 0.00 to 10.0 seconds:

$$\text{Rate}_{0\text{-}10} = \frac{1.000 \text{ mol}/\text{L} - 0.833 \text{ mol}/\text{L}}{10.0 \text{ s} - 0.00 \text{ s}} = \frac{0.167 \text{ mol}/\text{L}}{10.0 \text{ s}} = 0.0167 \frac{\text{mol}}{\text{L} \cdot \text{s}}$$

From 10.0 to 20.0 seconds:

$$\text{Rate}_{10\text{-}20} = \frac{0.833 \text{ mol/L} - 0.714 \text{ mol/L}}{20.0 \text{ s} - 10.0 \text{ s}} = \frac{0.119 \text{ mol/L}}{10.0 \text{ s}} = 0.0119 \frac{\text{mol}}{\text{L} \cdot \text{s}}$$

From 20.0 to 30.0 seconds:

$$\text{Rate}_{20\text{-}30} = \frac{0.714 \text{ mol/L} - 0.625 \text{ mol/L}}{30.0 \text{ s} - 20.0 \text{ s}} = \frac{0.089 \text{ mol/L}}{10.0 \text{ s}} = 0.0089 \frac{\text{mol}}{\text{L} \cdot \text{s}}$$

From 30.0 to 40.0 seconds:

$$\text{Rate}_{30\text{-}40} = \frac{0.625 \text{ mol/L} - 0.555 \text{ mol/L}}{30.0 \text{ s} - 20.0 \text{ s}} = \frac{0.070 \text{ mol/L}}{10.0 \text{ s}} = 0.0070 \frac{\text{mol}}{\text{L} \cdot \text{s}}$$

The rate of change of changes decreases because, as time passes, the concentration of reactant is decreasing.

Check your answers: According to the discussion of rate dependence on concentration of Section 13.2, it makes sense that the rate decreases as the reactant concentration decreases.

(b) As described in Section 13.1, since there are two B products formed whenever one A reactant undergoes reaction, that means that B is forming twice as fast as A is being depleted. Therefore, the rate of change of [B] is twice as fast as the rate of change of [A].

(c) In (a) above, the rate of depletion of A calculated for the 10 - 20 second range is 0.0119 mol/L·s. According to (b), the rate of production of B is twice that number:

$$2 \times (0.0119 \text{ mol/L·s}) = 0.0238 \text{ M/s}$$

17. *Define the problem*: Given data of concentrations at various times during the course of a chemical reaction, find the rate of change for each time interval. *NOTE: The 5.00-hour data is omitted from the text table. It should be: Time = 5.00 h [N$_2$O$_5$] = 0.196 mol/L*

Develop a plan: To calculate the average rate for each time interval, use the method described in Section 13.1 and in the answers to Questions 15 and 16:

Execute the plan:

(a) From 0 to 0.50 h:

$$\text{Rate}_{0\text{-}0.5} = \frac{0.849 \text{ mol/L} - 0.733 \text{ mol/L}}{0.50 \text{ h} - 0.00 \text{ h}} = \frac{0.116 \text{ mol/L}}{0.50 \text{ h}} = 0.23 \frac{\text{mol}}{\text{L} \cdot \text{h}}$$

(b) From 0.5 to 1.00 h:

$$\text{Rate}_{0.5\text{-}1.0} = \frac{0.733\ \text{mol}/L - 0.633\ \text{mol}/L}{1.00\ h - 0.50\ h} = \frac{0.100\ \text{mol}/L}{0.50\ h} = 0.20\ \frac{\text{mol}}{L \cdot h}$$

(c) From 1.00 to 2.00 h:

$$\text{Rate}_{1.0\text{-}2.0} = \frac{0.633\ \text{mol}/L - 0.472\ \text{mol}/L}{2.00\ h - 1.00\ h} = \frac{0.161\ \text{mol}/L}{1.00\ h} = 0.161\ \frac{\text{mol}}{L \cdot h}$$

(d) From 2.00 to 3.00 h:

$$\text{Rate}_{2.0\text{-}3.0} = \frac{0.472\ \text{mol}/L - 0.352\ \text{mol}/L}{3.00\ h - 2.00\ h} = \frac{0.120\ \text{mol}/L}{1.00\ h} = 0.120\ \frac{\text{mol}}{L \cdot h}$$

(e) From 3.00 to 4.00 h:

$$\text{Rate}_{3.0\text{-}4.0} = \frac{0.352\ \text{mol}/L - 0.262\ \text{mol}/L}{4.00\ h - 3.00\ h} = \frac{0.090\ \text{mol}/L}{1.00\ h} = 0.090\ \frac{\text{mol}}{L \cdot h}$$

(f) From 4.00 to 5.00 h:

$$\text{Rate}_{3.0\text{-}4.0} = \frac{0.262\ \text{mol}/L - 0.196\ \text{mol}/L}{5.00\ h - 4.00\ h} = \frac{0.066\ \text{mol}/L}{1.00\ h} = 0.066\ \frac{\text{mol}}{L \cdot h}$$

Check your answers: According to the discussion of rate dependence on concentration of Section 13.2, it makes sense that the rate decreases as the reactant concentration decreases.

18. *Define the problem*: Given data of concentrations at various times during the course of a chemical reaction and the average rate calculated over different time intervals, show that the reaction obeys a specific given rate law, evaluate the rate constant by averaging. *NOTE: The 5.00-hour data is omitted from the text table. It should be:*

$$\text{Time} = 5.00\ h \qquad [N_2O_5] = 0.196\ mol/L$$

Develop a plan: The average rates over each time interval were calculated in the answer to Question 17. Find the average concentration over that interval. If the rate is proportional to the concentration, then a graph of concentration vs. rate should be linear.

Execute the plan:

(a) From 0 to 0.50 h, the rate was $0.23\ \dfrac{\text{mol}}{L \cdot h}$.

$$\text{Average concentration}_{0\text{-}0.5} = \frac{0.849\ \text{mol}/L + 0.733\ \text{mol}/L}{2} = 0.791\ \frac{\text{mol}}{L}$$

From 0.5 to 1.00 h, the rate was $0.20 \frac{mol}{L \cdot h}$.

$$\text{Average concentration}_{0.5-1.0} = \frac{0.733 \ mol/L + 0.633 \ mol/L}{2} = 0.683 \frac{mol}{L}$$

From 1.00 to 2.00 h, the rate was $0.161 \frac{mol}{L \cdot h}$

$$\text{Average concentration}_{1.0-2.0} = \frac{0.633 \ mol/L + 0.472 \ mol/L}{2} = 0.553 \frac{mol}{L}$$

From 2.00 to 3.00 h, the rate was $0.120 \frac{mol}{L \cdot h}$

$$\text{Average concentration}_{2.0-3.0} = \frac{0.472 \ mol/L + 0.352 \ mol/L}{2} = 0.412 \frac{mol}{L}$$

From 3.00 to 4.00 h, the rate was $0.090 \frac{mol}{L \cdot h}$.

$$\text{Average concentration}_{3.0-4.0} = \frac{0.352 \ mol/L + 0.262 \ mol/L}{2} = 0.307 \frac{mol}{L}$$

From 4.00 to 5.00 h, the rate was $0.066 \frac{mol}{L \cdot h}$.

$$\text{Average concentration}_{4.0-5.0} = \frac{0.262 \ mol/L + 0.196 \ mol/L}{2} = 0.229 \frac{mol}{L}$$

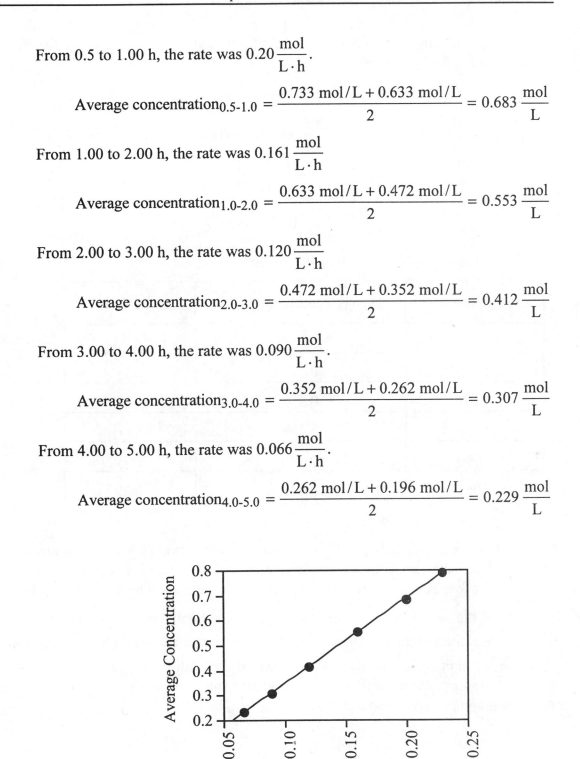

The linear relationship between rate and concentration proves the data satisfies the equation described in the problem:

$$Rate = k\,[N_2O_5]$$

(b) From 0 to 0.50 h, the rate is $0.23\,\dfrac{mol}{L\cdot h}$ and the average concentration is $0.791\,\dfrac{mol}{L}$.

$$k = \frac{Rate}{[N_2O_5]} = \frac{0.23\,\dfrac{mol}{L\cdot h}}{0.791\,\dfrac{mol}{L}} = 0.29h^{-1}$$

Similarly, for each of the other time increments:

Average rate $\dfrac{mol}{L\cdot h}$	Average concentration $\dfrac{mol}{L}$	$k\ (h^{-1})$
0.23	0.791	0.29
0.20	0.683	0.29
0.161	0.553	0.291
0.120	0.412	0.291
0.090	0.307	0.29
0.066	0.229	0.29

The average value of k is $0.29\ hr^{-1}$.

Check your answers: According to the discussion of rate dependence on concentration of Section 13.2, it makes sense that the rate decreases as the reactant concentration decreases. It also makes sense that the value of k is the same each time it was calculated.

22. Qualitatively, the concentration of the reactant will drop nonlinearly (first-order reaction) as the concentrations of the reactants increase. The rate of increase of O_2 concentration will be twice that of O_3 and O, due to their stoichiometric ratios being $2:1:1$. At late times, the concentrations of all species will level off to an unchanging value. (The dashed line on the graph is used to find the initial rate; see (a) below.)

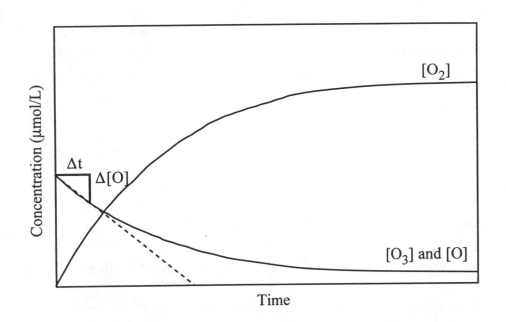

(a) As shown on the graph, the initial rate can be determined by taking a small time interval (Δt) near the initial time where the curve is still well approximated by a straight line (represented by the dashed line) and finding the slope of that line. For example, the initial rate of disappearance of O atoms is described this way:

$$\text{Rate} = \frac{-\Delta[O]}{\Delta t} = -\text{ slope of the straight line}$$

(b) As shown on the graph at very long times, the final rate will be zero because the concentration stops changing and the numerator of the rate expression is zero. That will happen for one of two reasons, either one or all of the reactants will have been used up or the reaction will reach an equilibrium state.

Effect of Concentration on Reaction Rates

23. *Define the problem*: Given the rate law of a reaction, determine what happens to the rate when the concentration is changed.

Develop a plan: Knowing the rate law, the rate increase can be determined by creating a ratio to provide a simple relationship between the rate changes and the concentration changes.

Execute the plan: The rate law is: rate $= k[A]^2$. Make a ratio of the rate law of the reaction under two different conditions: the initial condition (designated by a subscript "1", rate$_1 = k[A]_1^2$) compared to the final condition (designated by subscript "2", rate$_2 = k[A]_2^2$)

$$\text{rate change factor} = \frac{\text{rate}_2}{\text{rate}_1} = \frac{k[A]_2^2}{k[A]_1^2} = \left(\frac{[A]_2}{[A]_1}\right)^2$$

(a) When the concentration is tripled, the concentration ratio is $\dfrac{[A]_2}{[A]_1} = 3$,

so the rate change factor $= 3^2 = 9$. The rate increases by a factor of nine.

(b) When the concentration is halved, the concentration ratio is $\dfrac{[A]_2}{[A]_1} = \dfrac{1}{2}$,

so the rate change factor $= \left(\dfrac{1}{2}\right)^2 = \dfrac{1}{4}$. The rate will be one-fourth as fast.

Check your answers: It makes sense that the rate is faster when the concentration is larger and vice versa. It also makes sense that the rate change is larger than the concentration change, because the reaction is second-order.

25. Use the methods described in the answer to Question 23.

(a) The reaction is second-order in NO_2 and zeroth-order in CO, so the rate law has powers of 2 and zero on the concentrations of NO_2 and CO, respectively.

$$\text{rate} = k[NO_2]^2[CO]^0$$

Because any number raised to the zero power results in 1, that says the rate is unaffected by the concentration of CO.

$$\text{rate} = k[NO_2]^2$$

(b) A ratio gives the rate change factor:

$$\text{rate change factor} = \frac{\text{rate}_2}{\text{rate}_1} = \frac{k[NO_2]_2^2}{k[NO_2]_1^2} = \left(\frac{[NO_2]_2}{[NO_2]_1}\right)^2$$

When the concentration of NO_2 is halved, the concentration ratio is $\dfrac{[NO_2]_2}{[NO_2]_1} = \dfrac{1}{2}$, so the rate change factor $= \left(\dfrac{1}{2}\right)^2 = \dfrac{1}{4}$. The rate will be one fourth as fast.

(c) Because the rate law does not include concentration dependence for CO, the rate will be unchanged when the CO concentration is doubled.

27. (a) - (c) *Define the problem*: Given the rate law of a reaction, the value of the rate constant, and several concentrations, determine the instantaneous rates at each of those concentrations.

Develop a plan: Plug the concentration and the rate constant into the rate law.

Execute the plan: rate $= k[Pt(NH_3)_2Cl_2]$

(a) $rate_{0.010} = (0.090\ hr^{-1})(0.010\ M) = 9.0 \times 10^{-4}\ M/hr$

(b) $rate_{0.020} = (0.090\ hr^{-1})(0.020\ M) = 1.8 \times 10^{-3}\ M/hr$

(c) $rate_{0.040} = (0.090\ hr^{-1})(0.040\ M) = 3.6 \times 10^{-3}\ M/hr$

Check your answers: The rate is faster when the concentration is higher. This makes sense.

(d) If initial concentration of $Pt(NH_3)_2Cl_2$ is high, the initial rate of disappearance of $Pt(NH_3)_2Cl_2$ will be high. If initial concentration of $Pt(NH_3)_2Cl_2$ is low, the initial rate of disappearance of $Pt(NH_3)_2Cl_2$ will be small. The rate of disappearance of $Pt(NH_3)_2Cl_2$ is directly proportional to the concentration of $Pt(NH_3)_2Cl_2$.

(e) Rate law shows direct proportionality between rate and the concentration of $Pt(NH_3)_2Cl_2$.

(f) If the initial concentration of $Pt(NH_3)_2Cl_2$ is high, the rate of appearance of Cl^- will be high. If initial concentration is low, rate of appearance Cl^- will be small. The rate of appearance of Cl^- is directly proportional to the concentration of $Pt(NH_3)_2Cl_2$.

30. *Define the problem*: Given the initial concentrations and initial rates of a reaction for several different experimental conditions for the same chemical reaction, determine the rate law and the rate constant for the reaction.

Develop a plan: In Section 13.2, the method of finding the rate law from initial rates is described for getting the orders. However, comparing pairs of experiments where only one of the concentrations is different and relating that to the changes in the rate does not give consistent results. So, here we will derive a linear equation relating the concentrations and the rates and graph the results. Once the orders are determined, plug the data into the rate law to determine the value of k.

Execute the plan: Because I and II are reactants and are both varied in the different experiments we will seek a rate law looks like this: $rate = k[I]^i[II]^j$ where k, i, and j are currently unknown.

(a) Comparing experiments with [I] constant, the rate law can be simplified to:

$$rate = k'[II]^j \qquad\qquad where\ k' = k[I]^i$$

If we take the log of both sides of the equation, we can derive a linear relationship:

$$log(rate) = log(k') + log([II]^j)$$

$$log(rate) = log(k') + j\ log[II]$$

$$log(rate) = j\ log[II] + log(k')$$

Comparing to the equation of a line: $y = m\,x + b$, we see that if we plot

log(rate) against log[II], the slope of the line will be the reaction order, j.

Similarly comparing experiments with [II] constant, the rate law can be simplified to:

$$rate = k''[II]^I \qquad\qquad where\ k'' = k[II]^j$$

This gives a similar linear relationship:

$$log(rate) = i\ log[I] + log(k'')$$

Comparing to the equation of a line: $y = m\,x + b$, we see that if we plot log(rate) against log[I], the slope of the line will be the reaction order, i.

Four data sets have constant [I] and four data sets have [II].

log[II] with constant [I]	log rate with constant [I]	log[I] with constant [II]	log rate with constant [II]
-4.745	-8.509	-4.783	-8.824
-4.453	-8.201	-4.606	-8.569
-4.149	-7.951	-4.304	-8.345
-3.975	-7.752	-3.827	-7.752

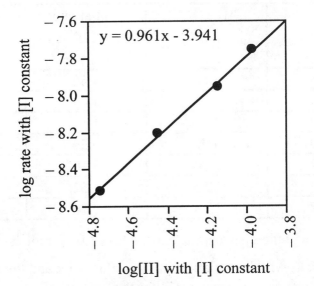

So, the reaction order of reactant II (j) is one.

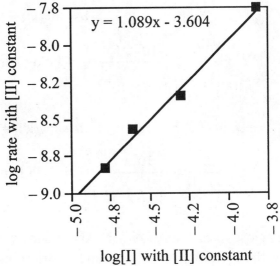

So, the reaction order of reactant I (i) is also one. So, rate = k[I][II]

(b) Solve the rate law for k:
$$k = \frac{\text{rate}}{[I][II]}$$

Plug in each experiment's data. Here is an example of the first Experiment's calculation:

$$k_1 = \frac{1.50 \times 10^{-9}\,\text{mol L}^{-1}\text{s}^{-1}}{(1.65 \times 10^{-5}\,\text{mol/L})(10.6 \times 10^{-5}\,\text{mol/L})} = 0.858\,\text{L mol}^{-1}\text{s}^{-1}$$

$[I] \times 10^5$ (mol/L)	$[II] \times 10^5$ (mol/L)	Initial rate $\times 10^9$ (mol $L^{-1}s^{-1}$)	Rate constant (L mol^{-1}s^{-1})
1.65	10.6	1.50	0.858
14.9	10.6	17.7	1.12
14.9	7.10	11.2	1.06
14.9	3.52	6.30	1.20
14.9	1.76	3.10	1.18
4.97	10.6	4.52	0.858
2.48	10.6	2.70	1.03

The average of these seven rate constants is 1.04 L mol^{-1}s^{-1}

Check your answers: The variations in the values of k and the deviation of the slopes from the exact integer values expected for the orders make these results unsatisfying. In addition, we can try to get the value of k from the y-intercepts of the graphs:

First Graph: $b = \log(k') = \log(k[I]^i)$

$$k = \frac{10^b}{[I]^i} = \frac{10^{-3.941}}{\left(14.9 \times 10^{-5}\right)^1} = 0.769 \text{ L mol}^{-1}\text{s}^{-1}$$

Second Graph: $b = \log(k'') = \log(k[II]^j)$

$$k = \frac{10^b}{[II]^j} = \frac{10^{-3.604}}{\left(10.6 \times 10^{-5}\right)^1} = 2.35 \text{ L mol}^{-1}\text{s}^{-1}$$

These are not very close to each other either. These wide variations might suggest a systematic error in the collection of the data under various conditions.

32. *Define the problem*: Given the initial concentrations and initial rates of a reaction for several different experimental conditions for the same chemical reaction, determine the reaction orders, the rate law, the rate constant and the relative rates of the reaction given specific rate.

Develop a plan: In Section 3.2, the method of finding the rate law from initial rates is described for getting the orders. Then compare pairs of experiments where only one of the concentrations is different and relate that to the changes in the rate. Once the orders are determined, plug the data into the rate law to determine the value of k. Use stoichiometric relationships between reactants and products, as described in Section 3.1, to relate the rates of reactants and products.

Execute the plan: The reactants are A and B. Data are available for the changes in each of these reactants' concentrations, so the rate law looks like this:

$$\text{rate} = k[A]^i[B]^j \quad \text{where } k, i \text{ and } j \text{ are currently unknown.}$$

(a) Looking at Experiments 1 and 2, the initial concentration of B doubles, the initial concentration of A stays constant, and the initial rate changes by a factor of two. The rate change is the same as the concentration change, which suggests that the rate is proportional to the concentration of B, and the order with respect to B is one.

(b) Looking at Experiments 3 and 4, the initial concentration of A doubles, the initial concentration of B stays constant, and the initial rate changes by a factor of four. The rate change is the square of the concentration change, which suggests that the rate is proportional to the square of the concentration of A, and the order with respect to A is two.

(c) The overall order is the sum of each reactant order. Here, the reaction is third-order.

(d) The rate law now looks like this: $\text{rate} = k[A]^2[B]^1$

(e) Solve the rate law for k: $$k = \frac{\text{rate}}{[A]^2[B]}$$

Plug in each experiment's data. Here is an example of the calculation for Experiment 1:

$$k_2 = \frac{0.012 \text{ mol } L^{-1}s^{-1}}{(6.0 \times 10^{-3} \text{ mol/L})^2(1.0 \times 10^{-3} \text{ mol/L})} = 3.3 \times 10^5 \text{ } L^2 \text{ mol}^{-2}s^{-1}$$

[A] (mol/L)	[B] (mol/L)	Initial rate (mol $L^{-1}s^{-1}$)	Rate constant (L^2 mol$^{-2}s^{-1}$)
6.0×10^{-3}	1.0×10^{-3}	0.012	3.3×10^5
6.0×10^{-3}	2.0×10^{-3}	0.024	3.3×10^5
2.0×10^{-3}	1.5×10^{-3}	0.0020	3.3×10^5
4.0×10^{-3}	1.5×10^{-3}	0.0080	3.3×10^5

The average of these four rate constants is 3.3×10^5 L^2 mol$^{-2}s^{-1}$

(f) The relative rates depend on the stoichiometric coefficients (Equation 13.3). Here, the stoichiometric relationship is 2A : 1C : 3D

$$-\frac{1}{2}\left(\frac{\Delta[A]}{\Delta t}\right) = \frac{1}{1}\left(\frac{\Delta[C]}{\Delta t}\right) = \frac{1}{3}\left(\frac{\Delta[D]}{\Delta t}\right)$$

$$\frac{\Delta[C]}{\Delta t} = \frac{1}{2}\left(-\frac{\Delta[A]}{\Delta t}\right) = \frac{1}{2}\left(0.034 \text{ mol L}^{-1}\text{s}^{-1}\right) = 0.017 \text{ mol L}^{-1}\text{s}^{-1}$$

$$\frac{\Delta[D]}{\Delta t} = \frac{3}{2}\left(-\frac{\Delta[A]}{\Delta t}\right) = \frac{3}{2}\left(0.034 \text{ mol L}^{-1}\text{s}^{-1}\right) = 0.051 \text{ mol L}^{-1}\text{s}^{-1}$$

Check your answers: The fact that every experiment gives the same rate constant makes it clear that the rate law determined properly describes the experimental data.

34. Follow the method described in the answer to Question 32.

The reactants are NO and O_2. Data are available for the changes in each of these reactants' concentrations, so the rate law looks like this:

$$\text{rate} = k[NO]^i[O_2]^j \quad \text{where k, i and j are currently unknown.}$$

(a) Looking at Experiments 1 and 2, the initial concentration of NO doubles, the initial concentration of O_2 stays constant, and the initial rate changes by a factor of four $(1.0 \times 10^{-4}/2.5 \times 10^{-5})$. The rate change is the square of the concentration change, which suggests that the rate is proportional to the square of the concentration of NO, and the order with respect to NO is two.

Looking at Experiments 1 and 3, the initial concentration of O_2 doubles, the initial concentration of NO stays constant, and the initial rate changes by a factor of two. The rate change is the same as the concentration change, which suggests that the rate is proportional to the concentration of O_2, and the order with respect to O_2 is one.

(b) The rate law (also called the rate equation) now looks like this: $\text{rate} = k[NO]^2[O_2]^1$

(c) Solve the rate law for k: $$k = \frac{\text{rate}}{[NO]^2[O_2]}$$

Plug in each experiment's data. Here is an example of the calculation for Experiment 1:

$$k_1 = \frac{2.5 \times 10^{-5} \text{ mol L}^{-1}\text{s}^{-1}}{(0.010 \text{ mol}/L)^2(0.010 \text{ mol}/L)} = 25 \text{ L}^2 \text{ mol}^{-2}\text{s}^{-1}$$

[NO] (mol/L)	[O$_2$] (mol/L)	Initial rate (mol L^{-1}s^{-1})	Rate constant (L^{2}mol^{-2}s^{-1})
0.010	0.010	2.5×10^{-5}	25
0.020	0.010	1.0×10^{-4}	25
0.010	0.020	5.0×10^{-5}	25

The average of these three rate constants is 25 L^{2}mol^{-2}s^{-1}.

(d) rate $= k[NO]^2[O_2]^1 = (25 \text{ L}^2\text{mol}^{-2}\text{s}^{-1})(0.025 \text{ mol/L})^2(0.050 \text{ mol/L})^1$

$$= 7.8 \times 10^{-4}\text{mol L}^{-1}\text{s}^{-1}$$

(e) The relative rates depend on the stoichiometric coefficients (Equation 13.3). Here, the stoichiometric relationship is 2NO : 1O$_2$: 2NO$_2$

$$-\frac{1}{2}\left(\frac{\Delta[NO]}{\Delta t}\right) = -\frac{1}{1}\left(\frac{\Delta[O_2]}{\Delta t}\right) = \frac{1}{2}\left(\frac{\Delta[NO_2]}{\Delta t}\right)$$

$$-\frac{\Delta[NO]}{\Delta t} = 2\left(-\frac{\Delta[O_2]}{\Delta t}\right) = 2\left(1.0\times10^{-4}\text{ mol L}^{-1}\text{s}^{-1}\right) = 2.0\times10^{-4}\text{ mol L}^{-1}\text{s}^{-1}$$

$$\frac{\Delta[NO_2]}{\Delta t} = 2\left(-\frac{\Delta[O_2]}{\Delta t}\right) = 2\left(1.0\times10^{-4}\text{ mol L}^{-1}\text{s}^{-1}\right) = 2.0\times10^{-4}\text{ mol L}^{-1}\text{s}^{-1}$$

Rate Law and Order of Reaction

36. Reaction order is the power to which the concentration of a component is raised. The overall order is the sum of all the individual orders.

(a) In the rate law: Rate $= k[A][B]^3$ the order with respect to A is one and the order with respect to B is three. The overall order is four.

(b) In the rate law: Rate $= k[A][B]$ the order with respect to A is one and the order with respect to B is one. The overall order is two.

(c) In the rate law: Rate $= k[A]$ the order with respect to A is one and the order with respect to B is zero. The overall order is one.

(d) In the rate law: Rate $= k[A]^3[B]$ the order with respect to A is three and the order with respect to B is one. The overall order is four.

39. *Define the problem*: Given concentration data as a function of time, find the rate law, the reaction order, the rate constant, and the rate at a specific concentration.

Develop a plan: In Section 3.3, the method of finding the rate law from linear graphs is described. Construct the three graphs described in Table 13.2, representing zero, first and second-order functions. Determine which one of these three graphs is the linear graph. The linear graph describes the order of the reaction. Once the order is determined, the slope of the straight line is used to determine the value of k.

Execute the plan:

(a) The reactants are phenyl acetate and water, and the problem tells us to assume that the concentration of water does not change. Data are available for the concentration changes with time for phenyl acetate, so the rate law looks like this:

$$\text{rate} = k[\text{phenyl acetate}]^i \quad \text{where k and i are currently unknown.}$$

Graph [phenyl acetate] vs. time, ln[phenyl acetate] vs. time, and 1/[phenyl acetate] vs. time, to see which one produces a straight line.

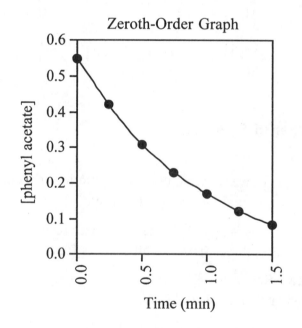

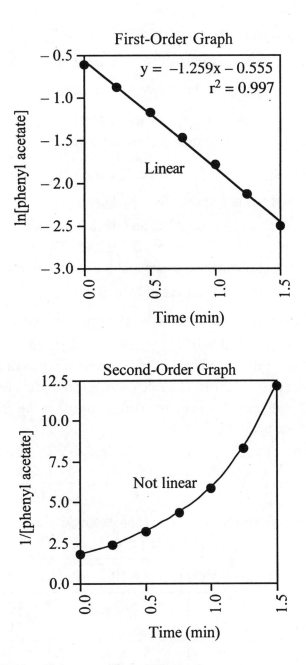

First-Order Graph

$y = -1.259x - 0.555$

$r^2 = 0.997$

Linear

ln[phenyl acetate]

Time (min)

Second-Order Graph

Not linear

1/[phenyl acetate]

Time (min)

Of the three graphs, the one that is the most linear is the ln[phenyl acetate] vs. time graph. That means the reaction is first-order.

$$rate = k[phenyl\ acetate]^1$$

(b) The order of the reaction with respect to phenyl acetate is one.

(c) The slope of the straight-line graph is equal to –k. So, here: $k = 1.259\ min^{-1}$

(d) rate = k[phenyl acetate]^1 = (1.259 min^{-1})(0.10 mol/L)^1 = 0.13 mol L^{-1}min^{-1}

Check your answers: It is satisfying to get one graph much more linear than the others.

42. Follow the method described in the answer to Question 32.

(a) The reactants are CO and NO$_2$. Data are available for the changes in each of these reactants' concentrations, so the rate law looks like this:

$$rate = k[CO]^i[NO_2]^j \quad \text{where k, i and j are currently unknown.}$$

(b) Looking at Experiments 1 and 4, the initial concentration of CO stays constant, the initial concentration of NO$_2$ doubles, and the initial rate changes by a factor of approximately two (1.0 × 10^{-3}/5.1 × 10^{-4} = 2.0). The rate change is the same as the concentration change, which suggests that the rate is proportional to the concentration of NO$_2$, and the order with respect to NO$_2$ is one.

Looking at Experiments 3 and 4, the initial concentration of CO approximately doubles (0.35 × 10^{-3}/0.18 × 10^{-4}=1.9), the initial concentration of NO$_2$ stays constant, and the initial rate changes by a factor of approximately two (1.0 × 10^{-3}/5.1 × 10^{-4} = 2.0). The rate change is the same as the concentration change, which suggests that the rate is proportional to the concentration of CO, and the order with respect to CO is one.

(c) The rate law now looks like this: $$rate = k[CO]^1[NO_2]^1$$

Solve the rate law for k: $$k = \frac{rate}{[CO][NO_2]}$$

Plug in each experiment's data. Here is an example using Experiment 1:

$$k_1 = \frac{5.1 \times 10^{-4} \text{ mol L}^{-1}\text{h}^{-1}}{(0.35 \times 10^{-4} \text{ mol/L})(3.4 \times 10^{-8} \text{ mol/L})} = 4.3 \times 10^8 \text{ L mol}^{-1}\text{h}^{-1}$$

[CO] (mol/L)	[NO$_2$] (mol/L)	Initial rate (mol L^{-1}h^{-1})	Rate constant (L mol^{-1}h^{-1})
0.35 × 10^{-4}	3.4 × 10^{-8}	5.1 × 10^{-4}	4.3 × 10^8
0.70 × 10^{-4}	1.7 × 10^{-8}	5.1 × 10^{-4}	4.3 × 10^8
0.18 × 10^{-4}	6.8 × 10^{-8}	5.1 × 10^{-4}	4.2 × 10^8
0.35 × 10^{-4}	6.8 × 10^{-8}	1.0 × 10^{-3}	4.2 × 10^8
0.35 × 10^{-4}	10.2 × 10^{-8}	1.5 × 10^{-3}	4.2 × 10^8

The average of these five rate constants is 4.2 × 10^8 L mol^{-1}h^{-1}.

43. Follow the method described in the answer to Question 32.

(a) The reactants are CH_3COCH_3, Br_2 and water. The reaction is catalyzed by acid, so the rate is affected by the concentration of H_3O^+. Data are available for the changes in the first two reactants' concentrations and the catalyst concentration, so we will assume the quantity of water is relatively unchanged in this reaction. Hence, the rate law looks like this:

$$rate = k[CH_3COCH_3]^i[Br_2]^j[H_3O^+]^h \quad \text{where k, i, j and h are currently unknown.}$$

Looking at Experiments 1 and 2, the initial concentration of CH_3COCH_3 stays constant, the initial concentration of Br_2 doubles, the initial concentration of H_3O^+ stays constant, and the initial rate does not change. The rate is independent of the concentration of Br_2, and the order with respect to Br_2 is zero.

Looking at Experiments 1 and 3, the initial concentration of CH_3COCH_3 stays constant, the initial concentration of Br_2 stays constant, the initial concentration of H_3O^+ doubles, and the initial rate changes by a factor of approximately two $(12.0 \times 10^{-5}/5.7 \times 10^{-5} = 2.1)$. The rate change is the same as the concentration change, which suggests that the rate is proportional to the concentration of H_3O^+, and the order with respect to H_3O^+ is one.

Looking at Experiments 1 and 5, the initial concentration of CH_3COCH_3 increases by a factor of 1.3 (0.40/0.30), the initial concentration of Br_2 stays constant, the initial concentration of H_3O^+ stays constant, and the initial rate changes by a factor of 1.3 $(7.6 \times 10^{-5}/5.7 \times 10^{-5})$. The rate change is the same as the concentration change, which suggests that the rate is proportional to the concentration of CH_3COCH_3, and the order with respect to CH_3COCH_3 is one.

The rate law now looks like this:

$$rate = k[CH_3COCH_3]^1[H_3O^+]^1$$

(b) Solve the rate law for k: $k = \dfrac{rate}{[CH_3COCH_3][H_3O^+]}$

Plug in each experiment's data. Here is an example of the calculation for Experiment 1:

$$k_1 = \frac{5.7 \times 10^{-5} \text{ mol L}^{-1}\text{s}^{-1}}{(0.30 \text{ mol}/\text{L})(0.05 \text{ mol}/\text{L})} = 4 \times 10^{-3} \text{ L mol}^{-1}\text{s}^{-1}$$

$[CH_3COCH_3]$ (mol/L)	$[H_3O^+]$ (mol/L)	Initial rate (mol L^{-1}s^{-1})	Rate constant (L mol^{-1}s^{-1})
0.30	0.05	5.7×10^{-5}	4×10^{-3}
0.30	0.10	12.0×10^{-5}	4.0×10^{-3}
0.40	0.20	31.0×10^{-5}	3.9×10^{-3}
0.40	0.05	7.6×10^{-5}	4×10^{-3}

The first two experiments have the same rate (represented in the first row of the chart above). The average of these five rate constants is 4×10^{-3} L mol^{-1}s^{-1}.

(c) rate $= k[CH_3COCH_3]^1[H_3O^+]^1$

$$= (4 \times 10^{-3} \text{ L mol}^{-1}\text{s}^{-1})(0.10 \text{ mol/L})^1(0.050 \text{ mol/L})^1 = 2 \times 10^{-5} \text{ mol L}^{-1}\text{s}^{-1}$$

45. Using the integrated rate law equations from Table 13.2, and setting $[A]_0 = 1.0$ mol/L and $k = 1.0$ in appropriate units, the following values were determined:

Time(t) (s)	$[A]_t = -kt + [A]_0$ $[A]_t$ (Zeroth-order)	$\ln[A]_t = -kt + \ln[A]_0$ $[A]_t$ (First-order)	$\frac{1}{[A]_t} = kt + \frac{1}{[A]_0}$ $[A]_t$ (Second-order)
0.0	1.0	1.0	1.0
1.0	0.0	0.37	0.50
2.0	0.0	0.14	0.33
3.0	0.0	0.050	0.25
4.0	0.0	0.018	0.20
5.0	0.0	0.0067	0.17

These concentrations can now be plotted as a function time for each order:

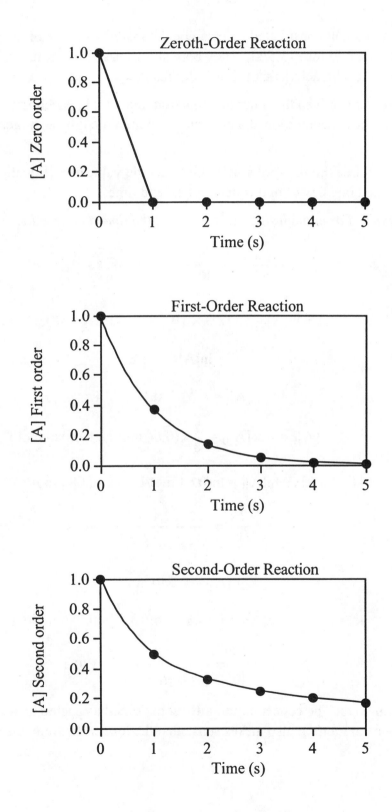

Only the second graph constructed here can be compared with one of the graphs in Figure 13.5 – in particular, Figure 13.5(a). They both show curved-down functional dependence of [A] verses time, characteristic of first-order reactions.

47. *Define the problem*: Given the order of a reaction, the initial concentration, and the half-life, determine the concentration at a new time and the time it takes to get to a new concentration.

Develop a plan: Use the first-order half-life to get the value of k (Equation 13.7). Then use the integrated rate law to find concentration and time.

Execute the plan: *For enough sig figs, assume all the time data is ± 1 s.*

$$k = \frac{\ln 2}{t_{1/2}} = \frac{\ln 2}{30.\ s} = 2.3 \times 10^{-2}\ s^{-1}$$

(a) $\ln[A]_t = -kt + \ln[A]_0 = -(2.3 \times 10^{-2}\ s^{-1})(60.\ s) + \ln(0.64\ mol/L)$

$$\ln[A]_t = -1.8$$

$$[A]_t = e^{-1.8} = 0.16\ mol/L$$

(b) $[A]_t = \frac{1}{8} \times [A]_0 = \frac{1}{8} \times (0.64\ mol/L) = 0.080\ mol/L$

$$kt = \ln[A]_0 - \ln[A]_t = \ln(0.64\ mol/L) - \ln(0.080\ mol/L) = 2.1$$

$$t = \frac{2.1}{k} = \frac{2.1}{2.3 \times 10^{-2}\ s^{-1}} = 90.\ s$$

(c) $[A]_t = 0.040\ mol/L$

$$kt = \ln[A]_0 - \ln[A]_t = \ln(0.64\ mol/L) - \ln(0.040\ mol/L) = 2.8$$

$$t = \frac{2.8}{k} = \frac{2.8}{2.3 \times 10^{-2}\ s^{-1}} = 120\ s$$

Check your answers: The concentration after some time has passed is always smaller than the initial concentration. It also makes sense that the longer the time, the smaller the concentration.

50. The correct answer for Question 49 provides an equation for a line, when plotting $\log[A]_t$ verses t. That equation looks like this:

$$\log[A]_t = \left(\frac{-\log 2}{t_{1/2}}\right) t + \log[A]_0$$

$$y \quad = \quad m\,x \quad + \quad b$$

Taking the logarithm of the *trans*-CHClCHCl concentrations and plotting them against time makes the following graph:

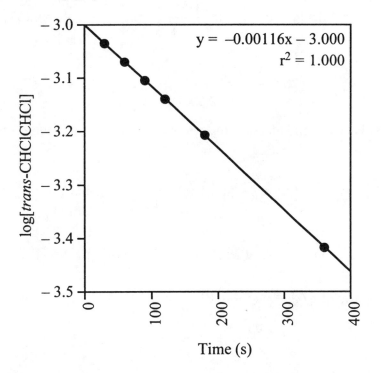

$$y = -0.00116x - 3.000$$
$$r^2 = 1.000$$

The slope of that line, m, is $\dfrac{-\log 2}{t_{1/2}}$, and the y-intercept, b, is $\log[\textit{trans}\text{-CHClCHCl}]_0$.

Use these relationships to find the half-life and the initial concentration.

$$m = \frac{-\log 2}{t_{1/2}} \qquad\qquad t_{1/2} = \frac{-\log 2}{m} = \frac{-\log 2}{-0.00116} = 260.\ s$$

$$b = \log[\textit{trans}\text{-CHClCHCl}]_0$$

$$[\textit{trans}\text{-CHClCHCl}]_0 = 10^b = 10^{-3.000} = 1.00 \times 10^{-3}\ \text{mol/L}$$

52. Use methods and equations similar to those described in the answer to Question 47.

(a)
$$t_{1/2} = \frac{\ln 2}{k} = \frac{\ln 2}{2.5 \times 10^{-4} \text{ s}^{-1}} = 2.8 \times 10^3 \text{ s}$$

(b)
$$[A]_t = \frac{1}{32} \times [A]_0$$

Note, the initial concentration is not known, but the ratio is all we need:

$$kt = \ln[A]_0 - \ln[A]_t = \ln\left(\frac{[A]_0}{[A]_t}\right) = \ln\left(\frac{[A]_0}{\frac{1}{32}[A]_0}\right) = \ln(32) = 3.47$$

$$t = \frac{3.47}{k} = \frac{3.47}{2.5 \times 10^{-4} \text{ s}^{-1}} = 1.4 \times 10^4 \text{ s}$$

Alternatively, we could also use the results derived in Question 49 to answer this question:

$$[A]_t = [A]_0 \left(\frac{1}{2}\right)^x \qquad\qquad x = \frac{t}{t_{1/2}}$$

Since $\left(\frac{1}{2}\right)^x = \frac{1}{32}$, when x = 5; therefore, t = 5 × (2.8 × 10³ s)= 1.4 × 10⁴ s.

(c) $kt = \ln[A]_0 - \ln[A]_t = \ln(3.4 \times 10^{-3} \text{ mol/L}) - \ln(2.3 \times 10^{-5} \text{ mol/L}) = 5.00$

$$t = \frac{5.00}{k} = \frac{5.00}{2.5 \times 10^{-4} \text{ s}^{-1}} = 2.0 \times 10^4 \text{ s}$$

A Nanoscale View: Elementary Reactions

55. A reaction with exactly one reactant is unimolecular and elementary. A reaction with exactly two reactants, either two of the same reactants or one of each of two different reactants, is bimolecular and elementary. A reaction that has more than two reactants, in any combination, is not elementary. We will assume that these reactions fit one of these three descriptions.

(a) The reaction has three reactants (one CH_4 and two O_2 molecules), so it is not elementary.

(b) The reaction has two reactants (one O_3 molecule and one O atom), so it is bimolecular and elementary.

(c) The reaction has three reactants (one Mg atom and two H_2O molecules), so it is not elementary.

(d) The reaction has one reactant (one O_3 molecule), so it is unimolecular and elementary.

57. The reaction of NO with O_3 will have a more important steric factor than the reaction of Cl with O_3, because NO is an asymmetrical molecule and Cl is a symmetrical atom. All collisions with a symmetrical atom could be effective, if they have enough energy. Collisions with the "wrong end" of the asymmetric molecule might be ineffective just because of which atoms came in contact during the collision.

Temperature and Reaction Rates

60. *Define the problem:* Given the activation energy of a reaction find the ratio of the rates at two different temperatures.

Develop a plan: A ratio of the rates when only the temperature changes must be equal to a ratio of the rate constants. Convert the temperatures to Kelvin, then use the equation derived in Problem-Solving Example 13.9 to calculate the rate constant ratio.

Execute the plan: Assume all the temperature data is $\pm$ 1 °C

E_a = 19 kJ/mol (from Problem-Solving Example 13.8).

$$T_1 = 50. \text{°C} + 273.15 = 323 \text{ K}, \qquad T_2 = 25 \text{°C} + 273.15 = 298 \text{ K}$$

$$\frac{rate_1}{rate_2} = \frac{k_1}{k_2} = e^{\frac{E_a}{R}\left(\frac{1}{T_2} - \frac{1}{T_1}\right)} = e^{\frac{(19 \text{ kJ}/\text{mol})}{(0.008314 \text{ kJ}/\text{mol·K})}\left(\frac{1}{(298 \text{ K})} - \frac{1}{(323 \text{ K})}\right)} = e^{0.6} = 1.8$$

Check your answer: The rate increases at a higher temperature. This makes sense.

62. As directed in Question 61, use the equation derived there (and also in Problem-Solving Example 13.9) to answer this question. We can use the method described in the answer to Question 60.

$$E_a = 76 \text{ kJ/mol}, \qquad T_1 = 50. \text{°C} + 273.15 = 323 \text{ K}, \qquad T_2 = 25 \text{°C} + 273.15 = 298 \text{ K}$$

$$\frac{rate_1}{rate_2} = \frac{k_1}{k_2} = e^{\frac{E_a}{R}\left(\frac{1}{T_2} - \frac{1}{T_1}\right)} = e^{\frac{(76 \text{ kJ/mol})}{(0.008314 \text{ kJ/mol·K})}\left(\frac{1}{(298 \text{ K})} - \frac{1}{(323 \text{ K})}\right)}$$

$$= e^{2.4} = 10.7 \text{ With strict sig figs, this should be } 1 \times 10^1 = 10$$

64. *Define the problem*: Given the values of the rate constant and several temperatures, determine the activation energy, the frequency factor and the rate constant at a new temperature.

Develop a plan: As described in Section 13.5, we can make a linear graph by taking the logarithm of the rate constant and the reciprocal of the absolute temperature. The slope is related to the activation energy and the y-intercept is related to the frequency factor. The Arrhenius equation can then be used to find the rate constant at a new temperature.

Execute the plan:

(a) *Assume all temperatures were measured to ±0.1 °C.* Convert temperatures to Kelvin:

T (°C)	T (K)	$\frac{1}{T}$ (K^{-1})	ln(k)	ln(k)
25.0	298.2	0.003354	−16.348	−16.348
30.0	303.2	0.003299	−15.255	−15.255
56.2	329.4	0.003036	−11.474	−11.474
78.2	351.4	0.002846	− 8.839	− 8.839

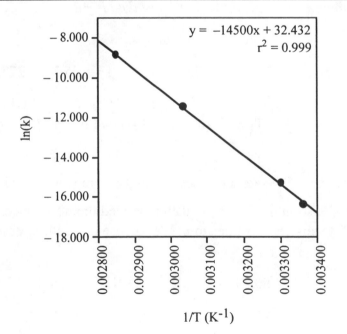

The slope, $m = -\dfrac{E_a}{R}$. Note that E_a has units of kJ/mol, using $R = 0.008314$ kJ/mol·K

$$E_a = -mR = -(-14500)(0.008314) = 120. \text{ kJ/mol}$$

The y-intercept, $b = \ln A$. Note that A has the same units as the rate constant, k.

$$A = e^b = e^{32.432} = 1.22 \times 10^{14} \text{ s}^{-1}$$

(b) Use the Arrhenius equation to estimate the rate constant.

$$T = 100.0 \text{ °C} + 273.15 \text{ K} = 373.2 \text{ K}$$

$$k = A e^{-E_a/RT} = (1.22 \times 10^{14} \text{ s}^{-1}) e^{-(120. \text{ kJ/mol})/(0.008314 \text{ kJ/mol·K})(373.2 \text{ K})}$$

$$= 1.7 \times 10^{-3} \text{ s}^{-1}$$

Check your answers: The linearity of the graph is satisfying. It also makes sense that the rate constant would be larger at a higher temperature.

66. Use the same method described in the answer to Question 64.

(a) Note that the k values must be multiplied by 10^{-5}, according to the table heading:

$\dfrac{1}{T}$ (K^{-1})	ln(k)
0.002393	11.626
0.002080	12.468
0.001923	12.889
0.001579	13.752
0.001500	13.955
0.001408	14.292
0.001355	14.433

Now make a graph of 1/T versus lnK:

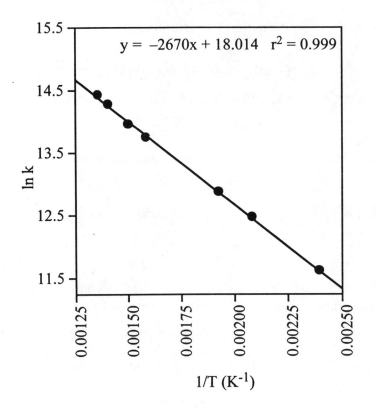

$$y = -2670x + 18.014 \quad r^2 = 0.999$$

The slope, $m = -\dfrac{E_a}{R}$. $E_a = -mR = -(-2670)(0.008314) = 22.2$ kJ/mol

The y-intercept, $b = \ln A$. $A = e^b = e^{18.014} = 6.66 \times 10^7$ $L^2 mol^{-2} s^{-1}$

(b) Use the Arrhenius Equation to estimate the rate constant.

$$k = A\,e^{-E_a/RT}$$

$$k = (6.66 \times 10^7\ L^2 mol^{-2} s^{-1})\,e^{-(22.2\,kJ/mol)/(0.008314\,kJ/mol\cdot K)(400.0\ K)}$$

$$= 8.39 \times 10^4\ L^2 mol^{-2} s^{-1}$$

68. *Define the problem*: Given the activation energy and several temperatures, determine the fraction of molecules whose energies would be energetic enough to react.

Develop a plan: In Equation 13.8 of Section 13.5, the exponential term, $e^{-E_a/RT}$, is described as the fraction of molecules with sufficient energy.

Execute the plan: *Assume all temperatures were measured to ±1 °C.* Convert temperatures to Kelvin: °C + 273.15 = Kelvin.

An example of the calculation is here for answer (a):

$$100. \,°C + 273.15 = 373 \text{ K}$$

$$\text{Fraction} = e^{-E_a/RT} = e^{-(139.7 \text{ kJ/mol})/(0.008314 \text{ kJ/mol·K})(373 \text{ K})} = e^{-45.0}$$

$$= 3 \times 10^{-20}$$

	T (K)	Fraction of sufficiently energetic molecules
(a)	373	$e^{-45.0} = 3 \times 10^{-20}$
(b)	473	$e^{-35.5} = 4 \times 10^{-16}$
(c)	773	$e^{-21.7} = 4 \times 10^{-10}$
(d)	1273	$e^{-13.20} = 1.9 \times 10^{-6}$

Check your answers: It makes sense that the fraction of molecules with enough energy to react increases with increasing temperature.

70. *Define the problem*: Given the activation energy, the concentration of the reactant, the frequency factor, and two temperatures, determine the rates of the reaction at those two temperatures.

Develop a plan: The units of the frequency factor are also the units of the rate constant. Use that information to determine the order of the reaction. Write the rate law for the reaction in terms of k and concentration. Use the Arrhenius equation to find the value of k, then plug it and the concentration into the rate law.

Execute the plan: The reaction is first-order since the units of A and therefore the units of k are time^{-1}. (Refer to Table 13.2)

$$\text{Rate} = k[CH_3CH_2I]^1$$

$$k = A \, e^{-E_a/RT}$$

Assume all temperatures were measured to ±1 °C. Convert temperatures to Kelvin.

An example of the calculation is here for answer (a):

$$400. \,°C + 273.15 = 673 \text{ K}$$

$$k = (1.2 \times 10^{14} \text{ s}^{-1}) \times e^{-(221 \text{ kJ/mol})/(0.008314 \text{ kJ/mol·K})(673 \text{ K})}$$

$$= (1.2 \times 10^{14} \text{ s}^{-1}) \times e^{-39.5} = 8 \times 10^{-4} \text{ s}^{-1}$$

$$\text{Rate} = (8 \times 10^{-4}\ s^{-1}) \times (0.012\ \text{mol/L})^1 = 1 \times 10^{-5}\ \text{mol}\ L^{-1}s^{-1}$$

	T (K)	k (s^{-1})	Rate (mol $L^{-1}s^{-1}$)
(a)	673	8×10^{-4}	1×10^{-5}
(b)	1073	2×10^{3}	3×10^{1}

Check your answers: It makes sense that the rate of the reaction increases with increasing temperature. The three significant figures of the activation energy and T limit the number of significant figures we can report.

72. *Define the problem:* Given two different temperatures and the rate constants at those temperatures, determine the activation energy of a reaction.

Develop a plan: Convert the temperatures to Kelvin, then use the equation derived in Problem-Solving Example 13.9 to calculate the activation energy:

$$\ln\left(\frac{k_1}{k_2}\right) = \frac{E_a}{R}\left(\frac{1}{T_2} - \frac{1}{T_1}\right)$$

Execute the plan: Assume all the temperature data is ± 1 °C.

$$25.\ °C + 273.15 = 298\ K \qquad\qquad 55.\ °C + 273.15 = 328\ K$$

$$\ln\left(\frac{3.46 \times 10^{-5}\ s^{-1}}{1.5 \times 10^{-3}\ s^{-1}}\right) = \frac{E_a}{R}\left(\frac{1}{328\ K} - \frac{1}{298\ K}\right)$$

$$\ln\left(2.3 \times 10^{-2}\right) = \frac{E_a}{(0.008134\text{kJ}/\text{mol} \cdot K)}\left(0.00305\ K^{-1} - 0.00336\ K^{-1}\right)$$

$$-3.77 = \frac{E_a}{(0.008134\text{kJ}/\text{mol} \cdot K)}\left(0.00031\ K^{-1}\right)$$

$$-3.77 = E_a\ (-.037\ \text{mol/kJ})$$

$$E_a = 1.0 \times 10^2\ \text{kJ/mol}$$

Check your answers: The calculated E_a has a reasonable size and sign, similar to those seen in previous problems. The result of the subtraction limits the number of significant figures we can report, suggesting that the temperature data should be kept more precisely.

Rate Laws for Elementary Reactions

75. *Define the problem*: Given the activation energy for the forward reaction of an elementary reaction, and the activation energy for the reverse of the same reaction, determine if the forward reaction is exothermic or endothermic.

Develop a plan: The difference between the forward and reverse activation energies is the ΔE for the reaction. If ΔE is positive, the forward reaction is endothermic, and if it is negative, the forward reaction is exothermic.

Execute the plan: $\Delta E = E_{a,forward} - E_{a,reverse} = 32\ kJ/mol - 58\ kJ/mol = -26\ kJ/mol$

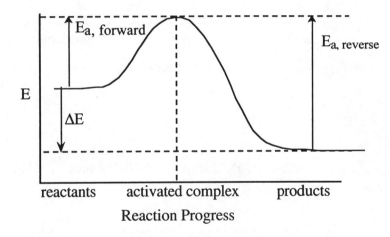

The reaction is exothermic.

Check your answer: The reverse reaction requires a higher activation energy than the forward reaction, suggesting that the reactants are higher in energy than the reactants. It makes sense that the reaction is exothermic.

77. Draw these diagrams using the information described in the answer to Question 75, and in Section 13.4. ΔH and ΔE are identical for these diagrams.

(a) $E_{a,reverse} = E_{a,forward} - \Delta H = (75 \text{ kJ mol}^{-1}) - (-145 \text{ kJ mol}^{-1}) = 220. \text{ kJ mol}^{-1}$

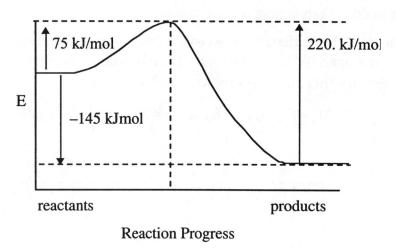

(b) $E_{a,reverse} = E_{a,forward} - \Delta H = (65 \text{ kJ mol}^{-1}) - (-70. \text{ kJ mol}^{-1}) = 135 \text{ kJ mol}^{-1}$

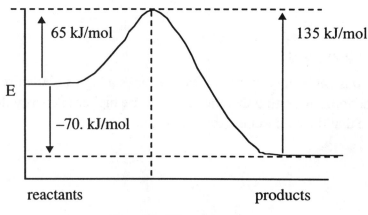

(c) $E_{a,reverse} = E_{a,forward} - \Delta H = (85 \text{ kJ mol}^{-1}) - (+70. \text{ kJ mol}^{-1}) = 15 \text{ kJ mol}^{-1}$

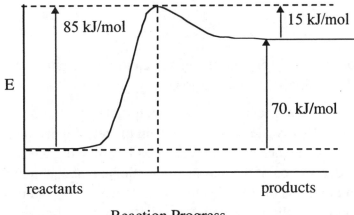

Reaction Progress

79. Assuming everything about the forward reactions in Questions 77 is identical except the activation energy, the reaction with the largest $E_{a,forward}$ will be the slowest, and that with the smallest $E_{a,forward}$ will be the fastest.

(a) Of the three, the smallest $E_{a,forward}$ is 65 kJ mol^{-1}. That is reaction (b).

(b) Of the three, the largest $E_{a,forward}$ is 85 kJ mol^{-1}. That is reaction (c).

81. Assuming everything about the reverse reactions in Questions 77 is identical except the activation energy, the reaction with the largest $E_{a,reverse}$ will be the slowest, and that with the smallest $E_{a,reverse}$ will be the fastest.

(a) Of the three, the smallest $E_{a,reverse}$ is 15 kJ mol^{-1}. That is the reverse of reaction (c).

(b) Of the three, the largest $E_{a,reverse}$ is 220 kJ mol^{-1}. That is the reverse of reaction (a).

83. *Define the problem*: Given the chemical equations of elementary reactions write their rate laws.

Develop a plan: As described in Section 13.6, the stoichiometric coefficient of a reactant in an elementary reaction is the reaction order for that reactant.

Execute the plan:

(a) $NO + NO_3 \longrightarrow$ products   rate = $k[NO][NO_3]$

(b) $O + O_3 \longrightarrow$ products   rate = $k[O][O_3]$

(c) $(CH_3)_3CBr \longrightarrow$ products   rate $= [(CH_3)_3CBr]$

(d) $2\ HI \longrightarrow$ products   rate $= k[HI]^2$

Reaction Mechanisms

87. (a) *Define the problem*: Given a mechanism, determine the rate law in terms of reactants (and possibly products') concentrations.

Develop a plan: This mechanism has a fast initial step. Following the description given in Section 13.7, write the rate law from the rate-determining step that includes an intermediate as a reactant. Set up an equation showing the steady state for the creation and destruction of the intermediate. Solve that equation for the intermediate concentration in terms of the concentrations of reactants (and possibly products). Use that equation to eliminate the intermediate concentration from the rate law.

Execute the plan: The slow step is step 2, so the rate of the reaction is equal to the rate of this rate-determining step. Because this reaction is elementary, its rate law is related to the stoichiometry of its reactants:

$$\text{rate of reaction} = \text{rate}_{\text{reaction 2}} = k_2[NOCl_2][NO]$$

$NOCl_2$ is an intermediate, so we need to eliminate it from the proposed mechanism. Look for all the ways this intermediate is created and destroyed during the mechanism. It is only created in the forward reaction of step 1. It is destroyed in the reverse reaction of step 1 and also in the reaction of step 2. Because all three of these reactions are elementary, use their reactants' stoichiometric coefficients to write rate laws for each of these three reactions:

$$\text{rate}_{\text{forward reaction 1}} = k_1[NO][Cl_2]$$

$$\text{rate}_{\text{reverse reaction 1}} = k_{-1}[NOCl_2]^1$$

$$\text{rate}_{\text{reaction 2}} = k_2[NOCl_2][NO]$$

The rate of $NOCl_2$ creation is equal to the rate of its destruction once the steady state condition is reached:

$$\text{rate}_{\text{forward reaction 1}} = \text{rate}_{\text{reverse reaction 1}} + \text{rate}_{\text{reaction 2}}$$

$$k_1[NO][Cl_2] = k_{-1}[NOCl_2]^1 + k_2[NOCl_2][NO]$$

Because the rate of step 2 is presumed to be much smaller than the rate of step 1, the second term is presumed to be negligibly small compared to the first term:

$$\text{rate}_{\text{forward reaction 1}} \cong \text{rate}_{\text{reverse reaction 1}}$$

$$k_1[NO][Cl_2] = k_{-1}[NOCl_2]^1$$

$$[NOCl_2] = \frac{k_1}{k_{-1}}[NO][Cl_2]$$

Substitute the equality just derived into the rate law: $\text{rate} = k_2[NOCl_2][NO]$

$$\text{rate} = k_2\left(\frac{k_1}{k_{-1}}[NO][Cl_2]\right)[NO]$$

We will define the new group of rate constants by one variable, $k' = k_2\dfrac{k_1}{k_{-1}}$.

$$\text{rate} = k'[NO][Cl_2][NO]$$

$$\text{rate} = k'[NO]^2[Cl_2]$$

Check your answer: The first step involves the reaction of NO and Cl_2 molecules in the formation of the steady state of the intermediate, so it makes sense that the concentration of Cl_2 is involved in the reaction's rate along with the concentration of NO.

(b) Suggesting mechanisms involves some creativity. We need to make sure that the rate law is satisfied, but we often must use our imaginations to determine what may gets formed during these reactions. Here is an example of a reaction that also satisfies the rate law: $\text{rate} = k'[NO]^2[Cl_2]$.

$$2\,NO \; \rightleftharpoons \; N_2O_2 \qquad \text{Fast}$$

$$N_2O_2 + Cl_2 \; \longrightarrow \; 2\,NOCl_2 \quad \text{Slow}$$

To think this one up, we just switched the roles of one NO molecule and the Cl_2 molecule, putting both NO molecules in the fast reaction and making an intermediate, then having the Cl_2 react with that intermediate.

To confirm that it qualifies, let's rederive the rate law for this new mechanism as described in (a):

$$rate\ of\ reaction = rate_{reaction\ 2} = k_2[N_2O_2][Cl_2]$$

Setting up the steady state for the N_2O_2 intermediate gives this equation:

$$k_1[NO]^2 = k_{-1}[N_2O_2]^1 + k_2[N_2O_2][Cl_2]$$

Again the rate of step 2 is presumed to be much smaller than the rate of step 1:

$$k_1[NO]^2 = k_{-1}[N_2O_2]^1$$

$$[N_2O_2] = \frac{k_1}{k_{-1}}[NO]^2$$

Substitute the equality just derived into the rate law: $rate = k_2[N_2O_2][Cl_2]$

$$rate = k_2\left(\frac{k_1}{k_{-1}}[NO]^2\right)[Cl_2]$$

We again define the new group of rate constants by one variable, $k' = k_2\dfrac{k_1}{k_{-1}}$.

$$rate = k'[NO]^2[Cl_2]$$

(c) Many mechanisms will not satisfy this rate law: $rate = k'[NO]^2[Cl_2]$. Here is an example of a reaction that does not satisfy the rate law: .

$$NO + Cl_2 \rightleftharpoons NOCl + Cl \quad Slow$$

$$NO + Cl \longrightarrow NOCl \quad Fast$$

This reaction has a rate law that looks like this: $rate = [NO][Cl_2]$

It is not the same as the observed rate law and cannot be the mechanism for this reaction.

89. *NOTE: There is a typo in this Question. The last $HOCH_3$ in the slow step of the mechanism should be " $+ H_3O^+$ ".*

(a) As we did in Chapter 6 with Hess's Law problems, cancel anything that ends up on both sides of the equation and add up the rest.

$$CH_3COOCH_3 + \cancel{H_3O^+} \longrightarrow \cancel{CH_3C(OH)OCH_3^+} + \cancel{H_2O}$$

$$+ \cancel{CH_3C(OH)OCH_3^+} + \cancel{2}\,H_2O \longrightarrow CH_3COOH + CH_3OH + \cancel{H_3O^+}$$

$$\overline{CH_3COOCH_3 + H_2O \longrightarrow CH_3COOH + CH_3OH}$$

(b) Follow the same method as described in the answer to Question 87.

The slow step is step 2, so the rate of the reaction is equal to the rate of this rate-determining step. Because this reaction is elementary, its rate law is related to the stoichiometry of its reactants:

$$\text{rate of reaction} = \text{rate}_{\text{reaction 2}} = k_2[CH_3C(OH)OCH_3{}^+][H_2O]^2$$

$CH_3C(OH)OCH_3{}^+$ is an intermediate, so we need to eliminate it from the proposed mechanism. Look for all the ways this intermediate is created and destroyed during the mechanism. It is only created in the forward reaction of step 1. It is destroyed in the reverse reaction of step 1 and also in the reaction of step 2. Because all three of these reactions are elementary, use their reactants' stoichiometric coefficients to write rate laws for each of these three reactions:

$$\text{rate}_{\text{forward reaction 1}} = k_1[CH_3COOCH_3][H_3O^+]$$

$$\text{rate}_{\text{reverse reaction 1}} = k_{-1}[CH_3C(OH)OCH_3{}^+][H_2O]$$

$$\text{rate}_{\text{reaction 2}} = k_2[CH_3C(OH)OCH_3{}^+][H_2O]^2$$

The rate of $CH_3C(OH)OCH_3{}^+$ creation is equal to the rate of its destruction once the steady state condition is reached:

$$\text{rate}_{\text{forward reaction 1}} = \text{rate}_{\text{reverse reaction 1}} + \text{rate}_{\text{reaction 2}}$$

$$k_1[CH_3COOCH_3][H_3O^+] = k_{-1}[CH_3C(OH)OCH_3{}^+][H_2O]$$
$$+ k_2[CH_3C(OH)OCH_3{}^+][H_2O]^2$$

Because the rate of step 2 is presumed to be much smaller than the rate of step 1, the second term is presumed to be negligibly small compared to the first term:

$$k_1[CH_3COOCH_3][H_3O^+] \cong k_{-1}[CH_3COHOCH_3{}^+][H_2O]$$

$$[CH_3C(OH)OCH_3{}^+] = \frac{k_1}{k_{-1}} \frac{[CH_3C(OH)OCH_3{}^+][H_3O^+]}{[H_2O]}$$

Substitute the equality just derived into the rate law:

$$\text{rate} = k_2[CH_3C(OH)OCH_3{}^+][H_2O]^2$$

$$\text{rate} = k_2\left(\frac{k_1}{k_{-1}} \frac{[CH_3C(OH)OCH_3{}^+][H_3O^+]}{[H_2O]}\right)[H_2O]^2$$

We will define the new group of rate constants by one variable, $k' = k_2 \dfrac{k_1}{k_{-1}}$.

$$\text{rate} = k'[CH_3COOCH_3][H_3O^+][H_2O]$$

In aqueous solutions, the concentration of the solvent water is essentially constant, so we can make a new rate constant, $k'' = k'[H_2O]$

$$\text{rate} = k''[CH_3COOCH_3][H_3O^+]$$

(c) A catalyst shows up in a mechanism as a reactant in an early step and then again as a product in a later step. There is a catalyst in this reaction. It is H_3O^+.

(d) An intermediate appears in a mechanism as a product in an early step and then again as a reactant in a later step. As described above, one intermediate is $H_3C(OH)OCH_3^+$. A molecule of H_2O is also created then destroyed in this reaction, but another H_2O molecule is consumed, water may be considered to be both a reactant and an intermediate.

Catalysts and Reaction Rate

92. (a) is true. The concentration of a homogeneous catalyst may appear in the rate law.

(b) is false. A catalyst may be consumed in a reaction, but must always be recreated.

(c) is false. A homogeneous catalyst is in the same phase as the reactants, but a heterogeneous catalyst is always in a different phase.

(d) is false. A catalyst can change the mechanism of a reaction and allow different intermediates to be produced, but it cannot change the products formed. A chemical that changes the products is not called a catalyst. Note that if a multiple set of reactions occurs simultaneously with the same reactants, a catalyst that speeds only one of these reactions would help produce one product in favor of some others. In this regard, the statement might be considered true, but it is probably not something students studying this chapter would think of.

94. *NOTE: There are typos in (b) and (c).*

A catalyst may appear in the rate law, though it will not appear in the overall equation representing the reaction. Any rate law with chemicals that are not reactants or products involve catalysts. If the catalyst has the same phase as the reactants, it is homogeneous. If the catalyst has a different phase from the reactants, it is heterogeneous.

(a) This aqueous reaction involves a homogeneous catalyst, $H_3O^+(aq)$.

(b) This reaction does not appear to involve a catalyst. Both of the concentrations in the rate law are of reactants.

(c) This gas phase reaction involves a heterogeneous catalyst, Pt(s), and apparently also a homogeneous catalyst gas-phase I_2.

In reality, there are probably typos in (b) and (c). They should be:

(b) $H_2(g) + I_2(g) \longrightarrow 2\,HI(g)$ *Rate = k[H₂][I₂](area of Pt surface)*

   with a heterogeneous catalyst, Pt(s)

(c) $2\,H_2(g) + O_2(g) \longrightarrow 2\,H_2O(g)$ *Rate = k[H₂][O₂]*

   with no catalyst.

(d) This reaction does not appear to involve a catalyst. Both of the concentrations in the rate law are of reactants.

96. Considering the two equations describing the different k values:

$$k_1 = A\,e^{-E_a/RT} \qquad\qquad k_2 = A\,e^{-E_a'/RT}$$

Divide the first equation by the second equation to eliminate A.

$$\frac{k_1}{k_2} = \frac{A e^{-E_a/RT}}{A e^{-E_a'/RT}} = e^{-E_a/RT} \times e^{+E_a'/RT} = e^{\frac{E_a'-E_a}{RT}}$$

97. The instructions in Question 96 were to use the equation derived there to solve this problem. This is the equation derived:

$$\frac{k_1}{k_2} = e^{\frac{E_a'-E_a}{RT}}$$

In Exercise 13.8, we are given $E_a = 10$ kJ/mol and T = 370 K. If $E_a' = 0$ kJ/mol, the rate increases because the E_a is lowered, the ratio of the rates is equal to the ration of the rate constants. The rate changes by a factor of:

$$\frac{rate_1}{rate_2} = e^{(10\text{ kJ/mol}-0\text{ kJ/mol})/(0.008314\text{ kJ/mol·K})(370\text{ K})} = 26$$

The reaction goes approximately 26 times faster. *(Strictly following the significant figures provided, the answer should really be $e^3 = 10^1$.)*

Enzymes: Biological Catalysts

100. We will use information described in Section 13.2, and methods similar to those in the answers to Questions 39 - 46.

The reaction rate is proportional to the concentration in a first-order reaction, rate = k[E]. That means the rate increase is related to the ratios:

$$\frac{\text{rate}_2}{\text{rate}_1} = \frac{[E]_2}{[E]_1} = \frac{4.5 \times 10^{-6} \text{ M}}{1.5 \times 10^{-6} \text{ M}} = 30.$$

The reaction rate is increased by a factor of 30. times.

Catalysts in Industry

103. Catalysts make possible the production of vital products. Without them many necessities and luxuries could not be made efficiently, if at all.

General Questions

106. We will use information described in Section 13.2, and methods similar to those in the answers to Questions 39 - 46.

The reactants are H_2 and NO, so the rate law looks like this:

$$\text{rate} = k[H_2]^i[NO]^j \quad \text{where k, i and j are currently unknown.}$$

When the initial concentration of H_2 is halved (presuming that the initial concentration of NO stays constant), the initial rate is halved. The rate change is the same as the concentration change, which suggests that the rate is proportional to the concentration of H_2, and the order with respect to H_2 is one.

When the initial concentration of NO increases by a factor of 3 (presuming that the initial concentration of H_2 stays constant), the initial rate changes by a factor of 9. The rate change is the square of the concentration change, which suggests that the rate is proportional to the square of the concentration of NO, and the order with respect to NO is two.

The rate equation, also called the rate law, now looks like this: $\text{rate} = k[H_2]^1[NO]^2$

This is different from the rate law derived in the answer to Question 33, the mechanism must be different for the reaction at this (undisclosed) temperature.

108. Follow methods similar to those used the answers to Questions 85 - 91.

(a) The observed rate law can be used to find the reaction orders of the reactants:

$$\text{rate} = k[HCrO_4^-][H_2O_2][H_3O^+]$$

Since all the concentrations are raised to the same power, all three of them are first-order. That is, the reaction is first-order in $HCrO_4^-$, first-order in H_2O_2, and first-order in H_3O^+.

(b) Cancel intermediates, H_2CrO_4 and $H_2CrO(O_2)_2$, and add the three reactions.

$$HCrO_4^- + H_3O^+ \rightleftharpoons H_2\cancel{CrO_4} + H_2O$$
$$H_2\cancel{CrO_4} + H_2O_2 \longrightarrow H_2\cancel{CrO(O_2)_2} + H_2O$$
$$+\ H_2\cancel{CrO(O_2)_2} + H_2O_2 \longrightarrow CrO(O_2)_2 + 2\ H_2O$$
$$\overline{HCrO_4^- + 2\ H_2O_2 + H_3O^+ \longrightarrow CrO(O_2)_2 + 4\ H_2O}$$

(c) It is clear, for two reasons, that the first step is not the rate limiting step. First, the double arrow used between the reactants and products indicates that the reaction is fast enough to reach a steady state. Second, the rate law that would be derived if the first step was slow is: rate = $k[HCrO_4^-][H_3O^+]$, which is incompatible with the observed rate law.

If we derive the rate law for the mechanism assuming that the second step were slow, we have a step that looks like this:

$$\text{rate} = k_2[H_2CrO_4][H_2O_2]$$

We set up a steady state condition for the intermediate, H_2CrO_4.

$$k_1[HCrO_4^-][H_3O^+] = k_{-1}[H_2CrO_4][H_2O] + k_2[H_2CrO_4][H_2O_2]$$

Assume that the second term is negligibly small since the rate of step 2 is small:

$$k_1[HCrO_4^-][H_3O^+] = k_{-1}[H_2CrO_4][H_2O]$$

Solve for the concentration of the intermediate:

$$[H_2CrO_4] = \frac{k_1}{k_{-1}} \frac{[HCrO_4^-][H_3O^+]}{[H_2O]}$$

Plug into the rate law:

$$\text{rate} = k_2 \frac{k_1}{k_{-1}} \frac{[HCrO_4^-][H_3O^+]}{[H_2O]}[H_2O_2]$$

Because the concentration of water in aqueous solutions is essentially constant, the observed rate constant is defined as $k = k_2 \dfrac{k_1}{k_{-1}[H_2O]}$, giving the derived rate law the same functional form as the observed rate law.

$$\text{rate} = k[HCrO_4^-][H_2O_2][H_3O^+]$$

Therefore, the second step is the rate-limiting step.

109. $Pt(NH_3)_2Cl_2$ is a reactant in the reaction and its concentration decreases as the reaction proceeds in the forward direction to form products, whereas Cl^- is a product in the reaction and its concentration increases as the reaction proceeds in the forward direction.

110. Most mechanisms do not specify relative rates for the sub-processes. A bicycle gear-changing mechanism and the mechanism of an elevator are similar to a chemical mechanism in that they all describe how something is accomplished (production of products, gear changed, elevator lifted or lowered).

111. Catalysts make possible the efficient production of many modern materials and substances. Without them, the cost of producing many of these products would be much higher (because of higher energy requirements, for example), or we might not be able to make them quickly enough to meet the demand.

112. The catalytic role of the chlorine atom (produced from the light decomposition of chlorofluorocarbons) in the mechanism for the destruction of ozone indicates that even small amounts of CFCs released into the atmosphere pose a serious risk.

113. A reaction is first-order in A only if the rate is proportional to the concentration of A. Looking that the three choices, (a) rate = $k[A][B]$, (b) rate = $k[A][B]^2$, and (c) rate = $k[B]^2$, the only one that doesn't show rate proportional to [A] is (c). Therefore, (c) cannot be correct.

Applying Concepts

114. This question uses methods described in the answers to Questions 15 - 17.

Line A represents the increase in the concentration of the product $H_2O(g)$ with time. Line B represents the increase in the concentration of product O_2 with time. The O_2 curve is half as steep as the H_2O line because the stoichiometric relationship between them is 1:2. Line C represents the decrease in the concentration of reactant $H_2O_2(g)$ with time.

116. Products form more quickly at higher temperatures, so (a) the snapshot with more HI molecules is the one corresponding to a higher temperature.

118. *NOTE: THE FIRST TWO OF THE PICTURES IN THIS QUESTION ARE INCORRECT. Both of the top figures should have one less C and one more A (i.e., In both pictures 1 and 2, one of the green balls should be orange).*

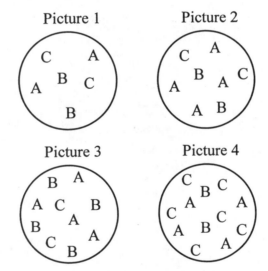

We will use the corrected information described above, the techniques described in Section 13.2, and methods similar to those in the answers to Questions 39 - 46.

The reactants are A, B and C, so the rate law looks like this:

$$\text{rate} = k[A]^i[B]^j[C]^h \quad \text{where k, i, j, and h are currently unknown.}$$

Between pictures 1 and 2, the initial concentration of A doubles (2 A → 4 A), and the initial concentrations of B and C stay constant (2 of each), the initial rate changes by a factor of (12/1.5 =) eight. The rate change is the cube of the concentration change $(8 = 2^3)$, which suggests that the order with respect to A is three.

Between pictures 2 and 3, the initial concentration of B doubles (2 B → 4 B), and the initial concentrations of A and C stay constant (4 A and 2 C), the initial rate changes by a factor of (23/12 =) two. The rate change is equal to the concentration change, which suggests that the order with respect to B is one.

Between pictures 2 and 4, the initial concentration of C triples (2 C → 6 C), and the initial concentrations of A and B stay constant (4A and 2B), the initial rate changes by a factor of nine. The rate change is the square of the concentration change($9 = 3^2$), which suggests that the the order with respect to C is two.

The rate equation, also called the rate law, now looks like this: rate = $k[A]^3[B][C]^2$

Chapter 14: Chemical Equilibrium

Characteristics of Chemical Equilibrium

9. Solid/liquid equilibria are discussed in Section 11.3

   (a) The temperature of an equilibrium mixture of ice and water must be at the melting point of water: 0 °C.

   (b) This is a dynamic equilibrium. Molecules are not smart enough to stay in a particular phase. Some molecules at the interface between the water and the ice detach from the ice and enter the liquid phase or attach to the solid phase leaving the liquid phase.

The Equilibrium Constant

11. Draw the reactant concentration vs. time as a downward curve to level off at the equilibrium concentration. Draw the product concentration vs. time curve as an upward curve to level off at the equilibrium concentration. The slope of [NO] increase is steeper than slope of [N_2] decrease, and the equilibrium concentration of NO is equal to half of the equilibrium concentration of N_2 or O_2.

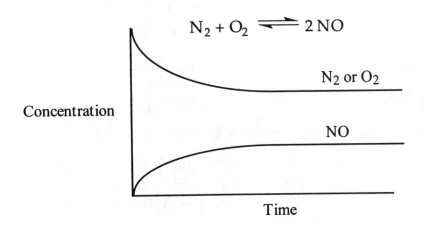

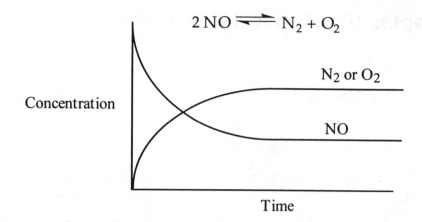

$$2\,NO \rightleftharpoons N_2 + O_2$$

13. For the expression of the equilibrium constant, use this form: $K_c = \dfrac{[\text{products}]}{[\text{reactants}]}$.

Remember that the equilibrium is not affected by the addition of any solids or liquids, so those materials do not show up in the equilibrium expression.

(a) $K_c = \dfrac{[H_2O]^2[O_2]}{[H_2O_2]^2}$

(b) $K_c = \dfrac{[PCl_5]}{[PCl_3][Cl_2]}$

(c) $K_c = [CO]^2$

(d) $K_c = \dfrac{[H_2S]}{[H_2]}$

15. Follow the same instructions given in the answer to Question 13.

(a) $K_c = [Cl_2]$

(b) $K_c = \dfrac{[Cu^{2+}][Cl^-]^4}{[CuCl_4^{2-}]}$

(c) $K_c = \dfrac{[CO_2][H_2]}{[CO][H_2O]}$

(d) $K_c = \dfrac{[Mn^{2+}][Cl_2]}{[H_3O^+]^4[Cl^-]^2}$

17. Follow the same instructions given in the answer to Question 13.

(a) $K_c = [H_2O]^2$

(b) $K_c = \dfrac{[HF]^4}{[SiF_4][H_2O]^2}$

(c) $K_c = \dfrac{[HCl]^2}{[H_2O]}$

19. A very large K value means that the reaction is product-favored; however, it does not indicate how fast the reaction is. It is clear from the identity of the reactants and products that a great deal of rearrangement in bonding must occur to make products from reactants. That suggests a complex mechanism and/or a high activation energy. Reactants of reactions with a high activation energy are kinetically stable.

21. The two equations are related so we will use the information in Section 14.2 to identify how their equilibrium constants are related.

Multiplying the first equation by a constant factor of 3 gives the second equation. This change means we need to raise the K from the first reaction to the power of 3. Therefore the second equation's equilibrium constant is related to the first equation's equilibrium constant as represented in (c), $K_p^3 = K_p'$.

23. The two equations are related so we will use the information in Section 14.2 to identify how their equilibrium constants are related.

Multiplying the first equation by a constant factor of $\frac{1}{2}$ gives the second equation. This change means we need to raise the K from the first reaction to the power of $\frac{1}{2}$. Therefore the second equation's equilibrium constant (K_{c_2}) is related to the first equation's equilibrium constant (K_{c_1}) this way:

$$K_{c_2} = K_{c_1}^{1/2} = (0.16)^{1/2} = 0.40$$

24. For the expression of the equilibrium constant, use this form: $K_P = \dfrac{P_{products}}{P_{reactants}}$.

Remember that the equilibrium is not affected by the addition of any solids or liquids, so those materials do not show up in the equilibrium expression.

(a) $K_P = \dfrac{P_{H_2O}^2 P_{O_2}}{P_{H_2O_2}^2}$

(b) $K_p = \dfrac{P_{PCl_5}}{P_{PCl_3} P_{Cl_2}}$

(c) $K_p = P_{CO}^2$

(d) $K_p = \dfrac{P_{H_2S}}{P_{H_2}}$

26. *Define the problem:* Given a phase change reaction and the vapor pressure of the gaseous product, find the value of K_c.

Develop a plan: Write the K_p expression for the reaction and plug in the known value. Then use the relationship between K_p and K_c (Equation 14.5, with $R = 0.08206 \dfrac{L \cdot atm}{mol \cdot K}$, and Δn = change in the number of moles of gas in the reaction) to get K_c.

Execute the plan: $H_2O(\ell) \rightleftharpoons H_2O(g)$ $K_p = P_{H_2O(g)} = 0.467$ atm

(Assume temperature is known within ± 1 °C.) $T = 80.\ °C + 273.15 = 353$ K

$$\Delta n = 1 \text{ mole } H_2O \text{ gas product} - 0 \text{ moles gas reactants} = 1$$

$$K_p = (RT)^{\Delta n} K_c \qquad\qquad (RT)^{-\Delta n} K_p = K_c$$

$$K_c = \left\{ \left(0.08206 \ \frac{L \cdot atm}{mol \cdot K}\right) \times (353 \text{ K}) \right\}^{-1} \times (0.467 \text{ atm}) = 0.0161$$

Check your answer: This $(RT)^{-1}$ factor makes K_c smaller than K_p. This answer makes sense.

Determining Equilibrium Constants

29. *Define the problem:* Given the equation for a dimerization reaction and the concentrations of the gaseous reactants and product at equilibrium, find the value of K_c.

Develop a plan: Write the K_c expression for the reaction and plug in the known values.

Execute the plan: $2\ A(g) \rightleftharpoons B(g)$ $K_c = \dfrac{[B]}{[A]^2}$

(a) $$K_c = \frac{0.74 \text{ mol}/L}{\left(0.74 \text{ mol}/L\right)^2} = 1.4$$

(b) $$K_c = \frac{2.0 \text{ mol}/L}{\left(2.0 \text{ mol}/L\right)^2} = 0.50$$

(c) $$K_c = \frac{0.01 \text{ mol}/L}{\left(0.01 \text{ mol}/L\right)^2} = 100$$

If the concentrations of the reactants and products are equal **and their stoichiometric coefficients are the same,** then the equilibrium constant is always 1.0. This is seen in the answers to Question 28. Here, the stoichiometric coefficients are not equal and the value of K is not 1.0.

Check your answers: Because the reactants' concentrations are squared, it makes sense that the value of K_c gets larger as the concentrations get smaller.

32. *Define the problem*: Given an equation for a reaction and the moles of the gaseous reactants and products at equilibrium in a known volume, find the value of K_p.

Develop a plan: Write the K_c expression for the reaction, calculate the concentrations of the gases, and plug them into the expression to get the value of K_c. Use the relationship between K_p and K_c (also described in the answer to Question 26) to get K_p.

Execute the plan:

$$H_2(g) + CO_2(g) \rightleftharpoons H_2O(g) + CO(g)$$

$$K_c = \frac{[H_2O][CO]}{[H_2][CO_2]} = \frac{\left(\dfrac{0.11 \text{ mol}}{1.0 \text{ L}}\right) \times \left(\dfrac{0.11 \text{ mol}}{1.0 \text{ L}}\right)}{\left(\dfrac{0.087 \text{ mol}}{1.0 \text{ L}}\right) \times \left(\dfrac{0.087 \text{ mol}}{1.0 \text{ L}}\right)} = 1.6$$

$$T = 986 \,^\circ C + 273.15 = 1259 \text{ K}$$

$$\Delta n = 1 \text{ mole } H_2O(g) + 1 \text{ mole } CO(g) - 1 \text{ mole } H_2(g) - 1 \text{ mol } CO_2(g) = 0$$

$$K_p = (RT)^{\Delta n} K_c$$

$$K_p = \left\{ (0.08206 \, \frac{L \cdot atm}{mol \cdot K}) \times (1259 \text{ K}) \right\}^0 \times (1.6) = 1.6$$

Check your answers: The slightly larger moles of products suggest this reaction is slightly product-favored, so it makes sense that the K_c is larger than 1. The equal moles of gas-phase reactants and gas-phase products, responsible for the power in the $(RT)^0$ factor, makes K_p the same as K_c. These answers make sense.

33. Use a method similar to that described in the answer to Question 32.

$$C(s) + CO_2(g) \rightleftharpoons 2 CO(g)$$

Remember that solids are omitted from the equilibrium expression.

$$K_c = \frac{[CO]^2}{[CO_2]} = \frac{\left(\dfrac{0.10 \text{ mol}}{2.0 \text{ L}}\right)^2}{\left(\dfrac{0.20 \text{ mol}}{2.0 \text{ L}}\right)} = 0.025$$

(Assume temperature is known within ± 1 °C.) $T = 700.$ °C $+ 273.15 = 973$ K

$$\Delta n = 2 \text{ mole } CO(g) - 1 \text{ mol } CO_2(g) = 1$$

$$K_p = (RT)^{\Delta n} K_c$$

$$K_p = \{(0.08206 \ \frac{L \cdot atm}{mol \cdot K}) \times (973 \text{ K})\}^1 \times (0.025) = 2.0$$

The smaller moles of products suggest this reaction is reactant-favored, so it makes sense that the K_c is smaller than 1. The larger moles of gas-phase reactants and gas-phase products, responsible for the power in the $(RT)^1$ factor, makes K_p larger than K_c. These answers make sense.

35. *Define the problem:* Given an equation for a reaction, the initial moles of the gaseous reactant and the equilibrium moles of a gaseous product in a known volume, find the value of K_c.

Develop a plan: Following the procedure given in Section 14.3, write the equation and construct a reaction table. Describe the stoichiometric changes in terms of one variable, x. Calculate the concentrations of the gases at equilibrium using what you know about the moles of product formed. Then write the K_c expression for the reaction, and plug the concentrations into the expression to get the value of K_c.

Execute the plan: (conc. NOCl) $= 2.00$ mol NOCl/1.00 L $= 2.00$ M

	2 NOCl(g) $\rightleftharpoons$	2 NO(g)	+ Cl$_2$(g)
conc. initial (M)	2.00	0	0
change conc. (M)	$-2x$	$+2x$	$+x$
equilibrium conc. (M)	$2.00 - 2x$	$2x$	x

$$[NO] = 0.66 \text{ mol NO}/1.00 \text{ L} = 0.66 \text{ M at equilibrium, so}$$

$$[NO] = 0.66 \text{ M} = 2x$$

$$x = 0.66 \text{ M}/2 = 0.33 \text{ M}$$

$$2.00 - 2x = 2.00 \text{ M} - 0.66 \text{ M} = 1.34 \text{ M}$$

$$K_c = \frac{[NO]^2[Cl_2]}{[NOCl]^2} = \frac{(2x)^2(x)}{(2.00-2x)^2} = \frac{(0.66)^2(0.33)}{(1.34)^2} = 0.080$$

Check your answer: The smaller product concentrations suggest this reaction is reactant-favored, so it makes sense that the K_c is smaller than 1.

36. Use a method similar to that described in the first part of the answer to Question 32.

$$CO(g) + Cl_2(g) \rightleftharpoons COCl_2(g) \qquad\qquad K_c = \frac{[COCl_2]}{[CO][Cl_2]}$$

Assume the volume is measured with a precision of ± 1 L.

$$K_c = \frac{\left(\dfrac{9.00 \text{ mol}}{50.\text{ L}}\right)}{\left(\dfrac{3.00 \text{ mol}}{50.\text{ L}}\right)\times\left(\dfrac{2.00 \text{ mol}}{50.\text{ L}}\right)} = 75$$

The larger moles of products suggest this reaction is slightly product-favored, so it makes sense that the K_c is larger than 1.

38. (a) Convert grams to moles, then use a method similar to that described in the first part of the answer to Question 32.

$$6.44 \text{ g NOBr} \times \frac{1 \text{ mol NOBr}}{109.9 \text{ g NOBr}} = 0.0586 \text{ mol NOBr}$$

$$3.15 \text{ g NO} \times \frac{1 \text{ mol NO}}{30.01 \text{ g NO}} = 0.105 \text{ mol NO}$$

$$8.38 \text{ g Br}_2 \times \frac{1 \text{ mol Br}_2}{159.8 \text{ g Br}_2} = 0.0524 \text{ mol Br}_2$$

$$2 \text{ NOBr}(g) \rightleftharpoons 2 \text{ NO}(g) + \text{Br}_2(g)$$

$$K_c = \frac{[NO]^2[Br_2]}{[NOBr]^2} = \frac{\left(\dfrac{0.105 \text{ mol}}{10.00 \text{ L}}\right)^2\left(\dfrac{0.0524 \text{ mol}}{10.00 \text{ L}}\right)}{\left(\dfrac{0.0586 \text{ mol}}{10.00 \text{ L}}\right)^2} = 0.0168$$

(b) Use a combination of Dalton's law of partial pressures (Section 10.8) and the ideal gas law, PV = nRT (Section 10.5), to calculate the total pressure.

$$n_{tot} = 0.0586 \text{ mol NOBr} + 0.105 \text{ mol NO} + 0.0524 \text{ mol Br}_2 = 0.216 \text{ mol total}$$

$$P_{tot}V = n_{tot}RT$$

$$T = 100. \text{ °C} + 273.15 = 373 \text{ K}$$

$$P_{tot} = \frac{n_{tot}RT}{V} = \frac{(0.261 \text{ mol}) \times \left(0.08206 \dfrac{\text{mol} \cdot \text{L}}{\text{K} \cdot \text{mol}}\right) \times (373 \text{ K})}{(10.00 \text{ L})} = 0.661 \text{ atm}$$

(c) Use a method similar to that described in the answer to Question 26.

$$\Delta n = 2 \text{ mole NO(g)} + 1 \text{ mole Br}_2\text{(g)} - 2 \text{ mol H}_2\text{(g)} = 1$$

$$K_p = (RT)^{\Delta n}K_c$$

$$K_p = \left\{(0.08206 \ \frac{\text{L} \cdot \text{atm}}{\text{mol} \cdot \text{K}}) \times (373 \text{ K})\right\}^1 \times (0.0168) = 0.514$$

This $(RT)^1$ factor makes K_p larger than K_c.

The Meaning of the Equilibrium Constant

40. *Define the problem:* Given a table of equations with values of K_c and K_p, order the members of a set or equations from most reactant-favored to most product-favored.

Develop a plan: Look up the given chemical equation or a related chemical equation, determine the size of its K_c, using techniques described at the end of Section 14.2; as needed. K_c values larger than 1 are product-favored. The smaller K_c is, the more reactant-favored the reaction is. The larger K_c is, the more product-favored the reaction is. Order the equations from smallest K_c to largest K_c.

Execute the plan:

(a) 2 NH$_3$(g) $\rightleftharpoons$ N$_2$(g) + 3 H$_2$(g) is the reverse of the third reaction in Table 14.1, so $K_{c,(a)} = 1/K_c = 1/(3.5 \times 10^8) = 2.9 \times 10^{-9}$. K_c is smaller than 1, so the reaction is not product-favored.

(b) $NH_4^+ + OH^-(aq) \rightleftharpoons NH_3(aq) + H_2O(\ell)$ is the reverse of the twelfth reaction in Table 14.1, so $K_{c,(b)} = 1/K_c = 1/(1.8 \times 10^{-5}) = 5.6 \times 10^4$. K_c is larger than 1, so the reaction is product-favored.

(c) $2 NO(g) \rightleftharpoons N_2(g) + O_2(g)$ is the reverse of the fourth reaction in Table 14.1, so $K_{c,(c)} = 1/K_c = 1/(4.5 \times 10^{-31}) = 2.2 \times 10^{30}$. K_c is larger than 1, so the reaction is product-favored.

Therefore, the order is (a) 2.9×10^{-9}, then (b) 5.6×10^4, then (c) 2.2×10^{30}.

Check your answer: Reversing a reaction takes what were the products and makes them the reactants and vice versa, so the reverse of a product-favored reaction will be a reactant-favored reaction and vice versa. The values of K_c make sense. Comparing the values of K to determine which is the most reactant- and product-favored among reaction with different stoichiometric relationships is not always legitimate. Here, while the denominators all have two concentration values, we find four concentration values in the numerator of (a), only two in (c), and only one in (b). These differences will affect the size of K_c dramatically. In general, one should compare the values of K_c only among reactions with the same stoichiometric relationships (i.e., with the same number of reactants and products.)

42. Dissolving ionic compounds produces aqueous solutions of ions as described in Chapters 2 and 3. Water is not explicitly included in this equation, because its only function during dissolving is as the solvent for the aqueous products. Write equilibrium expressions as was done in the answers to Question 13. Assess solubility by comparing the size of K_c as was done in the answer to Question 40.

(a) $Ag_2SO_4(s) \rightleftharpoons 2 Ag^+(aq) + SO_4^{2-}(aq)$ $K_c = [Ag^+]^2[SO_4^{2-}] = 1.7 \times 10^{-5}$

$Ag_2S(s) \rightleftharpoons 2 Ag^+(aq) + S^{2-}(aq)$ $K_c = [Ag^+]^2[S^{2-}] = 6 \times 10^{-30}$

(b) The compound that is more soluble has the larger K_c, and that is $Ag_2SO_4(s)$.

(c) The compound that is least soluble has the smaller K_c, and that is $Ag_2S(s)$.

Using Equilibrium Constants

44. *Define the problem:* Given an equation for a reaction, the initial concentration or moles of the gaseous reactant in a known volume, and the value of the equilibrium constant, determine the change in the concentrations and the equilibrium concentrations of the reactants and products.

Develop a plan: Follow the procedure given in Section 14.5. When necessary, find Q to determine direction of the reaction. Then write the equation and construct a reaction table. Describe the stoichiometric changes in terms of one variable, x. Write the K_c expression for the reaction in terms of x. Solve that equation to get x, and use it to calculate the concentrations of the gases at equilibrium.

Execute the plan: butane (g) $\rightleftharpoons$ 2-methylpropane (g)

(a)   (conc. butane) = (conc. 2-methylpropane) = 0.100 M

$$Q_c = \frac{(\text{conc. 2 - methylpropane})}{(\text{conc. butane})} = \frac{0.100 \text{ M}}{0.100 \text{ M}} = 1.00 < 2.5$$

$Q_c < K_c$, so reaction goes from reactants toward products:

	butane (g) $\rightleftharpoons$	2-methylpropane (g)
conc. initial (M)	0.100	0.100
change conc. (M)	$-x$	$+x$
equilibrium conc. (M)	$0.100 - x$	$0.100 + x$

(b) $$K_c = \frac{[2 - \text{methylpropane}]}{[\text{butane}]} = 2.5$$

At equilibrium $$\frac{0.100 + x}{0.100 - x} = 2.5$$

$$0.100 + x = 2.5(0.100 - x) = 0.25 - 2.5x$$

$$x + 2.5x = 0.25 - 0.100$$

$$3.5x = 0.15$$

$$x = 0.043$$

(c)   (conc. butane) = 0.017 mol butane/0.50 L = 0.034 M

	butane (g) $\rightleftharpoons$	2-methylpropane (g)
conc. initial (M)	0.034	0
change conc. (M)	$-x$	$+x$
equilibrium conc. (M)	$0.034 - x$	x

At equilibrium
$$\frac{x}{0.034 - x} = 2.5$$

$$x = 2.5(0.034 - x) = 0.085 - 2.5x$$

$$x + 2.5x = 0.085$$

$$3.5x = 0.085$$

$$x = 0.024 \text{ M} = [\text{2-methylpropane}]$$

$$[\text{butane}] = 0.034 \text{ M} - 0.024 \text{ M} = 0.010 \text{ M}$$

Check your answers: The equilibrium concentrations should combine to reproduce K_c:

Part (b), [butane] = 0.100 − 0.043 = 0.057 M, [2-methylpropane] = 0.100 + 0.043 =

0.143 M. $K_c = \dfrac{0.143}{0.057} = 2.5$. (c) $K_c = \dfrac{0.024}{0.010} = 2.4$. These are both right, within the

uncertainty of the data.

46. *Define the problem:* Given an equation for a reaction, the initial mass of the gaseous reactant in a known volume, and the value of the equilibrium constant, determine the mass of the reactant present at equilibrium.

Develop a plan: Convert grams to moles using the molar mass. Then follow the method described in the answer to Question 44 to get the equilibrium concentration of the reactant gas. Then use the volume and molar mass to determine the grams

Execute the plan: cyclohexane (g) $\rightleftharpoons$ methylcyclopentane (g)

$$(\text{conc. cyclohexane}) = \frac{3.79 \text{ g cyclohexane}}{2.80 \text{ L}} \times \frac{1 \text{ mol cyclohexane}}{84.15 \text{ g cyclohexane}} = 0.0161 \text{ M}$$

cyclohexane (g) $\rightleftharpoons$ methylcyclopentane (g)

	cyclohexane (g)	methylcyclopentane (g)
conc. initial (M)	0.0161	0
change conc. (M)	− x	+ x .
equilibrium conc. (M)	0.0161 − x	x

At equilibrium $K_c = \dfrac{[\text{methylcyclopentane}]}{[\text{cyclohexane}]} = 0.12$

$$\frac{x}{0.0161 - x} = 0.12$$

$$x = 0.12(0.0161 - x) = 0.0019 - 0.12x$$

$$x + 0.12x = 0.0019$$

$$1.12x = 0.0019$$

$$x = 0.0017 \text{ M} = [\text{methylcyclopentane}]$$

$$[\text{cyclohexane}] = 0.0161 \text{ M} - 0.0017 \text{ M} = 0.0144 \text{ M}$$

$$2.80 \text{ L} \times \frac{0.0144 \text{ mol cyclohexane}}{1 \text{ L}} \times \frac{84.15 \text{ g cyclohexane}}{1 \text{ mol cyclohexane}} = 3.39 \text{ g}$$

Check your answer: The quantity present at equilibrium is less than the initial quantity. The equilibrium concentrations can be used to recreate the value of the equilibrium constant: $K_c = \dfrac{0.0017}{0.0144} = 0.12$. These answers look right.

49. Use a method similar to that described in the answer to Question 44.

   (a)   (conc. I_2) = 1.00 mol/10.00 L = 0.100 M

                          (conc. H_2) = 3.00 mol/10.00 L = 0.300 M

	H_2 (g)	+ I_2 (g) $\rightleftharpoons$	2 HI (g)
conc. initial (M)	0.300	0.100	0
change conc. (M)	$-x$	$-x$	$+2x$
equilibrium conc. (M)	$0.300 - x$	$0.100 - x$	$2x$

At equilibrium $\quad K_c = \dfrac{[\text{HI}]^2}{[H_2][I_2]} = \dfrac{(2x)^2}{(0.300 - x)(0.100 - x)} = 50.0$

Set up to use the quadratic equation:

$$50.0\{(0.300)(0.100) - (0.300 + 0.100)x + x^2\} = 4x^2$$

$$1.50 - 20.0x + 50.0x^2 = 4x^2$$

$$1.50 - 20.0x + 46.0x^2 = 0$$

Plug into the quadratic equation:

$$x = \frac{-b \pm \sqrt{b^2 - 4ac}}{2a} = \frac{20.0 \pm \sqrt{(-20.0)^2 \pm 4(46.0)(1.50)}}{2(46.0)} = \frac{8.9}{92.0} = 0.097$$

$$2x = 0.194 \text{ M} = [\text{HI}]$$

$$10.00 \text{ L} \times \frac{0.194 \text{ mol}}{\text{L}} = 1.94 \text{ mol}$$

$$[H_2] = 0.300 - x = 0.300 \text{ M} - (0.097 \text{ M}) = 0.203 \text{ M}$$

$$[I_2] = 0.100 - x = 0.100 \text{ M} - (0.097 \text{ M}) = 0.003 \text{ M}$$

Check your answer: The larger concentration of product is consistent with a K greater than 1. The equilibrium concentrations should combine to reproduce K_c within the precision of the data:

$$K_c = \frac{(0.194)^2}{(0.203) \times (0.003)} = 6 \times 10^1 \cong 50.0 \text{ with only one significant figure. These}$$

results look right.

(b)   (conc. I_2) = 1.00 mol/5.00 L = 0.200 M

(conc. H_2) = 3.00 mol/5.00 L = 0.600 M

	H_2 (g)	+ I_2 (g) $\rightleftharpoons$	2 HI (g)
conc. initial (M)	0.600	0.200	0
change conc. (M)	$-x$	$-x$	$+2x$
equilibrium conc. (M)	$0.600 - x$	$0.200 - x$	$2x$

At equilibrium $K_c = \dfrac{[HI]^2}{[H_2][I_2]} = \dfrac{(2x)^2}{(0.600-x)(0.200-x)} = 50.0$

Set up to use the quadratic equation:

$$50.0\{(0.600)(0.200) - (0.600 + 0.200)x + x^2\} = 4x^2$$

$$6.00 - 40.0x + 46.0x^2 = 0$$

Plug into the quadratic equation:

$$x = \frac{-b \pm \sqrt{b^2 - 4ac}}{2a} = \frac{40.0 \pm \sqrt{(-40.0)^2 \pm 4(46.0)(6.00)}}{2(46.0)} = \frac{17.7}{92.0} = 0.192$$

$$2x = 0.384 \text{ M} = [HI]$$

$$5.00 \text{ L} \times \frac{0.384 \text{ mol}}{\text{L}} = 1.92 \text{ mol}$$

Check your answer: The larger concentration of product is consistent with a K greater than 1. The equilibrium concentrations should combine to reproduce K_c within the precision of the data:

$$[H_2] = 0.600 - x = 0.600 \text{ M} - (0.192 \text{ M}) = 0.408 \text{ M}$$

$$[I_2] = 0.200 - x = 0.200 \text{ M} - (0.192 \text{ M}) = 0.008 \text{ M}$$

$$K_c = \frac{(0.384)^2}{(0.408) \times (0.008)} = 5 \times 10^1 \cong 50.0 \text{ with only one significant figure.}$$ The answers for (a) and (b) should be the same, because the volume dependence is cancelled due to the fact that there are the same number of products as there are reactants. Within the cited uncertainty (a) and (b) give the same results.

(c)   (conc. I_2) = 1.00 mol/10.00 L = 0.100 M

(conc. H_2) = (3.00 mol + 3.00 mol)/10.00 L = 0.600 M

	H_2 (g)	+ I_2 (g) ⇌	2 HI (g)
conc. initial (M)	0.600	0.100	0
change conc. (M)	$-x$	$-x$	$+2x$
equilibrium conc. (M)	$0.600 - x$	$0.100 - x$	$2x$

At equilibrium $K_c = \dfrac{[HI]^2}{[H_2][I_2]} = \dfrac{(2x)^2}{(0.600 - x)(0.100 - x)} = 50.0$

Set up to use the quadratic equation:

$$50.0\{(0.600)(0.100) - (0.600 + 0.100)x + x^2\} = 4x^2$$

$$3.00 - 35.0x + 46.0x^2 = 0$$

Plug into the quadratic equation:

$$x = \frac{-b \pm \sqrt{b^2 - 4ac}}{2a} = \frac{35.0 \pm \sqrt{(-35.0)^2 \pm 4(46.0)(3.00)}}{2(46.0)} = \frac{9.1}{92.0} = 0.099$$

$$2x = 0.198 \text{ M} = [HI]$$

$$10.00 \text{ L} \times \frac{0.198 \text{ mol}}{\text{L}} = 1.98 \text{ mol}$$

Check your answer: The larger concentration of product is consistent with a K greater than 1. The equilibrium concentrations should combine to reproduce K_c within the precision of the data:

$$[H_2] = 0.300 - x = 0.600 \text{ M} - (0.099 \text{ M}) = 0.501 \text{ M}$$

$$[I_2] = 0.100 - x = 0.100 \text{ M} - (0.099 \text{ M}) = 0.001 \text{ M}$$

$K_c = \dfrac{(0.198)^2}{(0.501) \times (0.001)} = 7 \times 10^1 \cong 50.0.$ With only one significant figure round off

errors have started to dramatically affect the smallest number, $[I_2]$. These results still look reasonable.

51. Use a method similar to that described in the answer to Question 44.

(a)   (conc. CO) = 1.00 mol/1.00 L = 1.00 M

(conc. H_2O) = 1.00 mol/1.00 L = 1.00 M

	CO (g)	+	H_2O(g)	$\rightleftharpoons$	CO_2 (g)	+	H_2 (g)
conc. init. (M)	1.00		1.00		0		0
change conc. (M)	$-x$		$-x$		$+x$		$+x$
eq. conc. (M)	$1.00 - x$		$1.00 - x$		x		x

At equilibrium $K_c = \dfrac{[CO_2][H_2]}{[CO][H_2O]} = \dfrac{(x)(x)}{(1.00-x)(1.00-x)} = 4.00$

Take the square root of each side:

$$\dfrac{(x)}{(1.00-x)} = 2.00 \qquad x = 2.00 - 2x$$

$$x + 2x = 2.00 \qquad\qquad 3x = 2.00$$

$$x = 0.667 \text{ M} = [CO_2] = [H_2]$$

$$[CO] = [H_2O] = 1.00 \text{ M} - x = 1.00 \text{ M} - 0.667 \text{ M} = 0.33 \text{ M}$$

Check your answers: The larger concentration of products is consistent with a K greater than 1. The equilibrium concentrations should combine to reproduce K_c within the precision of the data: $K_c = \dfrac{(0.667) \times (0.667)}{(0.33) \times (0.33)} = 4.0 \cong 4.00.$ These results look right.

(b)   (conc. CO) = (1.00 mol + 1.00 mol)/1.00 L = 2.00 M

(conc. H_2O) = (1.00 mol + 1.00 mol)/1.00 L = 2.00 M

	CO (g)	+ H_2O(g)	⇌	CO_2 (g)	+	H_2 (g)
conc. init. (M)	2.00	2.00		0		0
change conc. (M)	$-x$	$-x$		$+x$		$+x$
eq. conc. (M)	$2.00-x$	$2.00-x$		x		x

At equilibrium $K_c = \dfrac{[CO_2][H_2]}{[CO][H_2O]} = \dfrac{(x)(x)}{(2.00-x)(2.00-x)} = 4.00$

Take the square root of each side:

$$\frac{(x)}{(2.00-x)} = 2.00$$

$x = 4.00 - 2x$ $x + 2x = 4.00$ $3x = 4.00$

$$x = 1.33 \text{ M} = [CO_2] = [H_2]$$

$$[CO] = [H_2O] = 1.00 \text{ M} - x = 2.00 \text{ M} - 1.33 \text{ M} = 0.67 \text{ M}$$

Check your answers: The larger concentration of products is consistent with a K greater than 1. The equilibrium concentrations should combine to reproduce K_c within the precision of the data: $K_c = \dfrac{(1.33)\times(1.33)}{(0.67)\times(0.67)} = 3.9 \cong 4.00$ within the uncertainty of the data. These results look right.

53. Use a method similar to that described in the answer to Question 44.

(conc. N_2) = (conc. H_2) = 1.00 mol/10.00 L = 0.100 M

	N_2 (g)	+	3 H_2 (g)	⇌	2 NH_3 (g)
conc. initial (M)	0.100		0.100		0
change conc. (M)	$-x$		$-3x$		$+2x$
equilibrium conc. (M)	$0.100-x$		$0.100-3x$		$2x$

At equilibrium $K_c = \dfrac{[NH_3]^2}{[N_2][H_2]^3} = \dfrac{(2x)^2}{(0.100-x)(0.100-3x)^3} = 5.97 \times 10^{-2}$

This value of x is not very small, so let's plug it back into the equation in the places where we ignored it, to obtain a more accurate value. (This method is described in Appendix A.7.) Repeat the procedure until the value of x stops changing:

$$x = \sqrt{\frac{5.97 \times 10^{-2}(0.100 - 0.00122)(0.100 - 3(0.00122))^3}{4}} = 0.00115$$

$$x = \sqrt{\frac{5.97 \times 10^{-2}(0.100 - 0.00115)(0.100 - 3(0.00115))^3}{4}} = 0.00115$$

$$\text{Percentage } N_2 \text{ converted} = \frac{\text{amount of } N_2 \text{ converted}}{\text{initial amount of } N_2} \times 100\%$$

$$= \frac{0.00115 \text{ M}}{0.100} \times 100\% = 1.15\%$$

Check your answers: The equilibrium concentrations should combine to reproduce K_c:

$$[N_2] = 0.100 - x = 0.099 \text{ M}$$

$$[H_2] = 0.100 - 3x = 0.097 \text{ M}$$

$$[NH_3] = 2x = 0.00230 \text{ M}$$

$$K_c = \frac{(0.00230)^2}{(0.099)(0.097)^3} = 5.9 \times 10^{-2}.$$

These results look right.

55. Use a method similar to that described in the answer to Question 44.

(a) (conc. PCl_5) = 0.75 mol PCl_5/5.00 L = 0.15 M

	PCl_5 (g) $\rightleftharpoons$	PCl_3 (g)	+ Cl_2 (g)
conc. initial (M)	0.15	0	0
change conc. (M)	$-x$	$+x$	$+x$
equilibrium conc. (M)	$0.15 - x$	x	x

At equilibrium $\qquad K_c = \dfrac{[PCl_3][Cl_2]}{[PCl_5]} = \dfrac{(x)(x)}{(0.15 - x)} = 3.30$

Set up to use the quadratic equation:

$$x^2 = 3.30(0.15 - x) = 0.50 - 3.30x$$

$$x^2 + 3.30x - 0.5 = 0$$

Plug into the quadratic equation:

$$x = \frac{-b \pm \sqrt{b^2 - 4ac}}{2a} = \frac{-3.30 \pm \sqrt{(3.30)^2 - 4(1)(-0.15)}}{2(1)} = \frac{0.29}{2} = 0.14$$

$$x = [PCl_5] = [Cl_2] = 0.14 \text{ M}$$

$$[PCl_5] = 0.15 - x = 0.15 \text{ M} - 0.14 \text{ M} = 0.01 \text{ M}$$

The reaction condition in (a) causes more than 90 % formation of products, so let's consider taking the reaction completely to the product's side, then bringing it back to equilibrium. This is a common tactic when reactants form products in the reaction with a K larger than 1. Use Chapter 4 stoichiometric "limiting reactant" procedure to accomplish this (Section 4.5).

	PCl$_5$ (g) $\rightleftharpoons$	PCl$_3$ (g)	+ Cl$_2$ (g)
conc. initial (M)	0.15	0	0
final conc. (M)	0	0.15	0.15
change conc. (M)	+x	−x	−x
equilibrium conc. (M)	x	0.15 − x	0.15 − x

At equilibrium $$K_c = \frac{[PCl_3][Cl_2]}{[PCl_5]} = \frac{(0.15 - x)(0.15 - x)}{(x)} = 3.30$$

Set up to use the quadratic equation:

$$(0.15)^2 - 2(0.15)x + x^2 = 3.30x$$

$$x^2 - 3.60x - 0.023 = 0$$

Plug into the quadratic equation:

$$x = \frac{-b \pm \sqrt{b^2 - 4ac}}{2a} = \frac{3.60 \pm \sqrt{(-3.60)^2 \pm 4(1)(0.023)}}{2(1)} = \frac{0.01}{2} = 0.005$$

$$[PCl_5] = 0.005 \text{ M}$$

$$x = [PCl_5] = [Cl_2] = 0.15 - 0.005 \text{ M} = 0.14 \text{ M}$$

This way we get slightly more reliable information about the smallest concentration, $[PCl_5]$.

(b)   (conc. PCl_5) = 0.75 mol PCl_5/5.00 L = 0.15 M

(conc. PCl_3) = 0.75 mol PCl_5/5.00 L = 0.15 M

	PCl_5 (g) $\rightleftharpoons$	PCl_3 (g) +	Cl_2 (g)
conc. initial (M)	0.15	0.15	0
change conc. (M)	$-x$	$+x$	$+x$
equilibrium conc. (M)	$0.15 - x$	$0.15 + x$	x

At equilibrium $K_c = \dfrac{\left[PCl_3\right]\left[Cl_2\right]}{\left[PCl_5\right]} = \dfrac{(0.15+x)(x)}{(0.15-x)} = 3.30$

Set up to use the quadratic equation:

$$0.15x + x^2 = 3.30(0.15 - x) = 0.50 - 3.30x$$

$$x^2 + 3.45x - 0.50 = 0$$

Plug into the quadratic equation:

$$x = \frac{-b \pm \sqrt{b^2 - 4ac}}{2a} = \frac{-3.45 \pm \sqrt{(3.45)^2 - 4(1)(0.50)}}{2(1)} = \frac{0.28}{2} = 0.14$$

$$x = [Cl_2] = 0.14 \text{ M}$$

$$[PCl_5] = 0.15 - x = 0.15 \text{ M} - 0.14 \text{ M} = 0.01 \text{ M}$$

$$[PCl_3] = 0.15 + x = 0.15 \text{ M} + 0.14 \text{ M} = 0.29 \text{ M}$$

The reaction condition in (b) also causes more than 90 % formation of products, but taking the reaction completely to the products side, as we did in (a), produces the same result, within the given significant figures, so it is not shown here.

Check your answers: The equilibrium concentrations should combine to reproduce K_c:

$$K_{c,(a)} = \frac{(0.14)(0.14)}{(0.005)} = 4, \; K_{c,(a)} = \frac{(0.29)(0.14)}{(0.01)} = 4.$$ Both are approximately equal to 3.30, within the uncertainty of the results (± 1).

Shifting a Chemical Equilibrium: Le Chatlier's Principle

58. As described in Section 14.6, the positive ΔH indicates that heat is a reactant in this reaction. Increasing the temperature will drive the reaction toward products and away from N_2O_4. That means the N_2O_4 concentration will decrease, and the answer is (b).

60. Use the methods described in Section 14.6 to answer this question. Changing concentrations or pressures of reactants and products or changing the available energy in the system will take the system out of equilibrium. The response to that change will be to shift away from an increase or shift towards a decrease. Changing the amounts of solids or liquids will not affect the equilibrium position. Because the reaction is endothermic, energy is a reactant.

$$16.1 \text{ kJ} + 2 \text{ NOBr(g)} \; \rightleftharpoons \; 2 \text{ NO(g)} + Br_2(\ell)$$

(a) The equilibrium is not affected by the addition of a pure liquid, so adding more liquid Br_2 will cause no change.

(b) Removing reactant, NOBr, shifts the reaction toward the reactants (left) to increase the [NOBr].

(c) Decreasing the temperature removes a reactant, energy. That shifts the reaction toward the reactants (left), to increase the energy.

62. Use methods similar to those described in the answers to Question 60 and Section 14.6.

$$2 \text{ NO(g)} + O_2(s) \; \rightleftharpoons \; 2 \text{ NO}_2(g) + \text{energy}$$

(a) Adding more reactant, $O_2(g)$, shifts the reaction toward the products (right) to decrease the [O_2].

(b) Adding more product, $NO_2(g)$, shifts the reaction toward the reactants (left) to decrease the [NO_2].

(c) Decreasing the temperature removes a product, energy. That shifts the reaction toward the products (right), to increase the energy.

63. (a) Use the methods described in Question 60 and Section 14.6.

$$PbCl_2(s) \rightleftharpoons Pb^{2+}(aq) + 2\ Cl^-(aq)$$

Adding the soluble ionic compound, NaCl, produces a solution with more of the product, Cl^-. That shifts the reaction toward the reactants (left) and decreases the $[Pb^{2+}]$.

(b) Each time NaCl is added, the chloride concentration rises, then the reaction shifts toward reactants producing more solid and dropping the concentrations of both reactants until a new equilibrium condition is achieved. The following qualitative sketch (similar to that of Figure 14.6) shows these changes visually.

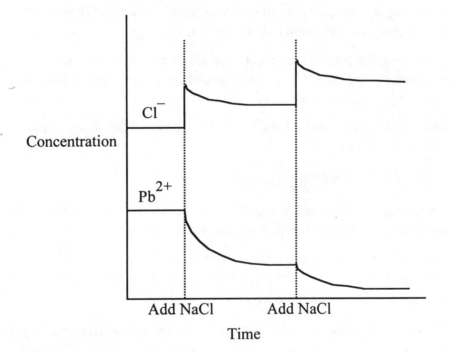

66. Use methods similar to those described in the answers to Question 60 and Section 14.6.

(a) $C(s) + H_2O(g) + 131.3\ kJ \rightleftharpoons CO(g) + 2\ H_2(g)$

   (i) Increasing the temperature adds more reactant, energy. That shifts the reaction toward the products (right), to decrease the energy.

   (ii) Decreasing the pressure affects the products (3 moles of gas) more than the reactants (1 mol gas). That results in a shift of the reaction toward the products (right), to increase the pressure.

   (iii) Increasing the amount of CO(g) increases the concentration of a product. That shifts the reaction toward the reactants (left), to decrease the [CO].

(b) $3 Fe(s) + 4 H_2O(g) \rightleftharpoons Fe_3O_4(s) + 4 H_2(g) + 149.9$ kJ

 (i) Increasing the temperature adds more product, energy. That shifts the reaction toward the reactants (left), to decrease the energy.

 (ii) Decreasing the pressure affects the both the products and the reactants equally (4 mol gas on each side). That results in no shift at all.

 (iii) Increasing the amount of $H_2O(g)$ increases the concentration of a reactant. That shifts the reaction toward the products (right), to decrease the $[H_2O]$.

(c) $C(s) + CO_2(g) + 172.5$ kJ $\rightleftharpoons 2 CO(g)$

 (i) Increasing the temperature adds more reactant, energy. That shifts the reaction toward the products (right), to decrease the energy.

 (ii) Decreasing the pressure affects the products (2 moles of gas) more than the reactants (1 mol gas). That results in a shift of the reaction toward the products (right), to increase the pressure.

 (iii) The equilibrium is not affected by the addition of any solid, so adding more C(s) will cause no shift.

(d) $N_2O_4(g) + 54.8$ kJ $\rightleftharpoons 2 NO_2(g)$

 (i) Increasing the temperature adds more reactant, energy. That shifts the reaction toward the products (right), to decrease the energy.

 (ii) Decreasing the pressure affects the products (2 moles of gas) more than the reactants (1 mol gas). That results in a shift of the reaction toward the products (right), to increase the pressure.

 (iii) Increasing the amount of $NO_2(g)$ increases the concentration of a product. That shifts the reaction toward the reactants (left), to decrease the $[NO_2]$.

Equilibrium at the Nanoscale

68. We will use methods described in Section 14.7 to answer this question.

The entropy effect is related to randomness and probability. The more random a system is, the more probable it is. In many of the chemical reactions we study, we can get a qualitative idea of the entropy change by looking at the physical phases of the reactants and products. In general, gases have more entropy than liquids, which have more entropy than solids. Aqueous substances have more entropy than solids, but less than gases.

The energy effect is related to the relative stability of the reactants and products. If the products are more stable than the reactants (exothermic, ΔH = negative), then the products are favored, energetically. If the reactants are more stable than the products (endothermic, ΔH = positive), then the reactants are favored, energetically.

(a) 4 mol gas $\rightleftharpoons$ 2 mol gas   Entropy effect favors reactants, not products.

Reaction is exothermic, so energy effect favors products.

Reaction (a) (in the forward direction) is favored only by the energy effect.

(b) 1 mol gas $\rightleftharpoons$ 2 mol gas   Entropy effect favors products.

Reaction is endothermic, so energy effect favors reactants, not products.

Reaction (b) (in the forward direction) is favored only by the entropy effect.

(c) 4 mol gas $\rightleftharpoons$ 2 mol gas   Entropy effect favors reactants, not products.

Reaction is endothermic, so energy effect favors reactants, not products.

Reaction (c) (in the forward direction) is not favored by either effect.

70. We will use methods described the answer to Question 68 and in Section 14.7. If the reaction is favored by both effects, it is product-favored and K will be greater than 1. If the reaction is favored by neither effect, it is reactant-favored and K will be less than 1. If the reaction is only favored by one of the two effects, we have insufficient information available.

(a) 3 mol gas $\rightleftharpoons$ 2 mol gas   Entropy effect favors reactants, not products.

Reaction is exothermic, so energy effect favors products.

We have insufficient information to judge the size of K for reaction (a) or which side of the reaction is favored at equilibrium.

(b) 2 mol gas $\rightleftharpoons$ 3 mol gas   Entropy effect favors products.

Reaction is exothermic, so energy effect favors products.

Reaction (b) (in the forward direction) is favored by both effects, so it is product-favored and its K will be greater than 1.

(c) 4 mol gas $\rightleftharpoons$ 2 mol gas   Entropy effect favors reactants, not products.

Reaction is endothermic, so energy effect favors reactants, not products.

Reaction (c) (in the forward direction) is not favored by either effect, so it is reactant-favored and K will be less than 1.

Controlling Chemical Reactions: The Haber-Bosch Process

73. (a) Follow the method described in the answer to Question 6.94 and Section 6.10.

Look up the ΔH_f° values in Appendix J.

Step 1:

$\Delta H^\circ = (1\ mol) \times \Delta H_f^\circ(SO_2) - (1\ mol) \times \Delta H_f^\circ(S) - (1\ mol) \times \Delta H_f^\circ(O_2)$

$\Delta H^\circ = (1\ mol) \times (-296.830\ kJ/mol) - (1\ mol) \times (0\ kJ/mol) - (1\ mol) \times (0\ kJ/mol)$

$$= -296.830\ kJ/mol$$

Step 2:

$\Delta H^\circ = (2\ mol) \times \Delta H_f^\circ(SO_3) - (2\ mol) \times \Delta H_f^\circ(SO_2) - (1\ mol) \times \Delta H_f^\circ(O_2)$

$\Delta H^\circ = (2\ mol) \times (-395.72\ kJ/mol)$

$$- (2\ mol) \times (-296.830\ kJ/mol) - (1\ mol) \times (0\ kJ/mol) = -197.78\ kJ/mol$$

Step 3:

$\Delta H^\circ = (1\ mol) \times \Delta H_f^\circ(H_2SO_4(\ell)) - (1\ mol) \times \Delta H_f^\circ(SO_3) - (1\ mol) \times \Delta H_f^\circ(H_2O(\ell))$

$\Delta H^\circ = (1\ mol) \times (-813.989\ kJ/mol) - (1\ mol) \times (-395.72\ kJ/mol)$

$$- (1\ mol) \times (-285.830\ kJ/mol) = -132.44\ kJ/mol$$

(b) All three reactions are exothermic (ΔH = negative)

(c) Follow the method described in the answer to Question 68.

Step 1:	1 mol gas $\rightleftharpoons$ 1 mol gas	No large entropy change.
Step 2:	3 mol gas $\rightleftharpoons$ 2 mol gas	Entropy decreases.
Step 3:	1 mol gas $\rightleftharpoons$ 0 mol gas	Entropy decreases.

(d) Heat is produced in all three steps. A low temperature in each step would serve to reduce the energy and drive the reaction toward products. All three steps would favor products more at lower temperatures.

General Questions

75. First Reaction: $K_c = [H^+][OH^-]$

Second Reaction: $K_c = \dfrac{[CH_3CO_2^-][H^+]}{[CH_3CO_2H]}$

Third Reaction:

$$K_c = \frac{[NH_3]^2}{[N_2][H_2]^3}$$

Fourth Reaction:

$$K_c = \frac{[O_2]^3}{[O_3]^2}$$

Fifth Reaction

$$K_c = \frac{[N_2O_4]}{[NO_2]^2}$$

Sixth Reaction:

$$K_c = \frac{[HCO_2H]}{[HCO_2^-][H^+]}$$

Seventh Reaction:

$$K_c = \frac{1}{[Ag^+][I^-]}$$

For each reaction, follow the directions given in the problem: Assume that all gases and solutes have initial concentrations of 1.0 mol/L. Then let the *first* reagent in each reaction changes its concentration by –x. *(Note: This is not always the best approach, but those are the instructions for this problem. Additional details are given afterward.)*

First reaction: $H_2O\ (\ell)$ ⇌ $H^+\ (aq)$ + $OH^-\ (aq)$

	$H_2O\ (\ell)$	$H^+\ (aq)$	$OH^-\ (aq)$
conc. initial (mol/L)		1.0	1.0
change conc. (mol/L)	– x	+ x	+ x
equilibrium conc. (mol/L)	– x	1.0 + x	1.0 + x

(a) $K_c = [H^+][OH^-] = (1.0 + x)(1.0 + x) = 1.0 \times 10^{-14}$

(b) This equation is quadratic.

(c) Not applicable.

Second reaction: $CH_3CO_2H\ (aq)$ ⇌ $CH_3CO_2^-\ (aq)$ + $H^+\ (aq)$

	$CH_3CO_2H\ (aq)$	$CH_3CO_2^-\ (aq)$	$H^+\ (aq)$
conc. initial (mol/L)	1.0	1.0	1.0
change conc. (mol/L)	– x	+ x	+ x
equilibrium conc. (mol/L)	1.0 – x	1.0 + x	1.0 + x

75. Second reaction: (continued)

(a) $K_c = \dfrac{[CH_3CO_2^-][H^+]}{[CH_3CO_2H]} = \dfrac{(1.0+x)(1.0+x)}{(1.0-x)} = 1.8 \times 10^{-5}$

(b) This equation is quadratic.

(c) Not applicable.

Note: We more often see the formulas CH_3CO_2H and $CH_3CO_2^-$ written as CH_3COOH and CH_3COO^-

Third reaction:

	$N_2\ (g)$ +	$3\ H_2\ (g)$ ⇌	$2\ NH_3\ (g)$
conc. initial (mol/L)	1.0	1.0	1.0
change conc. (mol/L)	$-x$	$-3x$	$+2x$
equilibrium conc. (mol/L)	$1.0 - x$	$1.0 - 3x$	$1.0 + 2x$

(a) $K_c = \dfrac{[NH_3]^2}{[N_2][H_2]^3} = \dfrac{(1.0+2x)^2}{(1.0-x)(1.0-3x)^3} = 3.5 \times 10^8$

(b) This equation is not quadratic.

(c) This K is so large, it might be worth running the reaction as far to the right as possible, using stoichiometry, then defining a new change variable, as was done in the answer to Question 55:

	$N_2\ (g)$ +	$3\ H_2\ (g)$ ⇌	$2\ NH_3\ (g)$
conc. initial (mol/L)	1.0	1.0	1.0
stoic. change conc. (mol/L)	-0.33	-1.0	$+0.67$
final conc. (mol/L)	0.7	0.0	1.7
equil. change conc. (mol/L)	$-x$	$-3x$	$+2x$
equilibrium conc. (mol/L)	$0.7 - x$	$-3x$	$1.7 + 2x$

Since this x is assumed to be very small compared to 0.7 or 1.7, we can ignore subtraction of x from 0.7 and addition of 2x from 1.7:

$K_c = \dfrac{[NH_3]^2}{[N_2][H_2]^3} = \dfrac{(1.7-2x)^2}{(0.7+x)(3x)^3} \cong \dfrac{(1.7)^2}{(0.7)(3x)^3} = 4 \times 10^8$, then solve for x.

$x = 0.0008$ mol/L $<<1.7$ or 0.7

Fourth reaction:	$2\ O_3\ (g)$ $\rightleftharpoons$	$3\ O_2\ (g)$
conc. initial (mol/L)	1.0	1.0
change conc. (mol/L)	$-x$	$+\dfrac{3}{2}x$
equilibrium conc. (mol/L)	$1.0-x$	$1.0+\dfrac{3}{2}x$

(a) $\quad K_c = \dfrac{[O_2]^3}{[O_3]^2} = \dfrac{\left(1.0+\dfrac{3}{2}x\right)^3}{\left(1.0-x\right)^2} = 7\times10^{56}$

(b) Not quadratic

(c) This K is so large, it might be worth running the reaction as far to the right as possible, using stoichiometry, then defining a new change variable, as was done in the answer to Question 55:

	$2\ O_3\ (g)$ $\rightleftharpoons$	$3\ O_2\ (g)$
conc. initial (mol/L)	1.0	1.0
conc. final (mol/L)	0	2.5
change conc. (mol/L)	$+x$	$-\dfrac{3}{2}x$
equilibrium conc. (mol/L)	x	$2.5-\dfrac{3}{2}x$

Since this x is assumed to be very small compared to 2.5, we can ignore $\dfrac{3}{2}x$ subtraction from 2.5

$$K_c = \frac{[O_2]^3}{[O_3]^2} = \frac{\left(2.5-\dfrac{3}{2}x\right)^3}{x^2} \cong \frac{(2.5)^3}{x^2} = 7\times10^{56}, \text{ then solve for x.}$$

$x = 1\times10^{-28}$ mol/L $\ll$ 2.5 mol/L

75. (continued)

Fifth reaction:

	2 NO$_2$ (g) $\rightleftharpoons$	N$_2$O$_4$ (g)
conc. initial (mol/L)	1.0	1.0
change conc. (mol/L)	$-x$	$+\dfrac{1}{2}x$
equilibrium conc. (mol/L)	$1.0 - x$	$1.0 + \dfrac{1}{2}x$

(a) $\quad K_c = \dfrac{[N_2O_4]}{[NO_2]^2} = \dfrac{\left(1.0 + \dfrac{1}{2}x\right)}{\left(1.0 - x\right)^2} = 1.7 \times 10^2$

(b) This equation is quadratic.

(c) Not applicable.

Sixth reaction:

	HCO$_2^-$ (aq) +	H$^+$ (aq) $\rightleftharpoons$	HCO$_2$H (aq)
conc. initial (mol/L)	1.0	1.0	1.0
change conc. (mol/L)	$-x$	$-x$	$+x$
equilibrium conc. (mol/L)	$1.0 - x$	$1.0 - x$	$1.0 + x$

(a) $\quad K_c = \dfrac{[HCO_2H]}{[HCO_2^-][H^+]} = \dfrac{(1.0 + x)}{(1.0 - x)(1.0 - x)} = 5.6 \times 10^3$

(b) This equation is quadratic.

(c) Not applicable.

Seventh reaction:

	Ag$^+$ (aq) +	I$^-$ (aq) $\rightleftharpoons$	AgI (s)
conc. initial (mol/L)	1.0	1.0	
change conc. (mol/L)	$-x$	$-x$	$+$
equilibrium conc. (mol/L)	$1.0 - x$	$1.0 - x$	

(a) $\quad K_c = \dfrac{1}{[Ag^+][I^-]} = \dfrac{1}{(1.0 - x)(1.0 - x)} = 6.7 \times 10^{15}$

(b) This equation is quadratic.

(c) Not applicable.

IMPORTANT NOTE FOR QUESTION 75: If we were actually doing this problem to get the answers, it would be a very good idea to calculate Q to determine the direction of reaction before deciding which side of the reaction loses ($-x$) and which side gains ($+x$). The losing side should get the $-x$, so that we can be certain that the value of x is a positive quantity. Doing that significantly simplifies the mathematics, especially when dealing with quadratic equations and other higher order polynomials. For the reactions in this Question, here are the Q calculations.

First reaction: $Q = (1.0M)(1.0M) = 1.0 > K_c = 1.0 \times 10^{-14}$

Reaction goes toward **reactants**, not products. The x defined three pages ago for the first reaction will be negative!

Second reaction: $Q = \dfrac{(1.0)(1.0)}{(1.0)} = 1.0 > K_c = 1.8 \times 10^{-5}$

Reaction goes toward **reactants**, not products. The x defined four pages ago for the second reaction will be negative!

Third reaction: $Q = \dfrac{(1.0)^2}{(1.0)(1.0)^3} = 1.0 < K_c = 3.5 \times 10^8$

Reaction goes toward products. The x defined above will be positive.

Fourth reaction: $Q = \dfrac{(1.0)^3}{(1.0)^2} = 1.0 < K_c = 7 \times 10^{56}$

Reaction goes toward products. The x defined above will be positive.

Fifth reaction: $Q = \dfrac{(1.0)}{(1.0)^2} = 1.0 < K_c = 1.7 \times 10^2$

Reaction goes toward products. The x defined above will be positive.

Sixth reaction: $Q = \dfrac{(1.0)}{(1.0)(1.0)} = 1.0 < K_c = 5.6 \times 10^3$

Reaction goes toward products. The x defined above will be positive.

Seventh reaction: $Q = \dfrac{1}{(1.0)(1.0)} = 1.0 < K_c = 6.7 \times 10^{15}$

Reaction goes toward products. The x defined above will be positive.

Note: It is also a good idea to take the fifth, sixth and seventh reactions to completion as was done in the answer to Question 55, since they also have very large K values and are product-favored.

76. Use a method similar to that described in the answer to Question 44.

(a)   (conc. E_2) = 1.00 mol E_2/1.0 L = 1.0 M

	E_2 (g) ⇌	2 E (g)
conc. initial (M)	1.0	0
change conc. (M)	$-x$	$+2x$
equilibrium conc. (M)	$1.0-x$	$2x$

At equilibrium $$K_c = \frac{[E]^2}{[E_2]} = \frac{(2x)^2}{1.0-x}$$

Set up to use the quadratic equation:

$$(2x)^2 = K_c(1.0 - x) = K_c - K_c x$$

$$4x^2 + K_c x - K_c = 0$$

Plug into the quadratic equation:

$$x = \frac{-b \pm \sqrt{b^2 - 4ac}}{2a} = \frac{-K_c \pm \sqrt{K_c^2 - 4(K_c)(-K_c)}}{2(4)}$$

$$[E] = 2x$$

Here is an example of the calculation for the first species: E = Br, $K_c = 8.9 \times 10^{-2}$.

$$x = \frac{-(8.9 \times 10^{-2}) \pm \sqrt{(8.9 \times 10^{-2})^2 - 4(8.9 \times 10^{-2})(-8.9 \times 10^{-2})}}{2(4)} = \frac{1.1}{8} = 0.14$$

$$[Br] = 2 \times (0.14\ M) = 0.28\ M$$

Similar calculations give these results:

E species	K_c	x (M)	[E] (M)
Br	8.9×10^{-2}	0.14	0.28
Cl	3.4×10^{-3}	0.029	0.057
F	7.4	0.72	1.44
I	1.5	0.45	0.90
H	3.1×10^{-10}	8.8×10^{-6}	1.76×10^{-5}
N	1×10^{-27}	2×10^{-14}	4×10^{-14}
O	1.6×10^{-11}	2.0×10^{-6}	4.0×10^{-6}

Check your answers: [E] is small when K is small. [E] is large when K is large. The equilibrium concentrations combine to reproduce $K_c = \dfrac{(2x)^2}{1.0 - x}$. These look right.

(b) At this temperature, the lowest bond energy is predicted from the reaction that gives the most products, so I_2 is predicted to have the lowest bond dissociation energy. Comparing these results to the bond energy values in Table 8.2, the product production decreases as the bond energy increases: 151kJ I_2, 156kJ F_2, 193kJ Br_2, 242kJ Cl_2, 436kJ H_2, 498kJ O_2, 946kJ N_2.

Lewis structures of I_2, F_2, Br_2, Cl_2, H_2 have a single bonds and more products in these calculations than O_2 with double bond and N_2 with a triple bond.

78. (a) Follow the method described in the answer to Question 35.

(Assume percentages are known within ± 1 %.)

(conc. *cis*-dichloroethene) = 100. %

	cis-dichloroethene $\rightleftharpoons$ *trans*-dichloroethene	
conc. initial (%)	100.	0
change conc. (%)	$-x$	$+x$
equilibrium conc. (%)	$100. - x$	x

Given [*trans*-dichloroethene] = 40. % at equilibrium, so 40. % = x

[*cis*-dichloroethene] = 100. % − 40. % = 60. % M

$$K_c = \frac{\left[trans\text{ - dichloroethane}\right]}{\left[cis\text{ - dichloroethane}\right]} = \frac{x}{100. - x} = \frac{40.\ \%}{60.\ \%} = 0.67$$

(b) If we started with 100 % *trans*-dichloroethene, the reaction would have run backwards producing the *cis*-isomer and depleting the *trans*-isomer until the concentration reached exactly the same equilibrium ratio as determined above.

81. Use some of the methods described in the answer to Question 44.

$$2\ HI(g) \rightleftharpoons H_2(g) + I_2(g) \qquad K_c = \frac{[H_2][I_2]}{[HI]^2} = 0.0200$$

Construct Q_c: $\qquad\qquad Q_c = \dfrac{(conc.\ H_2)(conc.\ I_2)}{(conc.\ HI)^2}$

Then compare to K_c.

Case a:
$$Q_c = \frac{\left(\dfrac{0.10 \text{ mol H}_2}{10.00 \text{ L}}\right)\left(\dfrac{0.10 \text{ mol I}_2}{10.00 \text{ L}}\right)}{\left(\dfrac{1.0 \text{ mol HI}}{10.00 \text{ L}}\right)^2} = 0.010 < 0.0200$$

Reaction goes toward products, and the concentration of HI decreases.

Case b:
$$Q_c = \frac{\left(\dfrac{1.0 \text{ mol H}_2}{10.00 \text{ L}}\right)\left(\dfrac{1.0 \text{ mol I}_2}{10.00 \text{ L}}\right)}{\left(\dfrac{10. \text{ mol HI}}{10.00 \text{ L}}\right)^2} = 0.010 < 0.0200$$

Reaction goes toward products, and the concentration of HI decreases.

Case c:
$$Q_c = \frac{\left(\dfrac{10. \text{ mol H}_2}{10.00 \text{ L}}\right)\left(\dfrac{1.0 \text{ mol I}_2}{10.00 \text{ L}}\right)}{\left(\dfrac{10. \text{ mol HI}}{10.00 \text{ L}}\right)^2} = 0.10 > 0.0200$$

Reaction goes toward reactants, and the concentration of HI increases.

Case d:
$$Q_c = \frac{\left(\dfrac{0.381 \text{ mol H}_2}{10.00 \text{ L}}\right)\left(\dfrac{1.75 \text{ mol I}_2}{10.00 \text{ L}}\right)}{\left(\dfrac{5.62 \text{ mol HI}}{10.00 \text{ L}}\right)^2} = 0.0211 < 0.0200$$

Reaction goes toward reactants a very small amount, and the concentration of HI increases a very small amount.

In conclusion, the HI concentration increases in case c and very slightly in case d. The HI concentration decreases in cases a and b.

Applying Concepts

82. For the system to be at equilibrium, it must contain some quantity of both reactants and products, the concentrations of all the reactants and products must be constant over time, and the same relative concentrations are established regardless of which direction the equilibrium was approached. The 42 % *cis* concentration was achieved in two different experiments; therefore, the system is at equilibrium. No further experiments are needed.

84. As described near the end of Section 14.6, an endothermic reaction will have an increased K_c at higher temperatures, and an exothermic reaction will have a decrease in K_c at higher temperatures. As temperature increases in the *cis-trans* isomerism, we find that K_c decreases; therefore, we can predict that the reaction is exothermic.

85. As described near the end of Section 14.6, an endothermic reaction will have an increased K_c at higher temperatures, and an exothermic reaction will have a decrease in K_c at higher temperatures. In addition, the larger the change, the more exothermic or endothermic the reaction is, respectively.

(a) When $R = CH_3$, the values of K_c for the *cis-trans* isomerism decrease at higher temperatures. This is the only data set that shows this trend, so (iii) the *cis-trans* isomerism for $R = CH_3$ is the most exothermic.

(b) When $R = F$ and $R = Cl$, the values of K_c for the *cis-trans* isomerism increase at higher temperatures. That means these are both endothermic *cis-trans* isomerism reactions. The $R = F$ isomerism has a larger change, so (i) the *cis-trans* isomerism for $R = F$ is the most endothermic.

87. In the warmer sample, the molecules would be moving faster, and more NO_2 molecules would be seen. In the cooler sample, the molecules would be moving slower, and fewer NO_2 molecules would be seen. In both samples, the molecules are moving very fast. The average speed of gas molecules is commonly hundreds of miles per hour. In both samples, I would see some N_2O_4 molecules decomposing and some NO_2 molecules reacting with each other, at equal rates.

90. The equilibrium constant expression for this reaction is: $K_c = \dfrac{[AB]^2}{[A_2][B_2]}$.

We need to find diagrams that fit this range: $10^2 > = \dfrac{[AB]^2}{[A_2][B_2]} > 0.1$

Diagram (a) has only reactants, A_2 and B_2. No AB molecules are found in the mixture. That means the value of $K_c = 0$. Diagram (e) has only product molecules, AB. No A_2 or B_2 are found in the mixture. That means the value of $K_c = \infty$. Neither of these K_c values is within the stated range.

In the other three diagrams, we note that the number of A_2 molecules and the number of B_2 molecules are equal.

$$K = \frac{[AB]^2}{[A_2][B_2]} = \frac{[(\text{Number AB}) \times N_0 / V]^2}{(\text{Number } A_2) \times N_0 / V \times (\text{Number } B_2) \times N_0 / V}$$

$$= \frac{(\text{Number AB})^2}{(\text{Number } A_2) \times (\text{Number } B_2)} = \frac{(\text{Number AB})^2}{(\text{Number } A_2)^2} = \left(\frac{\text{Number AB}}{\text{Number } A_2}\right)^2$$

That means we can simplify the search range by taking the square-root of the range variables and compare them to the ratio of the numbers:

$$\sqrt{10^2} = 10 > = \frac{\text{Number of AB}}{\text{Number of } A_2} > = \sqrt{0.1} = 0.3$$

Diagram (b) has 4 AB and 2 A_2 = 2 B_2 : $\dfrac{\text{Number of AB}}{\text{Number of } A_2} = \dfrac{4}{2} = 2$

Diagram (c) has 6 AB and 1 A_2 = 1 B_2 : $\dfrac{\text{Number of AB}}{\text{Number of } A_2} = \dfrac{6}{1} = 6$

Diagram (d) has 2 AB and 3 A_2 = 3 B_2 : $\dfrac{\text{Number of AB}}{\text{Number of } A_2} = \dfrac{2}{3} = 0.667$

Therefore, all three of these remaining diagrams, (b), (c), and (d), represent equilibrium mixtures where $10^2 > = K_c > 0.1$.

92. We derived a relationship between the expression for K and the number ratios derived in Question 90:

$$K = = \left(\frac{\text{Number AB}}{\text{Number } A_2}\right)^2$$

So we must square the ratios determined in the answer to Question 90 to get the K values:

(a) Diagram (d) has $K = \left(\dfrac{2}{3}\right)^2 = (0.667)^2 = 0.44$

(b) Diagram (b) has $K = \left(\dfrac{4}{2}\right)^2 = 2^2 = 4.0$

(c) Diagram (c) has $K = \left(\dfrac{6}{1}\right)^2 = 6^2 = 36$

94. Dynamic equilibria introduce D^+ ions in place of H^+ ions for the acidic hydrogen.

$$H_2O(\ell) \rightleftharpoons H^+(aq) + OH^-(aq)$$

$$D_2O(\ell) \rightleftharpoons D^+(aq) + OH^-(aq)$$

$$C_6H_5COOH\ (s) \rightleftharpoons C_6H_5COOH\ (aq) \rightleftharpoons C_6H_5COO^-(aq) + H^+(aq)$$

$$C_6H_5COOD\ (s) \rightleftharpoons C_6H_5COOD\ (aq) \rightleftharpoons C_6H_5COO^-(aq) + D^+(aq)$$

$$C_6H_5COO^-(aq) + D^+(aq) \rightleftharpoons C_6H_5COOD\ (aq) \rightleftharpoons C_6H_5COOD\ (s)$$

96. Dynamic equilibria representing the decomposition of the dimer $N_2O_4(g)$ produces $NO_2\ (g)$ and $N^*O_2\ (g)$, which will occasionally recombine into the mixed dimer, $O_2N^*–NO_2\ (g)$

$$O_2N–NO_2\ (g) \rightleftharpoons 2\ NO_2\ (g)$$

$$O_2N^*–N^*O_2\ (g) \rightleftharpoons 2\ N^*O_2\ (g)$$

$$O_2N^*–NO_2\ (g) \rightleftharpoons N^*O_2\ (g) + NO_2\ (g)$$

Chapter 15: The Chemistry of Solutes and Solutions

How Substances Dissolve

24. If the solid interacts with the solvent using similar (or stronger) intermolecular forces, it will dissolve readily. If the solute interacts with the solvent using different intermolecular forces than those experienced in the solvent, it will be almost insoluble. For example, consider dissolving an ionic solid in water and oil. The interactions between the ions in the solid and water are very strong, since ions would be attracted to the highly polar water molecule; hence the solid would have a relatively high solubility. However, the ions in the solid interact with each other much more strongly than the London dispersion forces experienced between the nonpolar hydrocarbons in the oil; hence, the solid would have a low solubility.

26. The dissolving process was endothermic, so the temperature dropped as more solute was added. The solubility of the solid at the lower temperature is lower, so some of the solid did not dissolve. As the solution warmed up, however, the solubility increased again. What remained of the solid dissolved. The solution was saturated at the lower temperature, but is no longer saturated at the current temperature.

28. When an organic acid has a large (nonpolar) piece, it interacts primarily using London dispersion intermolecular forces. Since water interacts via hydrogen bonding intermolecular forces, it would rather interact with itself than with the acid. Hence, the solubility of the large organic acids drops, and some are completely insoluble.

30. The positive side (H side) of the very polar water molecule interacts with the negative ions. The negative side of the very polar water molecule (O side) interacts with the positive ions.

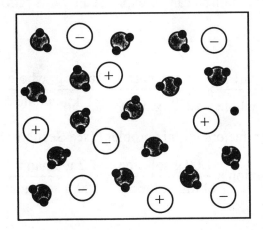

32. *Define the problem*: Given the partial pressure of a gas, the temperature, and the Henry's Law constant, determine the concentration of the gas dissolved in the water.

Develop a plan: Find the partial pressure of the gas in units of mmHg, then use Henry's law described in Section 15.5: $s_g = k_H P_g$ Where s_g is the solubility of the gas, k_H is the Henry's Law constant, and P_g is the partial pressure of a gas.

Execute the plan: 78 % by volume of gaseous N_2 in air is directly proportional to the percentage by mole (Avogadro's law, Section 10.4). The mole percentage is directly proportional to the pressure (Section 10.8), so we get 0.78 atm N_2 in every 1.00 atm air.

$$2.5 \text{ atm air } \times \frac{0.78 \text{ atm } N_2}{1 \text{ atm air}} \times \frac{760 \text{ mmHg } N_2}{1 \text{ atm } N_2} = 1.5 \times 10^3 \text{ mmHg } N_2$$

$$s_g = (8 \times 10^{-7} \text{ M/mmHg})(1.5 \times 10^3 \text{ mmHg}) = 1 \times 10^{-3} \text{ M}$$

Check your answer: Always double check to be sure that the gas does not react with the water, since Henry's law does not apply to those gases.

Concentration Units

34. The relationship between ppm and % was derived in the answer to Question 10.18: 1 % = 10,000 ppm.

$$73.2 \text{ ppm } \times \frac{1 \%}{10000 \text{ ppm}} = 7.32 \times 10^{-3} \%$$

36. The definition of parts per billion can be expressed by this ratio: $1\text{ppb} = \dfrac{1 \text{ g part}}{10^9 \text{ g whole}}$

Therefore,

$$1\text{ppb} = \frac{1 \text{ g part}}{10^9 \text{ g whole}} \times \frac{1 \text{ μg part}}{10^{-6} \text{ g part}} \times \frac{1000 \text{ g whole}}{1 \text{ kg whole}} = \frac{1 \text{ μg part}}{1 \text{ kg whole}}$$

38. Use standard conversion factors for this problem (see Section 15.6 for examples).

$$750 \text{ mL solution} \times \frac{1.00 \text{ g solution}}{1 \text{ mL solution}} \times \frac{12 \text{ g ethanol}}{100 \text{ g solution}} = 90. \text{ g ethanol}$$

40. Use standard conversion factors for this problem (see Section 15.6 for examples)

$$1.0\ cm^2 \times \left(\frac{1\ in}{2.54\ cm}\right)^2 \times \left(\frac{1\ ft}{12\ in}\right)^2 \times \frac{1\ gal\ paint}{500.\ ft^2}$$

$$\times \frac{8.0\ lb\ paint}{1\ gal\ paint} \times \frac{453.6\ g\ paint}{1\ lb\ paint} \times \frac{200.\ g\ Pb}{10^6\ g\ paint} = 1.6 \times 10^{-6}\ g\ Pb$$

42. Use standard conversion factors for this problem (see Section 15.6 for examples)

$$\frac{18\ mol\ H_2SO_4}{1\ L\ solution} \times \frac{1\ L\ solution}{1000\ cm^3\ solution} \times \frac{1\ cm^3\ solution}{1.84\ g\ solution} \times \frac{98.09\ g\ H_2SO_4}{1\ mol\ H_2SO_4} \times 100\ \%$$

$$= 96\ \%\ H_2SO_4$$

44. Use standard conversion factors for this problem (see Section 15.6 for examples)

$$950.\ g\ H_2O \times \frac{1\ kg\ H_2O}{1000\ g\ H_2O} \times \frac{1.0\ mol\ C_2H_4(OH)_2}{1\ kg\ H_2O} \times \frac{62.07\ g\ C_2H_4(OH)_2}{1\ mol\ C_2H_4(OH)_2}$$

$$= 59\ g\ C_2H_4(OH)_2$$

46. Use standard conversion factors for this problem (see Section 15.6 for examples).

$$\frac{25\ g\ lead}{10^9\ g\ solution} \times \frac{1.00\ g\ solution}{1\ mL\ solution} \times \frac{1000\ mL\ solution}{1\ L\ solution} \times \frac{1\ mol\ lead}{207.2\ g\ lead} = 1.2 \times 10^{-7}\ M\ lead$$

Colligative Properties

48. *Define the problem*: Given moles of solute, mass of solvent, and the normal boiling point of the solvent, determine the boiling point of the solution.

Develop a plan: Use method described in Section 15.7.

$$T_b(solution) = T_b(solvent) + \Delta T_b$$

$$\Delta T_b = K_b m_{solute}$$

$$m_{solute} = \frac{mol\ solute}{kg\ solvent}$$

Calculate the molality of the solute. Use molality to get ΔT_b with the molal boiling-point-elevation constant of the solvent, given in the text as $0.52 \, °C \, kg \, mol^{-1}$. Then add ΔT_b to the normal boiling point (exactly $100 \, °C$) to get the boiling point of the solution.

$$Execute \; the \; plan: \; m_{solute} = \frac{15.0 \, g \, (NH_2)_2 CO}{0.500 \, kg \, water} \times \frac{1 \, mol \, (NH_2)_2 CO}{60.06 \, g \, (NH_2)_2 CO} = 0.500 \, mol/kg$$

$$\Delta T_b = (0.52 \, °C \, kg \, mol^{-1}) \times (0.500 \, mol/kg) = 0.26 \, °C$$

$$T_b(solution) = 100.00 \, °C + 0.26 \, °C = 100.26 \, °C$$

Check your answer: The boiling point is elevated by the presence of the solute. The relative size and sign of the answers make sense.

50. In Section 15.7, we learn that the freezing point decreases as the molality of solute particles increases. As described near the end of Section 15.7, solutes that ionize provide a larger number of particles in the solution than solutes that do not ionize. So, we must calculate the concentration of particles in each solution and order the solutions from the smallest concentration to the largest concentration.

(a) One methanol (a covalent molecule) provides one particle. $= 0.10 \, mol/kg$ particles

(b) Two KCl ionizes into K^+ and Cl^-, two particles.

$$0.10 \, mol/kg \, KCl \times (2 \, particles/mol \, KCl) = 0.20 \, mol/kg \, particles$$

(c) One $BaCl_2$ ionizes into Ba^{2+} and $2 \, Cl^-$, three particles.

$$0.080 \, mol/kg \, BaCl_2 \times (3 \, particles/mol \, BaCl_2) = 0.24 \, mol/kg \, particles$$

(d) One Na_2SO_4 ionizes into $2 \, Na^+$ and SO_4^{2-}, three particles.

$$0.040 \, mol/kg \, Na_2SO_4 \times (3 \, particles/mol \, Na_2SO_4) = 0.12 \, mol/kg \, particles$$

Order of decreasing freezing points: (a) < (d) < (b) < (c)

52. Follow the method described in the answer to Question 48 for the boiling point calculation and a similar method for the freezing point calculation, as described in Section 15.7:

$$T_f(solution) = T_f(solvent) - \Delta T_f$$

$$\Delta T_f = K_f m_{solute}$$

The molal boiling-point-elevation and molal freezing-point-lowering constants for water are given in the text.

$$m_{solute} = \frac{4.00 \text{ g CO(NH}_2)_2}{75.0 \text{ g water}} \times \frac{1 \text{ mol CO(NH}_2)_2}{60.06 \text{ g CO(NH}_2)_2} \times \frac{1000 \text{ g water}}{1 \text{ kg water}} = 0.888 \text{ mol/kg}$$

$$\Delta T_b = (0.52 \text{ °C kg mol}^{-1}) \times (0.888 \text{ mol/kg}) = 0.46 \text{ °C}$$

$$T_b(\text{solution}) = 100.00 \text{ °C} + 0.54 \text{ °C} = 100.46 \text{ °C}$$

$$\Delta T_f = (1.86 \text{ °C kg mol}^{-1}) \times (0.888 \text{ mol/kg}) = 1.65 \text{ °C}$$

$$T_f(\text{solution}) = 0.00 \text{ °C} - 1.65 \text{ °C} = -1.65 \text{ °C}$$

54. *Define the problem*: Given the vapor pressure of the solvent at a given temperature, the vapor pressure of a solution at the same temperature and the mass of solvent, determine the mole fraction of the solvent and the mass of solute in the solution.

Develop a plan: Use Raoult's law, described in Section 15.7.

$$P_1 = X_1 P_1^0$$

Where P_1 is the vapor pressure over the solution, P_1^0 is the vapor pressure over the pure solvent, and X_1 is the mole fraction of the solvent (described in Section 10.8).

$$X_1 = \frac{n_{solvent}}{n_{tot}}$$

Calculate the mole fraction from the vapor pressures. Calculate moles of solvent from the grams, then use it and the mole fraction to find moles of solute. Then convert to mass.

Execute the plan:
$$X_{H_2O} = \frac{P_{H_2O}}{P_{H_2O}^0} = \frac{119.55 \text{ mmHg}}{149.44 \text{ mmHg}} = 0.79999$$

$$150. \text{ g H}_2O \times \frac{1 \text{ mol H}_2O}{18.02 \text{ g H}_2O} = 8.32 \text{ mol H}_2O$$

$$X_{H_2O} = \frac{n_{H_2O}}{n_{H_2O} + n_{sucrose}}$$

Solve for $n_{sucrose}$: $X_{H_2O}(n_{H_2O} + n_{sucrose}) = X_{H_2O}n_{H_2O} + X_{H_2O}n_{sucrose} = n_{H_2O}$

$$X_{H_2O}n_{sucrose} = n_{H_2O} - X_{H_2O}n_{H_2O}$$

$$n_{sucrose} = \frac{(1-X_{H_2O})n_{H_2O}}{X_{H_2O}} = \frac{(1-0.79999)(8.32 \text{ mol})}{0.79999} = \frac{(0.20001)(8.32 \text{ mol})}{(0.79999)} = 2.08 \text{ mol}$$

$$2.08 \text{ mol } C_{12}H_{22}O_{11} \times \frac{342.297 \text{ g } C_{12}H_{22}O_{11}}{1 \text{ mol } C_{12}H_{22}O_{11}} = 712 \text{ g } C_{12}H_{22}O_{11}$$

Check your answer: The mole fraction of sucrose = 2.08 mol/(8.32 mol + 2.08 mol) = 0.200. The sum of all the mole fractions (0.79999 + 0.200) is one. These answers make sense.

56. Adapt the method described in the answer to Question 48 to find moles of solute. *NOTE: Assume "approximate weight of the compound" means approximate molar mass of the compound.* Divide mass by moles to get molar mass.

$$\Delta T_b = 0.65 \text{ °C}$$

$$\Delta T_b = K_b m_{unknown}$$

$$m_{unknown} = \frac{\Delta T_b}{K_b} = \frac{0.65 \text{ °C}}{2.53 \text{ °C kg mol}^{-1}} = 0.26 \text{ mol/kg}$$

$$100. \text{ g benzene} \times \frac{1 \text{ kg benzene}}{1000 \text{ g benzene}} \times \frac{0.26 \text{ mol unknown}}{1 \text{ kg benzene}} = 0.026 \text{ mol unknown}$$

$$\text{Molar mass} = \frac{5.0 \text{ g}}{0.026 \text{ mol}} = 190 \text{ g/mol}$$

58. Adapt the method described in the answer to Question 48 to find moles of solute, then divide mass by moles for molar mass. Use methods described in Question 3.75 to find the molecular formula.

$$\Delta T_b = T_b(\text{solution}) - T_b(\text{solvent})$$

$$\Delta T_b = 80.34 \text{ °C} - 80.10 \text{ °C} = 0.24 \text{ °C}$$

$$\Delta T_b = K_b m_{C_7H_5}$$

$$m_{C_7H_5} = \frac{\Delta T_b}{K_b} = \frac{0.24 \text{ °C}}{2.53 \text{ °C kg mol}^{-1}} = 0.095 \text{ mol/kg}$$

$$30.0 \text{ g benzene} \times \frac{1 \text{ kg benzene}}{1000 \text{ g benzene}} \times \frac{0.095 \text{ mol } C_7H_5}{1 \text{ kg benzene}} = 0.0029 \text{ mol } C_7H_5$$

$$\text{Molar mass of the compound} = \frac{0.500\ \text{g}}{0.0029\ \text{mol}} = 1.8 \times 10^2\ \text{g/mol}$$

The molecular formula is a multiple of the empirical formula: $(C_7H_5)_n$

The molar mass of the empirical formula C_7H_5 is 89.117 g/mol

$$n = \frac{1.8 \times 10^2\ \text{g/mol}}{89.117\ \text{g/mol}} = 2.0 \cong 2$$

The molecular formula is: $C_{14}H_{10}$

Note: Anthracene is described in Question 8.65, and its formula is determined in the answer to Question 8.65 to be $C_{14}H_{10}$.

60. Adapt the method described in the answers to Questions 52 and 48:

(a) $$\Delta T_f = T_f(\text{solvent}) - T_f(\text{solution}) = 0.0\ °C - (-15.0\ °C) = 15.0\ °C$$

$$m_{C_2H_6O_2} = \frac{\Delta T_f}{K_f} = \frac{15.0\ °C}{1.86\ °C\ \text{kg mol}^{-1}} = 8.06\ \text{mol/kg}$$

$$5.0\ \text{kg water} \times \frac{8.06\ \text{mol}\ C_2H_6O_2}{1\ \text{kg water}} \times \frac{62.07\ \text{g}\ C_2H_6O_2}{1\ \text{mol}\ C_2H_6O_2} \times \frac{1\ \text{kg}\ C_2H_6O_2}{1000\ \text{g}\ C_2H_6O_2}$$

$$= 2.5\ \text{kg}\ C_2H_6O_2$$

You must add 2.5 kg of ethylene glycol to 5.0 kg of water for this much freezing protection.

(b) $$\Delta T_b = (0.52\ °C\ \text{kg mol}^{-1}) \times (8.06\ \text{mol/kg}) = 4.2\ °C$$

$$T_b(\text{solution}) = 100.00\ °C + 4.2\ °C = 104.2\ °C$$

62. *Define the problem*: Given the freezing point of a solution, the temperature, and an assumption that the density is the same as pure water, determine the osmotic pressure.

Develop a plan: Adapt the method described in the answers to Questions 52 to find the molality of the solution. With the molality and the density of the solution, determine the molar concentration. Then use the equation described in Section 15.7 to get the osmotic pressure:

$$\Pi = cRT$$

Π is the osmotic pressure, c is the concentration of the solute in the solution, R is the familiar gas constant with liter and atmosphere units, 0.08206 L·atm/mol·K, and T is the absolute temperature in kelvin units.

Execute the plan: $\Delta T_f = T_f(\text{solvent}) - T_f(\text{solution}) = 0.0\ °C - (-2.3\ °C) = 2.3\ °C$

$$m_{\text{solute}} = \frac{\Delta T_f}{K_f} = \frac{2.3\ °C}{1.86\ °C\ kg\ mol^{-1}} = 1.2\ mol/kg$$

Assume the mass of the solvent is essentially equal to the mass of the solution:

$$\frac{1.2\ mol\ solute}{1\ kg\ solvent} \times \frac{1\ kg\ solvent}{1\ kg\ solution} \times \frac{1.00\ kg\ solution}{1\ L\ solution} = 1.2\ mol/L$$

$$T = 20.0\ °C + 273.15 = 293.2\ K$$

$$\Pi = (1.2\ mol/L)(0.08206\ L\cdot atm/mol\cdot K)(293.2\ K) = 29\ atm$$

Check your answer: The solute concentration of seawater is fairly high, so it makes sense that the osmotic pressure is relatively high, also.

64. Adapt the methods described in the answers to Questions 62 and 58.

$$T = 25.00\ °C + 273.15 = 298.15\ K$$

$$c = \frac{\Pi}{RT} = \frac{7.6\ mmHg \times \left(\dfrac{1\ atm}{760\ mmHg}\right)}{\left(0.08206\ \dfrac{L\cdot atm}{mol\cdot K}\right)(298.15\ K)} = 4.1\times10^{-4}\ mol/L$$

Assuming that the addition of solute does not change the volume of the solution, the volume of the solvent is equal to the volume of the solution:

$$1.0\ L\ water \times \frac{1\ L\ solution}{1L\ water} \times \frac{4.1\times10^{-4}\ mol\ solute}{1\ L\ solution} = 4.1\times10^{-4}\ mol\ solute$$

$$\text{Molar mass of the compound} = \frac{5.0\ g}{4.1\times10^{-4}\ mol} = 1.2\times10^{4}\ g/mol$$

Water: Purification and Solutions

66. This is a standard conversion factor problem.

$$1\ week \times \frac{7\ days}{1\ week} \times \frac{6\ glasses}{1\ day} \times \frac{8\ oz}{1\ glass} \times \frac{1\ quart}{32\ oz} \times \frac{1\ L}{1.0567\ quart} \times \frac{1000\ mL}{1\ L}$$

$$\times \frac{1.00\ g\ water}{1\ mL} \times \frac{0.050\ g\ As}{10^{6}\ g\ water} = 5\times10^{-4}\ g\ As$$

68. The lime-soda process relies on the precipitation of insoluble compounds to remove the "hard water" ions. The ion exchange process relies on the high charge of the "hard water" ions to attract them to an ion exchange resin, thereby removing them from the water.

70. Soap forms insoluble curds in the presence of hard water ions.

72. Risk: There is a small risk of byproduct formation (THMs) that may be linked with liver cancer.

Benefit: Chlorine kills bacteria that pose a great risk to human health.

Benefit outweighs the risk.

74. Fish breathe oxygen by extracting it from the water, Plants "breathe" carbon dioxide by extracting it from the water. The concentrations of these gases in calm water drop, unless they are replenished. The concentration of dissolved gases in the water is replenished by bubbling air through the water in the aquarium.

General Questions

76. This is a standard percent problem: *Assume all masses are known ± 0.1 g.*

Total mass = 5.0 g + 0.2 g + 0.3 g = 5.5 g

$$\text{Weight \% of A} = \frac{0.2 \text{ g A}}{5.5 \text{ g total}} \times 100 \% = 4 \%$$

78. Water in the cells of the wood leaked out, since the osmotic pressure inside the cells was less than that of the seawater the wood was sitting in. See the discussion about hypertonic solutions and cells in Section 15.8.

80. Use standard conversion factors for this problem (see Section 15.6 for examples).

$$\frac{14.8 \text{ mol NH}_3}{1 \text{ L solution}} \times \frac{1 \text{ L solution}}{1000 \text{ cm}^3 \text{ solution}} \times \frac{1 \text{ cm}^3 \text{ solution}}{0.90 \text{ g solution}} \times \frac{17.03 \text{ g NH}_3}{1 \text{ mol NH}_3} \times 100 \%$$

$$= 28 \% \text{ NH}_3$$

82. Use standard conversion factors for this problem (see Section 15.6 for examples)

$$500. \text{ mL CH}_3\text{OH} \times \frac{0.7893 \text{ g CH}_3\text{OH}}{1 \text{ mL CH}_3\text{OH}} = 395. \text{ g CH}_3\text{OH}$$

$$\text{molality} = \frac{45.0 \text{ g C}_4\text{H}_8\text{N}_2\text{O}_2}{395 \text{ g ethanol}} \times \frac{1 \text{ mol C}_4\text{H}_8\text{N}_2\text{O}_2}{116.13 \text{ g C}_4\text{H}_8\text{N}_2\text{O}_2} \times \frac{1000 \text{ g}}{1 \text{ kg}} = 0.982 \text{ mol/kg}$$

$$\text{Total mass} = 45.0 \text{ g} + 395. \text{ g} = 440. \text{ g}$$

$$\text{Weight percent of } C_4H_8N_2O_2 = \frac{45.0 \text{ g } C_4H_8N_2O_2}{440. \text{ g total}} \times 100 \% = 10.2 \% \ C_4H_8N_2O_2$$

84. Use the method described in the answer to Question 50.

   (a) One ethylene glycol (a covalent molecule) provides one particle, so the actual concentration is 0.20 mol/kg particles.

   (b) One Na_2SO_4 ionizes into 2 Na^+ and SO_4^{2-}, three particles.

$$0.12 \text{ mol/kg } Na_2SO_4 \times (3 \text{ particles/mol } Na_2SO_4) = 0.36 \text{ mol/kg particles}$$

   (c) One NaBr ionizes into Na^+ and Br^-, two particles.

$$0.10 \text{ mol/kg NaBr} \times (2 \text{ particles/mol NaBr}) = 0.20 \text{ mol/kg particles}$$

   (d) One KI ionizes into K^+ and I^-, two particles.

$$0.12 \text{ mol/kg KI} \times (2 \text{ particles/mol KI}) = 0.24 \text{ mol/kg particles}$$

Order of decreasing freezing points: (a) = (c) < (d) < (b)

86. Adapt the method described in the answer to Question 56.

$$\Delta T_b = T_b(\text{solution}) - T_b(\text{solvent}) = 63.20 \text{ °C} - 61.70 \text{ °C} = 1.50 \text{ °C}$$

$$\Delta T_b = K_b m_{\text{solute}} i$$

One $[(C_4H_9)_4N][ClO_4]$ ionizes into $(C_4H_9)_4N^+$ and ClO_4^-, two particles, so i = 2.

$$m_{\text{solute}} = \frac{\Delta T_b}{K_b i} = \frac{1.50 \text{ °C}}{(3.63 \text{ °C kg mol}^{-1})(2)} = 0.207 \text{ mol}/\text{kg}$$

$$25.0 \text{ g chloroform} \times \frac{1 \text{ kg chloroform}}{1000 \text{ g chloroform}} \times \frac{0.207 \text{ mol } [(C_4H_9)_4N][ClO_4]}{1 \text{ kg chloroform}}$$

$$\times \frac{341.90 \text{ g }[(C_4H_9)_4N][ClO_4]}{1 \text{ mol}[(C_4H_9)_4N][ClO_4]} = 1.77 \text{ g }[(C_4H_9)_4N][ClO_4]$$

88. This is a combination of the problem-solving from several different chapters.

(a) From Chapter 3:

$$100.00 \text{ g compound} - 73.94 \text{ g C} - 8.27 \text{ g H} = 17.79 \text{ g Cr}$$

$$73.94 \text{ g C} \times \frac{1 \text{ mol C}}{12.011 \text{ g C}} = 6.156 \text{ mol C} \qquad 8.27 \text{ g H} \times \frac{1 \text{ mol H}}{1.0079 \text{ g H}} = 8.20 \text{ mol H}$$

$$17.79 \text{ g Cr} \times \frac{1 \text{ mol Cr}}{51.996 \text{ g Cr}} = 0.342 \text{ mol Cr}$$

Mole Ratio: 6.156 mol C : 8.20 mol H : 0.342 mol Cr

Simplify   18 C : 24 H : 1 Cr

Empirical Formula is $C_{18}H_{24}Cr$.

(b) Adapt the answer to Question 64.

$$T = 25 \text{ °C} + 273.15 = 298 \text{ K}$$

$$c = \frac{\Pi}{RT} = \frac{3.17 \text{ mmHg}\left(\dfrac{1 \text{ atm}}{760 \text{ mmHg}}\right)}{\left(0.08206 \dfrac{L \cdot atm}{mol \cdot K}\right)(298 \text{ K})} = 1.71 \times 10^{-4} \text{ mol/L}$$

Assuming that the addition of solute does not change the volume of the solution, so the volume of the solvent is equal to the volume of the solution:

$$100. \text{ mL chloroform} \times \frac{1 \text{ L chloroform}}{1000 \text{ mL chloroform}} \times \frac{1 \text{ L solution}}{1 \text{ L chloroform}}$$

$$\times \frac{1.71 \times 10^{-4} \text{ mol solute}}{1 \text{ L solution}} = 1.71 \times 10^{-5} \text{ mol solute}$$

$$\text{Molar mass of the compound} = \frac{5.00 \text{ mg}}{1.71 \times 10^{-5} \text{ mol}} \times \frac{1 \text{ g}}{1000 \text{ mg}} = 292 \text{ g/mol}$$

The molecular formula is a multiple of the empirical formula: $(C_{18}H_{24}Cr)_n$

The molar mass of the empirical formula $C_{18}H_{24}Cr$ is 292.37 g/mol, so the molecular formula is $C_{18}H_{24}Cr$.

NOTE: Non-bold Question 89 requires you to use methods and information from Chapter 17. If you are assigned non-bold Question 89 in homework or as practice before studying Chapter 17, ask your instructor about it.

90. (a) $AlCl_3(aq) + H_3PO_4(aq) \longrightarrow AlPO_4(s) + 3 HCl(aq)$

(b) $152. \text{ g } AlCl_3 \times \dfrac{1 \text{ mol } AlCl_3}{133.33 \text{ g } AlCl_3} \times \dfrac{1 \text{ mol } AlPO_4}{1 \text{ mol } AlCl_3} = 1.14 \text{ mol } AlPO_4$

$3.00 \text{ L solution} \times \dfrac{0.750 \text{ mol } H_3PO_4}{1 \text{ L solution}} \times \dfrac{1 \text{ mol } AlPO_4}{1 \text{ mol } AlCl_3} = 2.25 \text{ mol } AlPO_4$

Limiting reactant is $AlCl_3$.

$$1.14 \text{ mol } AlPO_4 \times \dfrac{121.95 \text{ g } AlPO_4}{1 \text{ mol } AlPO_4} = 139 \text{ g } AlPO_4$$

(c) We are told to assume that this solid ionizes completely and doesn't undergo hydrolysis. (*NOTE: That is completely unrealistic, since AlPO$_4$(s) is identified as* **insoluble** *in Table 5.1.*)

$$AlPO_4(s) \longrightarrow Al^{3+}(aq) + PO_4^{3-}(aq)$$

$$\dfrac{25.0 \text{ g } AlPO_4}{1 \text{ L}} \times \dfrac{1 \text{ mol } AlPO_4}{121.95 \text{ g } AlPO_4} \times \dfrac{1 \text{ mol ion}}{1 \text{ mol } AlPO_4} = 0.205 \text{ M ion}$$

$[Al^{3+}] = 0.0205$ M and $[PO_4^{3-}] = 0.0205$ M, with this assumption.

Applying Concepts

92. (a) Sugar and water interact with the same hydrogen bonding attractive forces, so they will commingle.

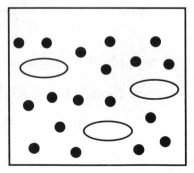

(b) Carbon tetrachloride and sugar interact with very different interactive forces, so they will remain separate phases.

94. Figure 15.10 gives curves showing solubility verses temperature. If the concentration is on curve at the given temperature, the solution is saturated. If the concentration is above the curve at the given temperature, the solution is supersaturated. If the concentration is below the curve at the given temperature, the solution is unsaturated.

(a) 120 g RbCl/100 g H_2O at 50 °C: This point is above the curve, so the solution is supersaturated.

(b) 50 g KCl/100 g H_2O at 70 °C: This point is below the curve, so the solution is unsaturated.

(c) 20 g NaCl/50 g H_2O at 60 °C, which is the same as 40 g NaCl/100 g H_2O at 60 °C: This point is above the curve, so the solution is supersaturated.

(d) 150 g CsCl/100 g H_2O at 10 °C: This point is below the curve, so the solution is unsaturated.

96. Mass fraction is calculated by dividing the mass of the substance by the total mass of the solution. Calculate weight percent by multiplying mass fraction by 100 %. Calculate ppm, using the equality 1 % = 10,000 ppm.

Compound	Mass of compound	Mass of water	Mass fraction	Weight percent	ppm of solute
Lye	75.0 g	125 g	0.375	37.5 %	3.75×10^5
Glycerol	33 g	200. g	0.14	14 %	1.4×10^5
Acetylene	0.0015 g	2×10^2 g	0.000009	0.0009 %	9

98. Molecules slow down and move less. The reduced motion prevents them from randomly translocating as they had in the liquid state. As a result, the intermolecular forces between one molecule and the next begin to organize them into a crystal form. The presence of a nonvolatile solute disrupts the formation of the crystal. Its size and shape will be different from that of the solute. Intermolecular forces between the solute and solvent are also different from those of solvent molecules with each other. To form the crystalline solid, the solute has to be excluded. If the ice in an iceberg is in regular crystalline form, the water will be pure. Only if the ice is crushed or dirty will other particles be included. So, melting an iceberg will produce relatively pure water.

100. (a) Sea water contains more dissolved solutes than fresh water. The presence of a solute lowers the freezing point. That means a lower temperature is required to freeze the sea water than to freeze fresh water.

     (b) Salt added to a mixture of ice and water will lower the freezing point of the water. If the ice cream is mixed at a lower temperature, its temperature will drop faster; hence, it will freeze faster.

Chapter 16: Acids and Bases

The Bronsted-Lowry Concept of Acids and Bases

12. (a) $HCO_3^-(aq) + H_2O(\ell) \rightleftharpoons CO_3^{2-}(aq) + H_3O^+(aq)$

(b) $HCl(aq) + H_2O(\ell) \rightleftharpoons Cl^-(aq) + H_3O^+(aq)$

(c) $CH_3COOH(aq) + H_2O(\ell) \rightleftharpoons CH_3COO^-(aq) + H_3O^+(aq)$

(d) $HCN(aq) + H_2O(\ell) \rightleftharpoons CN^-(aq) + H_3O^+(aq)$

14. (a) $HIO(aq) + H_2O(\ell) \rightleftharpoons IO^-(aq) + H_3O^+(aq)$

(b) $CH_3(CH_2)_4COOH(aq) + H_2O(\ell) \rightleftharpoons CH_3(CH_2)_4COO^-(aq) + H_3O^+(aq)$

(c) $HOOCCOOH(aq) + H_2O(\ell) \rightleftharpoons HOOCCOO^-(aq) + H_3O^+(aq)$

$HOOCCOO^-(aq) + H_2O(\ell) \rightleftharpoons {}^-OOCCOO^-(aq) + H_3O^+(aq)$

(d) $CH_3NH_3^+(aq) + H_2O(\ell) \rightleftharpoons CH_3NH_2(aq) + H_3O^+(aq)$

16. (a) $HSO_4^-(aq) + H_2O(\ell) \rightleftharpoons H_2SO_4(aq) + OH^-(aq)$

(b) $CH_3NH_2(aq) + H_2O(\ell) \rightleftharpoons CH_3NH_3^+(aq) + OH^-(aq)$

(c) $I^-(aq) + H_2O(\ell) \rightleftharpoons HI(aq) + OH^-(aq)$

(d) *There is a typo in this Question. Tthe charge on the reactant ion, $H_2PO_4^-$, in this problem is incorrect.*

$H_2PO_4^-(aq) + H_2O(\ell) \rightleftharpoons H_3PO_4(aq) + OH^-(aq)$

20. Two relationships between acids can be used to help us decide: If the central atoms are the same, then the acid with more O atoms is stronger.   If the numbers of O atoms are the same, then the acid with the more electronegative central atom is stronger. (Electronegativities are found in Figure 8.6.)

(a) H_2SO_4 is stronger than H_2SO_3 because it has more O atoms, and H_2SO_3 is stronger than H_2CO_3 because it has a more electronegative central atom ($EN_S > EN_C$), so H_2SO_4 is stronger than H_2CO_3.

(b) HNO_3 is stronger than HNO_2 because it has more O atoms.

(c) $HClO_4$ is stronger than H_2SO_4 because it has a more electronegative atom ($EN_{Cl} > EN_S$).

(d) $HClO_3$ is stronger than H_3PO_4 because it has a more electronegative atom ($EN_{Cl} > EN_P$).

(e) H_2SO_4 is stronger than H_2SO_3 because it has more O atoms.

22. Conjugate acid-base pairs differ by one H^+ ion.

(a) HI is a Bronsted acid. Its conjugate base is I^-, iodide ion.

(b) NO_3^- is a Bronsted base. Its conjugate acid is HNO_3, nitric acid.

(c) CO_3^{2-} is a Bronsted base. Its conjugate acid is HCO_3^-, hydrogen carbonate ion.

(d) H_2CO_3 is a Bronsted acid. Its conjugate base is HCO_3^-, hydrogen carbonate ion.

(e) HSO_4^- as a Bronsted base, has a conjugate acid of H_2SO_4, sulfuric acid. HSO_4^- as a Bronsted acid, has a conjugate of base SO_4^{2-}, sulfate ion.

(f) SO_3^{2-} is a Bronsted base. Its conjugate is HSO_3^-, hydrogen sulfite ion.

26. (a) $HS^-(aq)$ + $H_2O(\ell)$ $\rightleftharpoons$ $H_2S(aq)$ + $OH^-(aq)$

     acid   base   conj. acid   conj. base

(b) $S^{2-}(aq)$ + $NH_4^+(aq)$ $\rightleftharpoons$ $NH_3(aq)$ + $HS^-(aq)$

     base   acid   conj. base   conj. acid

(c) $HCO_3^-(aq)$ + $HSO_4^-(aq)$ $\rightleftharpoons$ $H_2CO_3(aq)$ + $SO_4^{2-}(aq)$

     base   acid   conj. acid   conj. base

(d) $NH_3(aq)$ + $NH_2^-(aq)$ $\rightleftharpoons$ $NH_2^-(aq)$ + $NH_3(aq)$

     acid   base   conj. base   conj. acid

28. (a) $CN^-(aq)$ $+$ $CH_3COOH(aq)$ $\rightleftharpoons$ $CH_3COO^-(aq)$ $+$ $HCN(aq)$

      base   acid   conj. base   conj. acid

(b) $O^{2-}(aq)$ $+$ $H_2O(\ell)$ $\rightleftharpoons$ $OH^-(aq)$ $+$ $OH^-(aq)$

      base   acid   conj. acid   conj. base

(c) $HCO_2^-(aq)$ $+$ $H_2O(\ell)$ $\rightleftharpoons$ $HCOOH(aq)$ $+$ $OH^-(aq)$

      base   acid   conj. acid   conj. base

    (Note: The first reactant is more often written with the formula $HCOO^-$.)

30. (a) $CO_3^{2-}(aq) + H_2O(\ell) \rightleftharpoons HCO_3^-(aq) + OH^-(aq)$

        $HCO_3^-(aq) + H_2O(\ell) \rightleftharpoons H_2CO_3(aq) + OH^-(aq)$

(b) $H_3AsO_4(aq) + H_2O(\ell) \rightleftharpoons H_2AsO_4^-(aq) + H_3O^+(aq)$

        $H_2AsO_4^-(aq) + H_2O(\ell) \rightleftharpoons HAsO_4^{2-}(aq) + H_3O^+(aq)$

        $HAsO_4^{2-}(aq) + H_2O(\ell) \rightleftharpoons AsO_4^{3-}(aq) + H_3O^+(aq)$

(c) $NH_2CH_3COO^-(aq) + H_2O(\ell) \rightleftharpoons NH_2CH_3COOH(aq) + OH^-(aq)$

        $NH_2CH_3COOH(aq) + H_2O(\ell) \rightleftharpoons NH_3CH_3COOH^+ + OH^-(aq)$

pH Calculations

32. *Define the problem*: Given the pH, determine the $[H_3O^+]$.

Develop a plan: Use the appropriate relationship between the concentration and the pH.

$$pH = -\log[H_3O^+], \text{ so } [H_3O^+] = 10^{-pH}$$

$$14.00 = pH + pOH$$

$$pOH = -\log[OH^-], \text{ so } [OH^-] = 10^{-pOH}$$

To get pH directly from $[OH^-]$, use the following relationship:

$$pH = 14.00 + \log[OH^-]$$

When pH < 7 and pOH > 7, the solution is acidic; when pH = 7 = pOH, the solution is neutral; when pH > 7 and pOH < 7, the solution is basic.

Execute the plan:

When pH = 10.5, $[H_3O^+] = 10^{-10.5}$ M = 3×10^{-11} M. The solution is basic.

Check your answers: The high pH gives a $[H_3O^+]$ that is relatively small ($< 10^{-7}$ M).
$Mg(OH)_2$, found in some antacids, is an ionic compound that puts hydroxide ions in the saturated solution; these answers make sense.

34. Follow the method described in the answer to Question 32.

    When pH = 2.44, $[H_3O^+] = 10^{-2.44}$ M = 3.6×10^{-3} M.

36. Follow the method described in the answer to Question 32.

    NaOH is a soluble hydroxide and a strong base. It completely ionizes to form Na^+ and
    OH^-: $[OH^-] = 0.025$ M, pH = $14.00 + \log[OH^-] = 12.40$

    $$pOH = -\log[OH^-] = 1.60$$

38. Follow the method described in the answer to Question 32.

    $$[H_3O^+] = 0.032 \text{ M} \qquad pH = -\log[H_3O^+] = 1.49$$

    $$pOH = 14.00 - pH = 14.00 - 1.49 = 12.51$$

42. Follow the methods described in the answer to Question 32.

	pH	$[H_3O^+]$ (M)	$[OH^-]$ (M)	acidic or basic
(a)	6.21	6.1×10^{-7}	1.6×10^{-8}	acidic
(b)	5.34	4.5×10^{-6}	2.2×10^{-9}	acidic
(c)	4.67	2.1×10^{-5}	4.7×10^{-10}	acidic
(d)	1.60	2.5×10^{-2}	4.0×10^{-13}	acidic
(e)	9.12	7.6×10^{-10}	1.3×10^{-5}	basic

44. (a) Black coffee has a pH about 5.0. $[H_3O^+] = 1 \times 10^{-5}$ M, about 100 times more acidic than neutral.

    (b) Household ammonia (assuming that is aqueous ammonia on the table) has a pH about 11.0. $[H_3O^+] = 1 \times 10^{-11}$ M, about 10,000 times more basic than neutral.

    (c) Baking soda (also known as sodium bicarbonate) has a pH about 8.2.
        $[H_3O^+] = 7 \times 10^{-9}$ M, about 16 times more basic than neutral.

(d) Vinegar has a pH about 2.7. $[H_3O^+] = 2 \times 10^{-3}$ M, about 20,000 times more basic than neutral.

Acid-Base Strengths

46. Acids react with water to make H_3O^+ and the conjugate base. Bases react with water to make hydroxide OH^- and the conjugate acid.

(a) $F^-(aq) + H_2O(\ell) \rightleftharpoons HF(aq) + OH^-(aq)$

$$K = \frac{[HF][OH^-]}{[F^-]}$$

(b) $NH_3(aq) + H_2O(\ell) \rightleftharpoons NH_4^+(aq) + OH^-(aq)$

$$K = \frac{[NH_4^+][OH^-]}{[NH_3]}$$

(c) $H_2CO_3(aq) + H_2O(\ell) \rightleftharpoons HCO_3^-(aq) + H_3O^+(aq)$

$$K = \frac{[HCO_3^-][H_3O^+]}{[H_2CO_3]}$$

(d) $H_3PO_4(aq) + H_2O(\ell) \rightleftharpoons H_2PO_4^-(aq) + H_3O^+(aq)$

$$K = \frac{[H_2PO_4^-][H_3O^+]}{[H_3PO_4]}$$

(e) $CH_3COO^-(aq) + H_2O(\ell) \rightleftharpoons CH_3COOH(aq) + OH^-(aq)$

$$K = \frac{[CH_3COOH][OH^-]}{[CH_3COO^-]}$$

(f) $S^{2-}(aq) + H_2O(\ell) \rightleftharpoons HS^-(aq) + OH^-(aq)$

$$K = \frac{[HS^-][OH^-]}{[S^{2-}]}$$

48. In all these comparisons, the concentrations are the same, so we can use the ionization constants from Table 16.2 to compare their strengths. The larger the K_b the more basic the solution. Ionic compounds provide cations or anions that may be bases, so watch for them.

(a) NH_3 is a base. ($K_b = 1.8 \times 10^{-5}$)

$$NaF \longrightarrow Na^+ + F^-, \text{ and } F^- \text{ is a base. } (K_b = 1.4 \times 10^{-11})$$

A solution of NH_3 is more basic than a solution of NaF.

(b) $K_2S \longrightarrow 2 K^+ + S^{2-}$, and S^{2-} is a base. $(K_b = 1 \times 10^5)$

$K_3PO_4 \longrightarrow 3 K^+ + PO_4^{3-}$, and PO_4^{3-} is a base. $(K_b = 2.8 \times 10^{-2})$

A solution of K_2S is more basic than a solution of K_3PO_4.

(c) $NaNO_3 \longrightarrow Na^+ + NO_3^-$, NO_3^- is base. $(K_b = 5 \times 10^{-16})$

$CH_3COONa \longrightarrow Na^+ + CH_3COO^-$, and CH_3COO^- is a base. $(K_b = 5.6 \times 10^{-10})$

A solution of CH_3COONa is more basic than a solution of $NaNO_3$.

(d) $KCN \longrightarrow K^+ + CN^-$, and CN^- is a base. $(K_b = 2.5 \times 10^{-5})$

A solution of KCN is more basic than a solution of NH_3.

50. Follow the procedure described in Section 16.6, adapting the methods described in the answer to Question 14.44. In all parts, the initial concentration of the solute is 0.10 M.

(a)

	$HNO_2(aq)$	$+ H_2O(\ell)$	$\rightleftharpoons$ $NO_2^-(aq)$	$+ H_3O^+(aq)$
conc. init. (M)	0.10	▱	0	0
change conc. (M)	$-x$	$^-$◖	$+x$	$+x$
eq. conc. (M)	$0.10 - x$	▱ $-$◖	x	x

At equilibrium $K_a = \dfrac{[NO_2^-][H_3O^+]}{[HNO_2]} = \dfrac{(x)(x)}{(0.10 - x)} = 4.5 \times 10^{-4}$

$$x^2 = (4.5 \times 10^{-4})(0.10 - x)$$

$$x^2 + 4.5 \times 10^{-4}x - 4.5 \times 10^{-5} = 0$$

Use the quadratic equation: (see answer to Question 14.49)

$$x = 6.5 \times 10^{-3} \text{ M} = [H_3O^+]$$

$$pH = -\log[H_3O^+] = -\log(6.5 \times 10^{-3}) = 2.19$$

If you did Question 49, compare this pH to your qualitative prediction. You should have predicted between 2 and 6.

(b)

	$NH_4^+(aq)$	$+$ $H_2O(\ell)$ $\rightleftharpoons$	$NH_3(aq)$	$+ H_3O^+(aq)$
conc. init. (M)	0.10	▬	0	0
change conc. (M)	$-x$	$^-$◊	$+x$	$+x$
eq. conc. (M)	$0.10-x$	▬ $^-$◊	x	x

At equilibrium $\quad K_a = \dfrac{[NH_3][H_3O^+]}{[NH_4^+]} = \dfrac{(x)(x)}{(0.10-x)} = 5.6 \times 10^{-10}$

The small-K, reactant-favored reaction will have a larger reactant concentration than product concentrations, so assume x is very small, such that subtraction from the reactant concentration is negligible: $0.10 - x \cong 0.10$.

$$x^2 = (5.6 \times 10^{-10})(0.10) \qquad\qquad x = 7.5 \times 10^{-6} \text{ M} = [H_3O^+]$$

$$pH = -\log[H_3O^+] = -\log(7.5 \times 10^{-6}) = 5.13$$

If you did Question 49, compare this pH to your qualitative prediction. You should have predicted between 2 and 6.

(c)

	$F^-(aq)$	$+$ $H_2O(\ell)$ $\rightleftharpoons$	$HF(aq)$	$+$ $OH^-(aq)$
conc. init. (M)	0.10	▬	0	0
change conc. (M)	$-x$	$^-$◊	$+x$	$+x$
eq. conc. (M)	$0.10-x$	▬ $^-$◊	x	x

At equilibrium $\quad K_b = \dfrac{[HF][OH^-]}{[F^-]} = \dfrac{(x)(x)}{(0.10-x)} = 1.4 \times 10^{-11}$

For the same reasons as in (b), assume x is very small and: $0.10 - x \cong 0.10$.

$$1.4 \times 10^{-11} = \frac{x^2}{(0.10)} \qquad\qquad x^2 = (1.4 \times 10^{-11})(0.10)$$

$$x = 1.2 \times 10^{-6} \text{ M} = [OH^-]$$

$$pH = 14.00 + \log[OH^-] = 14.00 + \log(1.2 \times 10^{-6}) = 8.07$$

If you did Question 49, compare this pH to your qualitative prediction. You should have predicted between pH 8 and 12.

(d) (conc. CH_3COO^-) = 0.10 M $Ba(CH_3COO)_2 \times \dfrac{2 \text{ mol } CH_3COO^-}{1 \text{ mol } Ba(CH_3COO)_2} = 0.20$ M

$$CH_3COO^-(aq) + H_2O(\ell) \rightleftharpoons CH_3COOH(aq) + OH^-(aq)$$

	CH_3COO^-	H_2O	CH_3COOH	OH^-
conc. init. (M)	0.20	▬	0	0
change conc. (M)	$-x$	$^-\lozenge$	$+x$	$+x$
eq. conc. (M)	$0.20 - x$	▬ $-\lozenge$	x	x

At equilibrium
$$K_b = \frac{[CH_3COOH][OH^-]}{[CH_3COO^-]} = \frac{(x)(x)}{(0.20 - x)} = 5.6 \times 10^{-10}$$

For the same reasons as in (b), assume x is very small and: $0.20 - x \cong 0.20$.

$$5.6 \times 10^{-10} = \frac{x^2}{(0.20)}$$

$$x^2 = (5.6 \times 10^{-10})(0.20)$$

$$x = 1.1 \times 10^{-5} \text{ M} = [OH^-]$$

$$pH = 14.00 + \log[OH^-] = 14.00 + \log(1.1 \times 10^{-5}) = 9.02$$

If you did Question 49, compare this pH to your qualitative prediction. You should have predicted between pH 8 and 12.

(e) $O^{2-}(aq) + H_2O(\ell) \longrightarrow 2\,OH^-(aq)$ This must have a very large K. 100 % reaction.

$$[OH^-] = 0.10 \text{ M } O^{2-} \times \frac{2 \text{ mol } OH^-}{1 \text{ mol } O^{2-}} = 0.20 \text{ M}$$

$$pH = 14.00 + \log[OH^-] = 14.00 + \log(0.20 \text{ M}) = 13.30$$

If you did Question 49, compare this pH to your qualitative prediction. You should have predicted pH 12 (or higher).

(f) $HSO_4^-(aq) + H_2O(aq) \rightleftharpoons SO_4^{2-}(aq) + H_3O^+$ $\qquad K_a = 1.2 \times 10^{-2}$

$HSO_4^-(aq) + H_2O(aq) \rightleftharpoons H_2SO_4(aq) + OH^-(aq)$ $\qquad K_b = $ very small

We will use the reaction with the largest K, since it produces more products:

50. (continued)

$$HSO_4^-(aq) \quad + \quad H_2O(\ell) \quad \rightleftharpoons \quad SO_4^{2-}(aq) \quad + \quad H_3O^+(aq)$$

	HSO_4^-	H_2O	SO_4^{2-}	H_3O^+
conc. init. (M)	0.10	🔲	0	0
change conc. (M)	−x	⁻◊	+x	+x
eq. conc. (M)	0.10 − x	🔲 −◊	x	x

At equilibrium
$$K_a = \frac{[SO_4^{2-}][H_3O^+]}{[HSO_4^-]} = \frac{(x)(x)}{(0.10 - x)} = 1.2 \times 10^{-2}$$

$$x^2 = (1.2 \times 10^{-2})(0.10 - x)$$

$$x^2 + 1.2 \times 10^{-2}x - 1.2 \times 10^{-3} = 0$$

Use the quadratic equation: (see answer to Question 14.49)

$$2.9 \times 10^{-2} \text{ M} = [H_3O^+] \qquad pH = -\log[H_3O^+] = -\log(2.9 \times 10^{-2}) = 1.54$$

If you did Question 49, compare this pH to your qualitative prediction. You should have predicted between 2 and 6. While this calculated pH doesn't fall in that range, it is still greater than the pH of 1 we would predict for a 0.10 M strong acid.

(g) $HCO_3^-(aq) + H_2O(aq) \rightleftharpoons CO_3^{2-}(aq) + H_3O^+ \qquad K_a = 4.8 \times 10^{-11}$

$HCO_3^-(aq) + H_2O(aq) \rightleftharpoons H_2CO_3(aq) + OH^-(aq) \qquad K_b = 2.4 \times 10^{-8}$

We will use the reaction with the largest K, since it produces more products:

$$HCO_3^-(aq) \quad + \quad H_2O(\ell) \quad \rightleftharpoons \quad H_2CO_3(aq) \quad + \quad OH^-(aq)$$

	HCO_3^-	H_2O	H_2CO_3	OH^-
conc. init. (M)	0.10	🔲	0	0
change conc. (M)	−x	⁻◊	+x	+x
eq. conc. (M)	0.10 − x	🔲 −◊	x	x

At equilibrium
$$K_b = \frac{[H_2CO_3][OH^-]}{[HCO_3^-]} = \frac{(x)(x)}{(0.10 - x)} = 2.4 \times 10^{-8}$$

For the same reasons as in (b), assume x is very small and: $0.10 - x \cong 0.10$.

$$2.4 \times 10^{-8} = \frac{x^2}{(0.10)} \qquad x^2 = (2.4 \times 10^{-8})(0.10)$$

$$x = 4.5 \times 10^{-5} \text{ M} = [OH^-]$$

$$pH = 14.00 + \log[OH^-] = 14.00 + \log(4.5 \times 10^{-5}) = 9.69$$

If you did Question 49, compare this pH to your qualitative prediction. You should have predicted between pH 8 and 12.

(h) $BaCl_2 \longrightarrow Ba^{2+} + 2\,Cl^-$ Neither of these ions affect the pH of a water solution, because Ba^{2+} is the cation of a soluble ionic hydroxide compound, $Ba(OH)_2$ and Cl^- is a weak base (K_b = very small), so pH 7.00 and no calculation needs to be done. If you did Question 49, you should have predicted pH 7.

52. Combine the methods described in Question 32 and 50 with those described in Section 16.6 and in the answer to Question 14.35.

We do not know the formula of butyric acid, so we will assume it is monoprotic and give it the symbol HBu.

	HBu(aq)	+ $H_2O(\ell)$	$\rightleftharpoons$	Bu^-(aq)	+ H_3O^+(aq)
conc. init. (M)	0.025	▬		0	0
change conc. (M)	$-x$	$^-$◊		$+x$	$+x$
eq. conc. (M)	$0.025 - x$	▬ $-$◊		x	x

At equilibrium, pH = 3.21, so $[H_3O^+] = 10^{-pH} = 10^{-3.21} = 6.2 \times 10^{-4}$ M = x

$$K_a = \frac{[Bu^-][H_3O^+]}{[HBu]} = \frac{(x)(x)}{(0.025 - x)} = \frac{(6.2 \times 10^{-4})(6.2 \times 10^{-4})}{(0.025 - 6.2 \times 10^{-4})} = 1.6 \times 10^{-5}$$

56. Combine the methods described in Question 32 and 50 with those described in Section 16.6 and in the answer to Question 14.35.

	$C_6H_5NH_2$(aq)	+ $H_2O(\ell)$	$\rightleftharpoons$	$C_6H_5NH_3^+$(aq)	+ OH^-(aq)
conc. init. (M)	0.12	▬		0	0
change conc. (M)	$-x$	$^-$◊		$+x$	$+x$
eq. conc. (M)	$0.12 - x$	▬ $-$◊		x	x

At equilibrium $\quad K_b = \dfrac{[C_6H_5NH_3^+][OH^-]}{[C_6H_5NH_2]} = \dfrac{(x)(x)}{(0.12-x)} = 4.2 \times 10^{-10}$

Assume x is very small and: $0.12 - x \cong 0.12$.

$$x^2 = (4.2 \times 10^{-10})(0.12)$$

$$x = 7.1 \times 10^{-6}\ M = [\,OH^-\,]$$

$$pH = 14.00 + \log[\,OH^-\,] = 14.00 + \log(7.1 \times 10^{-6}) = 8.85$$

58. First use methods from Chapters 3 and 4 to find the initial concentration of the $C_3H_6O_3$. Then combine the methods described in Question 32 and 50 with those described in Section 16.6 and in the answer to Question 14.35.

$$\dfrac{56\ mg\ C_3H_6O_3}{250\ mL\ soln} \times \dfrac{1\ g}{1000\ mg} \times \dfrac{1\ mol\ C_3H_6O_3}{90.08\ g\ C_3H_6O_3} \times \dfrac{1000\ mL}{1\ L} = 0.0025\ M$$

According to the discussion in Section 16.9, lactic acid is a monoprotic acid, so we'll write the formula as: $HC_3H_5O_3$.

	$HC_3H_5O_3$(aq)	$+ H_2O(\ell)$	$\rightleftharpoons$ $C_3H_5O_3^-$(aq)	$+ H_3O^+$(aq)
conc. init. (M)	0.0025	▬	0	0
change conc. (M)	$-x$	$^-$◊	$+x$	$+x$
eq. conc. (M)	$0.0025 - x$	▬$-$◊	x	x

At equilibrium $\quad K_a = \dfrac{[C_3H_5O_3^-][H_3O^+]}{[HC_3H_5O_3]} = \dfrac{(x)(x)}{(0.0025-x)} = 1.4 \times 10^{-4}$

$$x^2 = (1.4 \times 10^{-4})(0.0025 - x)$$

$$x^2 + 1.4 \times 10^{-4}x - 3.5 \times 10^{-7} = 0$$

Use the quadratic equation: (see answer to Question 14.49)

$$x = 5.2 \times 10^{-4}\ M = [H_3O^+]$$

$$pH = -\log[H_3O^+] = -\log(5.2 \times 10^{-4}) = 3.28$$

Acid-Base Reactions

60. Use the methods described in the answers to Questions 22, 26, and 48. Compare the reactant acid to the product acid and identify which is stronger and which is weaker. Do the same with the bases. Equilibrium favors the weaker species in the reaction.(aq)

(a) **$CN^-(aq)$**   + $HSO_4^-(aq)$ $\rightleftharpoons$ $HCN(aq)$ + $SO_4^{2-}(aq)$

    stronger base   stronger acid   weaker acid   weaker base

    The reaction is product-favored.

(b) $H_2S(aq)$ + $H_2O(\ell)$ $\rightleftharpoons$ $H_3O^+(aq)$ +   **$HS^-(aq)$**

    weaker acid   weaker base   stronger acid   stronger base

    The reaction is reactant-favored.

(c) $H^-(aq)$ + $H_2O(\ell)$ $\rightleftharpoons$ $OH^-(aq)$ +   **$H_2(g)$**

    stronger base   stronger acid   weaker base   weaker acid

    The reaction is product-favored. *(Note, the H_2 product is ultimately a gas, not aqueous as indicated in the problem.)*

62. Use the method described in the answer to Question 60.

(a) $NH_4^+(aq)$ + $HPO_4^{2-}(aq)$ $\rightleftharpoons$ **$NH_3(aq)$**   +   **$H_2PO_4^-(aq)$**

    weaker acid   weaker base   stronger base   stronger acid

    The reaction is reactant-favored.

(b) $CH_3COOH(aq)$ + $OH^-(aq)$ $\rightleftharpoons$ **$CH_3COO^-(aq)$**+ **$H_2O(\ell)$**

    stronger acid   stronger base   weaker base   weaker acid

    The reaction is product-favored.

(c) We choose to use HSO_4^- as the reactant acid, since it is a stronger acid than $H_2PO_4^-$

    $HSO_4^-(aq)$ + $H_2PO_4^-(aq)$ $\rightleftharpoons$ **$H_3PO_4(aq)$**   +   **$SO_4^{2-}(aq)$**

    stronger acid   stronger base   weaker acid   weaker base

    The reaction is product-favored.

(d) $CH_3COOH(aq) + F^-(aq) \rightleftharpoons CH_3COO^-(aq) + HF(\ell)$

weaker acid   weaker base   stronger base   stronger acid

The reaction is reactant-favored.

64. Adapt the method described in the answers to Questions 47 - 50.

(a) $AlCl_3 \longrightarrow Al^{3+} + 3\,Cl^-$ The Cl^- ions do not affect the pH of a water solution, because HCl is a strong acid. However, the hydrated aqueous aluminum ion is formed:

$$Al^{3+} + 6\,H_2O(aq) \longrightarrow Al(H_2O)_6^{3+}$$

It is a weak acid, so we predict pH less than 7.

(b) $Na_2S \longrightarrow 2\,Na^+ + S^{2-}$ The Na^+ ions do not affect the pH of a water solution. S^{2-} is a strong base, so we predict pH greater than 7.

(c) $NaNO_3 \longrightarrow Na^+ + NO_3^-$ Neither of these ions affects the pH of a water solution, because NaOH is a strong base and HNO_3 is a strong acid, so we predict pH equal to 7.

66. Adapt the method described in the answer to Question 64.

(a) $Na_2HPO_4 \longrightarrow 2\,Na^+ + HPO_4^{2-}$ The Na^+ ions do not affect the pH of a water solution. The HPO_4^{2-} ion is a weak acid and a weak base, so we must compare the size of K_a and K_b:

$$HPO_4^{2-}(aq) + H_2O(aq) \rightleftharpoons PO_4^{3-}(aq) + H_3O^+ \qquad K_a = 3.6 \times 10^{-13}$$

$$HPO_4^{2-}(aq) + H_2O(aq) \rightleftharpoons H_2PO_4^-(aq) + OH^- \qquad K_b = 1.7 \times 10^{-7}$$

$H_2PO_4^-$ is a stronger base than it is an acid, so we predict pH greater than 7.

(b) $(NH_4)_2S \longrightarrow 2\,NH_4^+ + S^{2-}$ The NH_4^+ ion is a weak acid, but the S^{2-} ion is a strong base, so we predict pH greater than 7.

(c) $KCH_3COO \longrightarrow K^+ + CH_3COO^-$ The K^+ ions do not affect the pH of a water solution. CH_3COO^- is a weak base, so we predict pH greater than 7.

68. All of these solids have the same cation, whose aqueous form, $Cu(H_2O)_6^{2+}$, is an acid with $K_a = 1.6 \times 10^{-7}$. The presence of extra H_3O^+ will not affect the cation concentration from one solution to the next. Therefore, we will compare the strength of the anionic bases to judge variations in the solubility in acid solutions. The solid will be more soluble at pH 2 than pH 7 if its anion reacts with H_3O^+ to form a weak conjugate acid.

(a) $Cu(OH)_2 \longrightarrow Cu^{2+} + 2\,OH^-$ The conjugate of OH^- is H_2O, a very weak acid. This salt would be more soluble at pH 2.

(b) $CuSO_4 \longrightarrow Cu^{2+} + SO_4^{2-}$ The conjugate of SO_4^{2-} is HSO_4^-, a slightly weak acid. This salt would be slightly more soluble at pH 2.

(c) $CuCO_3 \longrightarrow Cu^{2+} + CO_3^{2-}$ The conjugate of CO_3^{2-} is HCO_3^-, a weak acid. This salt would be more soluble at pH 2.

(d) $CuS \longrightarrow Cu^{2+} + S^{2-}$ The conjugate of S^{2-} is HS^-, a weak acid. This salt would be more soluble at pH 2.

(e) $CuS \longrightarrow 3\,Cu^{2+} + 2\,PO_4^{3-}$ The conjugate of PO_4^{3-} is HPO_4^{2-}, a weak acid. This salt would be more soluble at pH 2.

In conclusion, all of these salts would be more soluble at pH 2 than at pH 7.

Practical Acid-Base Chemistry

70. $Na_2CO_3 + 2\,CH_3(CH_2)_{16}COOH \longrightarrow 2\,CH_3(CH_2)_{16}COONa + H_2O + CO_2$

72. This question involves some interpretation and research. Of the three materials described, two are solutions containing acids: Vinegar contains acetic acid (CH_3COOH, $K_a = 1.8 \times 10^{-5}$) at a concentration of approximately 5 % and lemon juice contains triprotic citric acid ($H_3C_5H_5O_7$, Table 12.5) at a concentration of approximately 1 %. The third material is lactic acid ($C_3H_6O_3$, $K_a = 1.4 \times 10^{-4}$, from Question 16.58), a 100 % pure molecular acid that is sometimes found in milk.

The ion being neutralized is HCO_3^-(aq).

$$HCO_3^-(aq) + H_3O^+(aq) \rightleftharpoons H_2CO_3(aq) + H_2O(\ell)$$

$$K = \frac{1}{K_{a,H_2CO_3}} = \frac{1}{4.2 \times 10^{-7}} = 2.4 \times 10^6$$

$$HA(aq) + H_2O(\ell) \rightleftharpoons A^-(aq) + H_3O^+(aq) \qquad K_{a,HA}$$

Because lactic acid, acetic acid, and citric acid each presumably have $K_a > 10^{-7}$, the three neutralization reactions we are studying here will all be product-favored. $(K > 1)$. So, all we need to do is figure out which sample provides the largest number of H_3O^+ ions.

Using the molar mass of lactic acid, we can determine the moles of H_3O^+ ions available if one gram of lactic acid:

$$1 \text{ g C}_3\text{H}_6\text{O}_3 \times \frac{1 \text{ mol C}_3\text{H}_6\text{O}_3}{90.08 \text{ g C}_3\text{H}_6\text{O}_3} \times \frac{1 \text{ mol H}_3\text{O}^+}{1 \text{ mol C}_3\text{H}_6\text{O}_3} = 0.01 \text{ mol H}_3\text{O}^+$$

Using the percentages of acids in solution and the molar masses, determine the H_3O^+ ions available in the vinegar and lemon juice samples.

$$1 \text{ g vinegar} \times \frac{0.05 \text{ g CH}_3\text{COOH}}{1 \text{ g vinegar}} \times \frac{1 \text{ mol CH}_3\text{COOH}}{60.05 \text{ g CH}_3\text{COOH}} \times \frac{1 \text{ mol H}_3\text{O}^+}{1 \text{ mol CH}_3\text{COOH}} =$$

$$0.0008 \text{ mol H}_3\text{O}^+$$

$$1 \text{ g lemon juice} \times \frac{0.01 \text{ g H}_3\text{C}_5\text{H}_3\text{O}_7}{1 \text{ g lemon juice}} \times \frac{1 \text{ mol H}_3\text{C}_5\text{H}_3\text{O}_7}{178.1 \text{ g H}_3\text{C}_5\text{H}_3\text{O}_7} \times \frac{3 \text{ mol H}_3\text{O}^+}{1 \text{ mol H}_3\text{C}_5\text{H}_3\text{O}_7} =$$

$$0.0002 \text{ mol H}_3\text{O}^+$$

With this analysis, we find that the 1 gram sample of pure lactic acid will make more CO_2, even though the citric acid in the lemon juice is the strongest acid.

74. Dishwasher detergent is very basic, and should not be used to wash anything by hand, including a car. If it gets into the engine area, it can also dissolve automobile grease and oil, which could prevent the engine from running correctly.

Lewis Acids and Bases

76. The Lewis model focuses on what the electron pairs are doing. The substance capable of donating the electron pair to form a new bond is called a Lewis base. The substance capable of accepting an electron pair is a Lewis acid.

(a) O^{2-} has a lone pair of electrons that can form a new bond, so O^{2-} can be a Lewis base. It cannot accept any more electrons, so it is not a Lewis acid.

$$:\overset{\displaystyle ..}{\underset{\displaystyle ..}{O}}:$$

(b) CO_2 has a lone pair of electrons on the O atoms that can form a new bond, so CO_2 can be a Lewis base. Its central C atom can interact with lone pairs on other Lewis bases, so CO_2 can also be a Lewis acid.

(c) H^- has a lone pair of electrons that can form a new bond, so H^- can be a Lewis base. It cannot accept any more electrons, so it is not a Lewis acid.

$$H\mathbin{:}$$

78. Follow the method described in the answer to Question 76.

(a) Cr^{3+} can interact with lone pairs on Lewis bases, so Cr^{3+} can also be a Lewis acid. It has no valence electrons that can form a new bond, so it cannot be a Lewis base.

(b) SO_3 has lone pairs of electrons on the O atoms that can form new bonds, so SO_3 can be a Lewis base. Its S atoms could interact with lone pairs on other Lewis bases, so SO_3 can also be a Lewis acid.

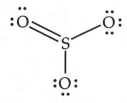

(c) CH_3NH_2 has a lone pair of electrons on N that can form a new bond, so CH_3NH_2 can be a Lewis base. The H atoms polar-covalently bonded to N could interact with other Lewis bases and be removed, so it could function as a Lewis acid; however, that reaction requires a very strong Lewis base.

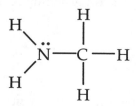

80. Identify which reactant is donating electrons and which is accepting them.

(a) The curved arrow shows how the lone pair on I^- becomes a new bond on the right I of I_2.

Therefore, I_2 is the Lewis acid and I^- is the Lewis base.

(b) The curved arrow shows how the lone pair on S becomes a new bond between the S atom and B atom.

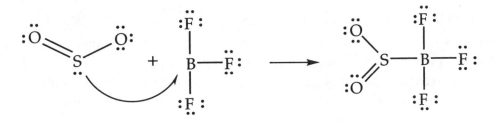

Therefore, BF_3 is the Lewis acid and SO_2 is the Lewis base.

(c) Au^+ metal ion has no bonds to start with and only two low-energy valence electrons. Electrons on CN^- make new bonds with Au^+, so Au^+ is Lewis acid and CN^- is Lewis base.

(d) A similar reaction is shown at the end of Section 16.10. The curved arrow shows how the lone pair on O becomes a new bond between C atom and O atom. (Note that after the initial Lewis acid-base reaction, one of the H atoms migrates to a nearby O atom to balance the positive and negative charges.)

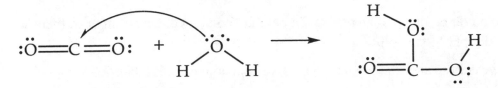

Therefore, CO_2 is the Lewis acid and H_2O is the Lewis base.

82. See the answer to Question 9.85 for the shape determinations. The shape of the ICl_3 molecule is T-shaped. The curved arrow shows how the lone pair on Cl^- becomes a new bond between the Cl atom and I atom.

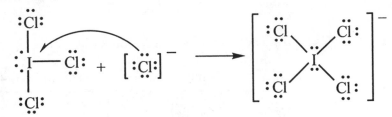

Therefore, ICl_3 is the Lewis acid in the reaction with chloride ion. The structure (i.e. shape) of the resulting ion, ICl_4^-, is square planar.

General Questions

84. Amphiprotic is defined in the margin note of page 730 in Section 16.1. In general, any species with an H can donate H^+ ions, though sometimes that will not happen in the common water solvent. In addition, any species with one or more lone pairs can be a base, since the electrons could make a bond to a proton. However, because this question asks for a judgement of "weak" and "strong," we will restrict our designations to the reactions of these species in water and use Table 16.2 to assist us.

(a) CH_3COOH is a weak acid.

(b) Na_2O, contains O^{2-}, which is a strong base.

(c) H_2SO_4 is a strong acid.

(d) NH_3 is a weak base.

(e) $Ba(OH)_2$, contains OH^-, which is a strong base.

(f) $H_2PO_4^-$ is amphiprotic.

86. If equal molar amounts of an acid and a base are combined, the stronger of them will dictate the pH of the solution.

(a) A weak base and a strong acid will have an acidic pH, less than 7.

(b) A strong base and a strong acid will have a neutral pH, equal to 7.

(c) A strong base and a weak acid will have a basic pH, greater than 7.

88. Follow the methods described in the answers to Questions 50 and 58.

$$\frac{5.0 \text{ mg } C_6H_8O_6}{1 \text{ mL soln}} \times \frac{1 \text{ g}}{1000 \text{ mg}} \times \frac{1 \text{ mol } C_6H_8O_6}{176 \text{ g } C_6H_8O_6} \times \frac{1000 \text{ mL}}{1 \text{ L}} = 0.028 \text{ M}$$

Ascorbic acid, $C_6H_8O_6$, is diprotic acid, so give it a formula of H_2A

	$H_2A(aq)$	$+ H_2O(\ell)$	$\rightleftharpoons$	$HA^-(aq)$	$+ H_3O^+(aq)$
conc. init. (M)	0.028	⬗		0	0
change conc. (M)	$-x$	$^-$⬗		$+x$	$+x$
eq. conc. (M)	$0.028 - x$	⬗ $^-$⬗		x	x

At equilibrium

$$K_a = \frac{[HA^-][H_3O^+]}{[H_2A]} = \frac{(x)(x)}{(0.028-x)} = 7.9 \times 10^{-5}$$

$$x^2 = (7.9 \times 10^{-5})(0.028 - x)$$

If we assume that x is small, and $0.028 - x \cong 0.028$

$$x^2 = (7.9 \times 10^{-5})(0.028)$$

$$x = 1.5 \times 10^{-3} \text{ M} = [H_3O^+]$$

$$pH = -\log[H_3O^+] = -\log(1.5 \times 10^{-3}) = 2.82$$

If we use the quadratic equation: (see answer to Question 14.49)

$$x^2 + 7.9 \times 10^{-5}x - 2.2 \times 10^{-6} = 0$$

$$x = 1.4 \times 10^{-3} \text{ M} = [H_3O^+]$$

$$pH = -\log[H_3O^+] = -\log(1.4 \times 10^{-32}) = 2.85$$

90. Adding acids to bases decreases the pH. Adding bases to acids increases the pH. Adding weak acids or bases to strong acids or bases, respectively, does not change the pH significantly. Use the methods described in Questions 47 - 50 and 84.

(a) $Na_2C_2O_4$ contains $C_2O_4^{2-}$, a weak base. The base is added to an acid, so the pH increases.

(b) NH_4Cl contains NH_4^+, a weak acid. The weak acid is added to a strong acid, so no change in pH will be noticed. pH stays the same.

(c) Adding a neutral salt, NaCl, will not change the pH, so the pH stays the same.

Applying Concepts

92. Use Table 16.1 for the values of K_w at 10 °C, at 25 °C, and at 50 °C.

$$2 \text{ H}_2O(\ell) \rightleftharpoons H_3O^+(aq) + OH^-(aq)$$

	$2 H_2O(\ell)$	$H_3O^+(aq)$	$OH^-(aq)$
conc. initial (M)	⬛	0	0
change conc. (M)	$-\delta -\delta$	$+x$	$+x$
equilibrium conc. (M)	⬛ $-\delta -\delta$	x	x

At equilibrium $\qquad K_w = [H_3O^+][OH^-] = x^2 \qquad x = \sqrt{K_w} = [H_3O^+]$

At 10 °C $\qquad [H_3O^+] = \sqrt{0.29 \times 10^{-14}} = 5.4 \times 10^{-8}$

$\qquad\qquad\qquad pH = -\log[H_3O^+] = -\log(5.4 \times 10^{-8}) = 7.27$

At 25 °C $\qquad [H_3O^+] = \sqrt{1.01 \times 10^{-14}} = 1.00 \times 10^{-7}$

$\qquad\qquad\qquad pH = -\log[H_3O^+] = -\log(1.00 \times 10^{-7}) = 6.998$

At 50 °C $\qquad [H_3O^+] = \sqrt{5.48 \times 10^{-14}} = 2.34 \times 10^{-7}$

$\qquad\qquad\qquad pH = -\log[H_3O^+] = -\log(2.34 \times 10^{-7}) = 6.631$

The solutions are neutral, since $x = [H_3O^+] = [OH^-]$ in each of them.

94. Use the methods described in the answer to Question 84.

(a) HCl is a strong acid, resulting in a solution of H_3O^+ and Cl^-. The solution contains the following molecules and ions in order of decreasing concentrations:

$$H_2O > H_3O^+ = Cl^- \gg OH^-$$

(b) $NaClO_4$ is a soluble ionic compound, resulting in a solution of Na^+ and ClO_4^-. Neither of these ions reacts with water. The solution contains the following molecules and ions in order of decreasing concentrations:

$$H_2O > Na^+ = ClO_4^- \gg H_3O^+ = OH^-$$

(c) HNO_2 is a weak acid, resulting in a solution of a large proportion of HNO_2 and a small proportion of H_3O^+ and NO_2^- The solution contains the following molecules and ions in order of decreasing concentrations:

$$H_2O > HNO_2 > H_3O^+ = NO_2^- \gg OH^-$$

(d) NaClO is a soluble ionic compound, resulting in a solution of Na^+ and ClO^-. The anion is a weak base and reacts with water to a small extent to make a small proportion of HClO and OH^-. The solution contains the following molecules and ions in order of decreasing concentrations:

$$H_2O > Na^+ \cong ClO^- > OH^- = HClO \gg H_3O^+$$

(e) NH_4Cl is a soluble ionic compound, resulting in a solution of NH_4^+ and Cl^-. The cation is a weak acid and reacts with water to a small extent to make a small proportion of NH_3 and H_3O^+. The solution contains the following molecules and ions in order of decreasing concentrations:

$$H_2O > NH_4^+ \cong Cl^- > H_3O^+ = NH_3 >> OH^-$$

(f) NaOH is a strong base, resulting in a solution of Na^+ and OH^-. The solution contains the following molecules and ions in order of decreasing concentrations:

$$H_2O > Na^+ = OH^- >> H_3O^+$$

96. Conjugates in an acid-base pair must differ by only one H^+ ion. The acids and bases were identified correctly, but the conjugates were not. HCO_3 is the conjugate of H_2CO_3 and HSO_4 is the conjugate of SO_4^{2-}.

98. The Arrhenius theory is described by the Lewis theory in the following way:

Electron pairs on the solvent water molecules (Lewis base) form a bond with the hydrogen ion (Lewis acid) producing aqueous H^+ ions. Electron pairs on the OH^- ions (Lewis base) form a bond with the hydrogen ion (Lewis acid) in the solvent water molecule, producing aqueous OH^- ions.

Bronsted Theory: The H^+ ion from the Bronsted acid is bonded to a Bronsted base using an electron pair on the base. The electron-pair acceptor, the H^+ ion, is the Lewis acid and the electron-pair donor is the Lewis base.

100. Stronger acids result in weaker conjugate bases. The stronger base has the larger pH. So, the salt solution with the highest pH contains the anion of the weakest acid.

$$\text{Smallest pH:}\quad NaM > NaQ > NaZ \quad \text{:Largest pH}$$

$$\text{Strongest:}\quad HM > HQ > HZ \quad \text{:Weakest}$$

The initial $[OH^-]$ of neutral water is 1×10^{-7} M. The initial concentration of the base is 0.1 M.

In general	A^-(aq)	+ H_2O(ℓ)	⇌	HA(aq)	+ OH^-(aq)
conc. init. (M)	0.1	▭		0	1×10^{-7}
change conc. (M)	$-x$	$-x$		$+x$	$+x$
eq. conc. (M)	$0.1 - x$	▭ $- x$		x	$10^{-7} + x$

At equilibrium
$$K_b = \frac{[HA][OH^-]}{[A^-]} = \frac{(x)(10^{-7} + x)}{(0.1 - x)}$$

$$1 \times 10^{-7} + x = [OH^-] = 10^{-pOH} = 10^{-(14.00 - pH)} = 10^{(pH - 14.00)}$$

$$x = 10^{(pH - 14.00)} - 1 \times 10^{-7}$$

For NaZ, A^- above is Z^- and pH = 9.0, so

$$x = 10^{(9.0 - 14.00)} - 1 \times 10^{-7} = 1 \times 10^{-5} - 1 \times 10^{-7} = 1 \times 10^{-5}, \text{ and}$$

$$K_b = \frac{(1 \times 10^{-5})(1 \times 10^{-7} + 1 \times 10^{-5})}{(0.1 - 1 \times 10^{-5})} = 1 \times 10^{-9}$$

For NaQ, A^- above is Q^- and pH = 8.0, so

$$x = 10^{(8.0 - 14.00)} - 1 \times 10^{-7} = 1 \times 10^{-6} - 1 \times 10^{-7} = 1 \times 10^{-6}, \text{ and}$$

$$K_b = \frac{(1 \times 10^{-6})(1 \times 10^{-7} + 1 \times 10^{-6})}{(0.1 - 1 \times 10^{-6})} = 1 \times 10^{-11}$$

For NaM, A^- above is M^- and pH = 7.0, so $x = 10^{(7.0 - 14.00)} - 1 \times 10^{-7}$

$x = 1 \times 10^{-7} - 1 \times 10^{-7} = 0$, suggesting that the reaction with water does not go toward products at all. That means K_b of M^- may be so small that the pH of the solution is accounted for exclusively by the ionization of water. If indeed the base does provide all the OH^- ions to make the solution's pH: $[HM] = [OH^-] = 10^{-7.0} = 1 \times 10^{-7}$:

$$K_b = \frac{(1 \times 10^{-7})(1 \times 10^{-7})}{(0.1)} = 1 \times 10^{-13}$$

A relationship between K_a and K_b is described at the end of Section 16.6.

For HZ, $K_a = K_w/K_b = (1.0 \times 10^{-14})/(1 \times 10^{-9}) = 1 \times 10^{-5}$

For HQ, $K_a = K_w/K_b = (1.0 \times 10^{-14})/(1 \times 10^{-11}) = 1 \times 10^{-3}$

For HM, $K_a = K_w/K_b = (1.0 \times 10^{-14})/(1 \times 10^{-13}) = 1 \times 10^{-1}$ or larger

Chapter 17: Additional Aqueous Equilibria

Buffer Solutions

16. To determine the pH of a buffer, look up the pK_a (Table 17.1) or look up the K_a (Table 16.2) and calculate the pK_a ($pK_a = -\log K_a$). The pK_a closest to the desired pH is the best buffer, since close to equal quantities of the acid and base would be used, giving the solution approximately equal ability to neutralize added acid or added base.

(a) The $CH_3COOH/NaCH_3COO$ buffer system has $pK_a = 4.74$.

(b) The acid in $HCl/NaCl$ is HCl. It has $K_a =$ very large. This is not a buffer.

(c) The NH_3/NH_4Cl buffer system has $pK_a = 9.25$.

The combination that would make the best pH 9 buffer system is (c) NH_3/NH_4Cl.

18. To answer this question quantitatively we need the value of pKa, since that is equal to the pH in an equimolar buffer solution. In some cases, that value is not in the textbook. Without doing any calculations, we must estimate the pK_a from the K_a. We will look at the size of the K_a value and estimate its power of ten. We do this by determining which two powers of ten the number is between, then we can also use the values provided in Table 17.1 to give us advice about the fractional part of the pK_a.

(a) The acid of the pair is H_3PO_4. It has $K_a = 7.5 \times 10^{-3}$. Here the K_a is between 10^{-2} and 10^{-3}. In Table 17.1, the dihydrogen phosphate buffer K_a has a slightly smaller number multiplying its power of ten, 6.2, and its pK_a has a fractional component of .21, so we will estimate the $HNO_2/NaNO_2$ buffer system will have a pH of about 2.1.

(b) The NaH_2PO_4/Na_2HPO_4 buffer system has $pK_a = 7.21$, so its pH is 7.21.

(c) The acid of the pair is HPO_4^{2-}, with $K_a = 3.6 \times 10^{-13}$. Here the K_a is between 10^{-12} and 10^{-13}. In Table 17.1, the hypochlorous acid K_a has a similar number multiplying its power of ten, 3.5, and its pK_a has a fractional component of .46, so we will estimate the Na_2HPO_4/Na_3PO_4 buffer system will have a pH of about 12.46.

To check the answers, we will disobey the explicit instructions and do the calculation. In this case, you can also compare the estimates to the calculations you may have already done in Question 15.

(a) $pK_a = -\log(7.5 \times 10^{-3}) = 2.12$

(c) $pK_a = -\log(3.6 \times 10^{-13}) = 12.44$

20. We will compare the pH to the values of pK_a in Table 17.1, since that is equal to the pH in an equimolar buffer solution. The pK_a closest to the desired pH is most suitable.

(a) pH = 3.45, needs a lactic acid/lactate buffer ($pK_a = 3.85$).

(b) pH = 5.48, needs an acetic acid/acetate buffer ($pK_a = 4.74$). Its pH is slightly closer to the desired pH than any of the other buffer systems in Table 17.1.

(c) pH = 8.32, needs a hypochlorous acid/hypochlorite buffer ($pK_a = 7.46$). Its pH is slightly closer to the desired pH than any of the other buffer systems.

(d) pH =10.15, needs a hydrogen carbonate/carbonate buffer ($pK_a = 10.32$).

22. This question uses methods learned in several previous chapters. First adapt the methods from the answer to Questions 16.50 - 16.58 to find the concentration of the NH_4^+ present in the equilibrium solution. Then adapt the method from the answer to Question 5.57 to find the mass of the salt.

	$NH_3(aq)$	$+ H_2O(\ell)$	$\rightleftharpoons$ $NH_4^+(aq)$	$+ OH^-(aq)$
conc. init. (M)	0.10	🪣	c	0
change conc. (M)	$-x$	$-$ 💧	$+x$	$+x$
eq. conc. (M)	$0.10 - x$	🪣 $-$ 💧	$c + x$	x

At equilibrium   pH = 9.00, so $[OH^-] = 10^{-pOH} = 10^{-14.00 + 9.00} = 1.0 \times 10^{-5} = x$

$$K_b = \frac{[NH_4^+][OH^-]}{[NH_3]} = \frac{(x)(c+x)}{(0.10-x)} = \frac{(1.0 \times 10^{-5})(c + 1.0 \times 10^{-5})}{(0.10 - 1.0 \times 10^{-5})} = 1.8 \times 10^{-5}$$

Solve for c: c = 0.18 M

$$500. \text{ mL} \times \frac{1 \text{ L}}{1000 \text{ mL}} \times \frac{0.18 \text{ mol NH}_4^+}{1 \text{ L}} \times \frac{1 \text{ mol NH}_4Cl}{1 \text{ mol NH}_4^+} \times \frac{53.49 \text{ g NH}_4Cl}{1 \text{ mol NH}_4Cl} = 4.8 \text{ g NH}_4Cl$$

24. Adapt the method described in the answer to Question 22.

$$(\text{conc. NH}_4^+) = \frac{5.15 \text{ g NH}_4NO_3}{0.10 \text{ L}} \times \frac{1 \text{ mol NH}_4NO_3}{80.05 \text{ g NH}_4NO_3} \times \frac{1 \text{ mol NH}_4^+}{1 \text{ mol NH}_4NO_3}$$

$$= 0.64 \text{ M NH}_4^+$$

	NH₃(aq)	+ H₂O(ℓ)	⇌ NH₄⁺(aq)	+ OH⁻(aq)
conc. init. (M)	0.15	▱	0.643	0
change conc. (M)	$-x$	$^-$▵	$+x$	$+x$
eq. conc. (M)	$0.10 - x$	▱ $-$▵	$0.643 + x$	x

At equilibrium $\quad K_b = \dfrac{[NH_4^+][OH^-]}{[NH_3]} = \dfrac{(x)(0.643 + x)}{(0.15 - x)} = 1.8 \times 10^{-5}$

Assume that x is small enough to ignore it in the sum and difference:

$$1.8 \times 10^{-5} = \frac{(x)(0.643)}{(0.15)}$$

$$x = 4.2 \times 10^{-6} \text{ M} = [OH^-]$$

$$pH = 14.00 + \log[OH^-] = 14.00 + \log(4.2 \times 10^{-6}) = 8.62$$

26. Adapt the method described in the answers to Question 22 and 24.

$$(\text{conc. NH}_4^+) = \frac{0.125 \text{ mol NH}_4\text{Cl}}{500. \text{ mL}} \times \frac{1000 \text{ mL}}{1 \text{ L}} \times \frac{1 \text{ mol NH}_4^+}{1 \text{ mol NH}_4\text{Cl}} = 0.250 \text{ M NH}_4^+$$

	NH₃(aq)	+ H₂O(ℓ)	⇌ NH₄⁺(aq)	+ OH⁻(aq)
conc. init. (M)	0.500	▱	0.250	0
change conc. (M)	$-x$	$^-$▵	$+x$	$+x$
eq. conc. (M)	$0.500 - x$	▱ $-$▵	$0.250 + x$	x

At equilibrium $\quad K_b = \dfrac{[NH_4^+][OH^-]}{[NH_3]} = \dfrac{(x)(0.250 + x)}{(0.500 - x)} = 1.8 \times 10^{-5}$

Assume that x is small enough to ignore it in the sum and difference:

$$1.8 \times 10^{-5} = \frac{(x)(0.250)}{(0.500)}$$

$$x = 3.6 \times 10^{-5} \text{ M} = [OH^-]$$

$$pH = 14.00 + \log[OH^-] = 14.00 + \log(3.6 \times 10^{-5}) = 9.56$$

Assume that all the HCl gas bubbled through the solution is actually dissolved in the solution, find the initial concentration of H_3O^+ ions from the ionization of HCl, after it is dissolved but before it reacts:

$$(\text{conc. } H_3O^+) = \frac{0.0100 \text{ mol HCl}}{500. \text{ mL}} \times \frac{1000 \text{ mL}}{1 \text{ L}} \times \frac{1 \text{ mol } H_3O^+}{1 \text{ mol HCl}} = 0.0200 \text{ M } H_3O^+$$

The acid neutralizes the strongest base in the solution, NH_3. Write the product-favored neutralization reaction and take it as far to the products as possible, using the method of limiting reactants.

	$H_3O^+(aq)$	+	$NH_3(aq)$	$\longrightarrow$	$NH_4^+(aq)$	+	$H_2O(\ell)$
conc. init. (M)	0.0200		0.500		0.250		⊟
change conc. (M)	− 0.0200		− 0.0200		+ 0.0200		+ ◊
final conc. (M)	0		0.480		0.270		⊟ + ◊

The solution is still a buffer solution, containing an acid-base conjugate pair, so we can find the pH using K_b of the base.

	$NH_3(aq)$	+	$H_2O(\ell)$	$\rightleftharpoons$	$NH_4^+(aq)$	+	$OH^-(aq)$
conc. init. (M)	0.480		⊟		0.270		0
change conc. (M)	− x		− ◊		+ x		+ x
eq. conc. (M)	0.480 − x		⊟ − ◊		0.270 + x		x

At equilibrium $K_b = \dfrac{[NH_4^+][OH^-]}{[NH_3]} = \dfrac{(x)(0.270 + x)}{(0.480 - x)} = 1.8 \times 10^{-5}$

Assume that x is small enough to ignore it in the sum and difference:

$$1.8 \times 10^{-5} = \frac{(x)(0.270)}{(0.480)}$$

$$x = 3.2 \times 10^{-5} \text{ M} = [OH^-]$$

$$pH = 14.00 + \log[OH^-] = 14.00 + \log(3.2 \times 10^{-5}) = 9.51$$

The original buffer solution has a pH of 9.56. After adding some quantity of strong acid, it makes sense that the pH is slightly more acidic.

28. A 1-L solution of 0.20 M NaOH provides a strong base. The added solution must provide a source of a weak acid is sufficient quantity for some of it to completely neutralize the strong base and some of it to remain in the solution

(a) Adding 0.10 mol CH_3COOH to the 1-L solution makes a 0.10 M CH_3COOH solution. The strong base neutralizes the acid. The product-favored neutralization runs towards the products until one of the reactants runs out.

$$OH^-(aq) \ + CH_3COOH(aq) \ \longrightarrow \ CH_3COO^-(aq) \ + H_2O(\ell)$$

	OH^-	CH_3COOH	CH_3COO^-	H_2O
conc. init. (M)	0.20	0.10	0	
change conc. (M)	− 0.10	− 0.10	+ 0.10	$^+$
final conc. (M)	0.10	0	0.10	$^+$

The solution produced is a not buffer solution. Too much base is present and all the CH_3COOH has been neutralized.

(b) Adding 0.30 mol CH_3COOH to the 1-L solution makes a 0.30 M CH_3COOH solution. The strong base neutralizes the acid. The product-favored neutralization runs towards the products until one of the reactants runs out.

$$OH^-(aq) \ + CH_3COOH(aq) \ \longrightarrow \ CH_3COO^-(aq) \ + H_2O(\ell)$$

	OH^-	CH_3COOH	CH_3COO^-	H_2O
conc. init. (M)	0.20	0.30	0	
change conc. (M)	− 0.20	− 0.20	+ 0.20	$^+$
final conc. (M)	0	0.10	0.20	$^+$

The solution produced is a buffer solution, containing an acid-base conjugate pair.

(c) Adding a strong acid to a strong base will not produce a buffer. A buffer needs a weak acid-base conjugate pair in solution and this solution has neither.

(d) Adding a weak base to a strong base will not produce a buffer. A buffer needs a weak acid-base conjugate pair in solution and this solution has no acid.

30. Calculate the initial pH by finding the pK_a. Calculate initial OH^- concentration using the method from the answer to Question 5.62. Then adapt the method used in the answers to Questions 28 and 14.55.

Equimolar buffer solutions have $pH = pK_a = - \log K_a$. So, $pH = - \log(1.8 \times 10^{-5}) = 4.74$

$$(\text{conc. } OH^-) = \frac{1.0 \text{ mL}}{\left(0.100 \text{ L} \times \dfrac{1000 \text{ mL}}{1 \text{ L}} + 1.0 \text{ mL}\right)} \times \frac{1.0 \text{ mol } OH^-}{1 \text{ L}} = 0.0099 \text{ M } OH^-$$

(a)

$$OH^-(aq) \ + CH_3COOH(aq) \ \longrightarrow \ CH_3COO^-(aq) \ + H_2O(\ell)$$

	OH^-	CH_3COOH	CH_3COO^-	H_2O
conc. init. (M)	0.0099	0.10	0.10	
change conc. (M)	− 0.0099	− 0.0099	+ 0.0099	+
final conc. (M)	0	0.09	0.11	+

$$CH_3COOH(aq) \ + H_2O(\ell) \ \rightleftharpoons \ CH_3COO^-(aq) \ + H_3O^+(aq)$$

	CH_3COOH	H_2O	CH_3COO^-	H_3O^+
conc. init. (M)	0.09		0.11	0
change conc. (M)	− x	−	+ x	+ x
eq. conc. (M)	0.09 − x	−	0.11 + x	x

At equilibrium $\quad K_a = \dfrac{[CH_3COO^-][H_3O^+]}{[CH_3COOH]} = \dfrac{(x)(0.11+x)}{(0.09-x)} = 1.8 \times 10^{-5}$

Assume x is very small and does not affect the sum or the difference.

$$1.8 \times 10^{-5} = \frac{(x)(0.11)}{(0.09)}$$

$$x = 1 \times 10^{-5} \text{ M} = [H_3O^+]$$

$$pH = -\log[H_3O^+] = -\log(1 \times 10^{-5}) = 4.8$$

$$\Delta pH = 4.8 - 4.74 = 0.1$$

(b)

$$OH^-(aq) \ + CH_3COOH(aq) \ \longrightarrow \ CH_3COO^-(aq) \ + H_2O(\ell)$$

	OH^-	CH_3COOH	CH_3COO^-	H_2O
conc. init. (M)	0.0099	0.010	0.010	
change conc. (M)	− 0.0099	− 0.0099	+ 0.0099	+
final conc. (M)	0	0	0.020	+

$$CH_3COO^-(aq) \; + \; H_2O(\ell) \; \rightleftharpoons \; CH_3COOH(aq) \; + \; OH^-(aq)$$

conc. init. (M)	0.020	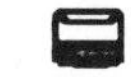 large	0	0
change conc. (M)	$-x$	$-\delta$	$+x$	$+x$
eq. conc. (M)	$0.020 - x$	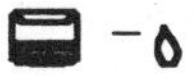 large $-\delta$	x	x

At equilibrium

$$K_b = \frac{[CH_3COOH][OH^-]}{[CH_3COO^-]} = \frac{(x)(x)}{(0.020 - x)} = 5.6 \times 10^{-10}$$

Assume x is very small and does not affect the sum or the difference.

$$5.6 \times 10^{-10} = \frac{x^2}{(0.020)}$$

$$x^2 = (5.6 \times 10^{-10})(0.20)$$

$$x = 3.3 \times 10^{-6} \, M = [OH^-]$$

$$pH = 14.00 + \log[OH^-] = 14.00 + \log(3.3 \times 10^{-6}) = 8.52$$

$$\Delta pH = 8.52 - 4.74 = 3.8$$

(c)

$$OH^- \; + CH_3COOH \; \longrightarrow \; CH_3COO^- \; + \; H_2O$$

conc. init. (M)	0.0099	0.0010	0.0010	large
change conc. (M)	-0.0010	-0.0010	$+0.0010$	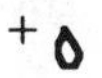 $+\delta$
final conc. (M)	0.0089	0	0.0020	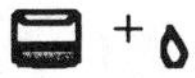 large $+\delta$

This time, some OH^- ions are left over. The minor amount of weak base ionization will not change this value, so:

$$pH = 14.00 + \log[OH^-] = 14.00 + \log(0.0089) = 11.95$$

$$\Delta pH = 11.95 - 4.74 = 7.25$$

32. Adapt the method described in the answers to Question 24, 28, and 30.

Use the abbreviation HProp for the monoprotic propanoic acid. The salt sodium propanoate contains the $Prop^-$ ion.

(a)

	HProp(aq) +	H₂O(ℓ) ⇌	Prop⁻(aq)	+ H₃O⁺(aq)
conc. init. (M)	0.20	🗆	0.30	0
change conc. (M)	$-x$	$^-$🌢	$+x$	$+x$
eq. conc. (M)	$0.20 - x$	🗆 $-$🌢	$0.30 + x$	x

At equilibrium $\quad K_a = \dfrac{[\text{Prop}^-][\text{H}_3\text{O}^+]}{[\text{HProp}]} = \dfrac{(x)(0.30 + x)}{(0.20 - x)} = 1.4 \times 10^{-5}$

Assume that x is small enough to ignore it in the sum and difference:

$$1.4 \times 10^{-5} = \frac{(x)(0.30)}{(0.20)} \qquad\qquad x = 9.3 \times 10^{-6}\ M = [\text{H}_3\text{O}^+]$$

$$\text{pH} = -\log[\text{H}_3\text{O}^+] = -\log(9.3 \times 10^{-6}) = 5.03$$

(b) Find the initial concentration of H_3O^+, upon adding 1.0 mL 0.10 M HCl to the 0.010-L solution:

$$(\text{conc. H}_3\text{O}^+) = \frac{1.0\ \text{mL}}{\left(0.010\ \text{L} \times \dfrac{1000\ \text{mL}}{1\ \text{L}} + 1.0\ \text{mL}\right)} \times \frac{0.10\ \text{mol H}_3\text{O}^+}{1\ \text{L}} = 0.0091\ \text{M H}_3\text{O}^+$$

The strong acid neutralizes the base. The product-favored neutralization runs towards the products until one of the reactants runs out.

	H₃O⁺ +	Prop⁻ ⟶	HProp	+ H₂O
conc. init. (M)	0.0091	0.30	0.20	🗆
change conc. (M)	-0.0091	-0.0091	$+0.0091$	$^+$🌢
final conc. (M)	0	0.29	0.21	🗆 $^+$🌢

	HProp(aq) +	H₂O(ℓ) ⇌	Prop⁻(aq)	+ H₃O⁺(aq)
conc. init. (M)	0.21	🗆	0.29	0
change conc. (M)	$-x$	$^-$🌢	$+x$	$+x$
eq. conc. (M)	$0.21 - x$	🗆 $-$🌢	$0.29 + x$	x

At equilibrium $\qquad K_a = \dfrac{[\text{Prop}^-][\text{H}_3\text{O}^+]}{[\text{HProp}]} = \dfrac{(x)(0.29 + x)}{(0.21 - x)} = 1.4 \times 10^{-5}$

Assume that x is small enough to ignore it in the sum and difference:

$$1.4 \times 10^{-5} = \dfrac{(x)(0.29)}{(0.19)} \qquad\qquad x = 1.0 \times 10^{-5}\ \text{M} = [\text{H}_3\text{O}^+]$$

$$\text{pH} = -\log[\text{H}_3\text{O}^+] = -\log(1.0 \times 10^{-5}) = 5.00$$

(c) Find the initial concentration of H_3O^+, upon adding 3.0 mL 1.0 M HCl to the 0.010 L solution:

$$(\text{conc. H}_3\text{O}^+) = \dfrac{3.0\ \text{mL}}{\left(0.010\ \text{L} \times \dfrac{1000\ \text{mL}}{1\ \text{L}} + 3.0\ \text{mL}\right)} \times \dfrac{1.0\ \text{mol H}_3\text{O}^+}{1\ \text{L}} = 0.23\ \text{M H}_3\text{O}^+$$

The strong acid neutralizes the base. The product-favored neutralization runs towards the products until one of the reactants runs out.

	H_3O^+	+	Prop^-	$\longrightarrow$	HProp	+	H_2O
conc. init. (M)	0.23		0.30		0.20		
change conc. (M)	− 0.23		− 0.23		+ 0.23		
final conc. (M)	0		0.07		0.43		

	HProp(aq)	+	$\text{H}_2\text{O}(\ell)$	$\rightleftharpoons$	$\text{Prop}^-\text{(aq)}$	+	$\text{H}_3\text{O}^+\text{(aq)}$
conc. init. (M)	0.43				0.07		0
change conc. (M)	−x				+x		+x
eq. conc. (M)	0.43 − x				0.07 + x		x

At equilibrium $\qquad K_a = \dfrac{[\text{Prop}^-][\text{H}_3\text{O}^+]}{[\text{HProp}]} = \dfrac{(x)(0.07 + x)}{(0.43 - x)} = 1.4 \times 10^{-5}$

Assume that x is small enough to ignore it in the sum and difference:

$$1.4 \times 10^{-5} = \dfrac{(x)(0.07)}{(0.43)} \qquad\qquad x = 9 \times 10^{-5}\ \text{M} = [\text{H}_3\text{O}^+]$$

$$\text{pH} = -\log[\text{H}_3\text{O}^+] = -\log(9 \times 10^{-5}) = 4.1$$

Titrations and Titration Curves

34. At the equivalence point, the solution contains the conjugate base of the weak acid. Weaker acids have stronger conjugate bases. Stronger bases have higher pH, so the titration of a weaker acid will have a more basic equivalence point.

36. The color change needs to be near the pH of the equivalence point.

(a) The strong base, NaOH, titrated with a strong acid, $HClO_4$, has a neutral equivalence point. It would be sensible to use bromthymol blue, which is shown changing color in Figure 17.5 at a pH near 7.

(b) The weak acid, CH_3COOH, titrated with a strong base, KOH, has a basic equivalence point, due to the presence of the weak base CH_3COO^- in the solution. It would be sensible to use phenolphthalein, which is shown changing color in Figure 17.5 at a pH near 9.

(c) The weak base, NH_3, titrated with a strong acid, HBr, has an acidic equivalence point, due to the presence of the weak acid NH_4^+ in the solution. It would be sensible to use methyl red, which is shown changing color in Figure 17.8 at a pH near 5.

(d) The strong base, KOH, titrated with a strong acid, HNO_3, has a neutral equivalence point. It would be sensible to use bromthymol blue, which is shown changing color in Figure 17.5 at a pH near 7.

38. Use the method described in the answer to Question 5.79.

$$\frac{22.6 \text{ mL Ba(OH)}_2}{25.00 \text{ mL HCl}} \times \frac{0.0140 \text{ mol Ba(OH)}_2}{1 \text{ L Ba(OH)}_2} \times \frac{2 \text{ mol OH}^-}{1 \text{ mol Ba(OH)}_2} \times \frac{1 \text{ mol HCl}}{1 \text{ mol OH}^-}$$

$$= 0.0253 \text{ M HCl}$$

40. Use the method described in the answers to Question 5.65 to 5.79.

$$24.4 \text{ mL NaOH} \times \frac{1 \text{ L}}{1000 \text{ mL}} \times \frac{0.110 \text{ mol NaOH}}{1 \text{ L NaOH}} \times \frac{1 \text{ mol C}_6H_8O_6}{1 \text{ mol NaOH}} \times \frac{176.12 \text{ g C}_6H_8O_6}{1 \text{ mol C}_6H_8O_6}$$

$$= 0.473 \text{ g C}_6H_8O_6$$

$$\frac{0.473 \text{ g C}_6H_8O_6}{0.505 \text{ g capsule}} \times 100 \% = 93.6 \%$$

42. *NOTE: There is a typo in part (d) of this question.*

Use the method described in the answers to Question 5.65 to 5.79.

(a) $25.0 \text{ mL KOH} \times \dfrac{0.175 \text{ mol KOH}}{1 \text{ L KOH}} \times \dfrac{1 \text{ mol HCl}}{1 \text{ mol KOH}} \times \dfrac{1 \text{ L HCl}}{0.150 \text{ mol HCl}} = 29.2 \text{ mL HCl}$

(b) $15.0 \text{ mL NH}_3 \times \dfrac{6.00 \text{ mol NH}_3}{1 \text{ L NH}_3} \times \dfrac{1 \text{ mol HCl}}{1 \text{ mol NH}_3} \times \dfrac{1 \text{ L HCl}}{0.150 \text{ mol HCl}} = 600. \text{ mL HCl}$

(c) $15.0 \text{ mL C}_3\text{H}_7\text{NH}_2 \times \dfrac{0.712 \text{ g C}_3\text{H}_7\text{NH}_2}{1 \text{ mL C}_3\text{H}_7\text{NH}_2} \times \dfrac{1 \text{ mol C}_3\text{H}_7\text{NH}_2}{58.10 \text{ g C}_3\text{H}_7\text{NH}_2}$

$\times \dfrac{1 \text{ mol HCl}}{1 \text{ mol C}_3\text{H}_7\text{NH}_2} \times \dfrac{1 \text{ L HCl}}{0.150 \text{ mol HCl}} = 1.23 \text{ L HCl}$

(d) *There is a typo in this part: It should read 40.0 mL of 0.0050 M Ba(OH)$_2$.*

$40.0 \text{ mL Ba(OH)}_2 \times \dfrac{0.0050 \text{ mol Ba(OH)}_2}{1 \text{ L Ba(OH)}_2} \times \dfrac{2 \text{ mol OH}^-}{1 \text{ mol Ba(OH)}_2}$

$\times \dfrac{1 \text{ mol HCl}}{1 \text{ mol OH}^-} \times \dfrac{1 \text{ L HCl}}{0.150 \text{ mol HCl}} = 2.7 \text{ mL HCl}$

44. Use a method similar to that used in Question 30. Use the abbreviation HBen for benzoic acid.

(a) $(\text{conc. OH}^-) = \dfrac{10.00 \text{ mL}}{\left(30.00 \text{ mL} + 10.00 \text{ mL}\right)} \times \dfrac{0.100 \text{ mol OH}^-}{1 \text{ L}} = 0.0250 \text{ M OH}^-$

$(\text{conc. HBen}) = \dfrac{30.00 \text{ mL}}{\left(30.00 \text{ mL} + 10.00 \text{ mL}\right)} \times \dfrac{0.100 \text{ mol HBen}}{1 \text{ L}} = 0.0750 \text{ M HBen}$

	OH$^-$	+ HBen $\longrightarrow$	Ben$^-$	+ H$_2$O
conc. init. (M)	0.0250	0.0750	0	⊟
change conc. (M)	− 0.0250	− 0.0250	+ 0.0250	+ ⬦
final conc. (M)	0	0.0500	0.0250	⊟ + ⬦

	HBen(aq)	+ H₂O(ℓ)	⇌	Ben⁻(aq)	+ H₃O⁺(aq)
conc. init. (M)	0.0500			0.0250	0
change conc. (M)	$-x$	$^-\diamond$		$+x$	$+x$
eq. conc. (M)	$0.0500 - x$	$\square - \diamond$		$0.0250 + x$	x

At equilibrium $\quad K_a = \dfrac{[\text{Ben}^-][\text{H}_3\text{O}^+]}{[\text{HBen}]} = \dfrac{(x)(0.0250 + x)}{(0.0500 - x)} = 6.3 \times 10^{-5}$

Assume x is very small and does not affect the sum or the difference.

$$6.3 \times 10^{-5} = \frac{(x)(0.0250)}{(0.0500)} \qquad\qquad x = 1.3 \times 10^{-4}\ \text{M} = [\text{H}_3\text{O}^+]$$

$$\text{pH} = -\log[\text{H}_3\text{O}^+] = -\log(1.3 \times 10^{-4}) = 3.90$$

(b) $(\text{conc. OH}^-) = \dfrac{30.00\ \text{mL}}{\left(30.00\ \text{mL} + 30.00\ \text{mL}\right)} \times \dfrac{0.100\ \text{mol OH}^-}{1\ \text{L}} = 0.0500\ \text{M OH}^-$

$(\text{conc. HBen}) = \dfrac{30.00\ \text{mL}}{\left(30.00\ \text{mL} + 30.00\ \text{mL}\right)} \times \dfrac{0.100\ \text{mol HBen}}{1\ \text{L}} = 0.0500\ \text{M HBen}$

	OH⁻	+ HBen	⟶	Ben⁻	+ H₂O
conc. init. (M)	0.0500	0.0500		0	
change conc. (M)	-0.0500	-0.0500		$+0.0500$	$^+\diamond$
final conc. (M)	0	0		0.0500	$\square + \diamond$

	Ben⁻(aq)	+ H₂O(ℓ)	⇌	HBen(aq)	+ OH⁻(aq)
conc. init. (M)	0.0500			0	0
change conc. (M)	$-x$	$^-\diamond$		$+x$	$+x$
eq. conc. (M)	$0.0500 - x$	$\square - \diamond$		x	x

At equilibrium $K_b = \dfrac{[HBen][OH^-]}{[Ben^-]} = \dfrac{(x)(x)}{(0.0500-x)} = 1.6 \times 10^{-10}$ (Table 16.2)

Assume x is very small and does not affect the sum or the difference.

$$1.6 \times 10^{-10} = \frac{x^2}{(0.0500)}$$

$$x^2 = (1.6 \times 10^{-10})(0.0500)$$

$$x = 2.8 \times 10^{-6} \, M = [OH^-]$$

$$pH = 14.00 + \log[OH^-] = 14.00 + \log(2.8 \times 10^{-6}) = 8.45$$

(c) $(\text{conc. } OH^-) = \dfrac{40.00 \text{ mL}}{(30.00 \text{ mL} + 40.00 \text{ mL})} \times \dfrac{0.100 \text{ mol } OH^-}{1 \text{ L}} = 0.0571 \text{ M } OH^-$

$(\text{conc. HBen}) = \dfrac{30.00 \text{ mL}}{(30.00 \text{ mL} + 40.00 \text{ mL})} \times \dfrac{0.100 \text{ mol HBen}}{1 \text{ L}} = 0.0429 \text{ M HBen}$

	OH^-	$+$ HBen $\longrightarrow$	Ben^-	$+$ H_2O
conc. init. (M)	0.0571	0.0429	0	▭
change conc. (M)	− 0.0429	− 0.0429	+ 0.0429	$^+$◊
final conc. (M)	0.0142	0	0.0429	▭ $+$ ◊

$$pH = 14.00 + \log[OH^-] = 14.00 + \log(0.0142) = 12.16$$

46. For each of the points, determine the limiting reactant and use the concentration of the excess reactant to calculate the pH.

(a) Solution contains only 0.150 M HCl, which ionizes to make 0.150 M H_3O^+, so

pH $= \log[H_3O^+] = - \log(0.150) = 0.824$

(b) $(\text{conc. } H_3O^+) =$

$$\frac{50.00 \text{ mL}}{(50.00 \text{ mL} + 25.00 \text{ mL})} \times \frac{0.150 \text{ mol HCl}}{1 \text{ L}} \times \frac{1 \text{ mol } H_3O^+}{1 \text{ mol HCl}} = 0.100 \text{ M } H_3O^+$$

$(\text{conc. } OH^-) = \dfrac{25.00 \text{ mL}}{(50.00 \text{ mL} + 25.00 \text{ mL})} \times \dfrac{0.150 \text{ mol } OH^-}{1 \text{ L}} = 0.0500 \text{ M } OH^-$

	H_3O^+	$+ OH^-$	$\longrightarrow$ 2 H_2O
conc. init. (M)	0.100	0.0500	
change conc. (M)	-0.0500	-0.0500	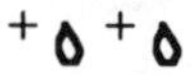
final conc. (M)	0.050	0	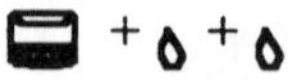

$$pH = -\log[H_3O^+] = -\log(0.050) = 1.30$$

(c) (conc. H_3O^+) =

$$\frac{50.00 \text{ mL}}{\left(50.00 \text{ mL} + 49.9 \text{ mL}\right)} \times \frac{0.150 \text{ mol HCl}}{1 \text{ L}} \times \frac{1 \text{ mol } H_3O^+}{1 \text{ mol HCl}} = 0.0751 \text{ M } H_3O^+$$

$$(\text{conc. } OH^-) = \frac{49.90 \text{ mL}}{\left(50.00 \text{ mL} + 49.9 \text{ mL}\right)} \times \frac{0.150 \text{ mol } OH^-}{1 \text{ L}} = 0.0749 \text{ M } OH^-$$

	H_3O^+	$+ OH^-$	$\longrightarrow$ 2 H_2O
conc. init. (M)	0.0751	0.0749	
change conc. (M)	-0.0749	-0.0749	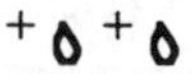
final conc. (M)	0.0002	0	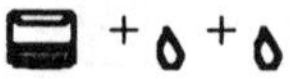

$$pH = -\log[H_3O^+] = -\log(0.0002) = 3.7$$

(d) (conc. H_3O^+) =

$$\frac{50.00 \text{ mL}}{\left(50.00 \text{ mL} + 50.0 \text{ mL}\right)} \times \frac{0.150 \text{ mol HCl}}{1 \text{ L}} \times \frac{1 \text{ mol } H_3O^+}{1 \text{ mol HCl}} = 0.0750 \text{ M } H_3O^+$$

$$(\text{conc. } OH^-) = \frac{50.00 \text{ mL}}{\left(50.00 \text{ mL} + 50.00 \text{ mL}\right)} \times \frac{0.150 \text{ mol } OH^-}{1 \text{ L}} = 0.0750 \text{ M } OH^-$$

	H_3O^+	$+ OH^-$	$\longrightarrow$ 2 H_2O
conc. init. (M)	0.0750	0.0750	
change conc. (M)	-0.0750	-0.0750	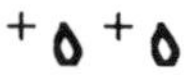
final conc. (M)	0	0	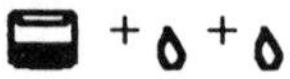

The reaction results in no excess acid or base, so the solution is neutral, and the pH = 7.00.

(e) (conc. H_3O^+) =

$$\frac{50.00 \text{ mL}}{\left(50.00 \text{ mL} + 50.1 \text{ mL}\right)} \times \frac{0.150 \text{ mol HCl}}{1 \text{ L}} \times \frac{1 \text{ mol } H_3O^+}{1 \text{ mol HCl}} = 0.0749 \text{ M } H_3O^+$$

(conc. OH^-) = $\dfrac{50.1 \text{ mL}}{\left(50.00 \text{ mL} + 50.1 \text{ mL}\right)} \times \dfrac{0.150 \text{ mol } OH^-}{1 \text{ L}} = 0.0751 \text{ M } OH^-$

	H_3O^+	+ OH^-	⟶ 2 H_2O
conc. init. (M)	0.0749	0.0751	🔋
change conc. (M)	− 0.0749	− 0.0749	+ ◊ + ◊
final conc. (M)	0	0.0002	🔋 + ◊ + ◊

$$pH = 14.00 + \log[OH^-] = 14.00 + \log(0.0002) = 10.3$$

(f) (conc. H_3O^+) =

$$\frac{50.00 \text{ mL}}{\left(50.00 \text{ mL} + 75.00 \text{ mL}\right)} \times \frac{0.150 \text{ mol HCl}}{1 \text{ L}} \times \frac{1 \text{ mol } H_3O^+}{1 \text{ mol HCl}} = 0.0600 \text{ M } H_3O^+$$

(conc. OH^-) = $\dfrac{75.00 \text{ mL}}{\left(50.00 \text{ mL} + 75.00 \text{ mL}\right)} \times \dfrac{0.150 \text{ mol } OH^-}{1 \text{ L}} = 0.0900 \text{ M } OH^-$

	H_3O^+	+ OH^-	⟶ 2 H_2O
conc. init. (M)	0.0600	0.0900	🔋
change conc. (M)	− 0.0600	− 0.0600	+ ◊ + ◊
final conc. (M)	0	0.0300	🔋 + ◊ + ◊

$$pH = 14.00 + \log[OH^-] = 14.00 + \log(0.0300) = 12.47$$

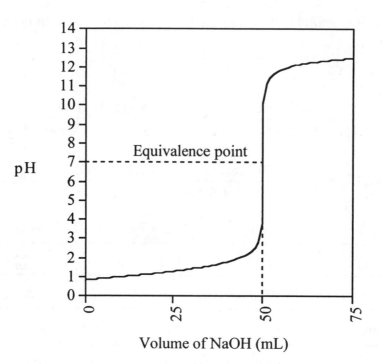

Volume of NaOH (mL)

Acid Rain

48. Two oxides that are key producers of acid rain are NO_2 and SO_3. NO_2 reacts with water in the air to make nitric acid and nitrous acid:

$$2\ NO_2(g) + H_2O(g) \longrightarrow HNO_3(g) + HNO_2(g)$$

SO_2 reacts with oxygen in the air to make SO_3 and SO_3 reacts water in the air to make sulfuric acid:

$$2\ SO_2(g) + O_2(g) \longrightarrow 2\ SO_3(g)$$

$$SO_3(g) + H_2O(g) \longrightarrow H_2SO_4(g)$$

50. Limestone is made of calcium carbonate. The anion of this compound is a base and reacts with acid in rainwater to neutralize it by a gas-forming exchange reaction (Section 5.2):

$$CaCO_3(s) + 2\ H^+(aq) \longrightarrow Ca^{2+}(aq) + CO_2(g) + H_2O(\ell)$$

Solubility Product

52. (a) $FeCO_3(s) \rightleftharpoons Fe^{2+}(aq) + CO_3^{2-}(aq)$ $K_{sp} = [Fe^{2+}][CO_3^{2-}]$

(b) $Ag_2SO_4(s) \rightleftharpoons 2\,Ag^+(aq) + SO_4^{2-}(aq)$ $K_{sp} = [Ag^+]^2[SO_4^{2-}]$

(c) $Ca_3(PO_4)_2(s) \rightleftharpoons 3\,Ca^{2+}(aq) + 2\,PO_4^{3-}(aq)$ $K_{sp} = [Ca^{2+}]^3[PO_4^{3-}]^2$

(d) $Mn(OH)_2(s) \rightleftharpoons Mn^{2+}(aq) + 2\,OH^-(aq)$ $K_{sp} = [Mn^{2+}][OH^-]^2$

54. Follow the method developed in the answers to Questions 14.32 - 14.37. Use the variable "s" to represent the molar solubility. We must keep in mind that the quantity of solids does not affect the position of equilibrium, so we will again use a small pile to represent the presence of solid. This is done primarily to remind us that those quantities do not belong in the equilibrium calculations.

	$CaSO_4(s) \rightleftharpoons$	$Ca^{2+}(aq)$ +	$SO_4^{2-}(aq)$
conc. initial (M)		0	0
change conc. (M)	$-$	$+s$	$+s$
equilibrium conc. (M)		s	s

At equilibrium $\dfrac{2.03 \text{ g } CaSO_4}{1\text{ L}} \times \dfrac{1 \text{ mol } CaSO_4}{136.14 \text{ g } CaSO_4} = 0.0149 \text{ M} = s$

$$K_{sp} = [Ca^{2+}][SO_4^{2-}] = (s)(s) = s^2 = (0.0149)^2 = 2.22 \times 10^{-4}$$

56. Follow the method developed in the answer to Question 54.

	$Ag_2CrO_4(s) \rightleftharpoons$	$2\,Ag^+(aq)$ +	$CrO_4^{2-}(aq)$
conc. initial (M)		0	0
change conc. (M)	$-$	$+2s$	$+s$
equilibrium conc. (M)		$2s$	s

At equilibrium, $\dfrac{2.7 \times 10^{-3} \text{ g } Ag_2CrO_4}{100. \text{ mL}} \times \dfrac{1 \text{ mol } Ag_2CrO_4}{331.8 \text{ g } Ag_2CrO_4} \times \dfrac{1000 \text{ mL}}{1 \text{ L}}$

$$= 8.1 \times 10^{-5} \text{ M} = s$$

$$K_{sp} = [Ag^+]^2[CrO_4^{2-}] = (2s)^2(s) = 4s^3 = 4 \times (8.1 \times 10^{-5})^3 = 2.2 \times 10^{-12}$$

58. Follow the method developed in the answer to Question 54.

	$PbCl_2(s)$	$\rightleftharpoons$	$Pb^{2+}(aq)$	+	$2\,Cl^-(aq)$
conc. initial (M)			0		0
change conc. (M)			+ s		+ 2s
equilibrium conc. (M)			s		2s

At equilibrium, $s = 1.62 \times 10^{-2}$ M

$$K_{sp} = [Pb^{2+}][Cl^-]^2 = (s)(2s)^2 = 4s^3 = 4 \times (1.62 \times 10^{-2})^3 = 1.70 \times 10^{-5}$$

Common Ion Effect

60. We will use the equilibrium expression for K_{sp} and the known values to determine the unknown value. We will get the value of K_{sp} from Table 17.2, Appendix H, or Question 61.

The soluble Na_2CO_3 salt produces carbonate ions in the solution. Neglecting the reaction of carbonate as a base*, $[CO_3^{2-}] = 0.25$ mol/L.

$$K_{sp} = [Zn^{2+}][CO_3^{2-}] = [Zn^{2+}] \times (0.25) = 1.5 \times 10^{-11}, [Zn^{2+}] = 6.0 \times 10^{-11} \text{ mol/L.}$$

Note, the value of K_{sp} for $ZnCO_3$ given in problem 61 is significantly different from that given in Appendix H. Using the K_{sp} from Appendix H:

$$K_{sp} = [Zn^{2+}][CO_3^{2-}] = [Zn^{2+}] \times (0.25) = 3 \times 10^{-8}, [Zn^{2+}] = 1 \times 10^{-7} \text{ mol/L.}$$

*The base reaction should be considered because its K is larger than the K_{sp} above:
$CO_3^{2-} + H_2O \rightleftharpoons HCO_3^- + OH^-$, some of the carbonate(–x) reacts to form hydrogen

carbonate (x) and hydroxide (x). $K_b = \dfrac{x^2}{0.25 - x} = 2.1 \times 10^{-4}$. Solving for x and

calculating the carbonate concentration, we find that it is slightly but measurably lower:

$$[CO_3^{2-}] = 0.25\ M - x = 0.25\ M - 0.01\ M = 0.24\ M$$

$$K_{sp} = [Zn^{2+}][CO_3^{2-}] = [Zn^{2+}] \times (0.24) = 1.5 \times 10^{-11}, [Zn^{2+}] = 6.2 \times 10^{-11}\ mol/L.$$

or

$$K_{sp} = [Zn^{2+}][CO_3^{2-}] = [Zn^{2+}] \times (0.24) = 3 \times 10^{-8}, [Zn^{2+}] = 1 \times 10^{-7}\ mol/L.$$

62. Follow the method developed in the answer to Questions 54 and 60.

The soluble Na_2SO_4 salt produces sulfate ions in the solution. Neglecting the reaction of sulfate as a base, (conc. SO_4^{2-}) = 0.010 mol/L.

	$SrSO_4(s) \rightleftharpoons$	$Sr^{2+}(aq)$ +	$SO_4^{2-}(aq)$
conc. initial (M)		0	0.010
change conc. (M)	–	+ s	+ s
equilibrium conc. (M)		s	0.010 + s

At equilibrium $K_{sp} = [Sr^{2+}][SO_4^{2-}] = (s)(0.010 + s) = 3.1 \times 10^{-7}$

Assuming s is small compared to 0.010, ignore its addition.

$$3.1 \times 10^{-7} = (s)(0.010) \qquad s = 3.1 \times 10^{-5} \text{ mol/L}$$

64. Follow the method developed in the answers to Questions 54 and 60.

(a)

	$Mg(OH)_2(s) \rightleftharpoons$	$Mg^{2+}(aq)$ +	$2\, OH^-(aq)$
conc. initial (M)		0	0
change conc. (M)	–	+ s	+ 2s
equilibrium conc. (M)		s	2s

At equilibrium $\dfrac{9 \text{ mg Mg(OH)}_2}{1 \text{ L}} \times \dfrac{1 \text{ g}}{1000 \text{ mg}} \times \dfrac{1 \text{ mol Mg(OH)}_2}{58.32 \text{ g Mg(OH)}_2}$

$$= 1.5 \times 10^{-4} \text{ M} \cong 2 \times 10^{-4} \text{ M} = s \text{ (rounded to 1 sig. fig.)}$$

$$K_{sp} = [Mg^{2+}][OH^-]^2 = (s)(2s)^2 = 4s^3 = 4 \times (1.5 \times 10^{-4})^3 = 1.47 \times 10^{-11}$$

$$K_{sp} \cong 1 \times 10^{-11} \text{ (rounded to 1 sig. fig.)}$$

(b) $[Mg^{2+}] = \dfrac{5.0 \text{ μg Mg}^{2+}}{1 \text{ L}} \times \dfrac{10^{-6} \text{ g}}{1 \text{ μg}} \times \dfrac{1 \text{ mol Mg}^{2+}}{24.305 \text{ g Mg}^{2+}} = 2.1 \times 10^{-7} \text{ M}$

$$K_{sp} = [Mg^{2+}][OH^-]^2$$

$$1.47 \times 10^{-11} = (2.1 \times 10^{-7})[OH^-]^2$$

$$[OH^-] = 8.45 \times 10^{-3} \text{ M} \cong 8 \times 10^{-3} \text{ M} \text{ (rounded to 1 sig. fig.)}$$

An equilibrium $[OH^-] = 8 \times 10^{-3}$ M or higher will keep $[Mg^{2+}]$ at or below 1.0 μg/L.

Factors Affecting the Solubility of Sparingly Soluble Solutes

66. Adapt the method developed in the answers to Questions 16.32 and 64. We will get the K_{sp} from Appendix H.

Given the pH, we can determine the concentration of the aqueous ion, OH^-. The precipitation of sparingly soluble $Zn(OH)_2$ will occur above a certain concentration of Zn^{2+}.

$$[OH^-] = 10^{-pOH} = 10^{pH-14.00} = 10^{pH-14.00} = 10^{10.00-14.00} = 1.0 \times 10^{-4} \text{ M}$$

$$K_{sp} = [Zn^{2+}][OH^-]^2$$

$$4.5 \times 10^{-17} = [Zn^{2+}](1.0 \times 10^{-4})^2$$

$$[Zn^{2+}] = 4.5 \times 10^{-9} \text{ M}$$

An equilibrium $[Zn^{2+}] = 4.5 \times 10^{-9}$ M or lower can exist in a solution with pH 10.00. Above that concentration, $Zn(OH)_2$ will precipitate.

68. Adapt the method developed in the answers to Questions 16.32 and 64. We will use the K_{sp} derived for $Mg(OH)_2$ in part (a) of Question 63: $K_{sp} = 1 \times 10^{-11}$.

The acid added affects the concentration of the hydroxide. So, if all the solid is dissolved it must be acidic enough to hold all the magnesium ions without precipitation. So, we will calculate the resulting magnesium ion concentration:

$$[Mg^{2+}] = \frac{5.00 \text{ g Mg(OH)}_2}{1 \text{ L}} \times \frac{1 \text{ mol Mg(OH)}_2}{58.32 \text{ g Mg(OH)}_2} \times \frac{1 \text{ mol Mg}^{2+}}{1 \text{ mol Mg(OH)}_2} = 0.0857 \text{ M}$$

$$K_{sp} = [Mg^{2+}][OH^-]^2$$

$$1 \times 10^{-11} = (0.0857)[OH^-]^2$$

$$[OH^-] = 1 \times 10^{-5} \text{ M}$$

$$pH = 14.00 + \log[OH^-] = 14.00 + \log(1 \times 10^{-5}) = 9.0$$

Enough acid must be added to drop the pH to 9, before all the solid will dissolve.

Complex Ion Formation

70. The charge of the reactant metal ion is determined by subtracting the Lewis base's charge(s), if any, from the complex ion charge.

(a) $Ag^+ + 2\ CN^- \rightleftharpoons Ag(CN)_2^-$ $\qquad K_f = \dfrac{[Ag(CN)_2^-]}{[Ag^+][CN^-]^2}$

(b) $Cd^{2+} + 4\ NH_3 \rightleftharpoons Cd(NH_3)_4^{2+}$ $\qquad K_f = \dfrac{[Cd(NH_3)_4^{2+}]}{[Cd^{2+}][NH_3]^4}$

72. Adapt the method described in the answer to Question 68, and Problem-Solving Example 17.12.

The $Na_2S_2O_3$ salt provides a source of $S_2O_3^{2-}$, a Lewis base capable of forming a complex ion with the silver ion. If all the solid is dissolved it must be have enough $S_2O_3^{2-}$ to complex enough the silver ions to prevent the precipitation of AgBr in the resulting Br^- solution. First, we get the balanced equation for the reaction from Problem-Solving Example 17.12 and determine the value of its equilibrium constant using Tables 17.2 and 17.3:

$$AgBr(s) + 2\ S_2O_3^{2-}(aq) \rightleftharpoons Ag(S_2O_3)_2^{3-}(aq) + Br^-(aq)$$

$$K = \frac{[Ag(S_2O_3)_2^{3-}][Br^-]}{[S_2O_3^{2-}]^2}$$

$$K = K_{sp} \times K_f = [Ag^+][Br^-] \times \frac{[Ag(S_2O_3)_2^{3-}]}{[Ag^+][S_2O_3^{2-}]^2} = \frac{[Ag(S_2O_3)_2^{3-}][Br^-]}{[S_2O_3^{2-}]^2}$$

$$K = K_{sp} \times K_f = (3.3 \times 10^{-13}) \times (2.0 \times 10^{-13}) = 6.6$$

Now, calculate the $[Br^-]$ and $[Ag(S_2O_3)_2^{3-}]$, once all the AgBr has dissolved:

$$[Br^-] = \frac{0.020\ mol\ AgBr}{1.0\ L} \times \frac{1\ mol\ Br^-}{1\ mol\ AgBr} = 0.020\ M,$$

Similarly, $[Ag(S_2O_3)_2^{3-}] = 0.020\ M$

Now, we can calculate the necessary $[S_2O_3{}^{2-}]$:

$$6.6 = \frac{(0.020)(0.020)}{[S_2O_3{}^{2-}]^2}$$

$$[S_2O_3{}^{2-}] = 7.8 \times 10^{-3}\ M$$

74. The hydroxide anion of the solid is neutralized in excess acid, and the aqueous complex ion is formed in excess base.

(a)
$$Zn(OH)_2(s) + 2\ H_3O^+(aq) \longrightarrow Zn^{2+}(aq) + 4\ H_2O(\ell)$$

$$Zn(OH)_2(s) + 2\ OH^-(aq) \longrightarrow Zn(OH)_4{}^{2-}(aq)$$

(b)
$$Sb(OH)_3(s) + 3\ H_3O^+(aq) \longrightarrow Sb^{3+}(aq) + 6\ H_2O(\ell)$$

$$Sb(OH)_3(s) + OH^-(aq) \longrightarrow Sb(OH)_4{}^-(aq)$$

General Questions

76. Adapt the methods used in the answers to Question 15 - 32.

(a) The solution contains an abundance of water, more sodium acetate than acetic acid, a small amount of H_3O^+ since the solution is an acidic buffer, and an even smaller amount of OH^-. Therefore, the ions and molecules in solution from the largest concentration to the smallest is:

$$H_2O,\ CH_3COO^-,\ Na^+,\ CH_3COOH,\ H_3O^+,\ OH^-$$

(b) $$\frac{4.95\ g\ NaCH_3COO}{250.\ mL} \times \frac{1\ mol\ NaCH_3COO}{82.03\ g\ NaCH_3COO} \times \frac{1000\ mL}{1\ L} \times \frac{1\ mol\ CH_3COO^-}{1\ mol\ NaCH_3COO}$$

$$= 0.241\ M\ CH_3COO^-$$

	$CH_3COOH(aq)$	$+ H_2O(\ell)$	$\rightleftharpoons$ $CH_3COO^-(aq)$	$+ H_3O^+(aq)$
conc. init. (M)	0.150	—	0.241	0
change conc. (M)	$-x$	—	$+x$	$+x$
eq. conc. (M)	$0.150 - x$	—	$0.241 + x$	x

At equilibrium $K_a = \dfrac{[CH_3COO^-][H_3O^+]}{[CH_3COOH]} = \dfrac{(x)(0.241+x)}{(0.150-x)} = 1.8 \times 10^{-5}$

Assume x is very small and does not affect the sum or the difference.

$1.8 \times 10^{-5} = \dfrac{(x)(0.241)}{(0.150)}$ $x = 1.1 \times 10^{-5} \, M = [H_3O^+]$

$$pH = -\log[H_3O^+] = -\log(1.1 \times 10^{-5}) = 4.95$$

(c) (conc. OH^-) =

$$\dfrac{80. \text{ mg NaOH}}{100. \text{ mL}} \times \dfrac{1000 \text{ mL}}{1 \text{ L}} \times \dfrac{1 \text{ g}}{1000 \text{ mg}} \times \dfrac{1 \text{ mol NaOH}}{40.00 \text{ g NaOH}} \times \dfrac{1 \text{ mol } OH^-}{1 \text{ mol NaOH}} = 0.020 \text{ M}$$

	OH^-(aq)	+ CH_3COOH(aq)	⟶ CH_3COO^-(aq)	+ $H_2O(\ell)$
conc. init. (M)	0.020	0.150	0.241	⬒
change conc. (M)	− 0.020	− 0.020	+ 0.020	⁺⬧
final conc. (M)	0	0.130	0.261	⬒ ⁺ ⬧

	CH_3COOH(aq)	+ $H_2O(\ell)$	⇌ CH_3COO^-(aq)	+ H_3O^+(aq)
conc. init. (M)	0.130	⬒	0.261	0
change conc. (M)	− x	⁻⬧	+ x	+ x
eq. conc. (M)	0.130 − x	⬒ ⁻ ⬧	0.261 + x	x

At equilibrium $K_a = \dfrac{[CH_3COO^-][H_3O^+]}{[CH_3COOH]} = \dfrac{(0.261+x)(x)}{(0.130-x)} = 1.8 \times 10^{-5}$

Assume x is very small and does not affect the sum or the difference.

$1.8 \times 10^{-5} = \dfrac{(0.261)(x)}{(0.130)}$ $x = 9.0 \times 10^{-6} \, M = [H_3O^+]$

$$pH = -\log[H_3O^+] = -\log(9.0 \times 10^{-6}) = 5.05$$

(d) CH_3COOH(aq) + $H_2O(\ell)$ ⇌ CH_3COO^-(aq) + H_3O^+(aq)

78. It is possible to adapt the methods used in the answers to Question 15 - 32; however, the Henderson-Hasselbalch equation is easily used when we know the pK_a of the conjugate acid.

$$pH = pK_a + \log\left(\frac{[o\text{-}ethylbenzoate]}{[o\text{-}ethylbenzoic acid]}\right)$$

At equilibrium, $pH = 4.0$ and $pK_a = 3.79$

$$pH - pK_a = \log\left(\frac{[o\text{-}ethylbenzoate]}{[o\text{-}ethylbenzoic acid]}\right)$$

$$4.0 - 3.79 = 0.2 = \log\left(\frac{[o\text{-}ethylbenzoate]}{[o\text{-}ethylbenzoic acid]}\right)$$

$$\frac{[o\text{-}ethylbenzoate]}{[o\text{-}ethylbenzoic acid]} = 10^{0.2} = 1.6 \cong 2 \text{ (must round to one sig. fig.)}$$

The [potassium o-ethylbenzoate] is approximately two times the [o-ethylbenzoic acid].

80. Adapt the methods used in the answers to Questions 15 - 32.

$$[NO_2^-] = \frac{7.50 \text{ g KNO}_2}{1 \text{ L}} \times \frac{1 \text{ mol KNO}_2}{85.11 \text{ g KNO}_2} \times \frac{1 \text{ mol NO}_2^-}{1 \text{ mol KNO}_2} = 8.81 \times 10^{-2}$$

At equilibrium, $pH = 4.00$, $[H_3O^+] = 10^{-pH} = 10^{-4.00} = 1.0 \times 10^{-4}$

$$K_a = \frac{[NO_2^-][H_3O^+]}{[HNO_2]} = 4.5 \times 10^{-4} = \frac{(8.81\times10^{-2})(1.0\times10^{-4})}{[HNO_2]}$$

$[HNO_2] = 0.020$ M (This could also be calculated using the Henderson-Hasselbalch equation.)

Add 0.020 mol HNO_2 to one liter of solution for this buffer.

82. Adapt the methods used in the answers to Question 15 - 32.

(a)

	$CH_3COOH(aq)$	$+ H_2O(\ell)$	$\rightleftharpoons$ $CH_3COO^-(aq)$	$+ H_3O^+(aq)$
conc. init. (M)	0.15	▬	0	0
change conc. (M)	$-x$	$-$▬	$+x$	$+x$
eq. conc. (M)	$0.15 - x$	▬ $-$▬	x	x

At equilibrium $K_a = \dfrac{[CH_3COO^-][H_3O^+]}{[CH_3COOH]} = \dfrac{(x)(x)}{(0.15-x)} = 1.8 \times 10^{-5}$

Assume x is very small and does not affect the difference.

$x^2 = (1.8 \times 10^{-5})(0.15)$ $\qquad\qquad$ $x = 1.6 \times 10^{-3} \text{ M} = [H_3O^+]$

$$pH = -\log[H_3O^+] = -\log(1.6 \times 10^{-3}) = 2.78$$

(b) $\dfrac{83 \text{ g NaCH}_3\text{COO}}{1.50 \text{ L}} \times \dfrac{1 \text{ mol NaCH}_3\text{COO}}{82.03 \text{ g NaCH}_3\text{COO}} \times \dfrac{1 \text{ mol CH}_3\text{COO}^-}{1 \text{ mol NaCH}_3\text{COO}}$

$$= 0.67 \text{ M CH}_3\text{COO}^-$$

	$CH_3COOH(aq)$	$+ H_2O(\ell)$	$\rightleftharpoons$ $CH_3COO^-(aq)$	$+ H_3O^+(aq)$
conc. init. (M)	0.15	🔲	0.67	0
change conc. (M)	$-x$	$-\delta$	$+x$	$+x$
eq. conc. (M)	$0.15 - x$	$🔲 - \delta$	$0.67 + x$	x

At equilibrium $K_a = \dfrac{[CH_3COO^-][H_3O^+]}{[CH_3COOH]} = \dfrac{(x)(0.67+x)}{(0.15-x)} = 1.8 \times 10^{-5}$

Assume x is very small and does not affect the sum or the difference.

$1.8 \times 10^{-5} = \dfrac{(x)(0.67)}{(0.15)}$ $\qquad\qquad$ $x = 4.0 \times 10^{-6} \text{ M} = [H_3O^+]$

$$pH = -\log[H_3O^+] = -\log(1.1 \times 10^{-5}) = 5.40$$

84. Adapt the methods described in the answers to Questions 33 - 46.

We will use the equivalence point data to find the initial concentration of the weak acid, HA, in the solution after 20.00 mL of titrant have been added.

$$35.00 \text{ mL NaOH} \times \dfrac{0.100 \text{ mmol NaOH}}{1 \text{ mL NaOH}} \times \dfrac{1 \text{ mmol HA}}{1 \text{ mmol NaOH}} = 3.5 \text{ mmol HA}$$

$$\dfrac{3.5 \text{ mmol HA}}{40.0 \text{ mL} + 20.00 \text{ mL}} = 0.0583 \text{ M HA}$$

Calculate the initial concentration of the OH⁻ ions in the solution after 20.00 mL of titrant have been added.

$$\frac{20.00 \text{ mL NaOH}}{40.0 \text{ mL} + 20.00 \text{ mL}} \times \frac{0.100 \text{ mol NaOH}}{1 \text{ L NaOH}} \times \frac{1 \text{ mol OH}^-}{1 \text{ mol NaOH}} = 0.0333 \text{ M OH}^-$$

	HA(aq)	+ OH⁻(aq) ⟶	A⁻(aq)	+ H₂O(ℓ)
conc. init. (M)	0.0583	0.0333	0	🝳
change conc. (M)	− 0.0333	− 0.0333	+ 0.0333	+ 💧
final conc. (M)	0.0250	0	0.0333	🝳 + 💧

	HA(aq)	+ H₂O(ℓ) ⇌	A⁻(aq)	+ H₃O⁺(aq)
conc. init. (M)	0.0250	🝳	0.0333	0
change conc. (M)	− x	− 💧	+ x	+ x
eq. conc. (M)	0.0250 − x	🝳 − 💧	0.0333 + x	x

At equilibrium   pH = 5.75, so $[H_3O^+] = 10^{-pH} = 10^{-5.75} = 1.8 \times 10^{-6} = x$

$$K_a = \frac{[A^-][H_3O^+]}{[HA]} = \frac{(x)(0.0333 + x)}{(0.0250 - x)} = \frac{(1.8 \times 10^{-6})(0.0333 + 1.8 \times 10^{-6})}{(0.0250 - 1.8 \times 10^{-6})} = 2.4 \times 10^{-6}$$

86. There is no effect on the equilibrium state if more solid is added, since the quantity of that solid present does not affect the equilibrium state.

Applying Concepts

90. The tiny amount of base (CH_3COO^-) present is insufficient to prevent the pH from changing dramatically if a strong acid is introduced into the solution.

92. When exactly half of the acid in the original solution has been neutralized, that means that equal quantities of the acid and base are present in the solution. As described in the answer to Question 18, when equimolar quantities are present, pH = pK_a.

$$pH = 3.64 = pK_a$$

$$K_a = 10^{-pK_a} = 10^{-3.64} = 2.3 \times 10^{-4}$$

94. Blood pH decreases because of the increase in H_2CO_3, which leads to acidosis, an acidification of the blood.

96. We must assume that any of these substances, if present, are present in "reasonable" concentrations.

Sample A: Phenolphthalein is colorless, suggesting the pH is below 8.3. The most acidic of the choices is $NaHCO_3$, and that is a likely guess for this sample's identity. (Note: any of these compounds in solution, even NaOH, can have a pH below 8.3, if it is sufficiently dilute, that's why we have to assume reasonable concentrations.)

Sample B: We interpret the evidence to say that the methyl orange changed to its acidic color as soon as it was added. That means the solution became acidic as soon as the phenolphthalein end point was reached. That suggests the titration of a strong base with a very dramatic decline to acidic pH values after the equivalence point. In addition, at very low pH values (below pH = 3.01), bubbles would have been observed as the carbonate reacted to form $CO_2(g)$. All of these interpretations lead us to believe that this sample is probably NaOH.

Sample C: Presumably, both indicators changed color rapidly, suggesting that they were true endpoints, and not just a result of pH changes. Because two different acidic end points were reached with different indicators, this sample must be a mixture of excess strong base, NaOH, and one or more of the salts, either Na_2CO_3 or $NaHCO_3$. It is not possible to distinguish which salt is present, due to the immediate reaction of the strong base with $NaHCO_3$ to make the same product Na_2CO_3.

Sample D: This sample has a two endpoints the second at exactly twice the volume of the first, suggesting that two neutralizations occurred. That means the sample may have been pure Na_2CO_3, which undergoes two sequential neutralization reactions during the titration. Alternative scenarios, equally plausible, would be a mixture of equimolar quantities of NaOH and $NaHCO_3$ or a mixture of NaOH, Na_2CO_3, and $NaHCO_3$ with the following proportions:

$$(\text{conc. NaOH}) = (\text{conc. NaHCO}_3) = \frac{1}{2}(\text{conc. Na}_2\text{CO}_3)$$

Other proportions are also possible. Without some more data, or other limits on the quantities, no more definite answers are possible.

Chapter 18: Thermodynamics: Directionality of Chemical Reactions

Reactant-Favored and Product-Favored Processes

16. Think about the process and whether you have observed it happening spontaneously. If it has, it will be classified as product-favored. If it has not, it will be classified as reactant-favored.

    (a) Water decomposing into its elements. $2 H_2O(\ell) \longrightarrow 2 H_2(g) + O_2(g)$.

    The decomposition of water has not been observed to occur spontaneously at normal temperature and pressure, thus this reaction is classified as reactant-favored.

    (b) Gasoline spills and evaporates: $C_8H_{18}(\ell) \longrightarrow C_8H_{18}(g)$

    The process has been observed to occur spontaneously at normal temperature and pressure, thus this reaction is classified as product-favored.

    (c) Dissolving sugar at room temperature: $C_{12}H_{22}O_{11}(s) \longrightarrow C_{12}H_{22}O_{11}(aq)$

    The process has been observed to occur spontaneously at normal temperature and pressure, thus this reaction is classified as product-favored.

Probability and Chemical Reactions

18. (a) "Heads" is one of two possible outcomes, so the probability is $\frac{1}{2}$.

    (b) "Tails" is one of two possible outcomes, so the probability is $\frac{1}{2}$.

    (c) If you flip a coin 100 times, the most likely number of "heads" and "tails" would be half of each, or 50 of each.

20. Start with all molecules in one flask or the other, then systematically move them to the other flask. Start with two molecules in each flask and alternate the identity of the pairs.

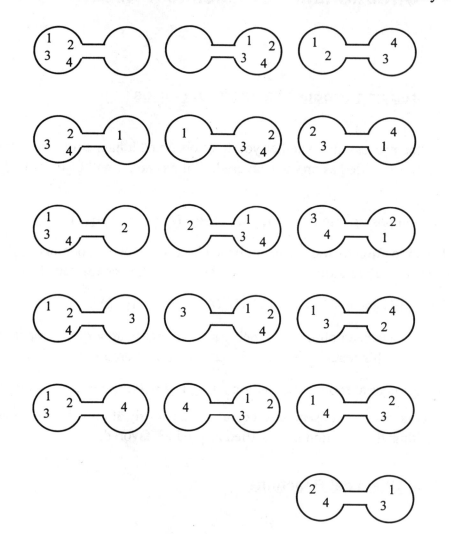

From these diagrams, we see that six of them have two molecules in each flask, two have no molecules in one flask. The most probably arrangement of molecules has two in each flask. Highest entropy arrangement of molecules has two in each flask.

Measuring Dispersal or Disorder: Entropy

21. Use the qualitative guidelines for entropy changes described in Section 18.3.

    (a) $H_2O(g) \longrightarrow H_2O(s)$ The solid product has lower entropy than the gas-phase reactant, so the entropy change is negative.

    (b) $CO_2(aq) \longrightarrow CO_2(g)$ The gas-phase product has higher entropy than the aqueous reactant, so the entropy change is positive.

    (c) glass(s) $\longrightarrow$ glass(ℓ)   The liquid product has higher entropy than the solid reactant, so the entropy change is positive.

23. Use the qualitative guidelines for entropy changes described in Section 18.3.

    (a) $2\,H_2O(\ell) \longrightarrow 2\,H_2(g) + O_2(g)$ The gas-phase products have higher entropy than the liquid reactant, so the entropy change is positive.

    (b) $C_8H_{18}(\ell) \longrightarrow C_8H_{18}(g)$ The gas-phase product has higher entropy than the liquid reactant, so the entropy change is positive.

    (c) $C_{12}H_{22}O_{11}(s) \longrightarrow C_{12}H_{22}O_{11}(aq)$ The aqueous product has higher entropy than the solid reactant, so the entropy change is positive.

25. Use the qualitative guidelines for entropy changes described in Section 18.3.

    (a) Item 2 has higher entropy since it is identical to item 1 except that its temperature is higher. Molecules at higher temperature have higher entropy.

    (b) Item 2, dissolved sugar, has higher entropy than item 1, solid sugar, because solute molecules are more random than those in a solid crystal.

    (c) Item 2, the mixture of water and alcohol together has higher entropy than item 1, water and alcohol separate. Mixing makes the molecules more random.

27. Use the qualitative guidelines for entropy changes described in Section 18.3.

    (a) Comparing NaCl and CaO, we find that the biggest difference between these two ionic solids is the attractions due to the charges on the ions. According to Coulomb's law, the Ca^{2+} and O^{2-} ions have greater interaction than the Na^+ and Cl^- ions. That means NaCl has a larger entropy per mole than CaO.

    (b) Molecules of P_4 have more atoms than molecules of Cl_2. That means P_4 has a larger entropy per mole than Cl_2.

(c) The solid NH_4NO_3 crystal is more ordered than the aqueous NH_4^+ and NO_3^- ions. That means the aqueous NH_4NO_3 has a larger entropy per mole than the solid NH_4NO_3.

29. Use the qualitative guidelines for entropy changes described in Section 18.3.

(a) Solid-state Ga atoms are more ordered than liquid Ga atoms. That means $Ga(\ell)$ has more entropy than $Ga(s)$.

(b) AsH_3 is a molecular gas and Kr is an atomic gas. Because they have similar mass, we look at the relative complexity. The molecule has more complexity. That means AsH_3 has a larger standard molar entropy than Kr.

(c) Comparing NaF and MgO, we find that the biggest difference between these two ionic solids is the attractions due to the charges on the ions. According to Coulomb's law, the Mg^{2+} and O^{2-} ions have greater interaction than the Na^+ and F^- ions. That means NaF has a larger standard molar entropy than MgO.

31. Use the qualitative guidelines for entropy changes described in Section 18.3.

(a) The reaction has more gas-phase reactants (2 mol) than gas-phase products (1 mol). That means the entropy change will be negative.

(b) The reaction has fewer gas-phase reactants ($\frac{3}{2}$ mol) than gas-phase products (3 mol). That means the entropy change will be positive.

(c) The reaction has more gas-phase reactants (4 mol) than gas-phase products (2 mol). That means the entropy change will be negative.

(d) The reaction has fewer gas-phase reactants (0 mol) than gas-phase products (1 mol). That means the entropy change will be positive.

33. Use the qualitative guidelines for entropy changes described in Section 18.3.

(a) The reaction has more gas-phase reactants (3 mol) than gas-phase products (2 mol). That means the entropy change will be negative.

(b) The reaction has more gas-phase reactants (3 mol) than gas-phase products (0 mol). That means the entropy change will be negative.

(c) The reaction has fewer gas-phase reactants (2 mol) than gas-phase products (3 mol). That means the entropy change will be positive.

Calculating Entropy Changes

35. Adapt the method described in Problem-Solving Example 18.6. Use Equation 18.4:

$$T = \frac{\Delta H^\circ}{\Delta S^\circ}$$

Use the enthalpy of vaporization and the normal boiling point:

$$\Delta S^\circ = \frac{\Delta H^\circ}{T} = \frac{\left(39.3 \ \frac{kJ}{mol}\right) \times \left(\frac{1000 \ J}{1 \ kJ}\right)}{(78.3 + 273.15)K} = 112 \ J \ K^{-1}mol^{-1}$$

37. Adapt the method described in Problem-Solving Example 18.1. Use the equation from the beginning of Section 18.3:

$$\Delta S = \frac{q_{rev}}{T}$$

For example, the calculation for (a) looks like this:

$$\Delta S = \frac{q_{rev}}{T} = \frac{(0.775 \ kJ) \times \left(\frac{1000 \ J}{1 \ kJ}\right)}{295 \ K} = 2.63 \ J \ K^{-1}$$

	q_{rev} (kJ)	T (K)	ΔS (J K^{-1})
(a)	0.775	295	2.63
(b)	500	500.	1000
(c)	2.45	1000.	2.45

39. (a) Adapt the method described in Problem-Solving Example 18.3. Write the balanced chemical equation and use the equation from the beginning of Section 18.4:

$$\Delta S^\circ = \Sigma \left[(\text{moles of product}) \times S^\circ (\text{product}) \right] - \Sigma \left[(\text{moles of reactant}) \times S^\circ (\text{reactant}) \right]$$

Vaporization equation: $CH_3OH(\ell) \longrightarrow CH_3OH(g)$

$$\Delta S^\circ = (1 \ \text{mole}) \times S^\circ(CH_3OH(g)) - (1 \ \text{mole}) \times S^\circ(CH_3OH(\ell))$$

$$\Delta S^\circ = (1 \ \text{mole}) \times (239.81 \ J \ K^{-1}mol^{-1}) - (1 \ \text{mole}) \times (126.8 \ J \ K^{-1}mol^{-1}) = 113.0 \ J \ K^{-1}$$

(b) Adapt the method described in the answer to Question 35.

$$\Delta H^\circ = T\Delta S^\circ = (64.6 + 273.15) \text{ K} \times (113.0 \text{ J K}^{-1}) \times \frac{1 \text{ kJ}}{1000 \text{ J}} = 38.17 \text{ kJ}$$

41. Adapt the method described in the answer to Question 39.

(a) $\Delta S^\circ = (1 \text{ mol}) \times S^\circ(C_2H_6) - (1 \text{ mol}) \times S^\circ(C_2H_4) - (1 \text{ mol}) \times S^\circ(H_2)$

$= (1 \text{ mol}) \times (229.60 \text{ J K}^{-1}\text{mol}^{-1}) - (1 \text{ mol}) \times (219.56 \text{ J K}^{-1}\text{mol}^{-1})$

$- (1 \text{ mol}) \times (130.684 \text{ J K}^{-1}\text{mol}^{-1}) = -120.64 \text{ J K}^{-1}$

This confirms the negative ΔS prediction from Question 31.

(b) $\Delta S^\circ = (1 \text{ mol}) \times S^\circ(CO_2) + (2 \text{ mol}) \times S^\circ(H_2O(g))$

$- (1 \text{ mol}) \times S^\circ(CH_3OH(\ell)) - (\frac{3}{2} \text{ mol}) \times S^\circ(O_2)$

$= (1 \text{ mol}) \times (213.74 \text{ J K}^{-1}\text{mol}^{-1}) + (2 \text{ mol}) \times (188.825 \text{ J K}^{-1}\text{mol}^{-1})$

$- (1 \text{ mol}) \times (126.8 \text{ J K}^{-1}\text{mol}^{-1}) - (\frac{3}{2} \text{ mol}) \times (205.138 \text{ J K}^{-1}\text{mol}^{-1}) = 156.9 \text{ J K}^{-1}$

This confirms the positive ΔS prediction from Question 31.

(c) $\Delta S^\circ = (2 \text{ mol}) \times S^\circ(NH_3) - (1 \text{ mol}) \times S^\circ(N_2) - (3 \text{ mol}) \times S^\circ(H_2)$

$= (2 \text{ mol}) \times (192.45 \text{ J K}^{-1}\text{mol}^{-1}) - (1 \text{ mol}) \times (191.61 \text{ J K}^{-1}\text{mol}^{-1})$

$- (3 \text{ mol}) \times (130.684 \text{ J K}^{-1}\text{mol}^{-1}) = -198.76 \text{ J K}^{-1}$

This confirms the negative ΔS prediction from Question 31.

(d) $\Delta S^\circ = (1 \text{ mol}) \times S^\circ(CaO) + (1 \text{ mol}) \times S^\circ(CO_2) - (1 \text{ mol}) \times S^\circ(CaCO_3)$

$= (1 \text{ mol}) \times (39.75 \text{ J K}^{-1}\text{mol}^{-1}) + (1 \text{ mol}) \times (213.74 \text{ J K}^{-1}\text{mol}^{-1})$

$- (1 \text{ mol}) \times (92.9 \text{ J K}^{-1}\text{mol}^{-1}) = 160.6 \text{ J K}^{-1}$

This confirms the positive ΔS prediction from Question 31.

43. Adapt the method described in the answer to Question 39.

(a) $\Delta S^\circ = (2 \text{ mol}) \times S^\circ(CO_2) - (2 \text{ mol}) \times S^\circ(CO) - (1 \text{ mol}) \times S^\circ(O_2)$

$= (2 \text{ mol}) \times (213.74 \text{ J K}^{-1}\text{mol}^{-1}) - (2 \text{ mol}) \times (197.674 \text{ J K}^{-1}\text{mol}^{-1})$

$- (1 \text{ mol}) \times (205.138 \text{ J K}^{-1}\text{mol}^{-1}) = -173.01 \text{ J K}^{-1}$

This confirms the negative ΔS prediction from Question 33.

(b) $\Delta S° = (2 \text{ mol}) \times S°(H_2O(\ell)) - (2 \text{ mol}) \times S°(H_2) - (1 \text{ mol}) \times S°(O_2)$

$= (2 \text{ mol}) \times (69.91 \text{ J K}^{-1}\text{mol}^{-1}) - (2 \text{ mol}) \times (130.684 \text{ J K}^{-1}\text{mol}^{-1})$

$- (1 \text{ mol}) \times (205.138 \text{ J K}^{-1}\text{mol}^{-1}) = -326.69 \text{ J K}^{-1}$

This confirms the negative ΔS prediction from Question 33.

(c) $\Delta S° = (3 \text{ mol}) \times S°(O_2) - (2 \text{ mol}) \times S°(O_3)$

$= (3 \text{ mol}) \times (205.138 \text{ J K}^{-1}\text{mol}^{-1}) - (2 \text{ mol}) \times (238.93 \text{ J K}^{-1}\text{mol}^{-1}) = 137.55 \text{ J K}^{-1}$

This confirms the positive ΔS prediction from Question 32.

46. Adapt the method described in the answer to Question 39. Formation is defined as producing one mole of a substance from standard state elements.

(a) $2 \text{ C(graphite)} + H_2(g) \longrightarrow C_2H_2(g)$

$\Delta S° = (1 \text{ mol}) \times S°(C_2H_2) - (2 \text{ mol}) \times S°(C) - (1 \text{ mol}) \times S°(H_2)$

$= (1 \text{ mol}) \times (200.94 \text{ J K}^{-1}\text{mol}^{-1}) - (2 \text{ mol}) \times (5.740 \text{ J K}^{-1}\text{mol}^{-1})$

$- (1 \text{ mol}) \times (130.684 \text{ J K}^{-1}\text{mol}^{-1}) = 58.78 \text{ J K}^{-1}$

(b) $2 \text{ C(graphite)} + 2 H_2(g) \longrightarrow C_2H_4(g)$

$\Delta S° = (1 \text{ mol}) \times S°(C_2H_4) - (2 \text{ mol}) \times S°(C) - (2 \text{ mol}) \times S°(H_2)$

$= (1 \text{ mol}) \times (219.56 \text{ J K}^{-1}\text{mol}^{-1}) - (2 \text{ mol}) \times (5.740 \text{ J K}^{-1}\text{mol}^{-1})$

$- (2 \text{ mol}) \times (130.684 \text{ J K}^{-1}\text{mol}^{-1}) = -53.29 \text{ J K}^{-1}$

(c) $2 \text{ C(graphite)} + 3 H_2(g) \longrightarrow C_2H_6(g)$

$\Delta S° = (1 \text{ mol}) \times S°(C_2H_6) - (2 \text{ mol}) \times S°(C) - (3 \text{ mol}) \times S°(H_2)$

$= (1 \text{ mol}) \times (229.60 \text{ J K}^{-1}\text{mol}^{-1}) - (2 \text{ mol}) \times (5.740 \text{ J K}^{-1}\text{mol}^{-1})$

$- (3 \text{ mol}) \times (130.684 \text{ J K}^{-1}\text{mol}^{-1}) = -173.93 \text{ J K}^{-1}$

Adding more H atoms decreases the ΔS (i.e., makes it more negative).

Entropy and the Second Law of Thermodynamics

47. Adapt the method described in the answer to Question 39.

$$\Delta S° = (1 \text{ mol}) \times S°(C_2H_5OH) - (1 \text{ mol}) \times S°(C_2H_4) - (1 \text{ mol}) \times S°(H_2O(g))$$

$$= (1 \text{ mol}) \times (160.7 \text{ J K}^{-1}\text{mol}^{-1}) - (1 \text{ mol}) \times (219.56 \text{ J K}^{-1}\text{mol}^{-1})$$

$$- (1 \text{ mol}) \times (188.825 \text{ J K}^{-1}\text{mol}^{-1}) = -247.7 \text{ J K}^{-1}$$

We cannot tell from the results of this calculation whether this reaction is product-favored. We need the $\Delta H°$, also.

Use Equation 6.9, to calculate $\Delta H°$

$$\Delta H° = \Sigma \left[(\text{moles of product}) \times \Delta H_f^° (\text{product}) \right] -$$

$$\Sigma \left[(\text{moles of reactant}) \times \Delta H_f^° (\text{reactant}) \right]$$

$$\Delta H° = (1 \text{ mol}) \times \Delta H_f^°(C_2H_5OH) - (1 \text{ mol}) \times \Delta H_f^°(C_2H_4) - (1 \text{ mol}) \times \Delta H_f^°(H_2O(g))$$

$$= (1 \text{ mol}) \times (-277.69 \text{ kJ/mol}) - (1 \text{ mol}) \times (52.26 \text{ kJ/mol})$$

$$- (1 \text{ mol}) \times (-241.818 \text{ kJ/mol}) = -88.13 \text{ kJ}$$

Since, both $\Delta H°$ and $\Delta S°$ are negative, it depends on the temperature whether the reaction is product-favored or not.

When $T = \dfrac{(-88.13 \text{ kJ}) \times \left(\dfrac{1000 \text{ J}}{1 \text{ kJ}}\right)}{-247.7 \text{ JK}^{-1}} = 355.8 \text{ K}$ or lower, the reaction is product-favored.

The temperature here is 25 °C + 273.15 = 298 K, so the reaction is product-favored.

49. Adapt the method described in the answer to Question 32, then consult Table 18.2.

The reaction has more gas-phase reactants (1 mol) than gas-phase products (0 mol). That means the entropy change will be negative. When $\Delta H°$ and $\Delta S°$ are both negative, the reaction is product-favored only at low temperatures. The exothermicity is sufficient to favor products, if the temperature is low enough to overcome the decrease in entropy.

51. Exothermic reactions with an increase in disorder, exhibited by a larger number of gas-phase products (7 mol) than gas-phase reactants (5 mol), never need help from the surroundings to favor products.

52. Exothermic reactions with an increase in disorder, exhibited by a larger number of gas-phase products (1 mol) than gas-phase reactants (0 mol), never need help from the surroundings to favor products.

54. The equation describing the chemical reaction is: $H_2O(\ell) \longrightarrow H_2(g) + \frac{1}{2} O_2(g)$

The reaction written here is the reverse of the highly exothermic reaction described in the first sentence of the Question. That means this reaction is highly endothermic. With more gas products ($1\frac{1}{2}$ mol) than gas reactants (0 mol), the entropy change is positive. Because we do not see water spontaneously decomposing, we will conclude that the reaction is reactant-favored. The entropy increase is insufficient to drive this highly endothermic reaction to form products without assistance from the surroundings at the temperature of 25 °C.

56. Adapt the method described in the answer to Questions 39 and 35.

$$\Delta H^\circ = (1 \text{ mol}) \times \Delta H_f^\circ(C_8H_{18}) - (1 \text{ mol}) \times \Delta H_f^\circ(C_8H_{16}) - (1 \text{ mol}) \times \Delta H_f^\circ(H_2)$$

$$= (1 \text{ mol}) \times (-208.45 \text{ kJ/mol}) - (1 \text{ mol}) \times (-82.93 \text{ kJ/mol}) - (1 \text{ mol}) \times (0 \text{ kJ/mol})$$

$$= -125.52 \text{ kJ}$$

We do not have $S^\circ(C_8H_{16})$, but we can assume that it is very close to that of $S^\circ(C_8H_{18})$. The flexibility and freedom of motion of the atoms in these large eight-carbon-chain molecules is probably about the same except near the one double bond in 1-octene. In support of this presumption, we compare $S^\circ(C_2H_6)$ and $S^\circ(C_2H_4)$, which are both larger than 200 J $K^{-1}mol^{-1}$ and differ by only about 10 J $K^{-1}mol^{-1}$.

$$\Delta S^\circ = (1 \text{ mol}) \times S^\circ(C_8H_{18}) - (1 \text{ mol}) \times S^\circ(C_8H_{16}) - (1 \text{ mol}) \times S^\circ(H_2)$$

$$\cong -(1 \text{ mol}) \times S^\circ(H_2) = -(1 \text{ mol}) \times (130.684 \text{ J } K^{-1}mol^{-1}) = -130.684 \text{ J } K^{-1}$$

We can determine the temperature at which this reaction becomes product-favored.

$$T \cong \frac{(-125.52 \text{ kJ}) \times \left(\frac{1000 \text{ J}}{1 \text{ kJ}}\right)}{-130.684 \text{ J} K^{-1}} = 960.48 \text{ K}$$

The given temperature of 25 °C (298K) is much lower, so the reaction is confirmed to be reactant-favored.

(With a complete thermodynamic data table, we could also have calculated the ΔG° for this reaction and proven that it is negative.)

57. Adapt the method described in the answer to Question 39 and use Table 18.2.

(a) $\Delta H^\circ = (2 \text{ mol}) \times \Delta H_f^\circ(\text{Fe}) + (1 \text{ mol}) \times \Delta H_f^\circ(\text{Al}_2\text{O}_3)$

$$- (1 \text{ mol}) \times \Delta H_f^\circ(\text{Fe}_2\text{O}_3) - (2 \text{ mol}) \times \Delta H_f^\circ(\text{Al})$$

$$= (2 \text{ mol}) \times (0 \text{ kJ/mol}) + (1 \text{ mol}) \times (- 1675.7 \text{ kJ/mol})$$

$$- (1 \text{ mol}) \times (- 824.2 \text{ kJ/mol}) - (2 \text{ mol}) \times (0 \text{ kJ/mol}) = -851.5 \text{ kJ}$$

$\Delta S^\circ = (2 \text{ mol}) \times S^\circ(\text{Fe}) + (1 \text{ mol}) \times S^\circ(\text{Al}_2\text{O}_3)$

$$- (1 \text{ mol}) \times S^\circ(\text{Fe}_2\text{O}_3) - (2 \text{ mol}) \times S^\circ(\text{Al})$$

$$= (2 \text{ mol}) \times (27.78 \text{ J K}^{-1}\text{mol}^{-1}) + (1 \text{ mol}) \times (50.92 \text{ J K}^{-1}\text{mol}^{-1})$$

$$- (1 \text{ mol}) \times (87.40 \text{ J K}^{-1}\text{mol}^{-1}) - (2 \text{ mol}) \times (28.3 \text{ J K}^{-1}\text{mol}^{-1}) = -37.52 \text{ J K}^{-1}$$

Reaction is exothermic, but its entropy change is negative, so it can only be product-favored at low temperatures. (Table 18.2)

(b) $\Delta H^\circ = (2 \text{ mol}) \times \Delta H_f^\circ(\text{NO}_2) - (1 \text{ mol}) \times \Delta H_f^\circ(\text{N}_2) - (2 \text{ mol}) \times \Delta H_f^\circ(\text{O}_2)$

$$= (2 \text{ mol}) \times (33.18 \text{ kJ/mol}) - (1 \text{ mol}) \times (0 \text{ kJ/mol}) - (2 \text{ mol}) \times (0 \text{ kJ/mol})$$

$$= 66.36 \text{ kJ}$$

$\Delta S^\circ = (2 \text{ mol}) \times S^\circ(\text{NO}_2) - (1 \text{ mol}) \times S^\circ(\text{N}_2) - (2 \text{ mol}) \times S^\circ(\text{O}_2)$

$$= (2 \text{ mol}) \times (240.06 \text{ J K}^{-1}\text{mol}^{-1}) - (1 \text{ mol}) \times (191.61 \text{ J K}^{-1}\text{mol}^{-1})$$

$$- (2 \text{ mol}) \times (205.138 \text{ J K}^{-1}\text{mol}^{-1}) = - 21.77 \text{ J K}^{-1}$$

Reaction is endothermic and its entropy change is negative, so it will never be product-favored. (Table 18.2)

Gibbs Free Energy

59. Adapt the method described in the answer to Question 40, then use three equations:

Equation 18.1 with the Kelvin temperature:

$$\Delta S^\circ_{\text{universe}} = \Delta S^\circ_{\text{universe}} = -\frac{\Delta H^\circ_{\text{system}}}{T} + \Delta S^\circ_{\text{system}}$$

$$T(K) = T(^\circ C) + 273.15 = 25 \,^\circ C + 273.15 = 298 \text{ K}$$

Equation 18.2 $\Delta G^{\circ}_{system} = \Delta H^{\circ}_{system} - T\Delta S^{\circ}_{system}$

and Equation 18.3, as described in Problem-Solving Example 18.5.

$$\Delta G^{\circ} = \sum \left[(\text{moles of product}) \times \Delta G^{\circ}_f (\text{product}) \right]$$

$$- \sum \left[(\text{moles of reactant}) \times \Delta G^{\circ}_f (\text{reactant}) \right]$$

(a) $\Delta H^{\circ}_{system} = (2 \text{ mol}) \times \Delta H^{\circ}_f (CO_2) + (3 \text{ mol}) \times \Delta H^{\circ}_f (H_2O(\ell))$

$$- (1 \text{ mol}) \times \Delta H^{\circ}_f (C_2H_6) - (\frac{7}{2} \text{ mol}) \times \Delta H^{\circ}_f (O_2)$$

$$= (2 \text{ mol}) \times (-393.509 \text{ kJ/mol}) + (3 \text{ mol}) \times (-285.83 \text{ kJ/mol})$$

$$- (1 \text{ mol}) \times (-84.68 \text{ kJ/mol}) - (\frac{7}{2} \text{ mol}) \times (0 \text{ kJ/mol}) = -1559.83 \text{ kJ}$$

$\Delta S^{\circ}_{system} = (2 \text{ mol}) \times S^{\circ}(CO_2) + (3 \text{ mol}) \times S^{\circ}(H_2O(\ell))$

$$- (1 \text{ mol}) \times S^{\circ}(C_2H_6) - (\frac{7}{2} \text{ mol}) \times S^{\circ}(O_2)$$

$$= (2 \text{ mol}) \times (213.74 \text{ J K}^{-1}\text{mol}^{-1}) + (3 \text{ mol}) \times (69.91 \text{ J K}^{-1}\text{mol}^{-1})$$

$$- (1 \text{ mol}) \times (229.60 \text{ J K}^{-1}\text{mol}^{-1}) - (\frac{7}{2} \text{ mol}) \times (205.138 \text{ J K}^{-1}\text{mol}^{-1}) = -310.37 \text{ J K}^{-1}$$

$$\Delta S^{\circ}_{universe} = -\frac{(-1559.83 \text{ kJ})}{(298 \text{ K})} \times \frac{1000 \text{ J}}{1 \text{ kJ}} + (-310.37 \text{ J K}^{-1}) = 4.92 \times 10^3 \text{ J K}^{-1}$$

(b) $\Delta G^{\circ}_{system} = \Delta H^{\circ}_{system} - T\Delta S^{\circ}_{system}$

$$= (-1559.83 \text{ kJ}) - (298 \text{ K})(-310.37 \text{ J K}^{-1}) \times \frac{1 \text{ kJ}}{1000 \text{ J}} = -1467.3 \text{ kJ}$$

Independently, we can get $\Delta G^{\circ}_{system}$ using Equation 18.3:

$\Delta G^{\circ}_{system} = (2 \text{ mol}) \times \Delta G^{\circ}_f (CO_2) + (3 \text{ mol}) \times \Delta G^{\circ}_f (H_2O(\ell))$

$$- (1 \text{ mol}) \times \Delta G^{\circ}_f (C_2H_6) - (\frac{7}{2} \text{ mol}) \times \Delta G^{\circ}_f (O_2)$$

$$= (2 \text{ mol}) \times (-394.359 \text{ kJ/mol}) + (3 \text{ mol}) \times (-237.129 \text{ kJ/mol})$$

$$- (1 \text{ mol}) \times (-32.82 \text{ kJ/mol}) - (\frac{7}{2} \text{ mol}) \times (0 \text{ kJ/mol}) = -1467.28 \text{ kJ}$$

A negative $\Delta G^{\circ}_{system}$ is consistent with a positive $\Delta S^{\circ}_{universe}$. In addition:

$$-T\Delta S^{\circ}_{universe} = (298 \text{ K})(4.92 \times 10^3 \text{ J K}^{-1}) \times \frac{1 \text{ kJ}}{1000 \text{ J}} = -1.47 \times 10^3 \text{ kJ} = \Delta G^{\circ}_{system}.$$

(c) Yes. Ethane is used as a fuel; hence, we would expect its combustion reaction to be product–favored.

62. Adapt the method described in the answer to Question 60 using Equation 18.2:

$$\Delta G^{\circ} = \Delta H^{\circ} - T\Delta S^{\circ}$$

Here: (sign of ΔH°) = negative, and (sign of ΔS°) = positive. The Kelvin temperature is always positive.

$$\Delta G^{\circ} = \text{(negative } \Delta H^{\circ}\text{)} - \text{(positive Kelvin temperature)(positive } \Delta S^{\circ}\text{)}$$

Take the absolute values of the each term and then include the resulting signs explicitly:

$$\Delta G^{\circ} = -|\Delta H^{\circ}| - |T\Delta S^{\circ}| = -(|\Delta H^{\circ}| + |T\Delta S^{\circ}|)$$

$$\text{(sign of } \Delta G^{\circ}\text{)} = -\text{(positive)} = \text{negative}$$

So, $\Delta G^{\circ} < 0$ for all temperature values.

65. Adapt the method described in the answer to Question 59.

$$\Delta G^{\circ} = (1 \text{ mol}) \times \Delta G^{\circ}_f(Si) + (1 \text{ mol}) \times \Delta G^{\circ}_f(CO_2)$$

$$- (1 \text{ mol}) \times \Delta G^{\circ}_f(SiO_2) - (1 \text{ mol}) \times \Delta G^{\circ}_f(C)$$

$$= (1 \text{ mol}) \times (0 \text{ kJ/mol}) + (1 \text{ mol}) \times (-394.359 \text{ kJ/mol})$$

$$- (1 \text{ mol}) \times (-856.64 \text{ kJ/mol}) - (1 \text{ mol}) \times (0 \text{ kJ/mol}) = 462.28 \text{ kJ}$$

The reaction is not product-favored, so this would not be a good way to make pure silicon.

68. Adapt the method described in the answer to Question 60.

$$T = 25.00 \text{ °C} + 273.15 = 298.15 \text{ K}$$

(a) $\Delta H^{\circ} = (1 \text{ mol}) \times \Delta H^{\circ}_f(C_2H_4) - (1 \text{ mol}) \times \Delta H^{\circ}_f(C_2H_2) - (1 \text{ mol}) \times \Delta H^{\circ}_f(H_2)$

$$= (1 \text{ mol}) \times (52.26 \text{ kJ/mol})$$

$$- (1 \text{ mol}) \times (226.73 \text{ kJ/mol}) - (1 \text{ mol}) \times (0 \text{ kJ/mol}) = -174.47 \text{ kJ}$$

$\Delta G^{\circ} = -141.04 \text{ kJ}$ (from Question 67)

$$\Delta S° = \frac{\Delta H° - \Delta G°}{T} = \frac{(-174.47 \text{ kJ}) - (-141.04 \text{ kJ})}{(298.15 \text{ K})} \times \frac{1000 \text{ J}}{1 \text{ kJ}} = -112.1 \text{ J K}^{-1}$$

$$\Delta S° = (1 \text{ mol}) \times S°(C_2H_4) - (1 \text{ mol}) \times S°(C_2H_2) - (1 \text{ mol}) \times S°(H_2)$$

$$S°(C_2H_2) = \frac{\Delta S°}{(1 \text{ mol})} - S°(C_2H_4) + S°(H_2)$$

$$= (-112.1 \text{ J K}^{-1}\text{mol}^{-1}) - (219.56 \text{ J K}^{-1}\text{mol}^{-1}) + (130.685 \text{ J K}^{-1}\text{mol}^{-1})$$

$$= 201.0 \text{ J K}^{-1}\text{mol}^{-1}$$

This is consistent with the value of 200.94 J K^{-1}mol^{-1} given in Appendix J.

(b) $\Delta H° = (2 \text{ mol}) \times \Delta H_f°(SO_2) + (1 \text{ mol}) \times \Delta H_f°(O_2) - (2 \text{ mol}) \times \Delta H_f°(SO_3)$

$$= (2 \text{ mol}) \times (-296.830 \text{ kJ/mol}) + (1 \text{ mol}) \times (0 \text{ kJ/mol})$$

$$- (2 \text{ mol}) \times (-395.72 \text{ kJ/mol}) = 197.78 \text{ kJ}$$

$\Delta G° = 141.73 \text{ kJ}$ (from Question 67)

$$\Delta S° = \frac{\Delta H° - \Delta G°}{T} = \frac{(197.78 \text{ kJ}) - (141.73 \text{ kJ})}{(298.15 \text{ K})} \times \frac{1000 \text{ J}}{1 \text{ kJ}} = 188.0 \text{ J K}^{-1}$$

$$\Delta S° = (2 \text{ mol}) \times S°(SO_2) + (1 \text{ mol}) \times S°(O_2) - (2 \text{ mol}) \times S°(SO_3)$$

$$S°(SO_3) = -\frac{\Delta S°}{(2 \text{ mol})} + S°(SO_2) + \frac{1}{2} S°(O_2)$$

$$= -\frac{1}{2}(188.0 \text{ J K}^{-1}\text{mol}^{-1}) + (248.22 \text{ J K}^{-1}\text{mol}^{-1}) + \frac{1}{2}(205.138 \text{ J}^{-1}\text{mol}^{-1})$$

$$= 256.8 \text{ J K}^{-1}\text{mol}^{-1}$$

This is consistent with the value of 256.76 J K^{-1}mol^{-1} given in Appendix J.

(c) $\Delta H° = (4 \text{ mol}) \times \Delta H_f°(NO) + (6 \text{ mol}) \times \Delta H_f°(H_2O(g))$

$$- (4 \text{ mol}) \times \Delta H_f°(NH_3) - (5 \text{ mol}) \times \Delta H_f°(O_2)$$

$$= (4 \text{ mol}) \times (90.25 \text{ kJ/mol}) + (6 \text{ mol}) \times (-241.818 \text{ kJ/mol})$$

$$- (4 \text{ mol}) \times (-46.11 \text{ kJ/mol}) - (5 \text{ mol}) \times (0 \text{ kJ/mol}) = -905.47 \text{ kJ}$$

$\Delta G° = -959.43 \text{ kJ}$ (from Question 67)

$$\Delta S^\circ = \frac{\Delta H^\circ - \Delta G^\circ}{T} = \frac{(-905.47\ kJ)-(-959.43\ kJ)}{(298.15\ K)} \times \frac{1000\ J}{1\ kJ} = 181.0\ J\ K^{-1}$$

$$\Delta S^\circ = (4\ mol) \times S^\circ(NO) + (6\ mol) \times S^\circ(H_2O(g)) - (4\ mol) \times S^\circ(NH_3) - (5\ mol) \times S^\circ(O_2)$$

$$S^\circ(NO) = \frac{\Delta S^\circ}{(4\ mol)} - \frac{6}{4} S^\circ(H_2O(g)) + S^\circ(NH_3) + \frac{5}{4} S^\circ(O_2)$$

$$= \frac{1}{4}(181.0\ J\ K^{-1}mol^{-1}) - \frac{6}{4}(188.825\ J\ K^{-1}mol^{-1})$$

$$+ (192.45\ J\ K^{-1}mol^{-1}) + \frac{5}{4}(205.138\ J\ K^{-1}mol^{-1}) = 210.9\ J\ K^{-1}mol^{-1}$$

This is consistent with the value of $210.76\ J\ K^{-1}mol^{-1}$ given in Appendix J.

69. Adapt the method described in the answer to Question 39.

(a) $\Delta H^\circ = (1\ mol) \times \Delta H^\circ_f(CH_3OH) + (1\ mol) \times \Delta H^\circ_f(CO) - (2\ mol) \times \Delta H^\circ_f(H_2)$

$$= (1\ mol) \times (-238.66\ kJ/mol) + (1\ mol) \times (-110.525\ kJ/mol)$$

$$- (2\ mol) \times (0\ kJ/mol) = -128.14\ kJ$$

$\Delta S^\circ = (1\ mol) \times S^\circ(CH_3OH) + (1\ mol) \times S^\circ(CO) - (2\ mol) \times S^\circ(H_2)$

$$= (1\ mol) \times (126.8\ J\ K^{-1}mol^{-1}) + (1\ mol) \times (197.674\ J\ K^{-1}mol^{-1})$$

$$- (2\ mol) \times (130.684\ J\ K^{-1}mol^{-1}) = -332.2\ J\ K^{-1}$$

$$T = \frac{(-128.14\ kJ) \times \left(\dfrac{1000\ J}{1\ kJ}\right)}{-332.2\ J\ K^{-1}} = 385.7\ K$$

(b) $\Delta H^\circ = (4\ mol) \times \Delta H^\circ_f(Fe) + (3\ mol) \times \Delta H^\circ_f(CO_2)$

$$- (2\ mol) \times \Delta H^\circ_f(Fe_2O_3) - (3\ mol) \times \Delta H^\circ_f(C)$$

$$= (4\ mol) \times (0\ kJ/mol) + (3\ mol) \times (-393.509\ kJ/mol)$$

$$- (2\ mol) \times (-824.2\ kJ/mol) - (3\ mol) \times (0\ kJ/mol) = 467.9\ kJ$$

$\Delta S^\circ = (4\ mol) \times S^\circ(Fe) + (3\ mol) \times S^\circ(CO_2) - (2\ mol) \times S^\circ(Fe_2O_3) - (3\ mol) \times S^\circ(C)$

$$= (4\ mol) \times (27.78\ J\ K^{-1}mol^{-1}) + (3\ mol) \times (213.74\ J\ K^{-1}mol^{-1})$$

$$- (2\ mol) \times (87.40\ J\ K^{-1}mol^{-1}) - (3\ mol) \times (5.740\ J\ K^{-1}mol^{-1}) = 560.32\ J\ K^{-1}$$

$$T = \frac{\left(467.9 \text{ kJ}\right) \times \left(\dfrac{1000 \text{ J}}{1 \text{ kJ}}\right)}{560.32 \text{ JK}^{-1}} = 835.1 \text{ K}$$

72. Adapt methods described in the answer to Question 47 and 57.

(a) $\Delta H^\circ = (1 \text{ mol}) \times \Delta H^\circ_f(CaO) + (1 \text{ mol}) \times \Delta H^\circ_f(CO_2) - (1 \text{ mol}) \times \Delta H^\circ_f(CaCO_3)$

$= (1 \text{ mol}) \times (-635.09 \text{ kJ/mol}) + (1 \text{ mol}) \times (-393.509 \text{ kJ/mol})$

$- (1 \text{ mol}) \times (-1206.92 \text{ kJ/mol}) = 178.32 \text{ kJ}$

$\Delta S^\circ = (1 \text{ mol}) \times S^\circ(CaO) + (1 \text{ mol}) \times S^\circ(CO_2) - (1 \text{ mol}) \times S^\circ(CaCO_3)$

$= (1 \text{ mol}) \times (39.75 \text{ J K}^{-1}\text{mol}^{-1}) + (1 \text{ mol}) \times (213.74 \text{ J K}^{-1}\text{mol}^{-1})$

$- (1 \text{ mol}) \times (92.9 \text{ J K}^{-1}\text{mol}^{-1}) = 160.6 \text{ J K}^{-1}$

$\Delta G^\circ = \Delta H^\circ - T\Delta S^\circ = (178.32 \text{ kJ}) - (298 \text{ K})(160.6 \text{ J K}^{-1}) \times \dfrac{1 \text{ kJ}}{1000 \text{ J}} = 130.5 \text{ kJ}$

(b) The change in the Gibbs free energy is positive. That means the reaction is reactant-favored.

(c) Since both ΔH° and ΔS° are positive, the reaction is only product-favored at high temperatures.

(d)

$$T = \frac{\left(178.32 \text{ kJ}\right) \times \left(\dfrac{1000 \text{ J}}{1 \text{ kJ}}\right)}{160.6 \text{ JK}^{-1}} = 1110. \text{ K}$$

74. Adapt the method described in Question 68.

$\Delta G^\circ = (1 \text{ mol}) \times \Delta G^\circ_f(C_2H_2) + (1 \text{ mol}) \times \Delta G^\circ_f(Ca(OH)_2)$

$- (1 \text{ mol}) \times \Delta G^\circ_f(CaC_2) - (2 \text{ mol}) \times \Delta G^\circ_f(H_2O(\ell))$

$\Delta G^\circ_f(Ca(OH)_2) = -\dfrac{\Delta G^\circ}{(1 \text{ mol})} - \Delta G^\circ_f(C_2H_2) + \Delta G^\circ_f(CaC_2) + 2 \times \Delta G^\circ_f(H_2O(\ell))$

$= (-119.282 \text{ kJ/mol}) - (209.20 \text{ kJ/mol}) - (-64.9 \text{ kJ/mol}) - 2 \times (-237.129 \text{ kJ/mol})$

$= -867.8 \text{ kJ/mol}$

76. Adapt methods described in the answer to Question 39.

(a)
$$Br_2(\ell) \longrightarrow Br_2(g)$$

$\Delta H° = (1 \text{ mol}) \times \Delta H_f^°(Br_2(g)) - (1 \text{ mol}) \times \Delta H_f^°(Br_2(\ell))$

$\qquad = (1 \text{ mol}) \times (30.907 \text{ kJ/mol}) - (1 \text{ mol}) \times (0 \text{ kJ/mol}) = 30.907 \text{ kJ}$

$\Delta S° = (1 \text{ mol}) \times S°(Br_2(g)) - (1 \text{ mol}) \times S°(Br_2(\ell))$

$\qquad = (1 \text{ mol}) \times (245.463 \text{ J K}^{-1}\text{mol}^{-1}) - (1 \text{ mol}) \times (152.231 \text{ K}^{-1}\text{mol}^{-1}) = 93.232 \text{ J K}^{-1}$

$$T = \frac{(30.907 \text{ kJ}) \times \left(\dfrac{1000 \text{ J}}{1 \text{ kJ}}\right)}{93.232 \text{ J K}^{-1}} = 331.51 \text{ K}$$

(b)
$$SnCl_4(\ell) \longrightarrow SnCl_4(g)$$

$\Delta H° \quad = (1 \text{ mol}) \times \Delta H_f^°(SnCl_4(g)) - (1 \text{ mol}) \times \Delta H_f^°(SnCl_4(\ell))$

$\qquad = (1 \text{ mol}) \times (-471.5 \text{ kJ/mol}) - (1 \text{ mol}) \times (-511.3 \text{ kJ/mol}) = 39.8 \text{ kJ}$

$\Delta S° = (1 \text{ mol}) \times S°(SnCl_4(g)) - (1 \text{ mol}) \times S°(SnCl_4(\ell))$

$\qquad = (1 \text{ mol}) \times (365.8 \text{ J K}^{-1}\text{mol}^{-1}) - (1 \text{ mol}) \times (258.6 \text{ K}^{-1}\text{mol}^{-1}) = 107.2 \text{ J K}^{-1}$

$$T = \frac{(39.8 \text{ kJ}) \times \left(\dfrac{1000 \text{ J}}{1 \text{ kJ}}\right)}{107.2 \text{ J K}^{-1}} = 371 \text{ K}$$

Gibbs Free Energy Changes and Equilibrium Constants

78. Use Equation 18.2 ($\Delta G° = \Delta H° - T\Delta S°$) and Equation 18.7 ($\Delta G° = -RT\ln K°$) to relate equilibrium constant to thermodynamic values.

(a) It is false to say that a chemical reaction with K = 1.0 has $\Delta H° = 0$ (except in the trivial cases where the reaction has the same reactants as products).

(b) It is false to say that a chemical reaction with K = 1.0 has $\Delta S° = 0$ (except in the trivial cases where the reaction has the same reactants as products).

(c) It is true to say that a chemical reaction with K = 1.0 has $\Delta G° = 0$, because ln(1.0) = 0.0.

(d) It is true to say that $\Delta H°$ and $\Delta S°$ have equal sign, since $\Delta H°$ must have the same sign as $T\Delta S°$ so that their difference gives $\Delta G° = 0$.

(e) It is true to say that $\Delta H°/T = \Delta S°$ at the temperature T. The relationship is derived when we solve this equation: $\Delta G° = \Delta H° - T\Delta S°$ for $\Delta S°$, then plug in $\Delta G° = 0$.

79. Use methods described in the answer to Question 60, then apply Equation 18.7:

$$\Delta G° = - RTlnK_p$$

(a) $\Delta G° = (1\ mol) \times \Delta G_f^°(H_2) + (1\ mol) \times \Delta G_f^°(Cl_2) - (2\ mol) \times \Delta G_f^°(HCl)$

$= (1\ mol) \times (0\ kJ/mol) + (1\ mol) \times (0\ kJ/mol) - (2\ mol) \times (-95.299\ kJ/mol) = 190.598\ kJ$

$$K_p = e^{(-\Delta G°/RT)} = e^{-\left(\frac{190.598\ kJ}{(0.008314\ kJ/mol\cdot K)\times(298\ K)}\right)} = e^{-76.9} = 4 \times 10^{-34}$$

(b) $\Delta G° = (1\ mol) \times \Delta G_f^°(NO) - (1\ mol) \times \Delta G_f^°(N_2) - (2\ mol) \times \Delta G_f^°(O_2)$

$= (2\ mol) \times (86.55\ kJ/mol) - (1\ mol) \times (0\ kJ/mol) - (1\ mol) \times (0\ kJ/mol) = 173.10\ kJ$

$$K_p = e^{(-\Delta G°/RT)} = e^{-\left(\frac{173.10\ kJ}{(0.008314\ kJ/mol\cdot K)\times(298\ K)}\right)} = e^{-69.9} = 5 \times 10^{-31}$$

81. Use methods described in the answer to Question 79:

$$\Delta G° = - RTlnK_p$$

(a) $\Delta G° = (1\ mol) \times \Delta G_f^°(H_3C–CH_3) - (1\ mol) \times \Delta G_f^°(H_2C=CH_2) - (1\ mol) \times$

$\Delta G_f^°(H_2)$

$= (1\ mol) \times (-32.82\ kJ/mol) - (1\ mol) \times (-68.15\ kJ/mol) - (1\ mol) \times (0\ kJ/mol) = -100.97$ kJ

The negative sign of $\Delta G°$ indicates that the reaction is product-favored.

(b) $K_p = e^{(-\Delta G°/RT)} = e^{-\left(\frac{-100.97\ kJ}{(0.008314\ kJ/mol\cdot K)\times(298\ K)}\right)} = e^{+40.8} = 5 \times 10^{17}$

When $\Delta G°$ is positive, K is less than 1. When $\Delta G°$ is negative, K is greater than 1.

83. Adapt the methods described in the answers to Questions 47 and 79.

(a) $\Delta H° = (2 \text{ mol}) \times \Delta H_f^°(NO_2) - (2 \text{ mol}) \times \Delta H_f^°(NO) - (1 \text{ mol}) \times \Delta H_f^°(O_2)$

$= (2 \text{ mol}) \times (33.18 \text{ kJ/mol}) - (2 \text{ mol}) \times (90.25 \text{ kJ/mol}) - (1 \text{ mol}) \times (0 \text{ kJ/mol})$

$= -114.14 \text{ kJ}$

$\Delta S° = (2 \text{ mol}) \times S°(NO_2) - (2 \text{ mol}) \times S°(NO) - (1 \text{ mol}) \times S°(O_2)$

$= (2 \text{ mol}) \times (240.06 \text{ J K}^{-1}\text{mol}^{-1}) - (2 \text{ mol}) \times (210.76 \text{ J K}^{-1}\text{mol}^{-1})$

$- (1 \text{ mol}) \times (205.138 \text{ J K}^{-1}\text{mol}^{-1}) = -146.54 \text{ J K}^{-1}$

$\Delta G_{298}^° = \Delta H° - T\Delta S° = (-114.14 \text{ kJ}) - (298 \text{ K}) \times (-146.54 \text{ J K}^{-1}) \times \dfrac{1 \text{ kJ}}{1000 \text{ J}} = -70.5 \text{ kJ}$

$K_{298}^° = e^{(-\Delta G°/RT)} = e^{-\left(\dfrac{-70.5 \text{ kJ}}{(0.008314 \text{ kJ}/\text{mol·K}) \times (298 \text{ K})}\right)} = e^{+28.4} = 2 \times 10^{12}$

$\Delta G_{1000}^° = \Delta H° - T\Delta S° = (-114.14 \text{ kJ}) - (1000. \text{ K}) \times (-146.54 \text{ J K}^{-1}) \times \dfrac{1 \text{ kJ}}{1000 \text{ J}} = 32.4 \text{ kJ}$

$K_{1000}^° = e^{(-\Delta G°/RT)} = e^{-\left(\dfrac{32.4 \text{ kJ}}{(0.008314 \text{ kJ}/\text{mol·K}) \times (1000. \text{ K})}\right)} = e^{-3.89} = 2.0 \times 10^{-2}$

(b) $\Delta H° = (2 \text{ mol}) \times \Delta H_f^°(NaCl) - (2 \text{ mol}) \times \Delta H_f^°(Na) - (1 \text{ mol}) \times \Delta H_f^°(Cl_2)$

$= (2 \text{ mol}) \times (-411.153 \text{ kJ/mol}) - (2 \text{ mol}) \times (0 \text{ kJ/mol}) - (1 \text{ mol}) \times (0 \text{ kJ/mol})$

$= -822.306 \text{ kJ}$

$\Delta S° = (2 \text{ mol}) \times S°(NaCl) - (2 \text{ mol}) \times S°(Na) - (1 \text{ mol}) \times S°(Cl_2)$

$= (2 \text{ mol}) \times (72.13 \text{ J K}^{-1}\text{mol}^{-1}) - (2 \text{ mol}) \times (51.21 \text{ J K}^{-1}\text{mol}^{-1})$

$- (1 \text{ mol}) \times (223.066 \text{ J K}^{-1}\text{mol}^{-1}) = -181.23 \text{ J K}^{-1}$

$\Delta G_{298}^° = \Delta H° - T\Delta S° = (-822.306 \text{ kJ}) - (298 \text{ K}) \times (-181.23 \text{ J K}^{-1}) \times \dfrac{1 \text{ kJ}}{1000 \text{ J}} = -768.3 \text{ kJ}$

$K_{298}^° = e^{(-\Delta G°/RT)} = e^{-\left(\dfrac{-768.3 \text{ kJ}}{(0.008314 \text{ kJ}/\text{mol·K}) \times (298 \text{ K})}\right)} = e^{+310.} = 10^{135}$

$$\Delta G^{\circ}_{1000} = \Delta H^{\circ} - T\Delta S^{\circ} = (-822.306 \text{ kJ}) - (1000. \text{ K}) \times (-181.23 \text{ J K}^{-1}) \times \frac{1 \text{ kJ}}{1000 \text{ J}}$$

$$= -641.08 \text{ kJ}$$

$$K^{\circ}_{1000} = e^{(-\Delta G^{\circ}/RT)} = e^{-\left(\frac{-641.1 \text{ kJ}}{(0.008314 \text{ kJ/mol·K}) \times (1000. \text{ K})}\right)} = e^{77.1} = 3 \times 10^{33}$$

86. Use methods described in the answer to Question 79. In part (c), we will also use Equation 14.5 as in the answer to Question 14.26. Use the appropriate value of R in each equation.

(a) $\Delta G^{\circ} = -RT\ln K_p = -(0.008314 \text{ kJ/mol·K}) \times (298 \text{ K}) \times \ln(4.4 \times 10^{18}) = -106$ kJ/mol

(b) $\Delta G^{\circ} = -RT\ln K_p = -(0.008314 \text{ kJ/mol·K}) \times (298 \text{ K}) \times \ln(3.17 \times 10^{-2}) = 8.55 \text{ kJ/mol}$

(c) Get K_p from K_c: $K_p = (RT)^{\Delta n} K_c$

$\Delta n = 2$ mol product gas $- 4$ mol reactant gas $= -2$

$$\Delta G^{\circ} = -RT\ln K_p = -RT\ln[(RT)^{-2} K_c] = -(0.008314 \text{ kJ/mol·K}) \times (298 \text{ K}) \times$$

$$\ln[(0.08206 \text{ L·atm/mol·K} \times 298 \text{ K})^{-2} \times (3.5 \times 10^8)] = -33.8 \text{ kJ/mol}$$

88. Use methods described in the answer to Questions 79 and 86.

(a) $\Delta G^{\circ} = (2 \text{ mol}) \times \Delta G^{\circ}_f(\text{NOCl}) + (2 \text{ mol}) \times \Delta G^{\circ}_f(\text{NO}) - (1 \text{ mol}) \times \Delta G^{\circ}_f(\text{Cl}_2)$

$= (2 \text{ mol}) \times (66.08 \text{ kJ/mol}) + (2 \text{ mol}) \times (86.55 \text{ kJ/mol})$

$- (1 \text{ mol}) \times (0 \text{ kJ/mol}) = -40.94 \text{ kJ}$

$$K_p = e^{(-\Delta G^{\circ}/RT)} = e^{-\left(\frac{-40.94 \text{ kJ}}{(0.008314 \text{ kJ/mol·K}) \times (298 \text{ K})}\right)} = e^{16.5} = 10^{7.16} = 1.5 \times 10^7$$

(b) The negative Gibbs free energy change indicates that the reaction is product-favored.

(c) Get K_c from K_p: $K_p = (RT)^{\Delta n} K_c$

$\Delta n = 2$ mol product gas $- 3$ mol reactant gas $= -1$

$$K_c = (RT)^{-\Delta n} K_p = (0.08206 \text{ L·atm/mol·K} \times 298 \text{ K})^{-(-1)} (1.5 \times 10^7) = 3.7 \times 10^8$$

Gibbs Free Energy, Maximum Work, and Energy Resources

90. In Section 10.1 (page 410), we were told that for a gaseous substance the standard thermodynamic properties are given for a gas pressure of 1 bar. If we also assume that 25 °C is actually 25.00 °C, or 298.15 K, then we can use the standard thermodynamic table in Appendix J. We will calculate the $\Delta G°$, the Gibbs free energy change, which is a measure of maximum useful work. If it is negative, the reaction can be harnessed to do useful work. If it is positive, the reaction cannot be harnessed to do useful work.

(a) $\Delta G° = (12 \text{ mol}) \times \Delta G_f°(CO_2) + (6 \text{ mol}) \times \Delta G_f°(H_2O(g))$

$$- (2 \text{ mol}) \times \Delta G_f°(C_6H_6(\ell)) - (15 \text{ mol}) \times \Delta G_f°(O_2)$$

$$= (12 \text{ mol}) \times (-394.359 \text{ kJ/mol}) + (6 \text{ mol}) \times (-228.572 \text{ kJ/mol})$$

$$- (2 \text{ mol}) \times (124.5 \text{ kJ/mol}) - (15 \text{ mol}) \times (0 \text{ kJ/mol}) = -6352.7 \text{ kJ}$$

The reaction can be harnessed to do useful work

(b) $\Delta G° = (1 \text{ mol}) \times \Delta G_f°(N_2) + (3 \text{ mol}) \times \Delta G_f°(F_2) - (2 \text{ mol}) \times \Delta G_f°(NF_3)$

$$= (1 \text{ mol}) \times (0 \text{ kJ/mol}) + (3 \text{ mol}) \times (0 \text{ kJ/mol}) - (2 \text{ mol}) \times (-83.2 \text{ kJ/mol}) = 166.4 \text{ kJ}$$

The reaction requires work to be done.

(c) $\Delta G° = (1 \text{ mol}) \times \Delta G_f°(Ti) + (3 \text{ mol}) \times \Delta G_f°(O_2) - (2 \text{ mol}) \times \Delta G_f°(TiO_2)$

$$= (1 \text{ mol}) \times (0 \text{ kJ/mol}) + (3 \text{ mol}) \times (0 \text{ kJ/mol})$$

$$- (1 \text{ mol}) \times (-884.5 \text{ kJ/mol}) = 884.5 \text{ kJ}$$

The reaction requires work to be done.

93. Balance the equations, then use the method described in the answer to Question 59.

(a) $\qquad 7 \text{ C(s)} + 6 \text{ H}_2\text{O(g)} \longrightarrow 2 \text{ C}_2\text{H}_6\text{(g)} + 3 \text{ CO}_2\text{(g)}$

$\qquad\qquad 5 \text{ C(s)} + 4 \text{ H}_2\text{O(g)} \longrightarrow \text{C}_3\text{H}_8\text{(g)} + 2 \text{ CO}_2\text{(g)}$

$\qquad\qquad 3 \text{ C(s)} + 4 \text{ H}_2\text{O(g)} \longrightarrow 2 \text{ CH}_3\text{OH}(\ell) + \text{CO}_2\text{(g)}$

(b) C_2H_6 reaction:

$$\Delta X° = (2 \text{ mol}) \times X°(C_2H_6) + (3 \text{ mol}) \times X°(CO_2)$$

$$- (7 \text{ mol}) \times X°(C) - (6 \text{ mol}) \times X°(H_2O(g)) \qquad (X = \Delta H_f, S, \text{ or } \Delta G_f)$$

$\Delta H° \quad = (2\ \text{mol}) \times (-84.68\ \text{kJ/mol}) + (3\ \text{mol}) \times (-393.509\ \text{kJ/mol})$

$\qquad - (7\ \text{mol}) \times (0\ \text{kJ/mol}) - (6\ \text{mol}) \times (-241.818\ \text{kJ/mol}) = 101.02\ \text{kJ}$

$\Delta S° = (2\ \text{mol}) \times (229.60\ \text{J K}^{-1}\text{mol}^{-1}) + (3\ \text{mol}) \times (213.74\ \text{J K}^{-1}\text{mol}^{-1})$

$\qquad - (7\ \text{mol}) \times (5.740\ \text{J K}^{-1}\text{mol}^{-1}) - (6\ \text{mol}) \times (188.825\ \text{J K}^{-1}\text{mol}^{-1}) = -72.71\ \text{J K}^{-1}$

$\Delta G° \quad = (2\ \text{mol}) \times (-32.82\ \text{kJ/mol}) + (3\ \text{mol}) \times (-394.359\ \text{kJ/mol})$

$\qquad - (7\ \text{mol}) \times (0\ \text{kJ/mol}) - (6\ \text{mol}) \times (-228.572\ \text{kJ/mol}) = 122.72\ \text{kJ}$

C_3H_8 reaction:

$\Delta X° = (1\ \text{mol}) \times X°(C_3H_8) + (2\ \text{mol}) \times X°(CO_2)$

$\qquad - (5\ \text{mol}) \times X°(C) - (4\ \text{mol}) \times X°(H_2O(g)) \qquad (X = \Delta H_f,\ S,\ \text{or}\ \Delta G_f)$

$\Delta H° = (1\ \text{mol}) \times (-103.8\ \text{kJ/mol}) + (2\ \text{mol}) \times (-393.509\ \text{kJ/mol})$

$\qquad - (5\ \text{mol}) \times (0\ \text{kJ/mol}) - (4\ \text{mol}) \times (-241.818\ \text{kJ/mol}) = 76.5\ \text{kJ}$

$\Delta S° = (1\ \text{mol}) \times (269.9\ \text{J K}^{-1}\text{mol}^{-1}) + (2\ \text{mol}) \times (213.74\ \text{J K}^{-1}\text{mol}^{-1})$

$\qquad - (5\ \text{mol}) \times (5.740\ \text{J K}^{-1}\text{mol}^{-1}) - (4\ \text{mol}) \times (188.825\ \text{J K}^{-1}\text{mol}^{-1}) = -86.6\ \text{J K}^{-1}$

$\Delta G° \quad = (1\ \text{mol}) \times (-23.49\ \text{kJ/mol}) + (2\ \text{mol}) \times (-394.359\ \text{kJ/mol})$

$\qquad - (5\ \text{mol}) \times (0\ \text{kJ/mol}) - (4\ \text{mol}) \times (-228.572\ \text{kJ/mol}) = 102.08\ \text{kJ}$

CH_3OH reaction:

$\Delta X° = (2\ \text{mol}) \times X°(CH_3OH) + (1\ \text{mol}) \times X°(CO_2)$

$\qquad - (3\ \text{mol}) \times X°(C) - (4\ \text{mol}) \times X°(H_2O(g)) \qquad (X = \Delta H_f,\ S,\ \text{or}\ \Delta G_f)$

$\Delta H° = (2\ \text{mol}) \times (-238.66\ \text{kJ/mol}) + (1\ \text{mol}) \times (-393.509\ \text{kJ/mol})$

$\qquad - (3\ \text{mol}) \times (0\ \text{kJ/mol}) - (4\ \text{mol}) \times (-241.818\ \text{kJ/mol}) = 96.44\ \text{kJ}$

$\Delta S° = (2\ \text{mol}) \times (126.8\ \text{J K}^{-1}\text{mol}^{-1}) + (1\ \text{mol}) \times (213.74\ \text{J K}^{-1}\text{mol}^{-1})$

$\qquad - (3\ \text{mol}) \times (5.740\ \text{J K}^{-1}\text{mol}^{-1}) - (4\ \text{mol}) \times (188.825\ \text{J K}^{-1}\text{mol}^{-1}) = -305.2\ \text{J K}^{-1}$

$\Delta G° \quad = (2\ \text{mol}) \times (-166.27\ \text{kJ/mol}) + (1\ \text{mol}) \times (-394.359\ \text{kJ/mol})$

$\qquad - (3\ \text{mol}) \times (0\ \text{kJ/mol}) - (4\ \text{mol}) \times (-228.572\ \text{kJ/mol}) = 187.39\ \text{kJ}$

None of these is feasible. $\Delta G°$ is positive. In addition, $\Delta H°$ is positive and $\Delta S°$ is negative, suggesting that there is no temperature at which the products would be favored.

95. Balance the equations, then use the method described in the answer to Question 59.

(a)

$$2\ CuO(s) \longrightarrow 2\ Cu(s) + O_2(g)$$
$$CuO(s) + C(graphite) \longrightarrow Cu(s) + CO(g)$$

$$\Delta G° = (1\ mol) \times \Delta G_f°(Cu) + (1\ mol) \times \Delta G_f°(CO)$$
$$- (1\ mol) \times \Delta G_f°(CuO) - (1\ mol) \times \Delta G_f°(C)$$
$$= (1\ mol) \times (0\ kJ/mol) + (1\ mol) \times (-137.168\ kJ/mol)$$
$$- (1\ mol) \times (-129.7\ kJ/mol) - (1\ mol) \times (0\ kJ/mol) = -7.5\ kJ$$

(b)

$$2\ Ag_2O(s) \longrightarrow 4\ Ag(s) + O_2(g)$$
$$Ag_2O(s) + C(graphite) \longrightarrow 2\ Ag(s) + CO(g)$$

$$\Delta G° = (2\ mol) \times \Delta G_f°(Ag) + (1\ mol) \times \Delta G_f°(CO)$$
$$- (1\ mol) \times \Delta G_f°(Ag_2O) - (1\ mol) \times \Delta G_f°(C)$$
$$= (2\ mol) \times (0\ kJ/mol) + (1\ mol) \times (-137.168\ kJ/mol)$$
$$- (1\ mol) \times (-11.20\ kJ/mol) - (1\ mol) \times (0\ kJ/mol) = -125.97\ kJ$$

(c)

$$2\ HgO(s) \longrightarrow 2\ Hg(\ell) + O_2(g)$$

$$HgO(s) + C(graphite) \longrightarrow Hg(\ell) + CO(g)$$

$$\Delta G° = (1\ mol) \times \Delta G_f°(Hg) + (1\ mol) \times \Delta G_f°(CO)$$
$$- (1\ mol) \times \Delta G_f°(HgO) - (1\ mol) \times \Delta G_f°(C)$$
$$= (1\ mol) \times (0\ kJ/mol) + (1\ mol) \times (-137.168\ kJ/mol)$$
$$- (1\ mol) \times (-58.539\ kJ/mol) - (1\ mol) \times (0\ kJ/mol) = -78.63\ kJ$$

(d)

$$2\ MgO(s) \longrightarrow 2\ Mg(s) + O_2(g)$$

$$MgO(s) + C(graphite) \longrightarrow Mg(s) + CO(g)$$

$$\Delta G° = (1\ mol) \times \Delta G_f°(Mg) + (1\ mol) \times \Delta G_f°(CO)$$

$$- (1\ mol) \times \Delta G_f°(MgO) - (1\ mol) \times \Delta G_f°(C)$$

$$= (1 \text{ mol}) \times (0 \text{ kJ/mol}) + (1 \text{ mol}) \times (-137.168 \text{ kJ/mol})$$

$$- (1 \text{ mol}) \times (-569.43 \text{ kJ/mol}) - (1 \text{ mol}) \times (0 \text{ kJ/mol}) = 432.26 \text{ kJ}$$

(e) $2 \text{ PbO(s)} \longrightarrow 2 \text{ Pb(s)} + O_2(g)$

$$\text{PbO(s)} + \text{C(graphite)} \longrightarrow \text{Pb(s)} + \text{CO(g)}$$

$$\Delta G° = (1 \text{ mol}) \times \Delta G_f°(\text{Pb}) + (1 \text{ mol}) \times \Delta G_f°(\text{CO})$$

$$- (1 \text{ mol}) \times \Delta G_f°(\text{PbO}) - (1 \text{ mol}) \times \Delta G_f°(\text{C})$$

$$= (1 \text{ mol}) \times (0 \text{ kJ/mol}) + (1 \text{ mol}) \times (-137.168 \text{ kJ/mol})$$

$$- (1 \text{ mol}) \times (-187.89 \text{ kJ/mol}) - (1 \text{ mol}) \times (0 \text{ kJ/mol}) = 50.72 \text{ kJ}$$

The coupled reactions that have negative $\Delta G°$ can be used to produce the respective metals, so Cu, Ag, and Hg can be obtained by this method at 25 °C.

Gibbs Free Energy in Biological Systems

99. Adapt the method described in Section 8.6 as described in the answer to Question 8.40.

(a) Looking at the given ball-and-stick model structure (refer also to Section 1.1 for a description of the model) we see that we must break five moles of O–H bonds, seven moles of C–O bonds, seven moles of C–H bonds, and five moles of C–C bonds in glucose. We must also break six moles of O=O bonds. Two C=O bonds in each of six moles of CO_2 and two O–H bonds in each of six moles of H_2O are formed.

The enthalpy of the reaction is approximated by adding the energy required to break one mole of bonds of each type of the bonds broken, described by the bond energies (D), to the energy required to form one mole of bonds of each type of the bonds produced in the reactants, described by the negative of the bond energies (–D).

$\Delta H°$ $\cong$ (bonds broken in glucose) + (bonds broken in O_2)

$$+ \text{(bonds formed in } CO_2) + \text{(bonds formed in } H_2O)$$

$$= (5 \text{ mol} \times D_{O–H} + 7 \text{ mol} \times D_{C–O} + 7 \text{ mol} \times D_{C–H} + 5 \text{ mol} \times D_{C–C})$$

$$+ 6 \text{ mol} \times D_{O=O} - 6 \times (2 \text{ mol} \times D_{C=O}) - 6 \times (2 \text{ mol} \times D_{O–H})$$

Note that the number of moles multiplied by the energy per mole (in kJ/mol) gives the result for each term in kJ.

$$= 5\,D_{O-H} + 7\,D_{C-O} + 7\,D_{C-H} + 5\,D_{C-C} + 6\,D_{O=O} - 12\,D_{C=O} - 12\,D_{O-H}$$

$$= 7\,D_{C-O} + 7\,D_{C-H} + 5\,D_{C-C} + 6\,D_{O=O} - 12\,D_{C=O} - 7\,D_{O-H}$$

Now use the data in Table 8.2.

$$= 7 \times (336\text{ kJ}) + 7 \times (416\text{ kJ}) + 5 \times (356\text{ kJ}) + 6 \times (498\text{ kJ})$$

$$- 12 \times (803\text{kJ}) - 7 \times (467\text{ kJ}) = -2873\text{ kJ} \cong \Delta H°$$

(b) The actual $\Delta H°$ (−2816 kJ) is close to the estimated value in (a). Interactive forces in condensed phases (solid glucose and liquid water) are being neglected in this calculation, which could explain the discrepancy.

101. (a) This is a conversion factor problem:

$$\frac{197\text{ kJ produced}}{1\text{ mol glucose}} \times \frac{1\text{ mol ATP}}{30.5\text{ kJ needed}} = 6.46\text{ mol ATP per mol of glucose}$$

(b) This is a conversion factor problem:

$$\frac{3\text{ mol ATP produced}}{1\text{ mol glucose}} \times \frac{30.5\text{ kJ needed}}{1\text{ mol ATP}} = 91.5\text{ kJ per mol of glucose}$$

The actual reaction must be less exergonic than the given reaction to produce fewer ATP.

$$\Delta G°_{\text{overall reaction}} = \Delta G°_{\text{conversion}} + \Delta G°_{\text{ATP}}$$

$$-197\text{ kJ} + 91.5\text{ kJ} = -106\text{ kJ}$$

(c) The overall reaction in part (b) has a negative Gibbs free energy change, so it is product-favored.

Conservation of Gibbs Free Energy

103. Food we eat provides us with a supply of Gibbs free energy. Coal, petroleum, and natural gas are the most common fuel sources used to supply Gibbs free energy by combustion. We also use solar and nuclear energy, as well as the kinetic energy of wind and water.

Thermodynamic and Kinetic Stability

105. Balance the equation, when not provided, and adapt the methods from the answer to Question 59.

(a) $\Delta G° = (1 \text{ mol}) \times \Delta G_f°(CH_3COOH) - (1 \text{ mol}) \times \Delta G_f°(CH_3OH) - (1 \text{ mol}) \times \Delta G_f°(CO)$

$= (1 \text{ mol}) \times (-389.9 \text{ kJ/mol})$

$- (1 \text{ mol}) \times (-166.27 \text{ kJ/mol}) - (1 \text{ mol}) \times (-137.168 \text{ kJ/mol}) = -86.5 \text{ kJ}$

(b) $CH_3COOH(\ell) + 2\ O_2(g) \longrightarrow 2\ CO_2(g) + 2\ H_2O(\ell)$

$\Delta G° = (2 \text{ mol}) \times \Delta G_f°(CO_2) + (2 \text{ mol}) \times \Delta G_f°(H_2O(\ell))$

$- (1 \text{ mol}) \times \Delta G_f°(CH_3COOH) - (2 \text{ mol}) \times \Delta G_f°(O_2)$

$= (2 \text{ mol}) \times (-394.359 \text{ kJ/mol}) + (2 \text{ mol}) \times (-237.129 \text{ kJ/mol})$

$- (1 \text{ mol}) \times (-389.9 \text{ kJ/mol}) - (2 \text{ mol}) \times (0 \text{ kJ/mol}) = -873.1 \text{ kJ}$

(c) Based on the answer to (b), the products of oxidation are more stable, so acetic acid is not thermodynamically stable.

(d) Acetic acid can be kept both in liquid form and in solution form if stored properly. In the presence of air, it does not explode, so we will classify it as kinetically stable.

107. Kinetic stability relates to the difficulties in the conversion of the reactants to products. Even very stable products may be difficult to form from some reactants. The kinetically stable materials usually require extensive reorganization of bonds, which likely involve very high-energy activated complexes and intermediates.

110. Balance the equations, and adapt the methods from the answer to Question 60.

$$CH_4(g) + 2\ O_2(g) \longrightarrow CO_2(g) + 2\ H_2O(\ell)$$

$\Delta G° = (1 \text{ mol}) \times \Delta G_f°(CO_2) + (2 \text{ mol}) \times \Delta G_f°(H_2O(\ell))$

$- (1 \text{ mol}) \times \Delta G_f°(CH_4) - (2 \text{ mol}) \times \Delta G_f°(O_2)$

$= (1 \text{ mol}) \times (-394.359 \text{ kJ/mol}) + (2 \text{ mol}) \times (-237.129 \text{ kJ/mol})$

$- (1 \text{ mol}) \times (-50.72 \text{ kJ/mol}) - (2 \text{ mol}) \times (0 \text{ kJ/mol}) = -817.90 \text{ kJ}$

$$C_6H_6(g) + \frac{15}{2} O_2(g) \longrightarrow 6 CO_2(g) + 3 H_2O(\ell)$$

$$\Delta G° = (6 \text{ mol}) \times \Delta G\,°_f(CO_2) + (3 \text{ mol}) \times \Delta G\,°_f(H_2O(\ell))$$

$$- (1 \text{ mol}) \times \Delta G\,°_f(CH_4) - (\frac{15}{2} \text{ mol}) \times \Delta G\,°_f(O_2)$$

$$= (6 \text{ mol}) \times (-394.359 \text{ kJ/mol}) + (3 \text{ mol}) \times (-237.129 \text{ kJ/mol})$$

$$- (1 \text{ mol}) \times (124.5 \text{ kJ/mol}) - (\frac{15}{2} \text{ mol}) \times (0 \text{ kJ/mol}) = -3202.0 \text{ kJ}$$

$$CH_3OH(\ell) + \frac{3}{2} O_2(g) \longrightarrow CO_2(g) + 2 H_2O(\ell)$$

$$\Delta G° = (1 \text{ mol}) \times \Delta G\,°_f(CO_2) + (2 \text{ mol}) \times \Delta G\,°_f(H_2O(\ell))$$

$$- (1 \text{ mol}) \times \Delta G\,°_f(CH_3OH) - (\frac{3}{2} \text{ mol}) \times \Delta G\,°_f(O_2)$$

$$= (1 \text{ mol}) \times (-394.359 \text{ kJ/mol}) + (2 \text{ mol}) \times (-237.129 \text{ kJ/mol})$$

$$- (1 \text{ mol}) \times (-166.27 \text{ kJ/mol}) - (\frac{3}{2} \text{ mol}) \times (0 \text{ kJ/mol}) = -702.34 \text{ kJ}$$

Organic compounds are complex molecular systems that require significant rearrangement of atoms and bonds to undergo combustion. This makes them likely candidates for being kinetically stable (See the answer to Question 107 from this chapter.)

General Questions

111. Use standard conversion factors to answer this question.

Agriculture, mining and construction industries are represented on the graph with approximately 5.2 quadrillion BTUs per year. One quadrillion is a thousand times more than a trillion, or 10^{15}. One BTU is defined in Appendix B, in Table B.4., as 1055.06 J.

(a)
$$\frac{5.2 \times 10^{15} \text{ BTU}}{\text{yr}} \times \frac{1055.06 \text{ J}}{1 \text{ BTU}} = \frac{5.5 \times 10^{18} \text{ J}}{\text{yr}}$$

(b)
$$\frac{5.5 \times 10^{18} \text{ J}}{\text{yr}} \times \frac{1 \text{ year}}{365 \text{ day}} = \frac{1.5 \times 10^{16} \text{ J}}{\text{day}}$$

(c)
$$\frac{1.5 \times 10^{16} \text{ J}}{\text{day}} \times \frac{1 \text{ day}}{24 \text{ hr}} \times \frac{1 \text{ hr}}{3600 \text{ s}} = \frac{1.7 \times 10^{11} \text{ J}}{\text{s}}$$

(d)
$$\frac{1.7 \times 10^{11} \text{ J}}{\text{s}} \times \frac{1 \text{ W}}{1 \text{ J}/\text{s}} = 1.7 \times 10^{11} \text{ W}$$

(e)
$$\frac{1.7 \times 10^{11} \text{ W}}{300 \times 10^{6} \text{ persons}} = 6 \times 10^{2} \frac{\text{W}}{\text{person}}$$

112. Use standard conversion factors to answer this question.

(a)
$$\frac{1.5 \times 10^{16} \text{ J}}{\text{day}} \times \frac{1 \text{ kJ}}{1000 \text{ J}} \times \frac{1 \text{ mol glucose}}{16.7 \text{ kJ}} \times \frac{180.16 \text{ g glucose}}{1 \text{ mol glucose}} \times \frac{1 \text{ kg}}{1000 \text{ g}}$$

$$= \frac{1.6 \times 10^{11} \text{ kg glucose}}{\text{day}}$$

(b) According to the answer in Question 111.(d), I must generate 1.7×10^{11} W, when a sprinter manages to muster up only 900 W during a sprint. That means I must generate energy two hundred million times faster than a sprinter does while sprinting. I cannot uphold this contract.

114. Adapt the method described in the answer to Question 86(c) and concepts described in Sections 14.2 and 14.6.

(a) Because of the relationship: $K_p = (RT)^{\Delta n} K_c$, all we need to do is check Δn for each reaction. K_p is larger than K_c when Δn is positive, as long as RT >1 (RT <1 only below T = 12 K!)

Reaction	Δn	$K_p > K_c$
1	2 mol product gas – 2 mol reactant gas = 0	No
2	1 mol product gas – 0 mol reactant gas = 1	Yes
3	2 mol product gas – 2 mol reactant gas = 0	No
4	1 mol product gas – 3 mol reactant gas = –2	No
5	2 mol product gas – 2 mol reactant gas = 0	No

Only reaction 2 has K_p larger than K_c.

(b) Now we actually need the value of K_p, because $K_p = K° > 1$ for a gas-phase reaction that is product-favored.

$$(RT) = (0.08206 \text{ L atm K}^{-1}\text{mol}^{-1})(298 \text{ K}) = 24.5$$

$$K_p = (RT)^{\Delta n}K_c$$

Reaction	K_c	$(RT)^{\Delta n}$	K_p	Product-favored
1	3.6×10^{20}	1	3.6×10^{20}	Yes
2	1.24×10^{-5}	$(24.5)^1$	3.03×10^{-4}	No
3	9.5×10^{-13}	1	9.5×10^{-13}	No
4	3.76	$(24.5)^{-2}$	6.29×10^{-3}	No
5	2×10^9	1	2×10^9	Yes

Only reactions 1 and 5 are product-favored.

(c) Every gas contributes a concentration to the K_c expression, as described in Section 14.2. Only one reaction given has just one gas-phase component: Reaction 2.

(d) The product concentrations get higher when the value of K_c increases. In Section 14.6, we learned that the K_c increases with an increase in temperature when the reaction is endothermic. So, reactions 2 and 3 will have an increase in the product concentrations when the temperature increases.

(e) Only two reactions (1 and 2) have a product that is H_2O. The difference between $\Delta G_f°(H_2O(\ell))$ and $\Delta G_f°(H_2O(g))$ is 8.557 kJ/mol, so the switch would make the reaction $\Delta G°$ less negative by 8.557 kJ. If these reactions have positive $\Delta G°$ between 8.557 kJ and zero, then the sign would switch. If these reactions have negative $\Delta G°$ values with gas-phase water, then the $\Delta G°$ will still be negative with liquid water.

Reaction	K_p	$\Delta G° = -RT\ln K°$	$\Delta G°$ sign affected by (ℓ) to (g)?
1	3.6×10^{20}	-117	No
2	3.03×10^{-4}	20.1	No

None of these has a $\Delta G°$ whose sign is affected by a switch from $H_2O(g)$ to $H_2O(\ell)$.

116.
$$2\ SO_3(g) \longrightarrow 2\ SO_2(g) + O_2(g)$$

(a) $\Delta G° = (2\ mol) \times \Delta G_f°(SO_2) + (1\ mol) \times \Delta G_f°(O_2) - (2\ mol) \times \Delta G_f°(SO_3)$

$= (2\ mol) \times (-300.194\ kJ/mol) + (1\ mol) \times (0\ kJ/mol)$

$- (2\ mol) \times (-371.06\ kJ/mol) = 141.73\ kJ$

(b) The positive Gibbs free energy change indicates that the reaction is not product-favored at 25 °C.

(c) $\Delta H° = (2\ mol) \times \Delta H_f°(SO_2) + (1\ mol) \times \Delta H_f°(O_2) - (2\ mol) \times \Delta H_f°(SO_3)$

$= (2\ mol) \times (-296.830\ kJ/mol) + (1\ mol) \times (0\ kJ/mol)$

$- (2\ mol) \times (-395.72\ kJ/mol) = 197.78\ kJ$

$\Delta S° = (2\ mol) \times S°(SO_2) + (1\ mol) \times S°(O_2) - (2\ mol) \times S°(SO_3)$

$= (2\ mol) \times (248.22\ J\ K^{-1}mol^{-1}) + (1\ mol) \times (205.138\ J\ K^{-1}mol^{-1})$

$- (2\ mol) \times (256.76\ J\ K^{-1}mol^{-1}) = 188.06\ J\ K^{-1}$

With both $\Delta H°$ and $\Delta S°$ positive, the reaction will be product-favored at some high temperature, one higher than 25 °C.

(d) $T = 1500\ °C + 273.15 = 1773\ K = 1.8 \times 10^3\ K$ *(round to hundreds place)*

$$\Delta G_{1773}° = \Delta H° - T\Delta S° = (197.78\ kJ) - (1.8 \times 10^3\ K) \times (188.06\ J\ K^{-1}) \times \frac{1\ kJ}{1000\ J}$$

$$= -1.4 \times 10^2\ kJ$$

$$K_{1773}° = e^{(-\Delta G°/RT)} = e^{-\left(\frac{-1.4 \times 10^2\ kJ}{(0.008314\ kJ/mol \cdot K) \times (1.8 \times 10^3\ K)}\right)} = e^{+9.2} = 1 \times 10^4$$

(e) $$K_c = (RT)^{-\Delta n} K_p \qquad \Delta n = 3\ mol - 2\ mol = 1\ mol$$

$$K_c = [\ (0.08206\ L\ atm\ K^{-1}mol^{-1}) \times (1.8 \times 10^3\ K)]^{-1} \times (1 \times 10^4) = 7 \times 10^1$$

118. Use Equation 18.7 and Equation 18.2:

$$\Delta G° = -RT\ln K = \Delta H° - T\Delta S°$$

Divide everything by $-RT$:

$$\ln K = -\frac{\Delta H°}{RT} + \frac{\Delta S°}{R}$$

If lnK is plotted against 1/T, the slope would be $-\Delta H°/R$, and the y-intercept would be $\Delta S°/R$. A straight line on this graph proves that the quantities of $\Delta H°$ and $\Delta S°$ are independent of temperature.

120. Use the method described in the answer to Question 79. Then use the equation derived in the answer to Question 119.

(a) $\Delta G° = (1 \text{ mol}) \times \Delta G_f^°(Hg(g)) - (1 \text{ mol}) \times \Delta G_f^°(Hg(\ell))$

$$= (1 \text{ mol}) \times (31.8 \text{ kJ/mol}) - (1 \text{ mol}) \times (0 \text{ kJ/mol}) = 31.8 \text{ kJ}$$

(b) $K° = K_p = P_{Hg(\ell)}$

(c) $K° = e^{(-\Delta G°/RT)} = e^{-\left(\dfrac{31.8 \text{ kJ/mol}}{(0.008314 \text{ kJ/mol·K}) \times (298 \text{ K})}\right)} = e^{-12.8} = 10^{-5.56}$

$$= 2.7 \times 10^{-6}$$

(d) $K° = K_p = P_{Hg(\ell)} = 2.7 \times 10^{-6} \text{ atm}$

(e) $K°_1 = 2.7 \times 10^{-6} \text{ atm at } T_1 = 298 \text{ K}, K°_2 = P_{Hg(\ell)} = 10 \text{ mmHg at } T_2$

$$\Delta H° = (1 \text{ mol}) \times \Delta H_f^°(Hg(g)) - (1 \text{ mol}) \times \Delta H_f^°(Hg(\ell))$$

$$= (1 \text{ mol}) \times (61.4 \text{ kJ/mol}) - (1 \text{ mol}) \times (0 \text{ kJ/mol}) = 61.4 \text{ kJ}$$

$$\ln\left(\frac{K_1^°}{K_2^°}\right) = \frac{\Delta H°}{R}\left(\frac{1}{T_2} - \frac{1}{T_1}\right)$$

$$\ln\left(\frac{2.7 \times 10^{-6} \text{ atm}}{10 \text{ mmHg} \times \dfrac{1 \text{ atm}}{760 \text{ mmHg}}}\right) = \frac{61.4 \text{ kJ/mol}}{0.008314 \text{ kJ/mol·K}}\left(\frac{1}{T_2} - \frac{1}{298 \text{ K}}\right)$$

Solve for T_2: 　　　　　　　　　$T_2 = 450 \text{ K}$

Applying Concepts

122. A scrambled egg is a very disordered state for an egg. The second law of thermodynamics says that the more disordered state is the more probable state. Putting the delicate tissues and fluids back where they were before the scrambling occurred would take a great deal of energy. Humpty Dumpty is a fictional character who was also an egg. He fell off a wall. A very probable result of that fall is for an egg to become scrambled. The story goes on to tell that all the energy of the king's horses and men were not sufficient to put Humpty together again.

125. Many of the oxides have negative enthalpies of reaction, which means their oxidations are exothermic. These are probably product-favored reactions.

127. Determine the qualitative changes in $\Delta H°$ and $\Delta S°$, then refer to Table 18.2

(a) Using the hint at the end of the problem, we look up bond energies in Chapter 6. We find that forming a bond is exothermic ($\Delta H°$ = negative). Two moles of gas-phase reactants form one mole of gas-phase products, so the entropy decreases ($\Delta S°$ = negative). That means this reaction is (ii) product-favored at low temperatures but not at high temperatures.

(b) A combustion reaction for a hydrocarbon is exothermic ($\Delta H°$ = negative). Nine moles of gas-phase reactants form eleven moles of gas-phase products, so the entropy increases ($\Delta S°$ = positive). That means this reaction is (i) always product-favored.

(c) Going farther with the hint at the end of the problem, we can look up actual bond energies in Chapter 8 (Table 8.2). We conclude that forming very strong P–F bonds (490 kJ/mol) produces more energy than is used breaking the weak P–P (226 kJ/mol) and F–F (158 kJ/mol) bonds. Hence, we will predict that the reaction is exothermic ($\Delta H°$ = negative). Eleven moles of gas-phase reactants form four moles of gas-phase products, so the entropy decreases ($\Delta S°$ = negative). That means this reaction is (ii) product-favored at low temperatures but not at high temperatures.

129. NaCl, in an orderly crystal structure, and pure water, with only O–H hydrogen bonding interactions in the liquid state, are far more ordered than the dispersed hydrated sodium and chloride ions interacting with the water molecules.

131. $\Delta G < 0$ means products are favored; however, the equilibrium state will always have some reactants present, too. To get all the reactants to go away requires the removal of the products from the reactants, so that the reaction continues forward.

Chapter 19: Electrochemistry and Its Applications

Redox Reactions

6. Follow the methods described in the answers to Questions 5.35 to 5.44

	Substance oxidized	Substance reduced	Oxidizing agent	Reducing agent	Reactant oxidation numbers	Product oxidation numbers
(a)	Al	Cl in Cl_2	Cl_2	Al	Ox. # Al = 0 Ox. # Cl = 0	Ox. # Al = +3 Ox. # Cl = –1
(b)	Fe^{2+}	Mn in MnO_4^-	MnO_4^-	Fe^{2+}	Ox. # H = +1 Ox. # O = –2 Ox. # Mn = +7 Ox. # Fe = +2	Ox. # H = +1 Ox. # O = –2 Ox. # Mn = +2 Ox. # Fe = +3
(c)	Fe and S in FeS	N in NO_3^-	NO_3^-	FeS	Ox. # Fe = +2 Ox. # S = –2 Ox. # N = +5 Ox. # O = –2 Ox. # H = +1	Ox. # Fe = +3 Ox. # S = +6 Ox. # N = +2 Ox. # O = –2 Ox. # H = +1

Using Half-Reactions to Understand Redox Reactions

10. Follow the methods described in Section 5.3 and Problem-Solving Example 19.3.

(a) Balance Zn atoms, then balance charge with electrons:

$$Zn(s) \longrightarrow Zn^{2+}(aq) + 2\ e^-$$ (Check: 1 Zn, zero net charge)

(b) Balance H atoms, then balance O atoms with H_2O, then balance charge with electrons:

$$2\ H_3O^+(aq) + 2\ e^- \longrightarrow 2\ H_2O(\ell) + H_2(g)$$ (Check: 6 H, 2 O, zero net charge)

(c) Balance Sn atoms, then balance charge with electrons:

$$Sn^{4+}(aq) + 2\ e^- \longrightarrow Sn^{2+}(aq)$$ (Check: 1 Sn, +2 net charge)

(d) Balance Cl atoms, then balance charge with electrons:

$$Cl_2(g) + 2\ e^- \longrightarrow 2\ Cl^-(aq)$$ (Check: 2 Cl, –2 net charge)

(e) Balance S atoms, then balance O atoms with H_2O, then balance H atoms with H^+, then balance charge with electrons:

$$2\ H_2O(\ell) + SO_2(g) \longrightarrow SO_4^{2-}(aq) + 4\ H^+(aq) + 2\ e^-$$

Then add four water molecules to each side to convert the H^+ ions into H_3O^+ ions:

$$6\ H_2O(\ell) + SO_2(g) \longrightarrow SO_4^{2-}(aq) + 4\ H_3O^+(aq) + 2\ e^-$$

(Check: 12 H, 8 O, 1 S, zero net charge)

12. Follow the methods described in the answer to Question 10, Section 5.3, and Problem-Solving Example 19.3.

The chemicals for oxidation all go in one half-reaction and the chemicals for reduction all go in the other half-reaction. We do not have to figure out where the H_3O^+ ions and H_2O molecules go; they will show up where they belong as we balance the half-reactions.

(a) Put Al and Al^{3+} in one half-reaction. Put Cl_2 and Cl^- in the other. Balance as described in Questions 10.

$$Al(s) \longrightarrow Al^{3+}(aq) + 3\ e^-$$

$$Cl_2(g) + 2\ e^- \longrightarrow 2\ Cl^-(aq)$$

(b) Put Fe^{2+} and Fe^{3+} in one half-reaction. Put MnO_4^- and Mn^{2+} in the other. Balance as described in Questions 10.

$$Fe^{2+}(aq) \longrightarrow Fe^{3+}(aq) + e^-$$

$$MnO_4^-(aq) + 8\ H_3O^+(aq) + 5\ e^- \longrightarrow Mn^{2+}(aq) + 12\ H_2O(\ell)$$

(Check: 1 Mn, 12 O, 24 H, +2 net charge)

(c) Put FeS, $Fe^{3+}(aq)$ and SO_4^{2-} in one half-reaction. Put NO_3^- and NO in the other. Balance as described in Questions 10.

$$FeS(s) + 12\ H_2O(\ell) \longrightarrow Fe^{3+}(aq) + SO_4^{2-}(aq) + 8\ H_3O^+(aq) + 9\ e^-$$

(Check: 1 Fe, 1 S, 12 O, 24 H, zero net charge)

$$NO_3^-(aq) + 4\ H_3O^+(aq) + 3\ e^- \longrightarrow NO(g) + 6\ H_2O(\ell)$$

(Check: 1 N, 7 O, 12 H, zero net charge)

14. Follow the methods described in the answer to Questions 6, 10, 12, Section 5.3, and Problem-Solving Example 19.3. After the half-reactions are separated and balanced, equalize the electrons with appropriate multipliers and add the two half-reactions. The conversion of the H^+ ions into H_3O^+ ions can wait until the redox reaction is balanced.

(a) Put CO and CO_2 in one half-reaction. Put O_3 in the other.

$$3\ H_2O(\ell) + 3\ CO(g) \longrightarrow 3\ CO_2(g) + 6\ H^+(aq) + 6\ e^-$$

$$\underline{6\ e^- + 6\ H^+(aq) + O_3(g) \longrightarrow 3\ H_2O(\ell)}$$

$$3\ CO(g) + O_3(g) \longrightarrow 3\ CO_2(g)$$

O_3 is the oxidizing agent. CO is the reducing agent.

(b) This reaction is trivial to balance the old-fashioned way:

$$H_2(g) + Cl_2(g) \longrightarrow 2\ HCl(g)$$

Cl_2 is the oxidizing agent. H_2 is the reducing agent.

(c) Put H_2O_2 in one half-reaction. Put Ti^{2+} and Ti^{4+} in the other.

$$2\ e^- + 2\ H^+(aq) + H_2O_2(aq) \longrightarrow 2\ H_2O(\ell)$$

$$\underline{Ti^{2+}(aq) \longrightarrow Ti^{4+}(aq) + 2\ e^-}$$

$$2\ H^+(aq) + H_2O_2(aq) + Ti^{2+}(aq) \longrightarrow 2\ H_2O(\ell) + Ti^{4+}(aq)$$

Then add two water molecules to each side to convert the H^+ ions into H_3O^+ ions:

$$2\ H_3O^+(aq) + H_2O_2(aq) + Ti^{2+}(aq) \longrightarrow 4\ H_2O(\ell) + Ti^{4+}(aq)$$

(Check: 8 H, 4 O, 1 Ti, +4 net charge)

H_2O_2 is the oxidizing agent. Ti^{2+} is the reducing agent

(d) Put MnO_4^- and MnO_2 in one half-reaction. Put Cl^- and Cl_2 in the other.

$$3\ e^- + MnO_4^-(aq) + 4\ H^+(aq) \longrightarrow MnO_2(s) + 2\ H_2O(\ell)$$

$$2\ Cl^-(aq) \longrightarrow Cl_2(g) + 2\ e^-$$

Multiply each reaction by a constant to get the same number of electrons:

$$2 \times [3\ e^- + MnO_4^-(aq) + 4\ H^+(aq) \longrightarrow MnO_2(s) + 2\ H_2O(\ell)]$$

$$3 \times [2\ Cl^-(aq) \longrightarrow Cl_2(g) + 2\ e^-]$$

Now add them:

$$6\ e^- + 2\ MnO_4^-(aq) + 8\ H^+(aq) \longrightarrow 2\ MnO_2(s) + 4\ H_2O(\ell)$$

$$6\ Cl^-(aq) \longrightarrow 3\ Cl_2(g) + 6\ e^-$$

$$2\ MnO_4^-(aq) + 6\ Cl^-(aq) + 8\ H^+(aq) \longrightarrow 2\ MnO_2(s) + 3\ Cl_2(g) + 4\ H_2O(\ell)$$

Then add two water molecules to each side to convert the H^+ ions into H_3O^+ ions:

$$2\ MnO_4^-(aq) + 6\ Cl^-(aq) + 8\ H_3O^+(aq) \longrightarrow 2\ MnO_2(s) + 3\ Cl_2(g) + 12\ H_2O(\ell)$$

(Check: 2 Mn, 16 O, 24 H, zero net charge)

MnO_4^- is the oxidizing agent. Cl^- is the reducing agent.

(e) Put FeS_2, Fe_2O_3 and SO_2 in one half-reaction. Put O_2 in the other.

$$11\ H_2O(\ell) + 2\ FeS_2(s) \longrightarrow Fe_2O_3(s) + 4\ SO_2(g) + 22\ H^+(aq) + 22\ e^-$$

$$4\ e^- + 4\ H^+(aq) + O_2(g) \longrightarrow 2\ H_2O(\ell)$$

Multiply each reaction by a constant to get the same number of electrons:

$$2 \times [11\ H_2O(\ell) + 2\ FeS_2(s) \longrightarrow Fe_2O_3(s) + 4\ SO_2(g) + 22\ H^+(aq) + 22\ e^-]$$

$$11 \times [4\ e^- + 4\ H^+(aq) + O_2(g) \longrightarrow 2\ H_2O(\ell)]$$

Now add them:

$$22\ H_2O(\ell) + 4\ FeS_2(s) \longrightarrow 2\ Fe_2O_3(s) + 8\ SO_2(g) + 44\ H^+(aq) + 44\ e^-]$$

$$44\ e^- + 44\ H^+(aq) + 11\ O_2(g) \longrightarrow 22\ H_2O(\ell)$$

$$4\ FeS_2(s) + 11\ O_2(g) \longrightarrow 2\ Fe_2O_3(s) + 8\ SO_2(g)$$

(Check: 4 Fe, 8 S, 22 O, zero net charge)

O_2 is the oxidizing agent. FeS_2 is the reducing agent.

(f) This reaction is already balanced: $O_3(g) + NO(g) \longrightarrow O_2(g) + NO_2(g)$

O_3 is the oxidizing agent. NO is the reducing agent.

(g) This reaction is already balanced:

$$Zn(Hg)(amalgam) + HgO(s) \longrightarrow ZnO(s) + 2\ Hg(\ell)$$

HgO is the oxidizing agent. Zn(Hg) is the reducing agent.

Electrochemical Cells

16. The generation of electricity occurs when electrons are transmitted through a wire from the metal to the cation. Here, the transfer of electrons would occur directly from the metal to the cation and the electrons would not flow through any wire.

18. Conventionally, in chemistry, they are written as reduction reactions. Engineers and physicists may still use an oxidation-reaction convention.

20. (a) $Zn(s) + Pb^{2+}(aq) \longrightarrow Zn^{2+}(aq) + Pb(s)$

(b) Oxidation of zinc atoms occurs at the anode, which is metallic zinc:

$$Zn(s) \longrightarrow Zn^{2+}(aq) + 2\ e^-$$

The reduction of lead(II) ions occurs at the cathode, which is metallic lead:

$$2\ e^- + Pb^{2+}(aq) \longrightarrow Pb(s)$$

(c)

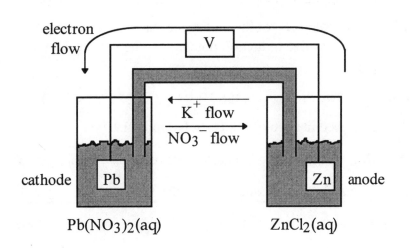

Electrochemical Cells and Voltage

22. This is a standard conversion factor problem.

$$\frac{1.0 \text{ hr}}{12 \text{ V}} \times \frac{60 \text{ min}}{1 \text{ hr}} \times \frac{60 \text{ s}}{1 \text{ min}} \times \frac{25 \text{ J}}{1 \text{ s}} \times \frac{1 \text{ C} \times 1 \text{ V}}{1 \text{ J}} = 7500 \text{ C}$$

24. (a) $Cu(s) \longrightarrow Cu^{2+}(aq) + 2 \text{ e}^-$

          $Ag^+(aq) + e^- \longrightarrow Ag(s)$

(b) The copper half-reaction is oxidation and it occurs in the anode compartment. The silver half-reaction is reduction and it occurs in the cathode compartment.

Using Standard Cell Potentials

26. On a chart (Table 19.1) where the standard reduction potentials are listed from most positive to most negative, the reactant of the first reaction is the strongest oxidizing agent, because it has the largest reduction potential. The reactant of the last reaction is the weakest oxidizing agent, because that reaction has the smallest reduction potential. The product of the first reaction is the weakest reducing agent, because that reaction, in reverse, has the smallest oxidation potential. The product of the last reaction is the strongest reducing agent, because that reaction, in reverse, has the largest oxidation potential. The Li is the strongest reducing agent and Li^+ is the weakest oxidizing agent. F_2 is the strongest oxidizing agent and F^- is the weakest reducing agent.

28. Look up the reduction potential for the half-reactions that have the given species as reactants. The largest positive reduction potential is the best at reducing and represents the best oxidizing agent.

$H_2O \ (-0.8277 \text{ V}) < PbSO_4 \ (-0.356 \text{ V}) < O_2 \ (+1.229 \text{ V}) < H_2O_2 \ (+1.77 \text{ V})$

30. Use the method described in Problem-Solving Example 19.7.

(a) $Mg(s) \longrightarrow Mg^{2+}(aq) + 2 \text{ e}^-$ oxidation half-reaction $E°_{ox} = -E°_{red} = +2.37 \text{ V}$

$I_2(s) + 2 \text{ e}^- \longrightarrow 2 \text{ I}^-(aq)$ reduction half-reaction $E°_{red} = +0.535 \text{ V}$

$I_2(s) + Mg(s) \longrightarrow Mg^{2+}(aq) + 2 \text{ I}^-(aq)$ $E°_{cell} = (+2.37 \text{ V}) + (+0.535 \text{ V}) = 2.91 \text{ V}$

This reaction is product-favored.

(b) $Ag(s) \longrightarrow Ag^+(aq) + e^-$ oxidation half-reaction $E°_{ox} = -E°_{red} = -0.80 \text{ V}$

$Fe^{3+}(aq) + e^- \longrightarrow Fe^{2+}(aq)$ reduction half-reaction $E°_{red} = +0.771 \text{ V}$

$Ag(s) + Fe^{3+}(aq) \longrightarrow Fe^{2+}(aq) + Ag^+(aq)$

$$E°_{cell} = (-0.80 \text{ V}) + (+0.771 \text{ V}) = -0.03 \text{ V}$$

This reaction is not product-favored.

(c) $Sn^{2+}(aq) \longrightarrow Sn^{4+}(aq) + 2 e^-$ oxidation half-reaction $E°_{ox} = -E°_{red} = -0.15 \text{ V}$

$2 Ag^+(aq) + 2 e^- \longrightarrow 2 Ag(s)$ reduction half-reaction $E°_{red} = +0.80 \text{ V}$

$Sn^{2+}(aq) + 2 Ag^+(aq) \longrightarrow Sn^{4+}(aq) + 2 Ag(s)$

$$E°_{cell} = (-0.15 \text{ V}) + (+0.80 \text{ V}) = 0.65 \text{ V}$$

This reaction is product-favored.

(d) $2 Zn(s) \longrightarrow 2 Zn^{2+}(aq) + 4 e^-$ oxidation half-reaction $E°_{ox} = -E°_{red} = +0.763 \text{ V}$

$O_2(s) + 2 H_2O(\ell) + 4 e^- \longrightarrow 4 OH^-(aq)$ reduction half-reaction $E°_{red} = +0.40 \text{ V}$

$2 Zn(s) + O_2(s) + 2 H_2O(\ell) \longrightarrow Zn^{2+}(aq) + 4 OH^-(aq)$

$$E°_{cell} = (+0.763 \text{ V}) + (+0.40 \text{ V}) = 1.16 \text{ V}$$

This reaction is product-favored.

32. Adapt the methods described in the answer to Question 26.

(a) The reactant of the last reaction is the weakest oxidizing agent, so the weakest oxidizing agent in this list is Al^{3+}.

(b) The reactant of the first reaction is the strongest oxidizing agent, so the strongest oxidizing agent is Ce^{4+}.

(c) The product of the last reaction is the strongest reducing agent, so the strongest reducing agent is Al.

(d) The product of the first reaction is the weakest reducing agent, so the weakest reducing agent is Ce^{3+}.

(e) $Sn(aq) \longrightarrow Sn^{2+}(aq) + 2 e^-$ oxidation half-reaction $E°_{ox} = -E°_{red} = +0.14$ V

   $2 Ag^+(aq) + 2 e^- \longrightarrow 2 Ag(s)$ reduction half-reaction $E°_{red} = +0.80$ V

   $Sn^{2+}(aq) + 2 Ag^+(aq) \longrightarrow Sn^{4+}(aq) + 2 Ag(s)$

$$E°_{cell} = (+0.14 \text{ V}) + (+0.80 \text{ V}) = 0.94 \text{ V}$$

   Yes, Sn(s) will reduce Ag^+ to Ag(s).

(f) $2 Hg(\ell) \longrightarrow Hg_2^{2+}(aq) + 2 e^-$ oxidation half-reaction $E°_{ox} = -E°_{red} = -0.79$ V

   $Sn^{2+}(aq) + 2 e^- \longrightarrow Sn(s)$ reduction half-reaction $E°_{red} = -0.14$ V

   $2 Hg(\ell) + Sn^{2+}(aq) \longrightarrow Hg_2^{2+}(aq) + Sn(s)$

$$E°_{cell} = (-0.79 \text{ V}) + (-0.14 \text{ V}) = -0.93 \text{ V}$$

   No, $Hg(\ell)$ will not reduce Sn^{2+} to Sn(s).

(g) Sn(s) can reduced any ion whose reduction potential is more positive than –0.14 V. In the table provided, Ce^{4+}, Ag^+, and Hg_2^{2+} can reduce Sn(s).

(h) $Ag^+(aq)$ can be oxidized any metal whose reduction potential is smaller than 0.80 V. In the table provided, Hg, Sn, Ni, and Al can oxidize $Ag^+(aq)$.

$E°$ and Gibbs Free Energy

34. In a product-favored chemical reaction, the standard cell potential, $E°$, is **greater** than zero and the Gibbs free energy, $\Delta G°$, is **less** than zero.

36. (a) Adapt the method described in Problem-Solving Example 19.8: $\Delta G° = -nFE°_{cell}$

   Two O atoms are going from Ox # = 0 to Ox # = –2. So, n = 4 mol.

$$E°_{cell} = \frac{-\Delta G°}{nF} = \frac{-(-598 \text{ kJ}) \times \left(\dfrac{1000 \text{ J}}{1 \text{ kJ}}\right) \times \left(\dfrac{1 \text{ C} \times 1 \text{ V}}{1 \text{ J}}\right)}{(4 \text{ mol})(96485 \text{ C/mol})} = 1.55 \text{ V}$$

   (b) When the reaction is written with all the coefficients doubled:

$$\Delta G°_{double} = 2 \times \Delta G°_{original} = 2 \times (-598 \text{ kJ}) = -1196 \text{ kJ}$$

   $E°_{cell}$ is independent of scale, so $E°_{cell} = 1.55$ V.

38. Adapt the method described in the answer to Question 36 and Problem-Solving Example 19.8.

$$Zn(s) + Cl_2(g) \longrightarrow Zn^{2+}(aq) + 2\,Cl^-(aq)$$

Zn is going from Ox. # = 0 to Ox. # = +2. So, n = 2 mol.

$$\Delta G° = -\,(2\text{ mol}) \times (96485\text{ C/mol}) \times (2.12\text{ V}) \times \frac{1\text{ J}}{1\text{ C} \times 1\text{ V}} \times \frac{1\text{ kJ}}{1000} = -409\text{ kJ}$$

40. Adapt the method described in the answer to Question 36 and Problem-Solving Example 19.8. *Note: Table 19.1 has a different value for the reduction potential of liquid bromine than Appendix I. Here we are using the value from Table 19.1.*

$$2\,Br^-(aq) \longrightarrow Br_2(\ell) + 2\,e^- \qquad\qquad E°_{ox} = -\,E°_{red} = -1.08\text{ V}$$

$$\underline{I_2(s) + 2\,e^- \longrightarrow 2\,I^-(aq) \qquad\qquad E°_{red} = +0.535\text{ V}}$$

$$I_2(s) + 2\,Br^-(aq) \longrightarrow 2\,I^-(aq) + Br_2(\ell) \quad E°_{cell} = (-1.08\text{ V}) + (+0.535\text{ V}) = -0.55\text{ V}$$

$$K° = 10^{\left(\frac{nE°_{cell}}{0.0592\text{ V}}\right)} = 10^{\left(\frac{(2\text{ mol})\times(-0.55\text{ V})}{0.0592\text{ V}}\right)} = 10^{-19}$$

$$\Delta G° = -\,(2\text{ mol}) \times (96485\text{ C/mol}) \times (-0.55\text{ V}) \times \frac{1\text{ J}}{1\text{ C} \times 1\text{ V}} \times \frac{1\text{ kJ}}{1000} = 1.1 \times 10^2\text{ kJ}$$

42. Adapt the method described in the answer to Question 36 and Problem-Solving Example 19.8.

$$2\,Br^-(aq) \longrightarrow Br_2(\ell) + 2\,e^- \quad\text{oxidation half-reaction} \qquad E°_{ox} = -\,E°_{red} = -1.08\text{ V}$$

$$\underline{Cl_2(g) + 2\,e^- \longrightarrow 2\,Cl^-(aq) \quad\text{reduction half-reaction} \qquad E°_{red} = +1.360\text{ V}}$$

$$Cl_2(g) + 2\,Br^-(aq) \longrightarrow Br_2(\ell) + 2\,Cl^-(aq) \quad E°_{cell} = (-1.08\text{ V}) + (+1.360\text{ V}) = 0.28\text{ V}$$

$$K° = 10^{\left(\frac{nE°_{cell}}{0.0592\text{ V}}\right)} = 10^{\left(\frac{(2\text{ mol})\times(0.28\text{ V})}{0.0592\text{ V}}\right)} = 10^{9.5} = 3 \times 10^9$$

$$\Delta G° = -\,(2\text{ mol}) \times (96485\text{ C/mol}) \times (0.28\text{ V}) \times \frac{1\text{ J}}{1\text{ C} \times 1\text{ V}} \times \frac{1\text{ kJ}}{1000} = -54\text{ kJ}$$

Effects of Concentration on Cell Potential

44. Adapt the method described in the answer to Question 30 and Problem-Solving Example 19.7, then use the Nernst Equation at T = 298 K, by adapting the method described in the Problem-Solving Example 19.10.

$$E_{cell} = E°_{cell} - \frac{0.0592 \text{ V}}{n} \log Q$$

Note: the bracket symbol, [X], which had been reserved exclusively for equilibrium concentrations, is sometimes used in this question. These should really be expressed as (conc X), since they are not equilibrium concentrations.

(a) $Cd(s) \longrightarrow Cd^{2+}(aq) + 2 e^-$ $E°_{ox} = -E°_{red} = +0.40 \text{ V}$

   $\underline{2 Ag^+(aq) + 2 e^- \longrightarrow 2 Ag(s) \qquad\qquad E°_{red} = +0.80 \text{ V}}$

   $Cd(s) + 2 Ag^+(aq) \longrightarrow Cd^{2+}(aq) + 2 Ag(s)$

$$E°_{cell} = (+0.40 \text{ V}) + (+0.80 \text{ V}) = 1.20 \text{ V}$$

For this reaction, $Q = \dfrac{(conc\ Cd^{2+})}{(conc\ Ag^+)^2}$ and n = 2

(b) $E_{cell} = E°_{cell} - \dfrac{0.0592 \text{ V}}{n} \log\left(\dfrac{(conc\ Cd^{2+})}{(conc\ Ag^+)^2}\right) = 1.20 \text{ V} - \dfrac{0.0592 \text{ V}}{2} \log\left(\dfrac{2.0 \text{ M}}{(0.25 \text{ M})^2}\right)$

$= 1.20 \text{ V} - \dfrac{0.0592 \text{ V}}{2} \log(32) = 1.20 \text{ V} - \dfrac{0.0592 \text{ V}}{2} \times 1.5 = 1.20 \text{ V} - 0.04 \text{ V}$

$$E_{cell} = 1.16 \text{ V}$$

(c) $\log\left(\dfrac{(conc\ Cd^{2+})}{(conc\ Ag^+)^2}\right) = \dfrac{n}{0.0592 \text{ V}}(E°_{cell} - E_{cell}) = \dfrac{2}{0.0592 \text{ V}} \times (1.20 \text{ V} - 1.25 \text{ V})$

$$= \dfrac{2}{0.0592 \text{ V}} \times (-0.05 \text{ V}) = -1.69 \cong -2 \text{ (round to 1 sig.fig.)}$$

$\log\left(\dfrac{0.100 \text{ M}}{(conc\ Ag^+)^2}\right) = -2$ $0.01 = \dfrac{0.100}{(conc\ Ag^+)^2}$

$$(conc\ Ag^+)^2 = 10$$

$$(conc\ Ag^+) = 3 \text{ M}$$

Common Batteries

48. Apply the methods described in the answers to Questions 20, 24, and 32.

(a) $Ni^{2+}(aq) + Cd(s) \longrightarrow Ni(s) + Cd^{2+}(aq)$

(b)

Substance oxidized	Substance reduced	Oxidizing agent	Reducing agent
Cd	Ni^{2+}	Ni^{2+}	Cd

(c) Metallic Cd is the anode and metallic Ni is the cathode.

(d) $Cd(s) \longrightarrow Cd^{2+}(aq) + 2\,e^{-}$ $E°_{ox} = -E°_{red} = +0.40$ V

$\underline{Ni^{2+}(aq) + 2\,e^{-} \longrightarrow Ni(s)\qquad E°_{red} = -0.25\text{ V}}$

$Cd(s) + Ni^{2+}(aq) \longrightarrow Cd^{2+}(aq) + Ni(s)\quad E°_{cell} = (+0.40\text{ V}) + (-0.25\text{ V}) = 0.15$ V

(e) The half-reactions above show that electrons flow spontaneously from the Cd electrode to the Ni electrode.

(f) The NO_3^{-} anions in the salt bridge flow toward the anode compartment to replenish the negative charges to neutralize the Cd^{2+} ions being formed.

Fuel Cells

50. A fuel cell has a continuous supply of reactants, and will be useable for as long as the reactants are supplied. A battery contains all the reactants of the reaction. Once the reactants are gone, the battery is no longer useable.

52. Adapt the methods described in the answer to Question 44, 4.55, and Problem-Solving Example 19.13.

(a) The N_2H_4 oxidation half-reaction occurs at the anode. The O_2 reduction half-reaction occurs at the cathode.

(b) Adding the two half-reactions gives: $N_2H_4(g) + O_2(g) \longrightarrow N_2(g) + 2H_2O(\ell)$

(c) $50.0 \text{ hr} \times 0.50 \text{ A} \times \dfrac{3600 \text{ s}}{1 \text{ hr}} \times \dfrac{1 \text{ C}}{1 \text{ A} \cdot 1 \text{ s}} \times \dfrac{1 \text{ mol e}^{-}}{96485 \text{ C}} \times \dfrac{1 \text{ mol N}_2\text{H}_4}{4 \text{ mol e}^{-}} \times \dfrac{32.05 \text{ g N}_2\text{H}_4}{1 \text{ mol N}_2\text{H}_4}$

$= 7.5 \text{ g N}_2\text{H}_4$

(d) $7.5 \text{ g N}_2\text{H}_4 \times \dfrac{1 \text{ mol N}_2\text{H}_4}{32.05 \text{ g N}_2\text{H}_4} \times \dfrac{1 \text{ mol O}_2}{1 \text{ mol N}_2\text{H}_4} \times \dfrac{32.00 \text{ g O}_2}{1 \text{ mol O}_2} = 7.5 \text{ g O}_2$

Electrolysis: Reactant-Favored Reactions

54. All the metals with a reduction potential more positive than the half-reaction with water as a reactant (– 0.8277 V) can be electrolyzed from their aqueous ions to the corresponding metals. The metals in Table 19.1 that qualify are:

$$\text{Au}^{3+}, \text{Hg}^{2+}, \text{Ag}^{+}, \text{Hg}_2{}^{2+}, \text{Fe}^{3+}, \text{Cu}^{2+}, \text{Sn}^{4+}, \text{Sn}^{2+}, \text{Ni}^{2+}, \text{Cd}^{2+}, \text{Fe}^{2+}, \text{Zn}^{2+}$$

56. Electrolysis of NaBr involves the oxidation of the anion Br⁻ to Br₂ and the reduction of the water.

Anode - oxidation: $2 \text{ Br}^-(\text{aq}) \longrightarrow \text{Br}_2(\ell) + 2 \text{ e}^-$

Cathode - reduction: $2 \text{ H}_2\text{O}(\ell) + 2 \text{ e}^- \longrightarrow \text{H}_2(\text{g}) + 2 \text{ OH}^-(\text{aq})$

$2 \text{ H}_2\text{O}(\ell) + 2 \text{ Br}^-(\ell) \longrightarrow \text{H}_2(\text{g}) + \text{Br}_2(\ell) + 2 \text{ OH}^-(\text{aq})$

H₂ and Br₂ are produced in a basic solution. After the reaction is complete, the solution contains Na⁺, OH⁻, a small amount of dissolved Br₂ (though it has low solubility in water), and a very small amount of H₃O⁺. H₂ is formed in the reduction reaction at the cathode. Br₂ is formed in the oxidation reaction at the anode.

Counting Electrons

58. Follow the method described in the answers to Question 52.

$$\text{Ag}^+(\text{aq}) + \text{e}^- \longrightarrow \text{Ag}(\text{s})$$

$$155 \text{ min} \times 0.015 \text{ A} \times \dfrac{60 \text{ s}}{1 \text{ min}} \times \dfrac{1 \text{ C}}{1 \text{ A} \cdot 1 \text{ s}} \times \dfrac{1 \text{ mol e}^-}{96485 \text{ C}} \times \dfrac{1 \text{ mol Ag}}{1 \text{ mol e}^-} \times \dfrac{107.9 \text{ g Ag}}{1 \text{ mol Ag}} = 0.16 \text{ g Ag}$$

60. Follow the method described in the answers to Question 52.

$$\text{Cu}^{2+}(\text{aq}) + 2 \text{ e}^- \longrightarrow \text{Cu}(\text{s})$$

$$2.00 \text{ hr} \times 2.50 \text{ A} \times \dfrac{3600 \text{ s}}{1 \text{ hr}} \times \dfrac{1 \text{ C}}{1 \text{ A} \cdot 1 \text{ s}} \times \dfrac{1 \text{ mol e}^-}{96485 \text{ C}} \times \dfrac{1 \text{ mol Cu}}{2 \text{ mol e}^-} \times \dfrac{63.55 \text{ g Cu}}{1 \text{ mol Cu}} = 5.93 \text{ g Cu}$$

62. Follow the method described in the answers to Question 52.

$$8.0 \text{ hr} \times (1 \times 10^5 \text{ A}) \times \frac{3600 \text{ s}}{1 \text{ hr}} \times \frac{1 \text{ C}}{1 \text{ A} \cdot 1 \text{ s}} \times \frac{1 \text{ mol e}^-}{96485 \text{ C}} \times \frac{1 \text{ mol Al}}{3 \text{ mol e}^-} \times \frac{26.98 \text{ g Al}}{1 \text{ mol Al}}$$

$$= 3 \times 10^5 \text{ g Al}$$

64. Follow the method described in the answers to Question 52. Equations for chemical reactions are found in Section 19.9.

$$50. \text{ hr} \times 1.0 \text{ A} \times \frac{3600 \text{ s}}{1 \text{ hr}} \times \frac{1 \text{ C}}{1 \text{ A} \cdot 1 \text{ s}} \times \frac{1 \text{ mol e}^-}{96485 \text{ C}} \times \frac{1 \text{ mol Pb}}{2 \text{ mol e}^-} \times \frac{207.2 \text{ g Pb}}{1 \text{ mol Pb}} = 190 \text{ g Pb}$$

66. Follow the method described in the answers to Question 52. Equations for chemical reactions are found in Section 19.9.

$$20 \text{ min} \times 250 \text{ mA} \times \frac{60 \text{ s}}{1 \text{ min}} \times \frac{10^{-3} \text{ A}}{1 \text{ mA}} \times \frac{1 \text{ C}}{1 \text{ A} \cdot 1 \text{ s}} \times \frac{1 \text{ mol e}^-}{96485 \text{ C}} \times \frac{1 \text{ mol Zn}}{2 \text{ mol e}^-} \times \frac{65.39 \text{ g Zn}}{1 \text{ mol Zn}}$$

$$= 0.1 \text{ g Zn}$$

68. Follow the method described in the answers to Question 52.

$$10. \text{ min} \times 1.0 \text{ A} \times \frac{60 \text{ s}}{1 \text{ min}} \times \frac{1 \text{ C}}{1 \text{ A} \cdot 1 \text{ s}} \times \frac{1 \text{ mol e}^-}{96485 \text{ C}} = 6.2 \times 10^{-3} \text{ mol e}^-$$

$$6.2 \times 10^{-3} \text{ mol e}^- \times \frac{1 \text{ mol Li}}{1 \text{ mol e}^-} \times \frac{6.941 \text{ g Li}}{1 \text{ mol Li}} = 0.043 \text{ g Li}$$

$$6.2 \times 10^{-3} \text{ mol e}^- \times \frac{1 \text{ mol Pb}}{2 \text{ mol e}^-} \times \frac{207.2 \text{ g Pb}}{1 \text{ mol Pb}} = 0.64 \text{ g Pb}$$

70. Perform a standard stoichiometry problem, then adapt the method described in the answers to Question 52 and at the end of section 19.12.

(a) $$9 \text{ metric tons} \times \frac{1000 \text{ kg}}{1 \text{ metric ton}} \times \frac{1000 \text{ g}}{1 \text{ kg}} \times \frac{1 \text{ mol F}_2}{38.00 \text{ g F}_2} \times \frac{2 \text{ mol HF}}{1 \text{ mol F}_2} \times \frac{20.01 \text{ g HF}}{1 \text{ mol HF}}$$

$$= 9 \times 10^6 \text{ g HF}$$

(b) $$24. \text{ hr} \times (6.0 \times 10^3 \text{ A}) \times \frac{3600 \text{ s}}{1 \text{ hr}} \times \frac{1 \text{ C}}{1 \text{ A} \cdot 1 \text{ s}} \times 12 \text{ V} \times \frac{1 \text{ J}}{1 \text{ C} \cdot 1 \text{ V}} \times \frac{1 \text{ kWh}}{3.60 \times 10^6 \text{ J}}$$

$$= 1.7 \times 10^3 \text{ kWh}$$

Corrosion – Product-Favored Reactions

72. One common metal that does not corrode readily under normal conditions is gold (Au), that is why it is used in fine jewelry;. There are others, but they are not as well known.

74. Galvanized iron has a thin coating of zinc, a more active metal. That metal coating corrodes instead of the iron.

General Questions

76. Adapt the method described in the answer to Question 58.

$$Ag^+(aq) + e^- \longrightarrow Ag(s), \quad 0.0234 \text{ g Ag} \times \frac{1 \text{ mol Ag}}{107.9 \text{ g Ag}} \times \frac{1 \text{ mol } e^-}{1 \text{ mol Ag}} = 2.17 \times 10^{-4} \text{ mol } e^-$$

$$Cu^{2+}(aq) + 2 e^- \longrightarrow Cu(s), \quad 2.17 \times 10^{-4} \text{ mol } e^- \times \frac{1 \text{ mol Cu}}{2 \text{ mol } e^-} \times \frac{63.55 \text{ g Cu}}{1 \text{ mol Cu}} = 0.00689 \text{ g Cu}$$

$$Al^{3+}(aq) + 3 e^- \longrightarrow Al(s), \quad 2.17 \times 10^{-4} \text{ mol } e^- \times \frac{1 \text{ mol Al}}{3 \text{ mol } e^-} \times \frac{26.98 \text{ g Al}}{1 \text{ mol Al}} = 0.00195 \text{ g Al}$$

Applying Concepts

78. The strongest reducing agent is the most reactive metal. From the information in (b), we find that metal C reacts with all the other metals' ions, so it will be last on the list. From the information in (a), we find that metals A and C are more reactive than the other metals, so A will precede C in the list. From the information in (c), we find that metal D reacts with the ions of metal B, so B will be first on the list.

$$B < D < A < C$$

80. The solution labeled "A^{2+}" is getting lighter and the solution labeled "B^{2+}" is getting darker. If we assume that means A^{2+} is getting less concentrated and that B^{2+} is getting more concentrated we can make the following conclusions:

(a) B(s) is being oxidized to B^{2+} and A^{2+} is being reduced to A(s).

(b) A^{2+} is the oxidizing agent and B(s) is the reducing agent.

(c) B(s) is the anode and A(s) is the cathode.

(d) $A^{2+} + 2 e^- \longrightarrow A(s)$ and $B(s) \longrightarrow B^{2+} + 2 e^-$

(e) The A metal gains mass.   (f) Electrons flow from B(s) to A(s).

(g) K^+ ions in the salt bridge will migrate towards the A^{2+} solution to replace the cations that plated out as A(s).

Chapter 20: Nuclear Chemistry

Nuclear Reactions

11. The mass numbers and atomic numbers must balance. Use that to determine the mass number and the atomic number of the missing entry. Use the periodic table and the atomic number to get the symbol of an element.

(a) $^{242}_{94}Pu \longrightarrow \, ^{4}_{2}He + \underline{\,^{238}_{92}U\,}$ $242 - 4 = 238, \; 94 - 2 = 92, \;$ Element 92 is U.

(b) $^{32}_{15}P \longrightarrow \, ^{32}_{16}S + \, ^{0}_{-1}e$ $32 + 0 = 32, \; 16 + (-1) = 15, \;$ Element 15 is P.

(c) $^{252}_{98}Cf + \underline{\,^{10}_{5}B\,} \longrightarrow 3 \, ^{1}_{0}n + \, ^{259}_{103}Lr$ $3 \times (1) - 259 - 252 = 10$

$3 \times (0) - 103 - 98 = 5, \;$ Element 5 is B.

(d) $^{55}_{26}Fe + \, ^{0}_{-1}e \longrightarrow \, ^{55}_{25}Mn$ $55 - 55 = 0, \; 25 - 26 = -1, \;$ electron captured $^{0}_{-1}e$.

(e) $^{15}_{8}O \longrightarrow \underline{\,^{15}_{7}N\,} + \, ^{0}_{+1}e$ $15 - 0 = 15, \; 8 - 1 = 7, \;$ Element 7 is N.

13. Interpret the statement by identifying the nuclear symbol(s) for the given reactant and/or product isotope(s) and identifying the details of the radioactive decay process. Then follow the balancing method described in the answer to Question 11.

(a) Magnesium-28 is $^{28}_{12}Mg$ and β emission is the production of $^{0}_{-1}e$.

$$^{28}_{12}Mg \longrightarrow \, ^{28}_{13}Al + \, ^{0}_{-1}e$$

(b) Uranium-238 is $^{238}_{92}U$, carbon-12 is $^{12}_{6}C$, and the neutron symbol is $^{1}_{0}n$.

$$^{238}_{92}U + \, ^{12}_{6}C \longrightarrow 4 \, ^{1}_{0}n + \, ^{246}_{98}Cf$$

(c) Hydrogen-2 is $^{2}_{1}H$, helium-3 is $^{3}_{2}He$, and helium-4 is $^{4}_{2}He$.

$$^{2}_{1}H + \, ^{3}_{2}He \longrightarrow \, ^{4}_{2}He + \, ^{1}_{1}H$$

(d) Argon-38 is $^{38}_{18}Ar$ and positron emission is the production of $^{0}_{+1}e$.

$$^{38}_{19}K \longrightarrow \, ^{38}_{18}Ar + \, ^{0}_{+1}e$$

(e) Platinum-175 is $^{175}_{78}$Pt, and osmium-171 is $^{171}_{76}$Os.

$$^{175}_{78}\text{Pt} \longrightarrow ^{4}_{2}\text{He} + ^{171}_{76}\text{Os}$$

15. The first five steps are: α, β, α, β, α. Therefore, start with uranium-235 undergoing an α decay reaction. Then take the radioisotope produced and make it the reactant of the second β decay reaction. Repeat this process for the remaining three steps undergoing α, then β, then α.

$$^{235}_{92}\text{U} \longrightarrow ^{4}_{2}\text{He} + ^{231}_{90}\text{Th}$$

$$^{231}_{90}\text{Th} \longrightarrow ^{0}_{-1}\text{e} + ^{231}_{91}\text{Pa}$$

$$^{231}_{91}\text{Pa} \longrightarrow ^{4}_{2}\text{He} + ^{227}_{89}\text{Ac}$$

$$^{227}_{89}\text{Ac} \longrightarrow ^{0}_{-1}\text{e} + ^{227}_{90}\text{Th}$$

$$^{227}_{90}\text{Th} \longrightarrow ^{4}_{2}\text{He} + ^{223}_{88}\text{Ra}$$

So, the radioisotopes produced in the first five steps are:

$$^{231}_{90}\text{Th},\ ^{231}_{91}\text{Pa},\ ^{227}_{89}\text{Ac},\ ^{227}_{90}\text{Th, and } ^{223}_{88}\text{Ra}$$

Nuclear Stability

17. Identify which type of radioactive decay is most likely for the isotope, by identifying the N/Z ratio and seeing where it falls on Figure 20.2.

(a) The neon-19 isotope has 9 neutrons and 10 protons, giving an N/Z ratio of 0.90. That means it has too few neutrons and undergoes positron emission or electron capture. Neon is relatively small, so we will predict that it undergoes positron emission:

$$^{19}_{10}\text{Ne} \longrightarrow ^{19}_{9}\text{F} + ^{0}_{+1}\text{e}$$

(b) The thorium-230 isotope has 140 neutrons and 90 protons, giving an N/Z ratio of 1.56. Isotopes in this range of the graph have an N/Z ratio of about 1.52. That means it has too many neutrons, so it undergoes β emission:

$$^{230}_{90}\text{Th} \longrightarrow ^{0}_{-1}\text{e} + ^{230}_{91}\text{Pa}$$

(c) The bromine-82 isotope has 47 neutrons and 35 protons, giving an N/Z ratio of 1.34. Isotopes in this range of the graph have an N/Z ratio of about 1.20. That means it has too many neutrons, so it undergoes β emission:

$$^{82}_{35}\text{Br} \longrightarrow \ ^{0}_{-1}\text{e} + \ ^{82}_{36}\text{Kr}$$

(d) The lead-212 isotope has 128 neutrons and 84 protons, more than the threshold limit of 126 neutrons and 83 protons above which all isotopes undergo α decay:

$$^{212}_{84}\text{Po} \longrightarrow \ ^{4}_{2}\text{He} + \ ^{208}_{82}\text{Pb}$$

19. Use the method described in Section 20.3. We want to compare the binding energy per nucleon. Calculate the mass defect when the nucleus is formed from the protons and neutrons. Use Einstein's equation: $E = (\Delta m)c^2$ to calculate the total energy generated, then divide that number by the number of nucleons to determine the binding energy per nucleon.

One mole of boron-10 nuclei has 5 moles of protons and 5 moles of neutrons, or a total of 10 moles of nucleons:

$$\Delta m = 10.01294 \text{ g } ^{10}\text{B}$$

$$- [(5 \text{ mol } ^{1}_{1}\text{p} \times 1.00783 \text{ g/mol } ^{1}_{1}\text{p}) + (5 \text{ mol } ^{1}_{0}\text{n} \times 1.00867 \text{ g/mol } ^{1}_{0}\text{n})] = -0.06956 \text{ g}$$

$$E = (\Delta m)c^2 = \left(-0.06956 \text{ g} \times \frac{1 \text{ kg}}{1000 \text{ g}}\right) \times (2.99792 \times 10^8 \text{ m/s})^2 \times \frac{1 \text{ J}}{1 \text{ kgm}^2 \text{ s}^{-2}} \times \frac{1 \text{ kJ}}{1000 \text{ J}}$$

$$= -6.252 \times 10^9 \text{ kJ}$$

$$E_b \text{ per mole nucleon} = \frac{6.252 \times 10^9 \text{ kJ}}{10 \text{ mol nucleons}} = 6.252 \times 10^8 \text{ kJ/mol nucleons}$$

We must go one step farther using Avogadro's number to find the actual energy per nucleon. *(We do not need to do this for comparing the two isotopes, but we will do it to get the actual E_b per nucleon. Note: The E_b per mole nucleon is incorrectly identified as E_b per nucleon in the answer to Exercise 20.5.)*

$$E_b \text{ per nucleon} = \frac{6.252 \times 10^8 \text{ kJ}}{1 \text{ mol nucleons}} \times \frac{1 \text{ mol nucleons}}{6.0221 \times 10^{23} \text{ nucleons}} = 1.038 \times 10^{-15} \text{ kJ/nucleon}$$

One mole of boron-11 nuclei has 5 moles of protons and 6 moles of neutrons, or a total of 11 moles of nucleons:

$\Delta m = 11.00931 \text{ g } ^{11}\text{B}$

$$- [(5 \text{ mol } ^1_1\text{p} \times 1.00783 \text{ g/mol } ^1_1\text{p}) + (6 \text{ mol } ^1_0\text{n} \times 1.00867 \text{ g/mol } ^1_0\text{n})] = -0.08186 \text{ g}$$

$$E = (\Delta m)c^2 = \left(-0.08186 \text{ g} \times \frac{1 \text{ kg}}{1000 \text{ g}}\right) \times (2.99792 \times 10^8 \text{ m/s})^2 \times \frac{1 \text{ J}}{1 \text{ kgm}^2 \text{ s}^{-2}} \times \frac{1 \text{ kJ}}{1000 \text{ J}}$$

$$= -7.357 \times 10^9 \text{ kJ}$$

$$E_b \text{ per nucleon} = \frac{7.357 \times 10^9 \text{ kJ}}{11 \text{ mol nucleons}} \times \frac{1 \text{ mol nucleons}}{6.0221 \times 10^{23} \text{ nucleons}}$$

$$= 1.111 \times 10^{-15} \text{ kJ/nucleon}$$

Boron-11 has a larger E_b per nucleon than boron-10, so ^{11}B is more stable than ^{10}B.

Rates of Disintegration Reactions

21. Adapt the methods described in Problem-Solving Examples 20.3 and 20.4 using these equations described in Section 20.4:

$$A = kN \qquad \ln\left(\frac{N}{N_0}\right) = -kt \qquad \ln\left(\frac{A}{A_0}\right) = -kt \qquad t_{1/2} = \frac{\ln 2}{k}$$

(Note: ln2 is used in the last equation instead of 0.693, for two reasons. First, it's faster to type into a calculator – requiring only two buttons be pushed instead of four, and it is more precise than its three sig. fig. approximation.)

Use the last equation to determine the value of k:

$$k = \frac{\ln 2}{t_{1/2}} = \frac{\ln 2}{15 \text{ h}} = 4.6 \times 10^{-3} \text{ h}^{-1}$$

The mass (m) of a sample of a pure substance is directly proportional to the number of atoms (N), so

$$\frac{N}{N_0} = \frac{m}{m_0}$$

Now, use the second equation to determine the new mass:

$$t = 1 \text{ d} \times \left(\frac{24 \text{ h}}{1 \text{ d}}\right) + 6 \text{ h} = 30. \text{ h} \quad \textit{(Assume in this context that 1 d is exact.)}$$

$$\ln\left(\frac{N}{N_0}\right) = \ln\left(\frac{m}{m_0}\right) = -kt = -(4.6 \times 10^{-3}\ h^{-1}) \times (30.\ h) = -1.4$$

$$m = m_0 e^{-1.4} = (20\ mg) \times e^{-1.4} = (20\ mg) \times (0.25) = 5\ mg$$

This answer makes sense, because 30 hours is two half-lives, and the sample mass is halved twice: $\frac{1}{2} \times \frac{1}{2} \times 20\ mg = 5\ mg$.

23. Adapt the methods described in the answers to Questions 13 and 21.

(a)
$$^{131}_{53}I \longrightarrow ^{\ \ 0}_{-1}e + ^{131}_{\ 54}Xe$$

(b)
$$k = \frac{\ln 2}{t_{1/2}} = \frac{\ln 2}{8.05\ d} = 8.61 \times 10^{-2}\ d^{-1}$$

The mass (m) of a sample of a compound is directly proportional to the number of radioactive atoms (N) because there is a fixed percentage of that atom in the compound, so:

$$\ln\left(\frac{m}{m_0}\right) = -kt = -(8.61 \times 10^{-2}\ d^{-1}) \times (32.2\ d) = -2.77$$

$$m = m_0 e^{-1.4} = (25.0\ mg) \times e^{-2.77} = (25.0\ mg) \times (0.063) = 1.6\ mg$$

This answer makes sense, because 32.2 days is four half-lives, and the sample mass is halved four times: $\frac{1}{2} \times \frac{1}{2} \times \frac{1}{2} \times \frac{1}{2} \times 25.0\ mg = 1.56\ mg$.

25. Adapt the methods described in the answer to Question 21.

$A_0 = 100.0\ \%$ and $A = 12.5\ \%$

$$\ln\left(\frac{A}{A_0}\right) = -kt \qquad \ln\left(\frac{12.5\ \%}{100.0\ \%}\right) = -2.079 = -k \times (12\ y)$$

$$k = 0.17\ y^{-1}$$

$$t_{1/2} = \frac{\ln 2}{k} = \frac{\ln 2}{0.17\ y^{-1}} = 4.0\ y$$

This answer makes sense, because 12.5 % is $\frac{1}{8}$ of 100 %, and: $\frac{1}{8} = \frac{1}{2} \times \frac{1}{2} \times \frac{1}{2}$, so 12 y is three half-lives and $\frac{1}{3} \times (12\ y) = 4.0\ y$

27. Adapt the methods described in the answers to Questions 21 and 23.

In Question 23, we calculated $k = 8.61 \times 10^{-2} \, d^{-1}$. $A_0 = 100.0 \, \%$ and $A = 5.0 \, \%$

$$\ln\left(\frac{A}{A_0}\right) = -kt \qquad \ln\left(\frac{5.0 \, \%}{100.0 \, \%}\right) = -3.00 = -(8.61 \times 10^{-2} \, d^{-1})t$$

$$t = 34.8 \, d$$

29. Adapt the methods described in Problem-Solving Example 20.5.

$$k = \frac{\ln 2}{t_{1/2}} = \frac{\ln 2}{5.73 \times 10^3 \, y} = 1.21 \times 10^{-4} \, y^{-1}$$

$A_0 = 15.3 \, d \, m^{-1} g^{-1}$ and $A = 11.2 \, d \, m^{-1} g^{-1}$ $\ln\left(\dfrac{A}{A_0}\right) = -kt$

$$\ln\left(\frac{11.2 \, d \, m^{-1} g^{-1}}{15.3 \, d \, m^{-1} g^{-1}}\right) = -0.312 = -(1.21 \times 10^{-4} \, y^{-1})t$$

$$t = 2.58 \times 10^3 \, y$$

Artificial Transmutation

31. Use the method described in the answer to Question 13.

$$^{239}_{94}Pu + 2 \, ^{1}_{0}n \longrightarrow ^{0}_{-1}e + ^{241}_{95}Am$$

33. Use the method described in the answer to Question 13.

$$^{238}_{92}U + ^{12}_{6}C \longrightarrow 4 \, ^{1}_{0}n + ^{246}_{98}Cf$$

Nuclear Fission and Fusion

35. Three components represent the fundamental parts of a nuclear fission reactor. Cadmium rods are used as a neutron absorber to control the rate of the fission reaction. Uranium rods are the source of fuel, since uranium is a reactant in the nuclear equation. Water is used for cooling by removing excess heat. It is also used in the form of steam in the steam/water cycle to produce the turning torque for the generator.

37. This is a standard dimensional analysis problem.

$$1.0 \text{ lb } {}^{235}\text{U} \times \frac{453.6 \text{ g}}{1 \text{ lb}} \times \frac{1 \text{ mol } {}^{235}\text{U}}{235 \text{ g U}} \times \frac{2.1 \times 10^{10} \text{ kJ}}{1 \text{ mol } {}^{235}\text{U}} \times \frac{1 \text{ ton coal}}{2.6 \times 10^7 \text{ kJ}}$$

$$= 1.6 \times 10^3 \text{ tons of coal}$$

Effects of Nuclear Radiation

39. The unit "rad" is the measure of the amount of radiation absorbed. The unit "rem" includes a quality factor that better describes the biological impact of a radiation dose. The unit rem would be more appropriate when talking about the effects of an atomic bomb on humans. The unit gray (Gy) is 100 rad.

41. Since most elements have some proportion of unstable isotopes that decay and we are composed of these elements (e.g., ^{14}C), our bodies emit radiation particles.

43. The gamma ray is a high energy photon. Its interaction with matter is most likely just going to be imparting large quantities of energy. The alpha and beta particles are charged particles of matter, which could interact and possibly react with the matter composing the food.

45. Adapt the methods described in the answer to Questions 21 and 5.61.

$$A = kN \propto \text{mol}$$

$$\textit{Molarity}(\text{conc}) = \text{mol/L} \propto A/L \propto A/\text{mL}$$

$$\textit{Molarity}(\text{conc}) \times V(\text{conc}) = \textit{Molarity}(\text{dil}) \times V(\text{dil})$$

$$A(\text{conc})/\text{mL} \times V(\text{conc}) = A(\text{dil})/\text{mL} \times V(\text{dil})$$

$$V(\text{dil}) = \frac{A/\text{mL}(\text{conc}) \times V(\text{conc})}{A/\text{mL}} =$$

$$= \frac{\left(2.0 \times 10^6 \text{ dps/mL conc}\right) \times (1.0 \text{ mL conc}) \times \dfrac{1 \text{ L}}{1000 \text{ mL}}}{(1.5 \times 10^4 \text{ dps/mL dil})} = 0.13 \text{ L}$$

General Questions

47. Follow the method described in the answers to Questions 11 and 13.

(a) $${}^{214}_{83}\text{Bi} \longrightarrow {}^{0}_{-1}\text{e} + {}^{214}_{84}\text{Po}$$

(b) $$4\,{}^{1}_{1}\text{H} \longrightarrow {}^{4}_{2}\text{He} + 2\,{}^{0}_{+1}\text{e}$$

(c) $${}^{249}_{99}\text{Es} + {}^{1}_{0}\text{n} \longrightarrow 2\,{}^{1}_{0}\text{n} + {}^{87}_{35}\text{Br} + {}^{161}_{64}\text{Gd}$$

(d) $${}^{220}_{86}\text{Rn} \longrightarrow {}^{216}_{84}\text{Po} + {}^{4}_{2}\text{He}$$

(e) $${}^{68}_{32}\text{Ge} + {}^{0}_{-1}\text{e} \longrightarrow {}^{68}_{31}\text{Ga}$$

49. Use the methods described in the answer to Question 21.

$$k = \frac{\ln 2}{t_{1/2}} = \frac{\ln 2}{10 \text{ min}} = 7 \times 10^{-2} \text{ min}^{-1}$$

$$\ln\left(\frac{m}{m_0}\right) = -kt = -(7 \times 10^{-2} \text{ min}^{-1}) \times (1 \text{ h}) \times \left(\frac{60 \text{ min}}{1 \text{ h}}\right) = -4$$

$$m = m_0 e^{-1.4} = (96 \text{ mg}) \times e^{-4} = (96 \text{ mg}) \times (0.02) = 1.5 \text{ mg} \cong 2 \text{ mg}$$

51. Adapt the methods described in the answer to Question 29.

$$k = \frac{\ln 2}{t_{1/2}} = \frac{\ln 2}{4.9 \times 10^{10} \text{ y}} = 1.4 \times 10^{-11} \text{ y}^{-1}$$

$$\frac{N}{N_0} = 0.951$$

$$\ln\left(\frac{N}{N_0}\right) = -kt$$

$$\ln(0.951) = -0.0502 = -(1.4 \times 10^{-11} \text{ y}^{-1})t$$

$$t = 3.6 \times 10^9 \text{ y}$$

53. Follow the method described in the answers to Questions 11 and 13.

(a) $$^{238}_{92}U + ^{14}_{7}N \longrightarrow ^{247}_{99}Es + 5\,^{1}_{0}n$$

(b) $$^{238}_{92}U + ^{16}_{8}O \longrightarrow ^{249}_{100}Fm + 5\,^{1}_{0}n$$

(c) $$^{253}_{99}Es + ^{4}_{2}He \longrightarrow ^{256}_{101}Md + ^{1}_{0}n$$

(d) $$^{246}_{96}Cm + ^{12}_{6}C \longrightarrow ^{254}_{102}No + 4\,^{1}_{0}n$$

(e) $$^{252}_{98}Cf + ^{10}_{5}B \longrightarrow ^{257}_{103}Lr + 5\,^{1}_{0}n$$

Applying Concepts

55. Alpha and beta radiation decay particles are charged ($^{4}_{2}He^{2+}$ and $^{0}_{-1}e^{-}$), so they are better able to interact with and ionize tissues, disrupting the function of the cancer cells. Gamma radiation, like X-rays, goes through soft tissue without much being absorbed. This is less likely to interfere with the cancerous cells.

57. The ^{20}Ne isotope is stable. The ^{17}Ne isotope is likely to decay by positron emission, to increase the ratio of neutrons to protons. The ^{23}Ne isotope is likely to decay by beta emission to decrease the ratio of neutrons to protons. For more details, refer to the discussion in Section 20.3 and to Figure 20.2.

59. A nuclear reaction occurred, making products. Therefore, some of the lost mass is found in the decay particles, if the decay is alpha or beta decay, and almost all the rest is found in the element produced by the reaction.

Chapter 21: The Chemistry of Selected Main Group Elements

Electrolytic Methods

20. This electrolysis process is described in Section 21.4

    (a) The magnesium ion is reduced to magnesium metal. This occurs at the cathode.

    (b) Chlorine gas, $Cl_2(g)$, is formed at the other electrode.

    (c) For each mole of Mg and Cl_2 produced, two moles of electrons are transferred. One mole of electrons is one Faraday. One faraday is equal to the charge of 96485 C. These are discussed in Chapter 19.

$$1000. \text{ kg MgCl}_2 \times \frac{1000 \text{ g}}{1 \text{ kg}} \times \frac{1 \text{ mol MgCl}_2}{95.210 \text{ g MgCl}_2} \times \frac{2 \text{ mol e}^-}{1 \text{ mol MgCl}_2} \times \frac{1 \text{ Faraday}}{1 \text{ mol e}^-}$$

$$= 2.101 \times 10^4 \text{ Faradays}$$

$$2.101 \times 10^4 \text{ Faradays} \times \frac{96485 \text{ C}}{1 \text{ Faraday}} = 2.027 \times 10^9 \text{ C}$$

    (d) The conversion factor 3.60×10^6 J/kWh was given in Question 19.70(b).

$$\frac{8.4 \text{ kwh}}{\text{lb Mg}} \times \frac{3.60 \times 10^6 \text{ J}}{1 \text{ kwh}} \times \frac{1 \text{ kJ}}{1000 \text{ J}} \times \frac{1 \text{ lb}}{453.6 \text{ g}} \times \frac{24.305 \text{ g Mg}}{1 \text{ mol Mg}} = 1.6 \times 10^3 \frac{\text{kJ}}{\text{mol}}$$

22. Adapt the methods described in the answer to Question 19.70.

$$1 \text{ ton Na} \times \frac{2000 \text{ lb}}{1 \text{ ton}} \times \frac{453.6 \text{ g}}{1 \text{ lb}} \times \frac{1 \text{ mol Na}}{22.99 \text{ g Na}} \times \frac{2 \text{ mol e}^-}{2 \text{ mol Na}} \times \frac{96485 \text{ C}}{1 \text{ mol e}^-}$$

$$\times 7.0 \text{ V} \times \frac{1 \text{ J}}{1 \text{ C} \cdot 1 \text{ V}} \times \frac{1 \text{ kWh}}{3.60 \times 10^6 \text{ J}} = 7 \times 10^3 \text{ kWh}$$

Note that it is not necessary to have or use the amperes given in the question.

24. Use the method described in the answer to Question 22.

$$2.00 \text{ hr} \times 100. \text{ A} \times \frac{3600 \text{ s}}{1 \text{ hr}} \times \frac{1 \text{ C}}{1 \text{ A} \cdot 1 \text{ s}} \times \frac{1 \text{ mol e}^-}{96485 \text{ C}} \times \frac{1 \text{ mol Al}}{3 \text{ mol e}^-} \times \frac{26.98 \text{ g Al}}{1 \text{ mol Al}} = 67.1 \text{ g Al}$$

General Questions

26. Use the methods described in Chapters 3 and 5.

Formula	Name	Oxidation state of phosphorus
P$_4$	Phosphorus	**0**
(NH$_4$)$_2$HPO$_4$	**Ammonium hydrogen phosphate**	**+5**
HNO$_2$	Phosphoric acid	**+5**
P$_5$O$_{10}$*	Pentaphosphorus decaoxide*	**+4***
Ca$_3$(PO$_4$)$_2$	**Calcium phosphate**	**+5**
Ca(H$_2$PO$_4$)$_2$	Calcium dihydrogen phosphate	**+5**

NOTE: There is a typo in this question. The fourth entry in the table should be tretraphosphorus decaoxide, P$_4$O$_{10}$, with the oxidation state on phosphorus of +5.

28. This is a standard stoichiometry and conversion factor problem:

$$5.0 \times 10^6 \text{ tons Al} \times \frac{2000 \text{ lb}}{\text{ton}} \times \frac{453.6 \text{ g}}{1 \text{ lb}} \times \frac{1 \text{ mol Al}}{26.98 \text{ g Al}} \times \frac{1 \text{ mol Al}_2\text{O}_3}{2 \text{ mol Al}}$$

$$\times \frac{101.96 \text{ g Al}_2\text{O}_3}{1 \text{ mol Al}_2\text{O}_3} \times \frac{100 \text{ g bauxite}}{55 \text{ g Al}_2\text{O}_3} \times \frac{1 \text{ lb}}{453.6 \text{ g}} \times \frac{1 \text{ ton}}{2000 \text{ lb}} = 1.7 \times 10^7 \text{ tons bauxite}$$

Note: The conversion from tons to grams and back to tons need not be done, if the mass relationship is done directly in tons.

30. Use the methods described in Chapter 8. In the answer to Question 8.16(d), we found that phosphate, PO_4^{3-}, has 4 O atoms single-bonded to the P atom. Use that pattern to develop a three-dimensional version of the P_4O_{10} molecule: Each P atom forms four single bonds to O atoms. Three of the four O atoms have single bonds to two different P atoms.

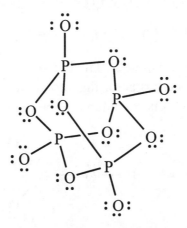

Other Lewis structures are possible, but the structure above is the most plausible one that follows the octet rule. If we look to reduce formal charges, the structure that allows the P atoms to expand their octet would be considered better:

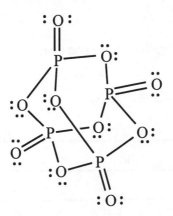

32. Use the methods described in the answer to Question 18.56. *NOTE: The units of $\Delta S°$ should be $J\,K^{-1}mol^{-1}$.*

$$T = \frac{(-17.6 \text{ kJ/mol}) \times \left(\dfrac{1000 \text{ J}}{1 \text{ kJ}}\right)}{-18.3 \text{ JK}^{-1}\text{mol}^{-1}} = 962 \text{ K}$$

34. The raw materials used in the synthesis of sulfuric acid are sulfur, water, oxygen, and catalyst (Pt or VO_5)

$$S_8(s) + 8\ O_2(g) \longrightarrow 8\ SO_2(g)$$

$$2\ SO_2(g) + O_2(g) \longrightarrow 2\ SO_3(g)$$

$$SO_3(g) + H_2SO_4(\ell) \longrightarrow H_2S_2O_7(\ell)$$

$$H_2S_2O_7(\ell) + H_2O(\ell) \longrightarrow 2\ H_2SO_4(aq)$$

36. Use the methods described in Chapter 8. The N atom in nitric acid has one double bond, so we can draw three resonance forms for this molecule. The third of these has the highest formal charges; hence, it is considered to be the worst structure of the three.

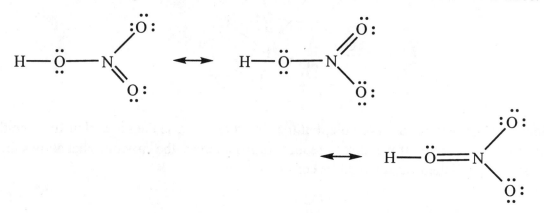

38. Use the methods described in in the answer to Question 13.35.

(conc. SO_2) = 1.00 mol SO_2/1.00 L = 1.00 M

(conc. O_2) = 5.00 mol O_2/1.00 L = 5.00 M

	2 SO_2(g) +	O_2(g) ⇌	2 SO_3(g)
conc. initial (M)	1.00	5.00	0
change conc. (M)	– 2x	– x	+ 2x
equilibrium conc. (M)	1.00 – 2x	5.00 – x	2x

(change conc. SO_2) = – 0.778 × (1.00 M) = – 0.778 M, so

$$0.778\ M = 2x$$

$$x = 0.389\ M$$

$$1.00 - 2x = 1.00\ M - 0.778\ M = 0.22\ M$$

$$5.00 - x = 5.00\ M - 0.389\ M = 4.61\ M$$

$$K_c = \frac{[SO_3]^2}{[SO_2]^2[O_2]} = \frac{(2x)^2}{(1.00-2x)^2(5.00-x)} = \frac{(0.778)^2}{(0.22)^2(4.61)} = 2.7$$

Applying Concepts

40. Use the method described in Problem-Solving Example 14.2.

$$Ca(OH)_2(s) \rightleftharpoons Ca^{2+}(aq) + 2\ OH^-(aq) \qquad K_{sp,1} = 7.9 \times 10^{-6}$$

$$Mg^{2+}(aq) + 2\ OH^-(aq) \rightleftharpoons Mg(OH)_2(s) \qquad K = 1/K_{sp,2} = 1/(1.5 \times 10^{-11})$$

$$Ca(OH)_2(s) + Mg^{2+}(aq) \rightleftharpoons Ca^{2+}(aq) + Mg(OH)_2(s) \qquad K_{net} = K_{sp,1} \times (1/K_{sp,2})$$

$$K_{net} = (7.9 \times 10^{-6}) \times [1/(1.5 \times 10^{-11})] = 5.3 \times 10^5$$

Putting sea water in the presence of $Ca(OH)_2$ will cause the precipitation of $Mg(OH)_2$. The solid can be isolated after it settles.

42. Use methods described in Chapter 8.

$$:\!N\!\!=\!\!N\!\!=\!\!N\!\!—\!\!N\!\!=\!\!\ddot{O}:$$

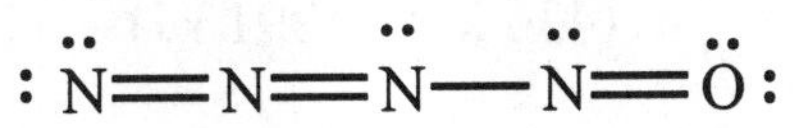

Other plausible resonance structures are also valid. The name helps confirm this structure: "Azide" is formed when three N atoms are bonded together, seen in this structure between the left-most three N atoms. "Nitrosyl" is the –NO functional group seen on the right side of the structure.

44. (a) Use methods described in Chapter 8.

$$H\!\!—\!\!\ddot{N}\!\!=\!\!N\!\!=\!\!\ddot{N}: \quad \longleftrightarrow \quad H\!\!—\!\!\ddot{N}\!\!—\!\!N\!\!\equiv\!\!N:$$

(b) Adapt the method described in the answer to Question 8.40.

Formation is described in Section 6.10 as the production of one mole of a compound from its standard state elements. That means we do not need to use the given ΔH_f for N and O.

Formation reaction for the first structure:

$$\frac{3}{2}(\ :N\equiv N:\ ) + \frac{1}{2}(\ H-H\ ) \longrightarrow H-\ddot{N}=N=\ddot{N}:$$

$$\Delta H_f \cong \frac{3}{2}\ mol \times D_{N\equiv N} + \frac{1}{2}\ mol \times D_{H-H} - 1\ mol \times D_{N-H} - 2\ mol \times D_{N=N}$$

Use the bond energies (D) in Table 8.2:

$$\Delta H_f \cong \frac{3}{2} \times (946\ kJ) + \frac{1}{2} \times (436\ kJ) - (391\ kJ) - 2 \times (418\ kJ) = 410.\ kJ$$

Formation reaction for the second structure:

$$\frac{3}{2}(\ :N\equiv N:\ ) + \frac{1}{2}(\ H-H\ ) \longrightarrow H-\ddot{N}-N\equiv N:$$

$$\Delta H_f \cong \frac{3}{2}\ mol \times D_{N\equiv N} + \frac{1}{2}\ mol \times D_{H-H}$$

$$- 1\ mol \times D_{N-H} - 1\ mol \times D_{N-N} - 1\ mol \times D_{N\equiv N}$$

Use the bond energies (D) in Table 8.2:

$$\Delta H_f \cong \frac{3}{2} \times (946\ kJ) + \frac{1}{2} \times (436\ kJ) - (391\ kJ) - (160\ kJ) - (946\ kJ) = 140\ kJ$$

46. (a)
$$NO_2(g) + NO(g) \longrightarrow N_2O_3(g)$$

(b) Use the methods described in Chapter 8.

(Note: There is a third resonance structure we could come up with, but two adjacent atoms have +1 formal charge, which eliminates that structure from being plausible.)

(c) Use the methods described in Chapter 9.

The O–N–O bond is on an N atom that fits the class AX_3. That means the bond angle is 120°. The N–N–O angles involving that same N atom as a central atom are also 120°. The N–N–O angle on the second N atom is slightly less than 120°, due to the repulsion of the lone pair on the central N atom.

48. Adapt the methods described in Section 10.7 *(Assume the temperature is known ± 1° C)*

$$\text{Molar mass} = \frac{dRT}{P} = \frac{(0.8012\,g/L)\times(0.08206\,L\,atm\,mol^{-1}K^{-1})\times(700.+273.15)K}{1.00\,atm}$$

$$= 64.0\ g/mol$$

The form of sulfur in the gas must be diatomic, $S_2(g)$, because the molar mass of the gas molecules is twice the molar mass of atomic sulfur.

50. Adapt the methods described in Chapter 18 and Chapter 19.

(a) $$T = 600\ ^{\circ}C + 273.15 = 9\times10^2\ K$$

$$\Delta G^{\circ} = \Delta H^{\circ} - T\Delta S^{\circ} = (820\ kJ) - (9\times10^2\ K)\times(180\ J\ K^{-1})\times\frac{1\,kJ}{1000\,J} = 7\times10^2\ kJ$$

(b) $$E^{\circ}{}_{cell} = -\frac{\Delta G}{nF} = -\frac{7\times10^2\ kJ}{(2)(96485\ C)}\times\frac{1000\,J}{1\,kJ}\times\frac{1\,C\times1\,V}{1\,J} = -3\ V$$

3 volts must be used to overcome this.

Chapter 22: Chemistry of Selected Transition Elements and Coordination Compounds

Transition Metals

14. Follow the method described in Sections 7.7 and 22.1.

The most common oxidation state of gold is zero. The $_{79}$Au electron configuration:

$[Xe]4f^{14}5d^{10}6s^1$ or $1s^22s^22p^63s^23p^63d^{10}4s^24p^64d^{10}4f^{14}5s^25p^65d^{10}6s^1$

Other oxidation states of gold are +1 and +3.

The $_{79}$Au$^+$ electron configuration: $[Xe]4f^{14}5d^{10}$

The $_{79}$Au^{3+} electron configuration: $[Xe]4f^{14}5d^8$

The most common oxidation states of silver are zero and +1.

The $_{47}$Ag electron configuration: $[Kr]4d^{10}5s^1$ or $1s^22s^22p^63s^23p^63d^{10}4s^24p^64d^{10}5s^1$

The $_{47}$Ag$^+$ electron configuration: $[Kr]4d^{10}$

16. The neutral Cr atom has the following orbital diagram, with six unpaired electrons:

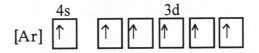

When Cr has lost all six electrons to gain an oxidation state of +6 it will have no unpaired electrons. All oxidation states between zero and +6 will apparently have unpaired electrons. Because the higher oxidation states of Cr^{4+} and Cr^{5+} usually have the chromium atom combined with other elements, which may change the number of unpaired electrons, we will predict that the question is asking for Cr^{2+} and Cr^{3+}, though all four of these oxidation states could be legitimate answers with the information we have.

Cr^{2+} Cr^{3+}

18. $Cr_2O_7^{2-}$ is the best oxidizing agent, then Cr^{3+}, then Cr^{2+}. The species with the more positive oxidation state of Cr has the greater the tendency to be reduced, hence acting as a stronger oxidizing agent. $Cr_2O_7^{2-}$ has Ox. # Cr = +6, Cr^{3+} has Ox. # Cr = +3, and Cr^{2+} has Ox. # Cr = +2.

20. (a) $Fe_2O_3(s) + 3\ CO(g) \longrightarrow Fe(s) + 3\ CO_2(g)$

(b) Use standard redox balancing methods shown in the answer to Question 19.14.

If Fe^{2+} is produced: $Fe(s) + 2\ H_3O^+(aq) \longrightarrow Fe^{2+}(aq) + H_2(g) + 2\ H_2O(\ell)$

If Fe^{3+} is produced: $2\ Fe(s) + 6\ H_3O^+(aq) \longrightarrow 2\ Fe^{3+}(aq) + 3\ H_2(g) + 6\ H_2O(\ell)$

22. Use standard redox balancing methods shown in the answer to Question 19.14.

$$3\ NO_3^-(aq) + 6\ H^+(aq) + 3\ e^- \longrightarrow 3\ NO_2(g) + 3\ H_2O(\ell)$$

$$\underline{\qquad Fe(aq) \longrightarrow Fe^{3+}(aq) + 3\ e^- \qquad}$$

$$3\ NO_3^-(aq) + 6\ H^+(aq) + Fe(aq) \longrightarrow Fe^{3+}(aq) + 3\ NO_2(g) + 3\ H_2O(\ell)$$

Then add two water molecules to each side to convert the H^+ ions into H_3O^+ ions:

$$3\ NO_3^-(aq) + 6\ H_3O^+(aq) + Fe(aq) \longrightarrow Fe^{3+}(aq) + 3\ NO_2(g) + 9\ H_2O(\ell)$$

(Check: 3 N, 15 O, 18 H, 1 Fe, +3 net charge)

Coordination Compounds

24. Cr^{3+} is bonded to two NH_3 ligands, three H_2O ligands, and one OH^-.

(a) The complex ion formed has a formula that looks like this:

$$[Cr(NH_3)_2(H_2O)_3(OH)]^{2+}$$

(b) The complex ion formed is a cation, so it will need a counter ion that is an anion.

26. (a) The complex ion $[Co(C_2O_4)_2Cl_2]^{3-}$ has two $C_2O_4^{2-}$ ligands, each with –2 charge and two Cl^- ions, each with –1 charge.

(b) The net charge is –3. The sum of the individual charges must add up to this number. So:

$$-3 = (\text{cobalt charge}) + 2 \times (-2) + 2 \times (-1)$$

The cobalt is in the form of Co^{3+}.

(c) $C_2O_4^{2-}$ is a bidentate ligand and NH_3 is a monodentate, so two $C_2O_4^{2-}$ must be replaced by four NH_3. Replacing two –2 ions with four molecules, reduces the net charge of the ion by 4, also. Therefore, the new complex ion has a formula that looks like this: $[Co(NH_3)_4Cl_2]^+$. It has a +1 charge.

28. For $Na_3[IrCl_6]$

(a) The complex ion has six Cl^- ions as ligands.

(b) $0 = + 3 \times (+1) + (\text{iridium charge}) + 6 \times (-1)$ The iridium is in the form of Ir^{3+} with a +3 charge.

(c) The formula of the complex ion is $[IrCl_6]^{3-}$ with a –3 charge.

(d) The ions that are not in the complex ion are the three Na^+ cations.

For $[Mo(CO)_4Br_2]$:

(a) The complex ion has four CO molecules and two Br^- ions as ligands.

(b) $0 = (\text{molybdenum charge}) + 4 \times (0) + 2 \times (-1)$ The molybdenum is in the form of Mo^{2+} with a +2 charge.

(c) As we have seen with Cr, we will assume that the Mo has a coordination number of 6. Therefore formula of the coordination complex must be $[Mo(CO)_4Br_2]$ with a zero charge. *This ought not be called a complex ion, since it does not have a net charge. If it were written as $[Mo(CO)_4]Br_2$ then there would be a complex ion with coordination number of 4.*

(d) All the ions are in the coordination complex. No ions are not in the complex.

30. *NOTE: There is a typo in the charge of the complex ion in (b). It should be 2– not 2+.*

Monodentates need one coordination site. Bidentates need two coordination sites. Thus, we need to add the number of monodentate ligands to $2 \times$ (number of bidentate ligands) to find the coordination number.

(a) In $[Pt(en)_2]^{2+}$, we have two bidentate ligands (2 en), so the coordination number is $2 \times (2) = 4$.

(b) In $[Cu(ox)_2]^{2-}$, we have two bidentate ligands (2 ox), so the coordination number is $2 \times (2) = 4$.

32. Use the method described in the answers to Questions 23 to 31.

   (a) $[Pt(NH_3)_2Br_2]$ Br^- and NH_3 are monodentate ligands.

   (b) $[Pt(en)(NO_2)_2]$ The ligand en is a bidentate ligand. NO_2^- is not described as a ligand in Chapter 22, but certainly has the requisite electron pairs needed to be monodentate.

   (c) $[PtClBr(NH_3)_2]$ Br^-, Cl^-, and NH_3 are monodentate ligands.

34. Use the method described in the answer to 30.

   (a) In $[FeCl_4]^-$, we have four monodentate ligands ($4\ Cl^-$), so the coordination number is $4 \times (1) = 4$.

   (b) In $[PtBr_4]^{2-}$, we have four monodentate ligands ($4\ Br^-$), so the coordination number is $4 \times (1) = 4$.

   (c) In $[Mn(en)_3]^{2+}$, we have three bidentate ligands (3 en), so the coordination number is $3 \times (2) = 6$.

   (d) In $[Cr(NH_3)_5H_2O]^{3+}$, we have six monodentate ligands ($5\ NH_3$ and $1\ H_2O$), so the coordination number is $6 \times (1) = 6$.

36. (a) $(CH_3)_3P$ has only one electron pair on the P atom, so it is a monodentate ligand.

   (b) $(^-OOC)-CH_2-N(COO^-)_2$ has lone pairs of electrons on every O atom as well as the N atom, so in theory it could be heptadentate. In practice, only one of the two O atoms on the carboxyl anion groups and the N atom probably have simultaneous access to the metal atom, due to geometric constraints, so this ligand will probably usually be tetradentate.

   (c) $H_2N-CH_2-CH_2-NH-CH_2-CH_2-NH_2$ has lone pairs of electrons on every N atom, so is a tridentate ligand.

   (d) H_2O has electron pairs on only one atom, so is a monodentate ligand.

38. There are many answers to this Question. We need a compound that is composed of a +3 cation, analogous to the $[Rh(en)]^{3+}$ ion, combined with three simple −1 anions. One example is $FeCl_3$.

Naming Complex Ions and Coordination Compounds

40. (a) Four chloride ligands coordinate-covalently bonded to a Mn^{2+} ion to make a complex anion with a −2 charge, called tetrachloromanganate(II).

(b) Three potassium counter ions combine with a complex anion composed of three oxalate anion ligands coordinate-covalently bonded to Fe^{3+} to form a neutral salt called potassium trioxalatoferrate(III).

(c) Two ammonia ligands (NH_3) and two cyanide ligands (CN^-) are coordinate-covalently bonded to a Pt^{2+} metal ion forming a neutral compound called diamminedicyanoplatinum(II).

(d) Five water ligands and one hydroxide ligand are coordinate-covalently bonded to Fe^{3+} to form a complex cation with a +2 charge called pentaquahydroxoiron(III).

(e) Hexacoordinated manganese(II) ion has two bidentate en ligands and two chloride ligands to form a neutral compound called diethylenediaminedichoromanganese(II).

Geometry of Coordination Complexes

42. (a) *cis*-$[Ni(H_2O)_2Cl_2]^{2-}$ has the H_2O molecules adjacent to each other in the square planar coordination arrangement:

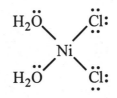

(b) *trans*-$[Cr(H_2O)_4Cl_2]^+$ has the Cl^- ligands across from each other in the octahedral coordination arrangement:

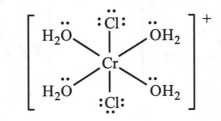

44. Fe(acac)$_3$ has the acetylacetonate ions in an octahedral coordination arrangement:

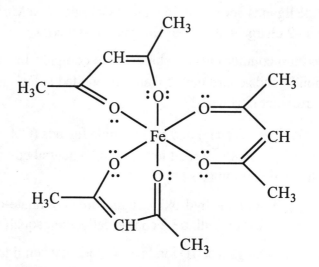

46. Both of these complexes can have geometric isomers:

(a) The triples could have an all-*cis*-orientation, or a pair of each triple could be *trans* while both are *cis* to the third:

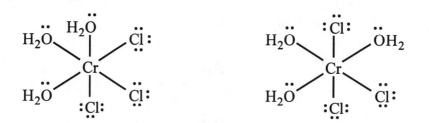

(b) The Cl$^-$ ligands can be *cis*- or *trans*-orientation:

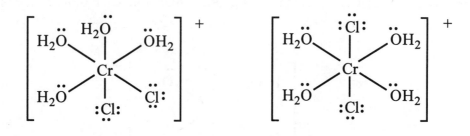

48. Adapt the methods used in the answers to Questions 41 to 47.

(a) The Cl^- ligands can be *cis*- or *trans*-orientation in the octahedral coordination arrangement:

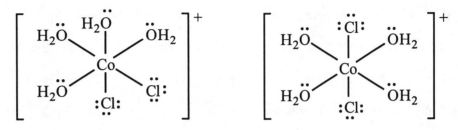

(b) This square planar structure has no isomers:

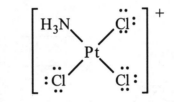

(c) The triples could have an all-*cis*-orientation, or a pair of each triple could be *trans* while both *cis* to the third in the octahedral coordination arrangement:

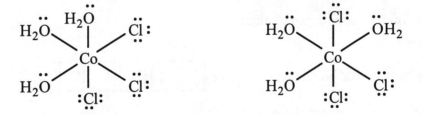

(d) The NH_3 ligands can be *cis*- or *trans*-orientation in the octahedral coordination arrangement:

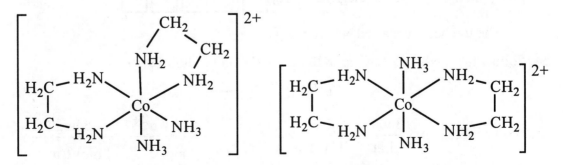

Therefore, (a), (c), and (d) have possible geometrical isomers.

General Questions

50. Follow the method described in Sections 7.7 and 22.1 and in the answer to Question 14

(a) $_{24}Cr^{2+}$ electron configuration: $[Ar]3d^4$ or $1s^22s^22p^63s^23p^63d^4$

(b) $_{30}Zn^{2+}$ electron configuration: $[Ar]3d^{10}$ or $1s^22s^22p^63s^23p^63d^{10}$

(c) $_{27}Co^{2+}$ electron configuration: $[Ar]3d^7$ or $1s^22s^22p^63s^23p^63d^7$

(d) $_{25}Mn^{4+}$ electron configuration: $[Ar]3d^3$ or $1s^22s^22p^63s^23p^63d^3$

52. Follow the method described in the answer to Question 7.78 using the answers from Question 50.

(a) $_{24}Cr^{2+}$ orbital box diagram: [Ar] ↑ ↑ ↑ ↑ ☐ (3d)

The ion has 4 unpaired electrons.

(b) $_{30}Zn^{2+}$ orbital box diagram: [Ar] ↑↓ ↑↓ ↑↓ ↑↓ ↑↓ (3d)

The ion has zero unpaired electrons.

(c) $_{27}Co^{2+}$ orbital box diagram: [Ar] ↑↓ ↑↓ ↑ ↑ ↑ (3d)

The ion has 3 unpaired electrons.

(d) $_{25}Mn^{4+}$ orbital box diagram: [Ar] ↑ ↑ ↑ ☐ ☐ (3d)

The ion has 3 unpaired electrons.

54. Follow the method described in the answers to Question 19.59.

$$Cu^{2+}(aq) + 2\,e^- \longrightarrow Cu(s)$$

$$5.00\ hr \times 2.50\ A \times \frac{3600\ s}{1\ hr} \times \frac{1\ C}{1\ A \cdot 1\ s} \times \frac{1\ mol\ e^-}{96485\ C} \times \frac{1\ mol\ Cu^{2+}}{2\ mol\ e^-} \times \frac{63.546\ g\ Cu}{1\ mol\ Cu^{2+}} = 14.8\ g\ Cu$$

56.
$$2 \, Cu_2S(s) + 3 \, O_2(g) \longrightarrow 2 \, Cu(\ell) + 2 \, SO_2(g)$$

$$1.0 \text{ ton CuS} \times \frac{2000 \text{ lb}}{1 \text{ ton}} \times \frac{453.6 \text{ g}}{1 \text{ lb}} \times \frac{1 \text{ mol } Cu_2S}{159.16 \text{ g } Cu_2S} \times \frac{2 \text{ mol } SO_2}{2 \text{ mol } Cu_2S}$$

$$\times \frac{64.06 \text{ g } SO_2}{1 \text{ mol } SO_2} \times \frac{1 \text{ lb}}{453.6 \text{ g}} \times \frac{1 \text{ ton}}{2000 \text{ lb}} = 0.40 \text{ ton } SO_2$$

58. Use the method described in the answers to Question 30.

(a) In $[Ni(en)Cl_2]^{2+}$, we have one bidentate ligand (en) and two monodentate ligands ($2 \, Cl^-$), so the coordination number is $2 + 2 \times (1) = 4$.

(b) In $[Mo(CO)_4Br_2]$, we have six monodentate ligands (4 CO and $2 \, Br^-$), so the coordination number is 6.

(c) In $[Cd(CN)_4]^{2-}$, we have four monodentate ligands ($4 \, CN^-$), so the coordination number is 4.

(d) In $[Co(CN)_5(OH)]^{3-}$, we have six monodentate ligands ($5 \, CN^-$ and $1 \, OH^-$), so the coordination number is 6.

60. Be systematic and remember that the complex ion has six ligands, so one of the ions is an uncomplexed counter ion.

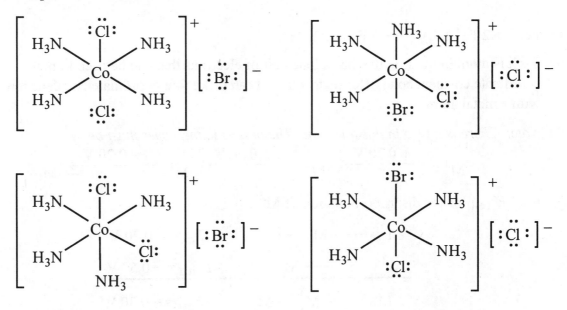

62. Use the methods described in the answers to Questions 23 - 34.

(a) The given statement is false. The coordination number of the Fe^{3+} ion in $[Fe(H_2O)_4(C_2O_4)]^+$ is six.

(b) The given statement is false. Cu^+ has no unpaired electrons.

(c) The given statement is false. The net charge of a coordination complex of Cr^{3+} with two NH_3 and one en is +3.

64. (a) The net charge = 0 = (platinum oxidation state) $+ 2 \times (0) + (-2)$. The platinum oxidation state is +2.

(b) The structural formula for $[Pt(NH_3)_2(C_2O_4)]$:

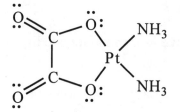

Applying Concepts

66. $Fe(s) + 2\ Fe^{3+}(aq) \longrightarrow 3\ Fe^{2+}(aq)$

68. Each N atom in the structure has a lone pair of electrons that can be used to make a coordinate covalent bond. Hence, this ligand can make four coordinate covalent bonds with a metal atom.

70. *Note: There is a typo in this question. The first M in the series must be M^{3+}.*

$$M^{3+} \xrightarrow{\ +\ 0.20\ V\ } M^{2+} \xrightarrow{\ +\ 0.50\ V\ } M^+ \xrightarrow{\ -\ 0.20\ V\ } M$$

Find $E°_{cell}$ for the disproportionation of M^{2+}:

$$
\begin{array}{ll}
M^{2+} \longrightarrow M^{3+} + e^- & E°_{ox} = -\ 0.20\ V \\
M^{2+} + e^- \longrightarrow M^+ & E°_{red} = +\ 0.50\ V \\
\hline
2\ M^{2+} \longrightarrow M^{3+} + M^+ & E°_{net} = +\ 0.30\ V
\end{array}
$$

Find $E°_{cell}$ for the disproportionation of M^+:

$$M^+ \longrightarrow M^{2+} + e^- \qquad E°_{ox} = -0.50 \text{ V}$$

$$M^+ + e^- \longrightarrow M \qquad E°_{red} = -0.20 \text{ V}$$

$$2\,M^{2+} \longrightarrow M^{3+} + M^+ \qquad E°_{net} = -0.70 \text{ V}$$

We conclude that only M^{2+} disproportionates spontaneously ($E°_{cell}$ = positive).

72. Use Table 19.1 and methods described in the answer to Question 19.20.

(a) Find a positive $E°_{cell}$ for the product-favored reaction in a cell containing Ni/Ni^{2+} and Ag/Ag$^+$.

$$Ni(s) \longrightarrow Ni^{2+}(aq) + 2\,e^- \qquad E°_{ox} = +0.25 \text{ V}$$

$$2\,Ag^+(aq) + 2\,e^- \longrightarrow 2\,Ag(s) \qquad E°_{red} = +0.80 \text{ V}$$

$$2\,Ag^+(aq) + Ni(s) \longrightarrow Ni^{2+}(aq) + 2\,Ag(s) \qquad E°_{cell} = +1.05 \text{ V}$$

(b) The cell potential is + 1.05 V.

(c)

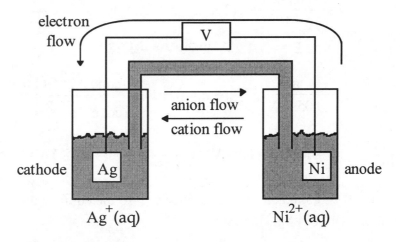

74. Balance the equation for the redox reaction as described in the answer to Question 19.14, then use conversion factor methods from Chapters 4 and 5 to determine the percent iron in the ore.

$$1 \times [5\,e^- + 8\,H^+(aq) + MnO_4^-(aq) \longrightarrow Mn^{2+}(aq) + 4\,H_2O(\ell)]$$

$$5 \times [Fe^{2+}(aq) + 1\,e^- \longrightarrow Fe^{3+}(aq)]$$

$$5\,Fe^{2+}(aq) + MnO_4^-(aq) + 8\,H^+(aq) \longrightarrow Mn^{2+}(s) + 5\,Fe^{3+}(aq) + 4\,H_2O(\ell)$$

$$\frac{18.6 \text{ mL}}{1.500 \text{ g ore}} \times \frac{1 \text{ L}}{1000 \text{ mL}} \times \frac{0.05012 \text{ mol MnO}_4^-}{1 \text{ L}} \times \frac{5 \text{ mol Fe}^{2+}}{1 \text{ mol MnO}_4^-}$$

$$\times \frac{1 \text{ mol Fe}}{1 \text{ mol Fe}^{2+}} \times \frac{55.847 \text{ g Fe}}{1 \text{ mol Fe}} \times 100 \% = 17.4 \%$$

76. Use the methods described in the answer to Question 19.44.

For this reaction, $Q = \dfrac{(\text{conc Ag}^{2+})}{(\text{conc Ag}^+)^2}$ and $n = 1$

(a)
$$E_{\text{cell}} = E^\circ{}_{\text{cell}} - \frac{0.0592 \text{ V}}{n} \log\left(\frac{(\text{conc Ag}^{2+})}{(\text{conc Ag}^+)^2}\right)$$

$$= -1.18 \text{ V} - \frac{0.0592 \text{ V}}{1} \log\left(\frac{\frac{1}{5}(1 \times 10^{-4} \text{ M})}{(1 \times 10^{-4} \text{ M})^2}\right)$$

$$= -1.18 \text{ V} - \frac{0.0592 \text{ V}}{1} \log(2 \times 10^3) = -1.18 \text{ V} - \frac{0.0592 \text{ V}}{1} \times 3.3$$

$$E = -1.18 \text{ V} - 0.20 \text{ V} = -1.38 \text{ V}$$

(b)
$$\frac{n}{0.0592 \text{ V}}(E^\circ{}_{\text{cell}} - E_{\text{cell}}) = \log\left(\frac{(\text{conc Ag}^{2+})}{(\text{conc Ag}^+)^2}\right)$$

$$\frac{1}{0.0592 \text{ V}} \times (-1.18 \text{ V} - 0.00 \text{ V}) = \frac{1}{0.0592 \text{ V}} \times (-1.18 \text{ V}) = -19.9$$

(When $E_{\text{cell}} = 0$ V, the reaction is at equilibrium, so the equilibrium concentration symbol $[\text{Ag}^{2+}]$ is used.)

$$-19.9 = \log\left(\frac{[\text{Ag}^{2+}]}{(1.0 \text{ M})^2}\right)$$

$$1 \times 10^{-20} = \frac{[\text{Ag}^{2+}]}{(1.0 \text{ M})^2}$$

$$[\text{Ag}^{2+}] = 1 \times 10^{-20} \text{ M}$$

78. Ions separate in aqueous solutions of ionic compounds, unless they are part of stable complex ions.

(a) [Pt(en)Cl$_2$] is a tetracoordinated coordination complex. When it is dissolved in water, the coordination complex becomes aqueous, but no ions are formed:

$$[Pt(en)Cl_2](s) \longrightarrow [Pt(en)Cl_2](aq)$$

(b) Na[Cr(en)$_2$(SO$_4$)$_2$] is a hexacoordinated complex anion with a sodium counter ion. When it is dissolved in water, it ionizes:

$$Na[Cr(en)_2(SO_4)_2](s) \longrightarrow Na^+(aq) + [Cr(en)_2(SO_4)_2]^-(aq)$$

According to this equation, when 1.00 mol of the solid dissolves, 2.00 mol of ions form.

(c) K$_3$[Au(CN)$_4$] is a tetracoordinated complex anion with a potassium counter ion. When it is dissolved in water, it ionizes:

$$K_3[Au(CN)_4](s) \longrightarrow 3\ K^+(aq) + [Au(CN)_4]^{3-}(aq)$$

According to this equation, when 1.00 mol of the solid dissolves, 4.00 mol of ions form.

(d) [Ni(H$_2$O)$_2$(NH$_3$)$_4$]Cl$_2$ is a hexacoordinated complex cation with a chloride counter ion. When it is dissolved in water, it ionizes:

$$[Ni(H_2O)_2(NH_3)_4]Cl_2(s) \longrightarrow [Ni(H_2O)_2(NH_3)_4]^{2+}(aq) + 2\ Cl^-(aq)$$

According to this equation, when 1.00 mol of the solid dissolves, 3.00 mol of ions form.

80. Use the percent by mass to determine the formula, using methods from Chapter 3. Arrange the atoms to form a complex ion with the cobalt such that there are four ions when one formula unit is dissolved in water. For this step, adapt the method described in the answer to Question 78.

(a) In 100.00 g sample of compound, we have 22.0 g Co, 31.4 g N, 6.78 g H, and 39.8 g Cl.

$$22.0\ \text{g Co} \times \frac{1\ \text{mol Co}}{58.933\ \text{g Co}} = 0.373\ \text{mol Co} \qquad 31.4\ \text{g N} \times \frac{1\ \text{mol N}}{14.01\ \text{g N}} = 2.24\ \text{mol N}$$

$$6.78\ \text{g H} \times \frac{1\ \text{mol H}}{1.0079\ \text{g H}} = 6.73\ \text{mol H} \qquad 39.8\ \text{g Cl} \times \frac{1\ \text{mol Cl}}{35.4527\ \text{g Cl}} = 1.12\ \text{mol Cl}$$

0.373 mol Co : 2.24 mol N : 6.73 mol H : 1.12 mol Cl

1 Co : 6 N : 18 H: 3 Cl

$CoN_6H_{18}Cl_3$

A typical ligand in complex ions is NH_3, so let's use the N atoms and H atoms to build NH_3 molecules. $Co(NH_3)_6Cl_3$. We can now arrange the components so that there is one hexacoordinated complex cation, $[Co(NH_3)_6]^{3+}$, with three Cl^- counter ions to make four ions. The formula is $[Co(NH_3)_6]Cl_3$.

(b) The dissociation equation to make an aqueous solution looks like this:

$$[Co(NH_3)_6]Cl_3(s) \longrightarrow [Co(NH_3)_6]^{3+}(aq) + 3\ Cl^-(aq)$$

82. Adapt the methods described in the answers to Questions 80 and 3.75.

The simplest formula, $PtN_2H_6Cl_2$, has a formula mass of 300.05 g/mol. The molar mass is twice that number, so the molecular formula must be $Pt_2N_4H_{12}Cl_4$. A typical ligand in complex ions is NH_3, so let's use the N atoms and H atoms to build NH_3 molecules. $Pt_2(NH_3)_4Cl_4$. We can now arrange the components so that there is one tetracoordinated platinum(II) complex cation and one tetracoordinated platinum(II) complex anion. We do this by putting the neutral NH_3 ligands on Pt^{2+} to form the cation, and the negative Cl^- ligands on Pt^{2+} to form the anion. Therefore, a logical formula for the compound is $[Pt(NH_3)_4]^{2+}[PtCl_4]^{2-}$. Its structural formula looks like this:

One other formula would also fit the given information: $[Pt(NH_3)_3Cl]^+[Pt(NH_3)Cl_4]^-$.

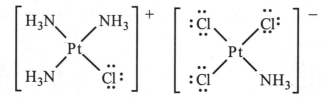